MODERN MATERIALS AND MANUFACTURING PROCESSES

MODERN MATERIALS AND MANUFACTURING PROCESSES

John E. Neely

Instructor, Machine Technology, Ret.
Lane Community College, Eugene, Oregon

Richard R. Kibbe

Director, RRK General Machining and Manufacturing
Instructor, Industrial Technology
Oxnard Community College, Oxnard, California

John Wiley & Sons

New York Chichester Brisbane Toronto Singapore

Cover photograph: Gabe Palmer/The Stock Market.

A number of illustrations have been reused with permission from Neely, White, Kibbe, and Meyer, *Machine Tools and Machining Practices*, Volumes I and II (copyright © 1977 John Wiley & Sons, Inc.) and John E. Neely, *Practical Metallurgy and Materials of Industry* 2 Ed. (copyright © 1984 John Wiley & Sons, Inc.).

Library of Congress Cataloging in Publication Data:

Neely, John E.
Modern materials and manufacturing processes.

Includes index.
1. Manufacturing processes. 2. Materials. 3. Metalwork. I. Kibbe, Richard R. II. Title.
TS183.N44 1987 670 86-15907
ISBN 0-471-81443-1

Printed in the United States of America

10 9 8 7 6 5 4 3 2 1

PREFACE

This textbook presents state-of-the art technology on materials and manufacturing processes in a style that can be understood by the typical technician or technology student. It is intended for vocational education students and technicians and for students in engineering technology.

The arrangement and format of chapters presents clear, sequential course material for any teaching system, whether competency based or traditional. Objectives at the beginning of each chapter clarify the goals of each chapter. Case studies and case problems are included where appropriate to promote thought and discussion on the part of the students. Questions and case problems may be used for review or examination purposes. Nearly 1000 illustrations and photographs complement the text.

This book is intended primarily for the survey course in materials and manufacturing processes. Rapidly growing technologies such as those found in the plastics and aerospace industries are emphasized. We not only discuss the specific materials and processes of manufacturing normally included in the survey course, but also, in a unique approach, show how those materials and processes are integrated into today's functioning manufacturing industry. However, this book does not provide a hands-on instructional system for the operation of machinery or the use of hand tools.

Part I, on materials and their applications, discusses frequently used materials such as metals, plastics, and rubbers, which are identified and classified. Their extraction from raw materials and their processing are presented in sufficient detail to enable the student to proceed to the study of product manufacturing from these materials.

Part II covers specific manufacturing processes. Conventional manufacturing processes including casting, cold and hot rolling, forming, forging, plastics molding, and joining of materials are discussed in detail. Newer processes and methods are presented where they apply.

Part III discusses the design, tooling, and production aspects of manufacturing. This part also introduces modern automation methods including CNC, CAD/CAM, industrial robotics, and Flexible Manufacturing systems (FMS), precursors to the totally automated factory of today and tomorrow. These topics include Computer Numerical Control (CNC), Computer-Aided Design (CAD) as a designer's tool, Computer-Aided Manufacturing (CAM), and the integration of these technologies as ICAM and CADAM.

In Part III we also discuss the way the manufacturing industry functions, job titles and general responsibilities, and how production manufacturing is accomplished. A separate chapter covers quality assurance and control aspects with particular emphasis on tolerances, production dimensional measurement, and calibration.

An extensive glossary is included at the end of the book. A complete instructor's manual is available from the publisher.

J.E.N.
Eugene, Oregon

R.R.K.
Oxnard, California

ACKNOWLEDGMENTS

The authors are most grateful to the companies and individuals who have contributed materials and help in preparing this text. Special thanks go to Eugene Aluminum & Brass Foundry, Hergert's Industry, Lane Community College, and Lockheed-California Company who granted access to their facilities for photography and for supplying illustrations and technical information.

Ajax Magnothermic Corporation, Warren, OH

Alewijnse Electrical Engineering, Environment System, Nijmegen, Holland

Allis Chalmers, Milwaukee, WI

Aluminum Company of America, Pittsburgh, PA

American Die Casting Institute, Des Plaines, IL

American Iron & Steel Institute, Washington, DC

American Motors, Southfield, MI

American Society for Metals, Metals Park, OH

Apex Broach & Machine Company, Detroit, MI

Armco Inc., Specialty Steels Division, Middletown, OH

Automated Finishing Incorporated, Menomonee Falls, WI

Baird Corporation, Bedford, MA

Barrington Automation, Fox River Grove, IL

Bethlehem Steel Corporation, Bethlehem, PA

Beuhler LTD, Lake Bluff, IL

Bird-Johnson Company, Walpole, MA

Boeing Commercial Airplane Company, Seattle, WA

Bracker Corporation, Pittsburgh, PA

Buck Chuck Company, Kalamazoo, MI

Buffalo Forge Company, Buffalo, NY

Bureau of Mines, Albany, OR

Burlington Industries, Greensboro, NC

Buss-Condux, Inc., Elk Grove Village, IL

CHR Industries, Inc., New Haven, CT

Chrysler Corporation, Detroit, MI

Cincinnati Milacron, Cincinnati, OH

Cincinnati Milacron Plastics Machinery Division, Batavia, OH

Clausing Machine Tools, Kalamazoo, MI

Coherent, Palo Alto, CA

Colt Industries, Crucible Specialty Metal Division, Syracuse, NY

Consolidated Metco, Inc., Clackamas, OR

Consulate General of the Netherlands, Economic Section, NY

Del Manufacturing, Oxnard, CA
Disogrin Industries Corporation, Manchester, NH
Dixon Industries Corporation, Bristol, RI
DoALL Company, Des Plaines, IL
Dover Publications, New York, NY
Ductr Mfg., Eugene, OR
DuPont Company, Engineering Polymers Division, Wilmington, DE

ESAB North America Incorporated, ESAB Robotic Welding Division, Fort Collins, CO
Ethan Allen, Danbury, CT
Eugene Aluminum & Brass Foundry, Eugene, OR
Exxon Company, Houston, TX

Federal Products, Providence, RI
Fellows Corporation, Springfield, VT
Ford Motor Company, Dearborn, MI

Gates Learjet, Phoenix, AZ
General Motors Corporation, Detroit, MI
Giddings and Lewis, Fond du Lac, WI
Gleason Machine Division, Rochester, NY
Globe Amerada Glass Company, Elk Grove Village, IL
Goodyear Tire and Rubber Company, Akron, OH

Hamilton Caster & Mfg. Co., Hamilton, OH
Heck Industries, Redford, MI
Hergert's Industries, Drain, OR

Illinois Gear, Chicago, IL
Intergraph Corporation, Huntsville, AL

Japax Incorporated, Kanagawa, Japan
John Wiley & Sons, Inc., New York, NY

Kearney and Trecker, Milwaukee, WI
Kennametal Inc., Raleigh, NC
Kevex Corporation, King of Prussia, PA

L.S. Starrett Company, Athol, MA
Landis Tool, Waynesboro, PA
Lane Community College, Eugene, OR
Laserdyne, Eden Prairie, MN
Laserfare, Johnston, RI
Leland Balber Associates, Pittsburgh, PA
Libby-Owens-Ford, Toledo, OH
Lidköping Mekaniska Verkstads AB, Sweden
Lockheed-California Company, Division of Lockheed Aircraft Corporation, Burbank, CA
Louis Levin & Son, Inc., Culver City, CA
LTV Steel Company, Cleveland, OH

McGraw-Edison Company, Power Systems Division, Pittsburgh, PA
Magnaflux Corporation, Chicago, IL
Martin Marietta Corporation, Bethesda, MD
The Minster Machine Company, Minster, OH
Miller Electric Manufacturing Company, Appleton, WI
Mittler Machine Corporation, Providence, RI
M. L. Sheldon Plastics Corporation, New York, NY
MTI Corporation, Paramus, NJ
Mueller-Phipps International Corporation, Poughkeepsie, NY

National Machinery Company, Tiffin, OH
Niagara Machine & Tool Works, Buffalo, NY
Nikon Incorporated, Instrument Division, Garden City, NY
Northrop Corporation, Hawthorne, CA

Oregon Portland Cement Company, Portland, OR

Pacific Machinery & Tool Steel Company, Portland, OR
Parker Hannifin Corporation, Cleveland, OH
Pennsylvania Pressed Metals, Inc., Emporium, PA
Pierce Corporation, Eugene, OR
Plastics Machinery Incorporated, Williamston, MI
Porter, Westbrook, ME
Positech Corporation, Laurens, IA
Pottstown Machine Company, Pottstown, PA

Republic Steel Corporation, Cleveland, OH
Rhino Robots, Champaign, IL

Sandusky Foundry & Machine Co., Sandusky, OH
Saunders Equipment, Inc., Cold Spring, NY
Siempelkamp Corporation, Buffalo Grove, IL
Siempelkamp Corporation, Marietta, GA
The Stanley Works, New Britain, CT
Storage Technology, Louisville, CO

Teledyne Wah Chang Albany, Albany, OR
Testing Machines Inc., Amityville, NY
Tinius Olsen Testing Machine Co., Inc., Willow Grove, PA
TRW Cutting Tools Division, Augusta, GA
TVT Die Casting & Mfg. Inc., Portland, OR

United States Steel Corporation, Pittsburgh, PA
Unitron Instruments, Inc., Plainview, NY

Warner & Swasey, Sheffield Measurement Division, Dayton, OH
Wellcraft Marine Corporation, Sarasota, FL
Wheaton Industries, Wheaton Glass Division, Millville, NJ
Whirlpool Corporation, Benton Harbor, MI

CONTENTS

CONTENTS

MODERN MATERIALS AND MANUFACTURING PROCESSES

MODERN MATERIALS AND MANUFACTURING PROCESSES

INTRODUCTION TO MATERIALS AND MANUFACTURING

Few human enterprises represent a more aggressive and dedicated use of time, energy, money, and material resources than the activities of extracting or otherwise making raw materials and then processing these materials into the endless line of products that make the modern life-style possible. Since the first appearance of humankind on this planet, **manufacturing,** the processes and methods of converting materials into different forms and products, has gone hand in hand with human development. Many examples of toolmaking by ancient humans have been discovered. With tools, ancient peoples were able to shape natural and ultimately synthetic materials for weapons, shelters, and other products that could be traded or sold, thus generating both commerce as well as demand for more, different, and improved types of materials and manufactured products. The desire and demand for exotic materials and products stimulated the explorer and the trader to travel to the far corners of the earth.

Such exploits created needs for higher technology in manufactured products. Navigational instruments and accurate maps are examples. These items were essential in order to make these voyages of discovery safer, more reliable, and to better ensure that the trader would return safely with the materials that were in such demand. Such factors stimulated the maximum efforts in design and materials application available at the time.

As ancient peoples gathered together to avail themselves of products, services, and materials, manufacturing societies began to develop and the structure for villages, towns, and ultimately cities was established. The increasing concentrations of people created needs for much improved manufacturing methods to supply demands. Specialty trades and crafts developed rapidly. This in turn created a need for faster methods of production, met by harnessing the power of animals, water, steam, and finally nuclear energy to power manufacturing machinery.

The lure of employment in manufacturing industries and the availability of manufactured products that would make life easier and more pleasant, contributed greatly to the industrialization and urbanization of much of the world. Great manufacturing centers developed, usually located close to fundamental sources of such raw materials as coal, oil, and iron ore, prime ingredients on which industrialization is built. In today's world, a vast international manufacturing society exists that aggressively makes use of all newly developed technology both in materials development and application, as well as in manufacturing processes.

In today's world of exotic materials and high technology, modern manufacturing technology is constantly changing and growing. New methods of accomplishing production and applying new materials become available almost daily. No matter how current information is in this area, it becomes obsolete rapidly as new technology comes on line. Although most of the well-established processes of manufacturing products will always be with us, the differences will come in how these processes are used. New applications of standard processes will, at the same time, create new processes. New materials will also create new processing methods. The manufacturing technologist of today and tomorrow must be well versed in materials application and in the standard processes but will always be looking forward toward new applications of these materials and processes as well as toward the development of entirely new processes. Manufacturing of the future will become more specialized, making it more difficult for the technologist to keep informed of this rapidly changing and expanding technology.

The central purpose of this textbook is to present an overview of materials science, a survey of traditional as well as high technology manufacturing processes, and a look at manufacturing systems in the computer high technology age.

THE MANUFACTURING ECONOMY —PAST, PRESENT, AND FUTURE

By the middle of the nineteenth century, a truly remarkable revolution, the Industrial Revolution, which had its beginning two centuries before, was well established. Goods formerly made by hand were now made by machine and the era of large-scale mass production had begun.

The effects of the Industrial Revolution have been far-reaching. Cheap production of goods in large quantities, now the order of the day, has raised the standard of living for many people in industrialized societies. The Industrial Revolution has formed the basis of the modern life-style that most of us take for granted. General industrialization has for the present also provided almost unlimited employment opportunities for anyone willing to participate in the working world.

In the latter half of the twentieth century, a new industrial revolution, one of computers and space-age high technology, is taking place. Like that of the past century, this new technology will have far-reaching effects. However, we are now more than 100 years further down the road of technological development. Thus, the effects of the current industrial revolution will be much more far reaching than anything even dreamed of in the past centuries.

Modern industrial technology has already given us the power and the tools to be far more productive with much less need for people to do the work of manufacturing. It is not to be denied that modern technology has made the life-style of the modern industrial society even more easy, convenient, and filled with incredible products, most of which are within the purchasing power of the average person living in that society. However it may have temporary negative effects as well—for those who are not prepared to work in a high technology, computerized, and automated manufacturing world.

In addition, the gradual industrialization of less technologically oriented societies throughout the world is moving ahead, perhaps at a slower pace but always in a steady forward direction. We of the established industrial world could, for much of the past 50 years, take little notice of this advance as our technology was always well in front of that in less developed countries.

As the wage demands of the old established industrial societies rose over time, manufacturers sought to take advantage of the lower wage structures generally prevalent in emerging countries. Thus, the great migration of manufacturing from the established industrial societies to the emerging countries began and still continues to the present day.

What have been the results of this migration? It is true that manufacturing industries that have moved abroad have for some years been able to operate under more favorable wage structures. However, the very in-

dustries that these manufacturers sought to staff with cheaper labor have, by a natural evolution, raised the standard of living for workers in the countries where the industries were located. More jobs have been created, and with jobs have come higher pay and the ability of industrial workers to buy not just necessities but goods that make life easier and more pleasant. This factor plus the often diligent attitude of foreign workers toward their jobs and the resultant high level of productivity and quality, have brought industrial revolutions similar to our own in emerging countries. This process is continuing unabated and now presents intense competition to our domestic employment and economy.

Manufacturers, seeing the same problems as before, have moved on to the remaining even less developed countries, thus initiating the process over again. The effects have been and will continue to be far-reaching. In those countries where the technology and industrialization occurred first, a diligent attitude toward productivity coupled with an almost limitless market in the established industrialized world for the products produced has for the first time, placed the technical leadership of the established industrial old world in question.

For example, the influx of high quality goods manufactured abroad and imported into the United States for sale at an often much cheaper price, has continued unabated for many years. The results are only too evident. Jobs in the domestic economy have been lost, and manufacturing industry has not always countered by keeping up with modern trends in automation, productivity, quality control, and worker efficiency. Foreign manufacturing, often operating in a very favorable unrestricted business environment, and often with the total support of local governments, has been able to take better advantage of modern technology and is engaged in an effective program of operating modern and efficiently run industries.

If the industrial structure of the old industrial societies is to remain intact and the standard of living is to remain at a high level, it is of paramount importance that the entire industrial and manufacturing community adopt new attitudes toward how manufacturing can be done efficiently and competitively. The worker, who is in the end responsible for industrial output, must be treated well, but, at the same time, quality work and high productivity must be maintained. Manufacturing industries must put to use all of the modern tools necessary to accomplish efficient quality production in the high technology age. The production worker of today will give way to the computerized robot of tomorrow and must therefore look ever forward to the future, or lack thereof, for his or her job skills.

It is a sure bet that product manufacturing in the future will be done on **a more automated** level of high technology and automation. Many of the products that will soon be commonplace have not even been designed yet, but as they are, more and more high technology manufacturing systems and methods will have to be created to produce them.

Automation trends, already at a high level in manufacturing, will become ever more the preferred method of production. **Flexible manufacturing systems (FMS)** (Figure 1), the forerunners of completely automated factories, already on line in many industries, will become commonplace as computer controlled automatic machines take over most of the mundane work heretofore done by fairly low-paid manual labor. The need for operating personnel will be minimized in this system, as robot operated transporters move workpieces automatically between manufacturing stations. **Industrial robots** (Figure 2) will perform the routine tasks of loading and off-loading parts into **computer numerically controlled machining centers (CNC).**

Figure 1
The flexible manufacturing system (FMS) with automated workpiece handling and transfer (Kearney and Trecker).

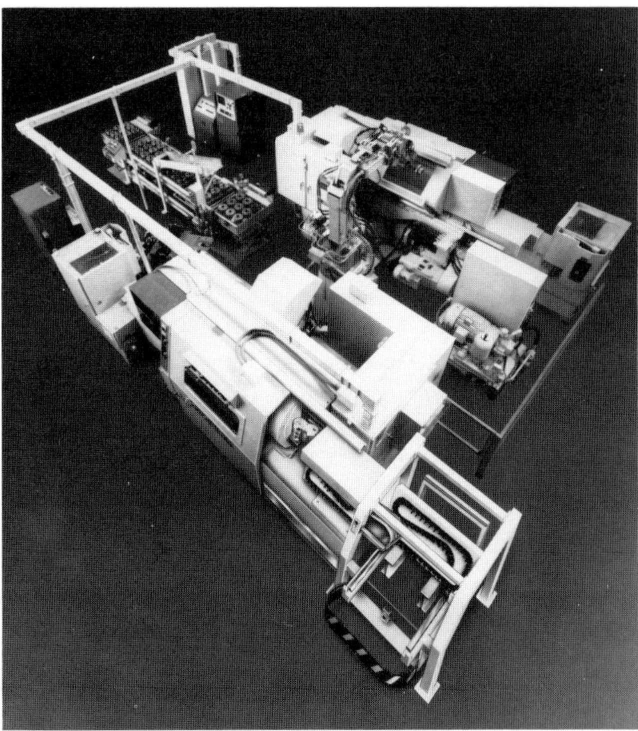

Figure 2
FMS work cell for CNC turning. The robot (center) loads and off loads parts for both machine tools (Cincinnati Milacron).

High technology is really not new. It has been with us since the first recorded manufacturing efforts, as each developmental age in manufacturing applied the latest available materials, tools, and methods to manufacturing requirements. Until now, however, most manufacturing industries still needed people to operate machines to turn out product units.

The computer, with its considerable ability to make decisions in process control applications, is already replacing people in the manufacturing process. As computer technology becomes more sophisticated, its decision-making powers will increase proportionally and thus its abilities for multitasking and decision making will increase as well. Artificial intelligence (AI) technology is presently under rapid development. This will give computers even more decision-making power, making them even more versatile and in more complete control of routine as well as sophisticated mass production.

THE COMPUTER INFLUENCE IN MANUFACTURING

Throughout the long and interesting history of technological development, many contributions have been made. First there was metallurgy, making available a wide selection of metals for toolmaking and manufactured products. The Industrial Revolution of the eighteenth and nineteenth centuries brought the age of machine-made mass produced goods. Electronics, especially solid state and integrated circuit (IC's or chips) technology, has made instant communication available to practically everyone on earth. Thus the message of what is available in manufactured goods has reached ever-widening markets.

All of these technological advances have transformed industry and society as a whole. However, the effect of the computer will be much more extensive than any technological device yet developed. This concept of computer control of processes is extremely significant in that it has and will continue to profoundly and forever change the way in which manufacturing is accomplished.

Does this mean that computers will control everything including the actions of people? The answer to this question is a definite no. However, the computer will aid those people who do direct the actions of people not just in manufacturing but in all phases of life.

The exact result of computers and computerized automation trends in manufacturing on the future worker is uncertain. However, it is sure that the factory worker of the future will have to become more of an electromechanical technician instead of being the simple machine tender of today's semi-automated factory.

The technician will have to learn broader skills, acquiring a more complete knowledge of electrical/electronic and hydraulic systems, transactional mechanisms, general mechanical skills, and very likely a working knowledge of computer systems and computer programming. For those not willing to learn new skills or accept retraining, future employment prospects at a suitable pay level may be limited. However, the skills and abilities of people with an open mind and a natural desire to understand how things work and how to operate, maintain, and repair complex electronic/mechanical machines are likely to be in demand.

MANUFACTURING AND YOU

For those who afford themselves of proper training, opportunities in manufacturing will abound. However, the future employee at any level in manufacturing will have to undergo broader and more complex training than at any time in the past.

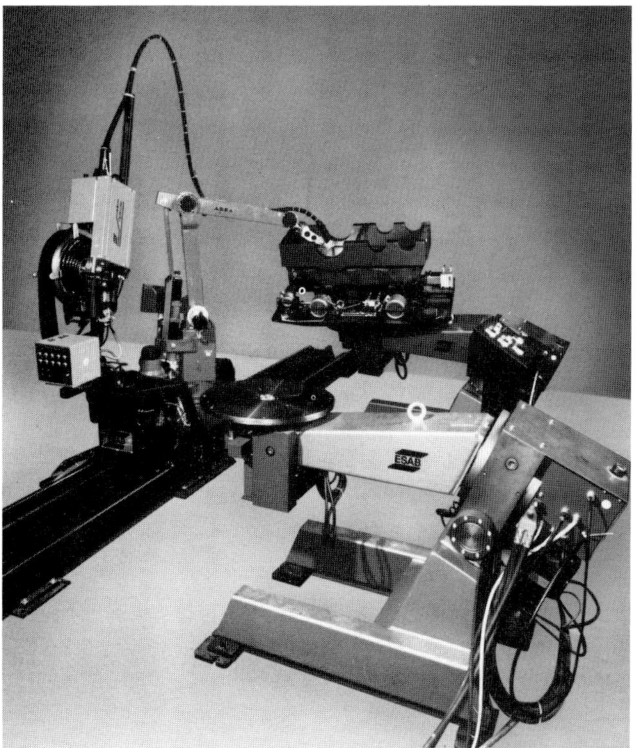

Figure 3
Robot welder welds a gear case mounted on one of two workpiece positioners (ESAB Robotic Welding Division).

Figure 4
Production technician assembles laser computer disks in an ultra-clean room environment (Storage Technology).

Trade Level Opportunities

At the trade level, many of the repetitive production jobs that now are done by minimum wage employees will be eliminated as automation such as computer numerical control **(CNC)** and industrial **robotics** take over. Robots will become more and more sophisticated and capable, and as they do they will move further into the areas of the skilled trades such as welding (Figure 3). The robot welder can, in many situations, actually out-perform its human counterpart in both quantity and quality on many welding tasks.

The higher level skilled trades, such as **tool and die** making, will always be in demand, but preparation for these jobs will require several years of diligent training. Much of the machine operator's work will be automated as numerical control of machine tools is more fully integrated with robotic systems. The **general machinist** will be needed, but only in sufficient numbers to produce one-of-a-kind prototypes and few-part batches and support machine repair and agriculture requirements.

The **production technician,** an individual who has broad training in electric/electronics, mechanical/hydraulic systems, and computers will probably have the brightest future in trade-level manufacturing jobs. Even though the capability of modern and future automated computerized manufacturing systems is truly amazing, the equipment and control systems are complex and will require constant and careful maintenance. The production equipment technician will have to be able to quickly diagnose problems and effect repairs rapidly and efficiently so that it can be placed back in production as soon as possible.

High technology production technicians will be specialists at their work. They will assemble and test delicate high precision equipment, often in controlled environments. In Figure 4 the technicians are assembling a laser computer disk with very large memory capacity. This precision equipment must be handled in an ultraclean room environment.

Middle Management Levels

The future in manufacturing for the college trained **industrial engineer (IE), industrial technologist (IT), manufacturing engineer,** and **applications engineer** will also be bright. As automated manufacturing systems are put in place, the industrial, manufacturing, and applications engineer will be needed for plant facilities, tool, jig, and fixture design (Figure 5), robot and numerical control programming, and implementation of all aspects of the production manufacturing system (Figure 6). As with the production technician, the industrial technologist and manufacturing engineer will need to have broad training in materials applications, com-

Figure 5
Automatic wing spar assembly tooling, new methods of tooling will challenge the industrial and manufacturing engineer (Boeing).

Figure 6
Integrated automation, machining, assembly, and testing in sequential automation (Giddings and Lewis).

puters, and manufacturing processes. Such individuals will also be heavily involved with all aspects of quality control and assurance as the manufacture of higher technology products becomes more commonplace.

Design Engineering

Creative **product designers** will always be needed as the capacity of high technology manufacturing makes possible production of a vast array of incredible new products, many of which have not even been designed as of the present. Tomorrow's designer will need to be thoroughly informed of the modern manufacturing methods and their capabilities. The designer of the future using the computer in a **Computer Aided Design (CAD)** system, will smoothly integrate new design ideas with new material and new processes as they become available. Preparation for jobs in this area will require college level education and on-the-job industrial training.

ON-GOING MANUFACTURING TECHNOLOGY

It is certain that modern manufacturing technology will always be in a continuing state of development. As new materials, methods, and computer aids come into play, the manufacturing engineer will apply these to solve production problems.

Materials

The product designer will continually seek new materials that are lighter, easier, and, especially, cheaper to process, while meeting design requirements for strength and durability. **Plastic** and **composite materials** will play a more important role in future products, thus permitting additional energy savings both in manufacturing processes and in powering road, air, and space vehicles (Figure 7).

Figure 7
High technology designs and materials for tomorrow's mach five high altitude aircraft. The leading edges of the air frame will glow red hot at a temperature of 1000° F (Lockheed–California Company).

Figure 8
The Computer Aided Design (CAD) engineering work station is fast becoming the preferred method of designing (Intergraph Corporation).

Design

The computer will play an ever-increasing role in initial product design and design evaluation (Figures 8 and 9). Computer Aided Design is fast becoming the preferred method of engineering design. To tomorrow's designer, CAD systems will be an absolutely essential design tool, and in most cases, actual manufacturing equipment control programs will be generated at the same time, thus forming **integrated Computer Aided Design/Computer Aided Manufacturing (CAD/CAM)** and **Computer Integrated Manufacturing (CIM) systems.** The day is fast approaching when design specifications entered at the CAD station will shortly appear as finished products at the shipping dock.

Figure 9
CAD used in vehicle designing (Ford Motor Company).

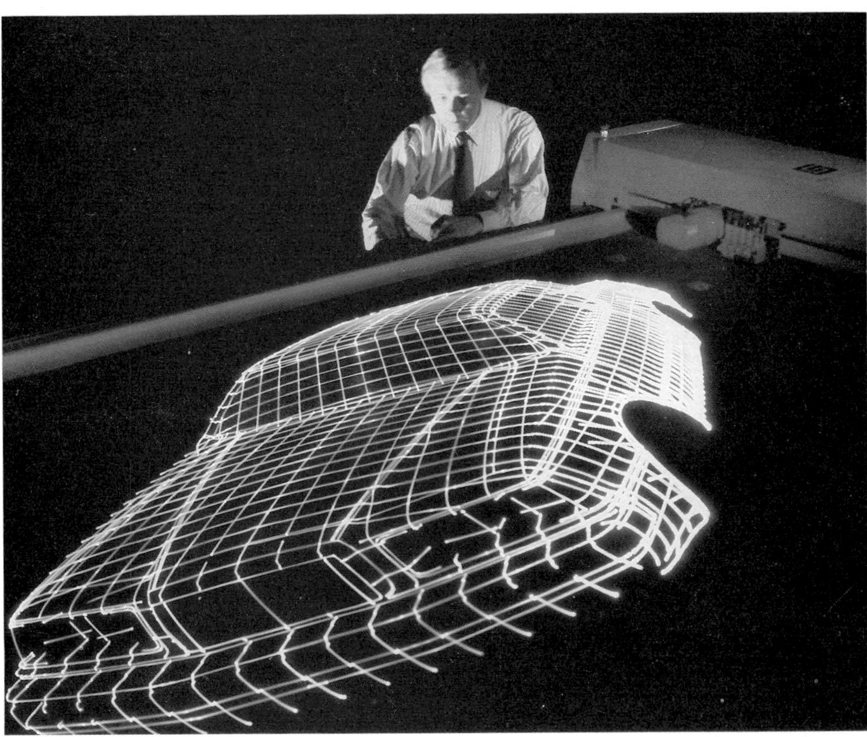

Figure 10
CAD for design evaluation of automotive aerodynamics (Ford Motor Company).

CAD will also become a more frequently chosen method of evaluating design, thus eliminating expensive prototyping. In Figure 10 complex compound contour shapes of today's aerodynamic automobile bodies can be studied and evaluated by these computer methods.

Figure 11
A new look at old and proven technology—high technology turbojet propellers (Lockheed–California Company).

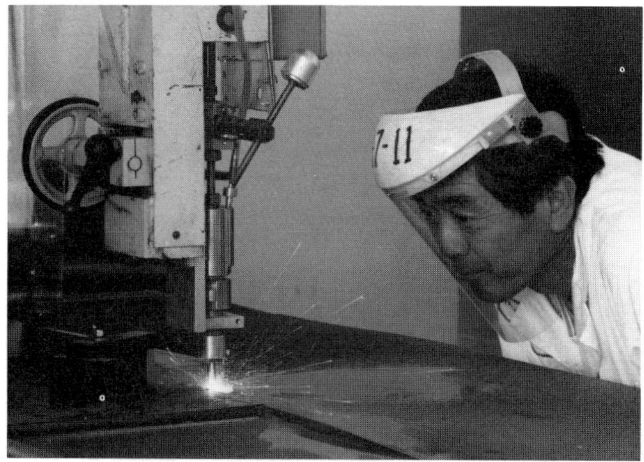

Figure 12
High pressure water containing an abrasive is used to cut titanium aircraft parts (Lockheed–California Company).

Proven designs of the past will be reevaluated with an eye toward energy and fuel consumption efficiency. Figure 11 shows a new high technology application of the airscrew. Driven by the efficient turboprop jet engine, the new propeller design can propel aircraft with the same speed and altitude capability of modern jets, but with a large saving in fuel consumption.

Manufacturing Processes

Manufacturers in all types of industries will search for and implement new manufacturing processes. Many of these will take on some unique and interesting forms. In Figure 12 a high pressure stream of water containing an abrasive material is used as a cutting tool for titanium aircraft parts. Since no heat is developed in this process, distortion of the part is eliminated.

Computer controlled automation using robotics will be the order of the day. Process control in this manner will be found in every aspect of manufacturing from electronic component insertion (Figure 13), robotic spot

Figure 13
Electronic components are automatically inserted onto circuit boards (Whirlpool Corporation).

Figure 14
Computer-driven spot welding robots spot weld front body assemblies with great speed and accuracy (Ford Motor Company).

welding (Figures 14 and 15), and precision measurement and inspection (Figure 16). Other high technology applications of available technology include the further use of **lasers** for many applications (Figure 17).

Figure 15
Underbody and two side panels are secured in a large fixture for production spot welding (Chrysler Corporation).

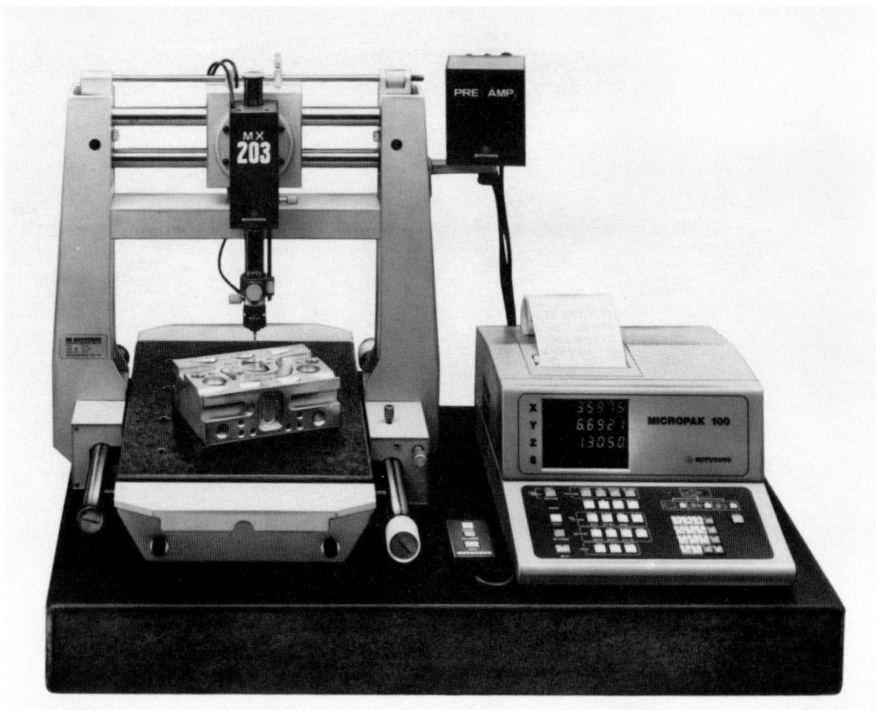

Figure 16
Computerization in precision measurement (MTI Corporation).

Figure 17
Laser used in precision measurement (Giddings and Lewis).

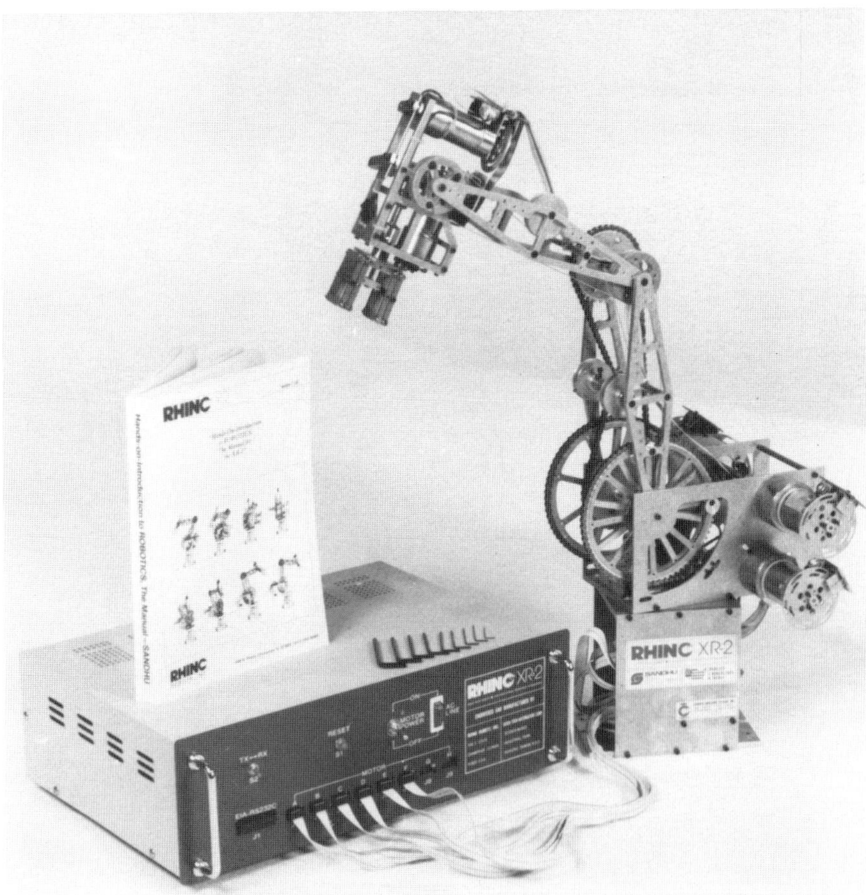

Figure 18
The Rhino XR industrial robot training system (Rhino Robots).

Figure 19
Students at Vincennes University, Vincennes, Indiana, study application and operational aspects of the Rhino XR industrial robot training system (Rhino Robots).

Figure 20
Students at Centennial College, Toronto, Canada, study programming and operation of the Rhino XR system (Rhino Robots).

THE PLACE OF TRAINING AND EDUCATION IN THE HIGH TECHNOLOGY WORLD

Whether one has a working knowledge of the newest skills in the world of high technology will determine to a great extent who will be employable at adequate pay levels. Progressive manufacturing industries, schools, colleges, and universities will be responsible for a continuing supply of adequately trained technicians and engineers to meet the needs of the high technology age. In manufacturing industries, in-service training programs in new methods will be a high priority.

Students of materials and manufacturing processes in the high technology world will need hands-on experience in the new systems and in their operations and applications. An example of the types of training system that can be implemented is the Rhino XR robotic system (Figure 18). This system, with its computer interfaces and available software, provides the user with an opportunity to learn industrial robotic programming, operation, and application (Figures 19 and 20). Note the open mechanism on this industrial robotic training system. This design enables the users to see how the robot functions mechanically under computer control.

For those who are properly prepared, the world of high technology manufacturing presents many opportunities. The technology of today and tomorrow will require many trained personnel to repair, service, and implement the manufacturing systems of an on-going high technology industrial revolution. New product designs and the manufacturing systems required to produce them, will be an ever-expanding field for those who seek the proper training. The production technician, manufacturing engineer, and designer will contribute greatly to maintaining a nation's international technological position. Will you be ready to meet these challenges? The following chapters in this book will aid you in answering this important question.

MATERIALS OF MANUFACTURE

PART I

Molten steel rushes from a tilted electric furnace into a huge ladle at Bethlehem Steel Corporation, Bethlehem, Pennsylvania, plant (Courtesy of Bethlehem Steel Corporation).

At the dawn of history, the only materials utilized by humans were taken directly from natural sources. Stones were used to erect crude shelters or structures; harder stones with cutting edges were used for arrow heads and spear points and for tools for cutting wood, bone, and leather. Clay was shaped into containers and building bricks. The discovery that firing these clay items made them stronger and water resistant was probably the first step in a long procession of the adaptation of natural materials to the needs and uses of humanity.

Today, there are literally thousands of materials derived from nature and many more synthetic materials that do not come directly from natural sources. All of these amazing developments have come about because of increasing knowledge and understanding of the properties of materials, especially in the last century—the arrangement of atoms and crystal structures and their behavior in various conditions. Plastics, elastomers, metals, ceramics, and composites are some of the important materials that are used today that will be studied in Part I.

CHAPTER 1 THE ATOMIC AND CRYSTALLINE STRUCTURE OF MATERIALS

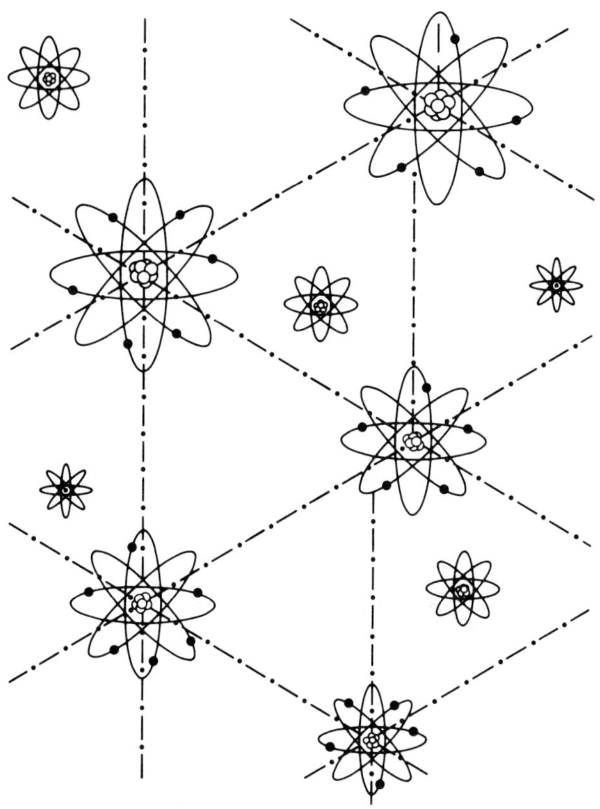

Much is presently known about the structure of the physical universe, and yet through the endeavors of the scientific community many more facts are constantly being revealed. The engineer uses these new discoveries to design new products, often with radically new concepts such as we have seen in the phenomenal growth in solid state electronics and high technology. The technician, therefore, must be able to understand some of the new terminology and technology of the scientist and engineer in order to communicate in the manufacturing field. It is useful to understand the basic structure of the atom prior to studying the properties of materials and their behavior under various conditions. Some of these principles and concepts are included in this chapter.

OBJECTIVES

This chapter should enable you to:

1. Explain a simplified model of the atom.
2. Understand several bonding arrangements and explain the role of valence in bonding.
3. Describe the crystalline structure of metals and how this affects their behavior.
4. Give an account of the structure of some high polymers for the manufacture of certain plastic materials and elastomers.
5. Describe the crystal structure of clays and other ceramics and their bonding arrangements.

Atoms are too small to be seen with the aid of ordinary light microscopes, but the outline of atoms can now be detected with such devices as the electron microscope and the field ion emission microscope. More than 100 elements have been discovered, 92 occur in nature, and the others are produced by atomic reactions. The **molecule** is defined as the smallest particle of any substance that can exist free and still exhibit all the chemical properties of that substance. A molecule

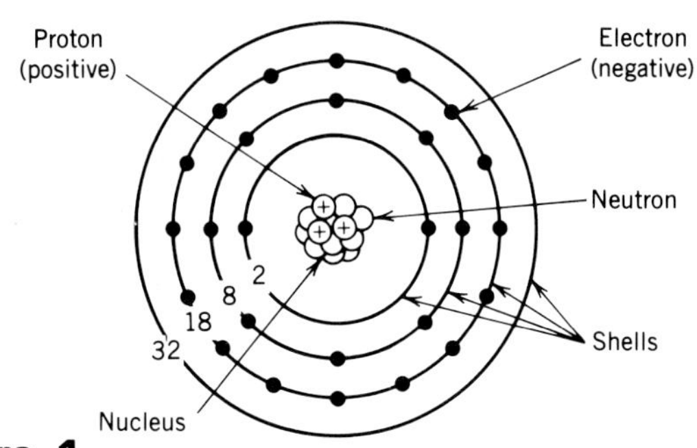

Figure 1
The basic parts of the atom.

can consist of one or more atoms of the same kind. For example, oxygen gas is found in nature as a molecule consisting of two oxygen atoms (O_2) or of three oxygen atoms, ozone (O_3). A **compound** is composed of two or more elements combined chemically; for example, hydrogen and oxygen combine to produce water (H_2O). A **mixture** consists of two or more elements or compounds that are physically, not chemically, combined.

Atoms, and in fact all materials, are composed of the same basic components: **protons, neutrons,** and **electrons** (Figure 1). The **nucleus** of the atom is composed of positively charged protons and neutrons. The latter have almost the same mass as protons but are without electrical charge. Surrounding the nucleus, orbiting it in what are called shells, somewhat like the orbits of planets in our solar system, are the electrons which have a negative charge equal and opposite to the charge of the proton in a neutral atom. It is this electrical charge that holds the atom together and which is responsible for much of the chemical and electrical behavior of the elements. Almost all of the atomic mass is contained in the nucleus; electrons have only about $\frac{1}{1839}$ the mass of the proton or neutron.

ATOMIC STRUCTURE

All matter is composed of one or more of the more than 100 chemical elements. Each element is a pure material, such as a metal or nonmetal, unlike any other; yet every element is composed of protons, neutrons, and electrons; only the number and arrangement differ.

Atomic Bonding

Valence is the capacity of an atom to combine with other atoms to form a molecule. There are positive and negative valences. When an atom has more or less electrons in the outer shell than it has in its uncombined or free state, it is called an ion and it possesses an electrical charge. The charge is negative when extra electrons are present and positive when some are missing. Metal atoms are easily stripped of their valence electrons and thus form positive ions. When the valence or outer shell has a full complement of electrons, it is said to have zero valence. The inert gases have zero valence and do not readily combine with other elements to form compounds.

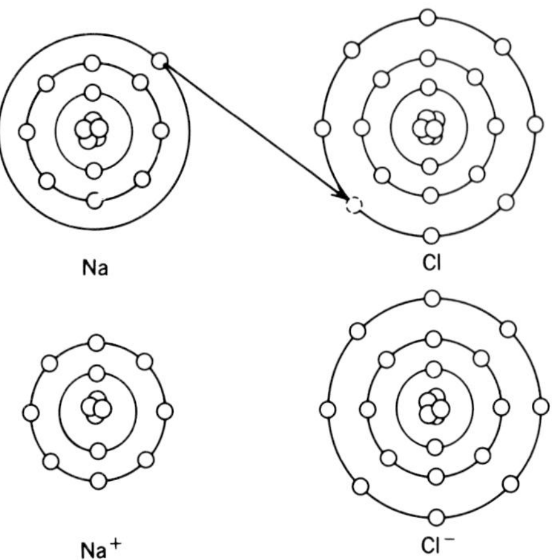

Figure 2
Ionic bonding of sodium chloride (White, Neely, Kibbe, Meyer, *Machine Tools and Machining Practices*, Vol. II, © 1977 John Wiley & Sons, Inc.).

There are four basic types of bonding arrangements that hold atoms together. They are **ionic** (also called electrovalent), **covalent** (polar or nonpolar shared electrons), **metallic**, and bonds made by **Van der Waals** forces. The simplest type of linking or bonding of atoms is the ionic bond. Ionic bonding links dissimilar atoms by the attraction of negative and positive ions. Sodium chloride (NaCl) is an example of ionic bonding (Figure 2) where a metal (sodium) loses its single valence electron to a nonmetal (chlorine) to complete its valence shell. The chlorine atom that became negatively charged and the sodium atom that became positively charged have combined to become a neutral compound (Figure 3). Ionic bonds tend to be somewhat weak because

Figure 3
Lattice structure of sodium chloride (White, Neely, Kibbe, Meyer, *Machine Tools and Machining Practices*, Vol. II, © 1977 John Wiley & Sons, Inc.).

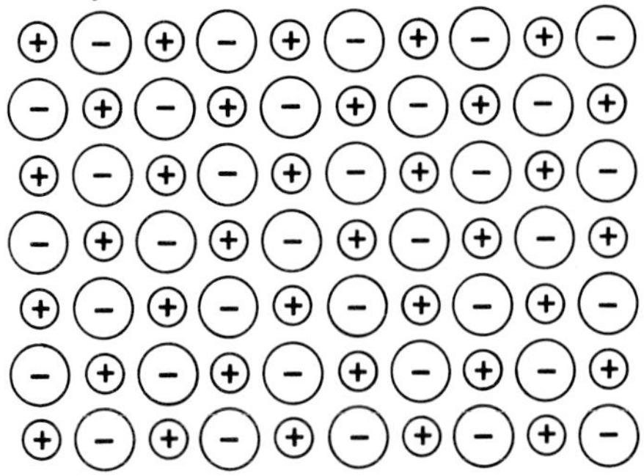

there are discontinuities in the lattice arrangement and there are unequal numbers of combining atoms so neutrality is not always maintained. These bonds therefore have the characteristics of brittleness, moderate strength, and high hardness, and they are good electrical insulators when they are in pure solid form because there is little opportunity for ion or electron movement. (It should be noted, however, that many ionic compounds, such as sodium chloride, are water soluble and that solutions of these compounds will conduct electricity very well.)

Covalent bonding, or the sharing of valence electrons, is a very strong atomic bond whose strength depends on the number of shared electrons. It is interesting to note, however, that a variety of covalent bonding patterns exist. In network covalent substances such as diamond and silicon carbide (carborundum) the sharing of electrons forms a large strongly bonded unit. On the other hand, molecular covalent substances such as sugars or waxes satisfy their sharing of electrons within the molecule and have very little attraction to other molecules surrounding them. Thus network covalent substances have high melting points and hardnesses, whereas molecular covalent substances have low melting points and hardnesses. Negative valences greater than four do not occur, but these atoms tend to assume an inert gas electronic structure of eight electrons in the outer shell (in the case of hydrogen it is one). Chlorine, having a complement of only seven electrons, therefore tends to join with any element that will fill its outer shell to eight electrons. As we have seen, it will join with sodium to obtain an extra electron and form an ionic substance; it will also join with another chlorine atom to gain an extra electron by sharing it to form a covalent substance (Figure 4). The element oxygen has the same tendency, but it shares two valence electrons (Figure 5).

The metallic bond is formed among metallic atoms when some electrons in the valence shell separate from their orbital path and exist in a free state as a cloud or gas, surrounding the atoms in a highly mobile state

Figure 4
Chlorine atoms form a covalent bond by sharing atoms.

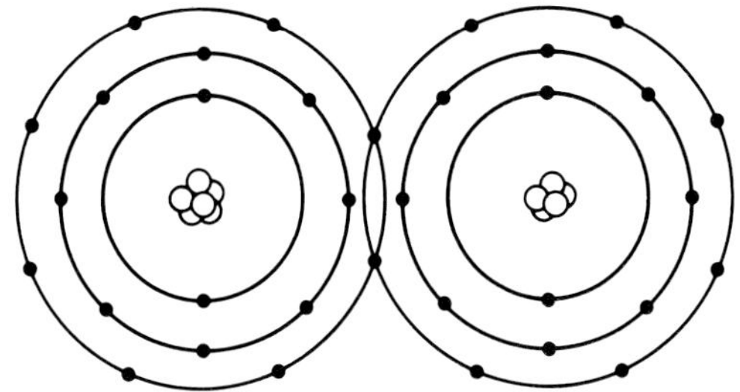

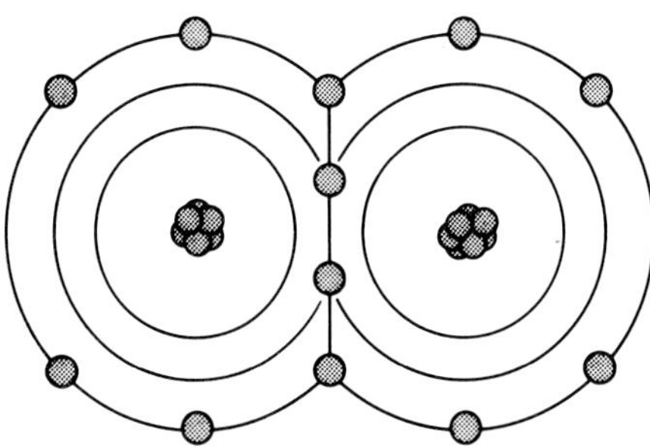

Figure 5
The covalent bond of the oxygen molecule (White, Neely, Kibbe, Meyer, *Machine Tools and Machining Practices*, Vol. II, © 1977 John Wiley & Sons, Inc.).

while being shared for bonding-like purposes (Figure 6). Because of this, bond strengths can vary, deformation is possible without rupture, and good heat and electrical conductivity is possible because of the easy movement of electrons from one atomic site to the next. This type of bonding makes possible some of the useful characteristics of metals such as plasticity and ductility, which will be discussed later in this chapter.

Van der Waals bonding is found between neutral atoms of elements such as the inert gases and between neutral molecules of covalent molecular compounds. This is only a very weak attractive force, and it is of importance in the study of metals only at very low temperatures.

CLASSIFICATION OF ATOMS

Atoms of the element hydrogen have in their normal state a single electron moving about a nucleus that has a positive charge equal to the negative charge of the electron. Likewise, the atom of the element helium has two electrons moving about its nucleus that has a positive charge equal to twice one electronic charge. An

Figure 6
The metallic bond.

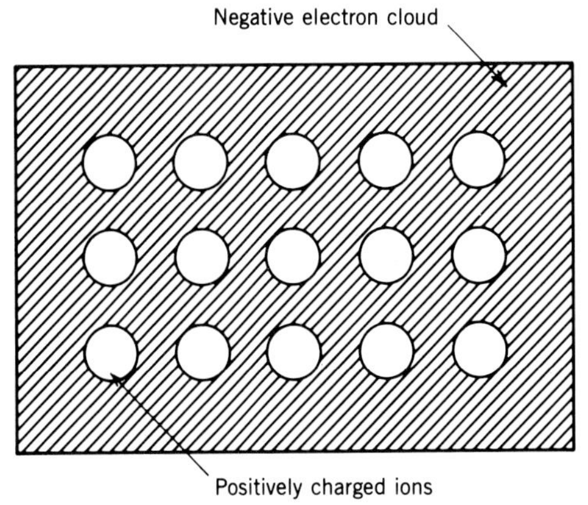

atom of the element lithium has three electrons and a positive nuclear charge equal to three times that of a hydrogen atom. This is the basis for the numbers on the periodic table (Table 1), called **atomic numbers.** These numbers range from 1 for hydrogen to 92 for uranium, and still higher for the transuranic elements. These numbers are important because they determine the chemical behavior of the atoms. The table of elements is called periodic because elements with similar properties appear at regular intervals in the arrangement of the elements. The horizontal rows of elements are known as **periods** and the vertical columns are known as **groups.** It is in the vertical columns that the greatest similarities exist. For example, copper (Cu), silver (Ag), and gold (Au) have similar crystal structures and readily combine as alloys. The vertical group of chromium (Cr), molybdenum (Mo), and tungsten (W) have similar characteristics as alloying elements in steel for promoting strength and hardenability. Also, sulfur (S), selenium (Se), and tellurium (Te) have all been used to promote machinability in steels. Besides the relationship of the elements in the vertical groups, there are resemblances in several horizontal triads: iron (Fe), cobalt (Co), and nickel (Ni) are all ferromagnetic, for example.

The six metals of the platinum group consist of two horizontal triads: ruthenium (Ru), rhodium (Rh), palladium (Pd), and osmium (Os), iridium (Ir), platinum (Pt). The first subgroup (Ru, Rh, Pd) has atomic weights and specific gravities that are about half those of the second subgroup (Os, Ir, Pt). Furthermore, each metal of the first group has a member in the second group that seems more closely related to it; for example, Ru and Os, Rh and Ir, and Pd and Pt may be considered as pairs. All of these metals are chemically resistant and have high melting points. These six metals along with gold and silver are called the precious metals group and are also called the noble metals because of their

Table 1
The Periodic Table of the Elements

Light Metals IA														Nonmetals			Inert gases 0
H 1	IIA	Heavy Metals										IIIA	IVA	VA	VIA	VIIA	He 2
Li 3	Be 4	←Brittle metals→					←Ductile metals→				Low melting	B 5	C 6	N 7	O 8	F 9	Ne 10
Na 11	Mg 12	IIIB	IVB	VB	VIB	VIIB	VIII			IB	IIB	Al 13	Si 14	P 15	S 16	Cl 17	Ar 18
K 19	Ca 20	Sc 21	Ti 22	V 23	Cr 24	Mn 25	Fe 26	Co 27	Ni 28	Cu 29	Zn 30	Ga 31	Ge 32	As 33	Se 34	Br 35	Kr 36
Rb 37	Sr 38	Y 39	Zr 40	Nb 41	Mo 42	Tc 43	Ru 44	Rh 45	Pd 46	Ag 47	Cd 48	In 49	Sn 50	Sb 51	Te 52	I 53	Xe 54
Cs 55	Ba 56	* 57–71	Hf 72	Ta 73	W 74	Re 75	Os 76	Ir 77	Pt 78	Au 79	Hg 80	Tl 81	Pb 82	Bi 83	Po 84	At 85	Rn 86
Fr 87	Ra 88	† 89															
	*	La 57	Ce 58	Pr 59	Nd 60	Pm 61	Sm 62	Eu 63	Gd 64	Tb 65	Dy 66	Ho 67	Er 68	Tm 69	Yb 70	Lu 71	
	†	Ac 89	Th 90	Pa 91	U 92	Np 93	Pu 94	Am 95	Cm 96	Bk 97	Cf 98	Es 99	Fm 100	Md 101	No 102	Lr 103	

In general, it can be stated that the properties of the elements are periodic functions of their atomic numbers. It is in the vertical columns that the greatest similarities between elements exist. There are also some similarities between the A and B groups on either side of the VIII groups. For example, the Scandium group, IIIB, is in some respects similar to the Boron group, IIIA. The most important properties in the study of metallurgy that vary periodically with the atomic numbers are crystal structure, atomic size, electrical and thermal conductivity, and possible oxidation states.

SOURCE John E. Neely, *Practical Metallurgy and Materials of Industry*, Second Edition, Wiley, New York, copyright © 1984.

resistance to atmospheric conditions and to most acids. In general, crystal structures, melting points, densities, and other properties of elements are functions of the atomic number.

In the foregoing discussion, the electrons were spoken of as "moving around the nucleus like planets in orbit around the sun." We know now by the **uncertainty principle** that this cannot be true of the electrons. The exact orbit of an electron about the nucleus has been replaced by a mathematical function called an **orbital.** These are electronic energy levels to which quantum numbers have been assigned in which electrons occupy certain **shells.** Each orbital electron contains discrete amounts of energy. These energy packets are called **photons** (that is, a photon is a quantum of radiant energy). If a group of electrons becomes excited, they will jump to a higher energy orbit. If an electron returns to its original energy orbit or shell, it will release a photon of energy (such as heat) that is characteristic of that jump. The first basic shell can contain no more than two electrons, the second shell contains eight, the third shell contains eighteen, and the fourth shell contains thirty-two. Electrons in the outer shell of metals are called valence electrons and can be easily stripped off in order to form compounds and bonding arrangements. These electrons are responsible for the electrical and chemical behavior of elements.

Metals and Nonmetals

As can be seen in Table 1, approximately three-quarters of all elements are considered to be metals. Chemical elements may be roughly classified into three groups: metals, nonmetals, and inert gases. Some nonmetals, such as carbon, germanium, and silicon, resemble metals in one way or another and are sometimes called **metalloids.** They can have a metallic luster and crystalline microstructure, and they can be semiconductors of electricity, but they cannot be deformed without fracturing because they are brittle, as are all nonmetals. Some of the properties that an element must exhibit to be considered a metal are:

1. Ability to donate electrons to form a positive ion.
2. Crystalline structure.
3. High thermal and electrical conductivity.
4. Ability to be deformed plastically.
5. Metallic luster or reflectivity.

Metals are the most important of the elements from a metallurgical standpoint, but plastics, rubbers, and other synthetic products are produced for the most part from nonmetals. Solid state electronic components are produced from metalloids such as germanium and silicon.

CRYSTALLINE STRUCTURE

When metals cool from the molten state and solidify, they form a crystalline structure called a **space lattice.** If imaginary lines are drawn connecting the centers of the atoms together, it is called a **lattice structure** (Figure 7). A recent development has created an exception to the foregoing statement. Metals can now be rapidly quenched from the liquid to the solid state so that there is no opportunity for a crystal structure to develop. These are appropriately called **metallic glasses** or amorphous metals. Only a small amount of this metal is being produced at present. Since it has very good electrical and magnetic properties, it is useful for making transformer cores. However, when metals cool at normal rates, they always form a space lattice which consists of orderly rows of atoms.

Interatomic Distances

Since similar electrical charges tend to repel and unlike charges attract, the electron-filled outer shells of the metallic ions tend to have a repulsive force. If it were not for the free valence electrons vibrating between the atoms to produce forces of attraction, metal atoms would fly apart. However, there is a distinct balance of forces that makes up metallic bonding. These atoms maintain a distance at a location where these two forces are in equilibrium. These interatomic distances can be altered by external forces, such as thermal energy and electrical or mechanical forces. This is the reason materials expand when heated and contract when cooled. The distance between atoms is different for different materials because it is related to the number of shells and the number of valence electrons. Atoms of different elements have different sizes. For example, an atom of

Figure 7
Lattice structure of atoms in solids.

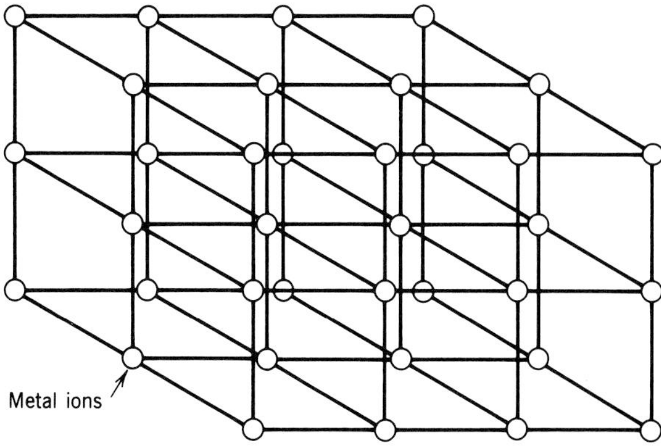

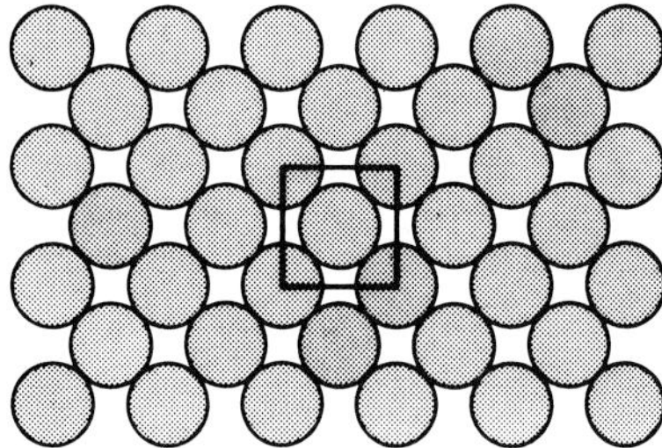

Figure 8
Body-centered cubic unit cell shown in the lattice (White, Neely, Kibbe, Meyer *Machine Tools and Machining Practices*, Vol. II, © 1977 John Wiley & Sons, Inc.).

iron is much larger than an atom of carbon. Removal of electrons from the outer shell, as is done in a metal ion, reduces its size.

Unit Cells

A space lattice can be described as having repeating unit building blocks throughout its structure. These are called **unit cells** (Figure 8). Metals solidify into 14 known crystal structures (Table 2), but most of the commercially important metals solidify into one of three types of lattice structures (Figure 9). They are body-centered cubic (BCC), face-centered cubic (FCC), and hexagonal close-packed (HCP).

If we assume that atoms are like tiny balls that almost touch their nearest neighbors, two common structures would look like that in Figure 10. The metallic bond, to have maximum plasticity and other needed properties, is best when the atoms are close-packed as in

Table 2
Crystal Structures of Some Common Metals

Symbol	Element	Crystal Structure
Al	Aluminum	FCC
Sb	Antimony	Rhombohedral
Be	Beryllium	HCP
Bi	Bismuth	Rhombohedral
Cd	Cadmium	HCP
C	Carbon (graphite)	Hexagonal
Cr	Chromium	BCC (above 26° C, 78.8° F)
Co	Cobalt	HCP
Cu	Copper	FCC
Au	Gold	FCC
Fe	Iron (alpha)	BCC
Pb	Lead	FCC
Mg	Magnesium	HCP
Mn	Manganese	Cubic
Mo	Molybdenum	BCC
Ni	Nickel	FCC
Nb	Niobium (columbium)	BCC
Pt	Platinum	FCC
Si	Silicon	Cubic, diamond
Ag	Silver	FCC
Ta	Tantalum	BCC
Sn	Tin	Tetragonal
Ti	Titanium	HCP
W	Tungsten	BCC
V	Vanadium	BCC
Zn	Zinc	HCP
Zr	Zirconium	HCP

SOURCE John E. Neely, *Practical Metallurgy and Materials of Industry*, Second Edition, Wiley, New York, copyright © 1984.

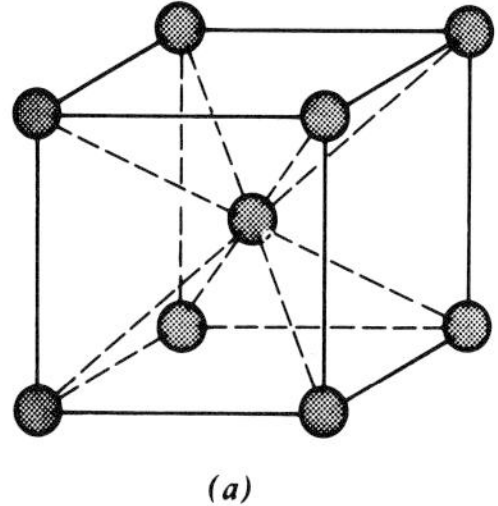

(a)

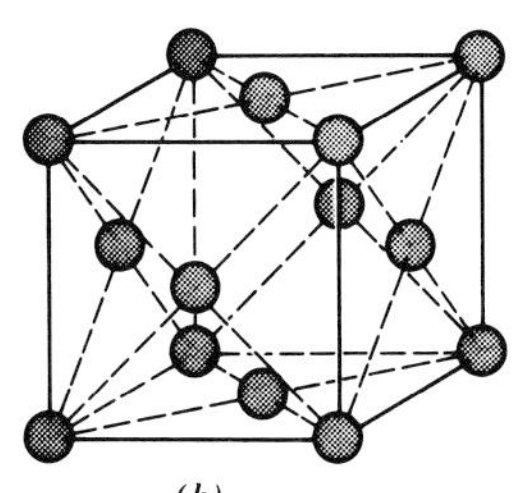

(b)

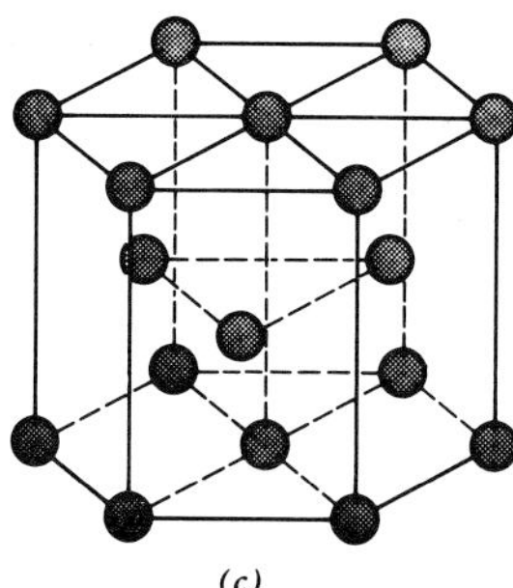

(c)

Figure 9
Three common unit cells of the lattice structure are (*a*) body-centered cubic, (*b*) face-centered cubic, and (*c*) hexagonal close-packed (John E. Neely, *Practical Metallurgy and Materials of Industry* 2 Ed., © 1984 John Wiley & Sons, Inc.).

(a)

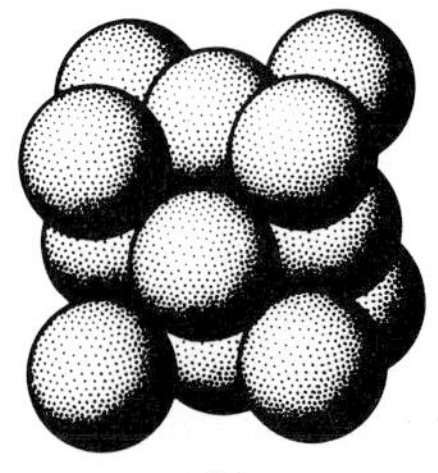

(b)

Figure 10
Model of atoms in unit cells. (*a*) Body-centered cubic (BCC). (*b*) Face-centered cubic (FCC) (White, Neely, Kibbe, Meyer, *Machine Tools and Machining Practices*, Vol. II, © 1977 John Wiley & Sons, Inc.).

body-centered cubic structures that form a cube with one atom in its center. Even better for plastic deformation is the face-centered cubic structure. The hexagonal close-packed unit structure does have some planes that are close-packed, but in another direction the atoms are spaced farther apart than in BCC and FCC, causing these metals to have poor formability.

Some materials have two or more of these structures under different conditions such as temperature change. Such a material is called **allotropic.** For example, the three allotropic forms of carbon are amorphous (charcoal, soot, and coal), graphite, and diamond. Iron also exists in three allotropic forms which are temperature dependent (Figure 11). (The term *iron* refers to elemental or pure iron unless specified as wrought or cast.) Below 1300° F (704° C) iron is BCC, called ferrite or alpha iron which is ferromagnetic. It becomes nonmagnetic at the Curie temperature (1440° F or 782° C), sometimes called beta iron. Between 1666° F (908° C) and 2554° F (1401° C) iron is FCC, called austenite or gamma iron. Between 2554° F (1401° C) and 2800° F (1538° C) iron is also BCC, called delta iron, which is of little importance in manufacturing. These three states of iron are called **phases;** the ferrite and austenite phases being of great importance in heat treating and manufacturing.

Note on the cooling curve in Figure 11 that the straight horizontal lines represent the time required to effect the transformation from one phase to another. Under cooling conditions heat is released during the transformation and, when being heated, heat energy is absorbed during the transformation.

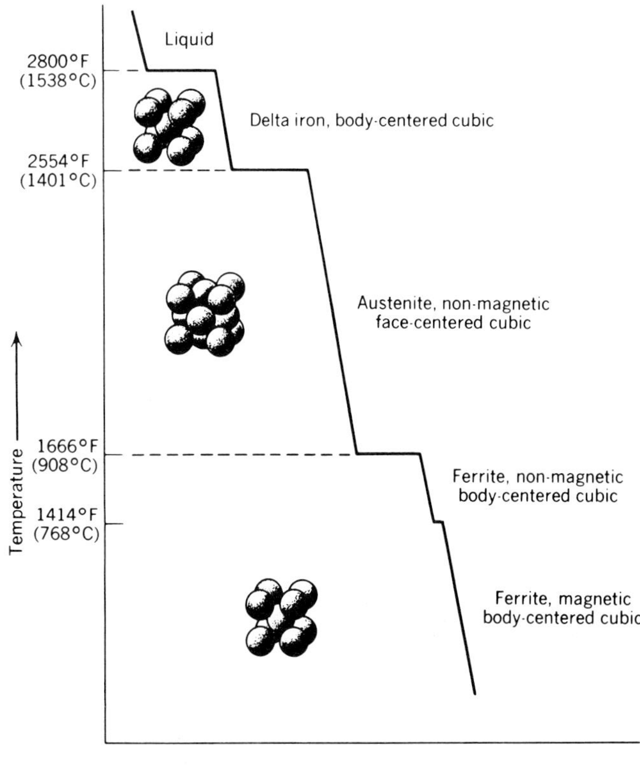

Figure 11
Cooling curve for pure iron. As iron is cooled slowly from the liquid phase (or heated from solid to liquid), it undergoes these allotropic transformations (White, Neely, Kibbe, Meyer, *Machine Tools and Machining Practices*, Vol. II, © 1977 John Wiley & Sons, Inc.).

Dendrite Formation

As solidification of the molten metal begins, the arrangement of the space lattice takes on the characteristic pattern for that metal. Each unit cell builds on another to form crystal growths that resemble pine trees. These structures, called dendrites (Figure 12), begin to grow from seed crystals or nuclei particles such as impurities. Aluminum is usually put into molten steel to make it fine grained and to eliminate dissolved oxygen in the molten metal. The aluminum oxide particles are so numerous that many crystals start growing from these nuclei and consequently the metal contains more grains than it normally would. Also, slow cooling promotes large grains and rapid cooling promotes smaller grains.

Figure 12
Dendrite growth in metal as shown in a photomicrograph.

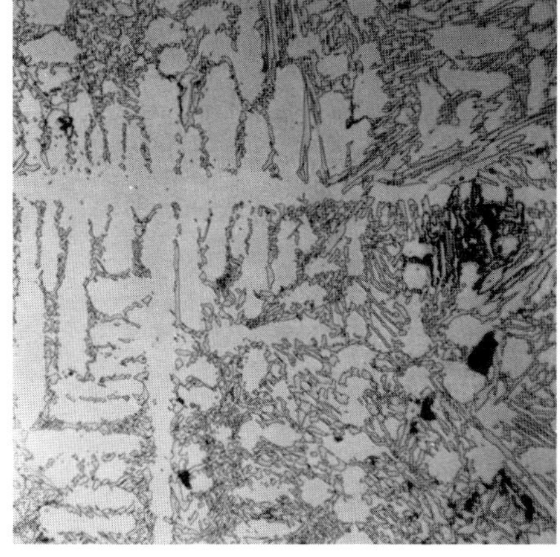

Grain Boundaries

The dendrites grow outward to form a single crystal or **grain** until it meets another dendrite crystal. The places where these grains meet are called **grain boun-**

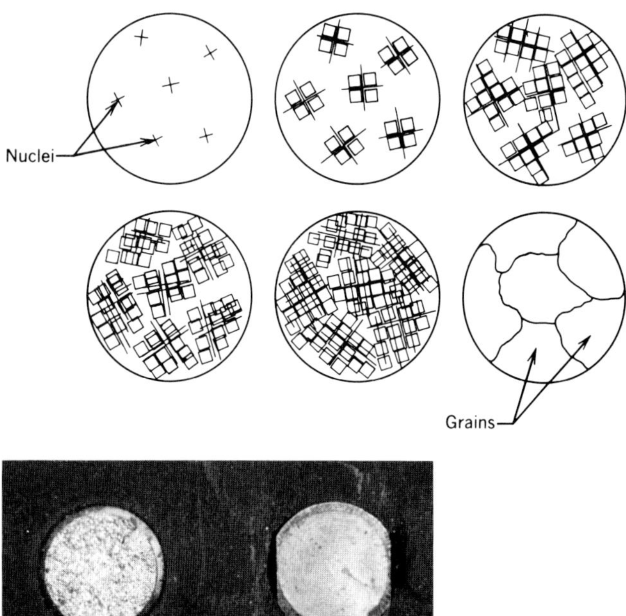

Figure 13
The formation of grains during solidification of molten metal (White, Neely, Kibbe, Meyer, *Machine Tools and Machining Practices*, Vol. II, © 1977 John Wiley & Sons, Inc.).

Figure 14
Broken sections of steel showing its crystalline structure. The one on the left is coarse and can be seen clearly. The one on the right is fine grained and requires magnification to see the grains.

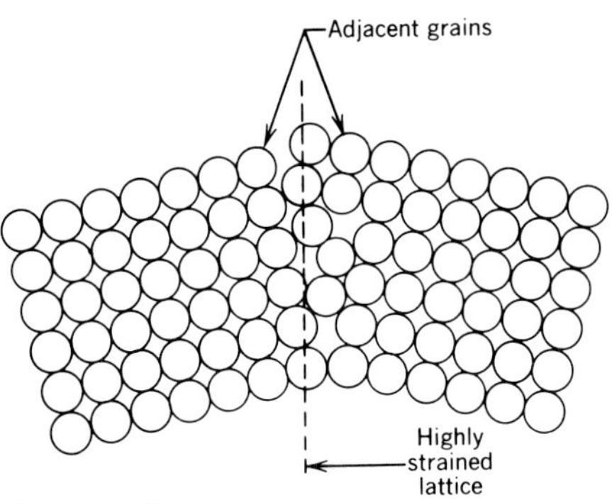

Figure 15
The grain boundary is in a highly strained condition (White, Neely, Kibbe, Meyer, *Machine Tools and Machining Practices*, Vol. II, © 1977 John Wiley & Sons, Inc.).

daries. Metals therefore become polycrystalline; that is, the axes of the lattice structures are oriented at random in each grain depending on how the dendrite began to grow (Figure 13). All metals when normally cooled are composed of these tiny grain structures which can be seen with the aid of a microscope when a specimen is cross-sectioned, polished, and etched. Often the grain structure can be seen with the naked eye as small crystals in a broken section of a piece of metal (Figure 14). Some metals, when slowly cooled, have extremely large grains. Large grains of zinc can often be seen on galvanized articles.

Since each crystal lattice or dendrite grows in different directions, it can be seen that the grain boundaries do not form a continuation of the lattice, but instead the atoms are jammed together in a misfit pattern (Figure 15). Although the nature of the atomic forces in the grain boundaries is not entirely understood, it is generally assumed to be an interlocking border in a highly strained condition. Grain boundaries are only about one or two atoms wide, but their strained condition causes them to etch darker than the less strained grain, allowing them to be seen with a microscope. Grain size in metals is very important in manufacturing products as will be explained later in this chapter.

PLASTIC DEFORMATION IN METALS

Two of the most important properties of metals for manufacturing purposes are **elastic deformation** and **plastic deformation. Stress** is the material's resistance to the applied load or force. When metals are placed under tensile, torsion, or compression stress, a slight elongation or compression takes place in the crystal lattice. This movement is called **strain.** The distortion of the crystal lattice is illustrated in Figure 16. This

Figure 16
Distortion of crystal lattice with various kinds of stresses. (*a*) Unloaded. (*b*) Tension. (*c*) Compression. (*d*) Shear.

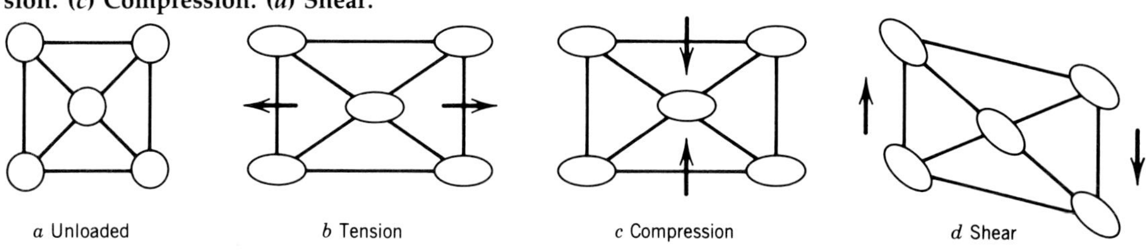

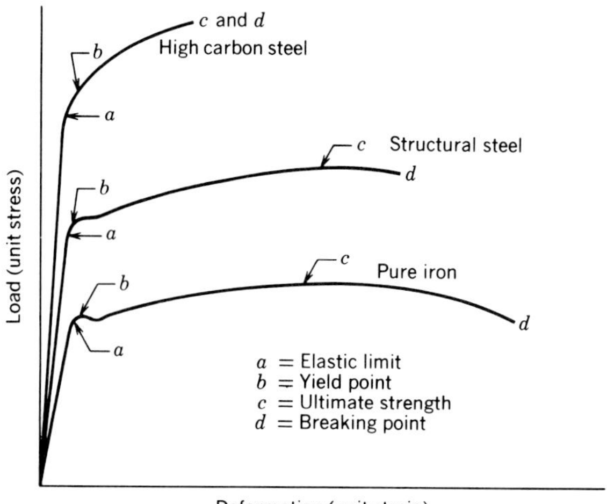

Figure 17
Load-deformation (stress–strain) diagram for a ductile steel (White, Neely, Kibbe, Meyer, *Machine Tools and Machining Practices*, Vol. II, © 1977 John Wiley & Sons, Inc.).

elastic deformation is not permanent. As soon as the stress is removed, the structure returns to its former shape. Also, this elongation or compression in one direction will produce an opposite change at 90 degrees to that force. That is, a piece of metal will become thinner if it is stretched lengthwise. This ratio of movement at right angles to the applied force is called **Poisson's ratio.**

Plastic deformation is one of the most useful characteristics of metals. Forging, drawing, forming, extruding, rolling, stamping, and pressing all involve plastic deformation. Plastic deformation in metals can take place only at a stress or load higher than the elastic limit (Figure 17). As the applied load is increased, the atoms must slide over each other to produce a permanent change in shape or break the atomic bonds, resulting in total rupture. The behavior of metal crystals under load depends on five things:

1. The spacing between planes of atoms.
2. Interatomic bond strength.
3. The density of atoms in various planes.
4. The lattice type.
5. Lattice irregularities (vacancies and discontinuities).

Plastic deformation tends to be favored along planes of highest atomic density and the greatest parallel separation (Figure 18). It can be seen in this figure that the more dense or closely packed the atoms are, the better chance there is for slip along planes or rows of atoms. Slip within grains is roughly analogous to a deck of cards, each card sliding a small amount (Figures 19 and 20). This movement of atoms tends to elongate the

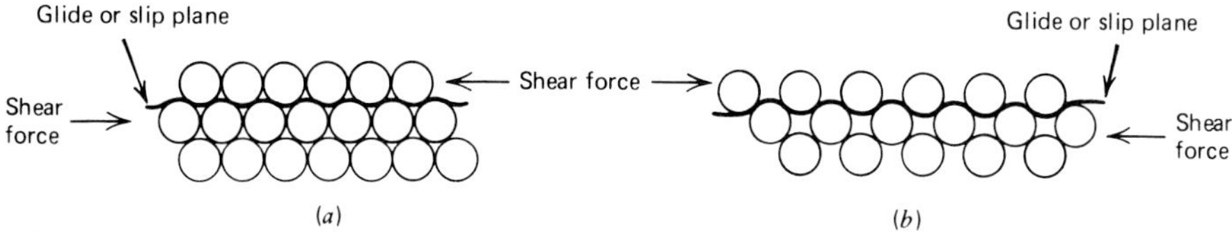

Figure 18
Rows of atoms can glide along slip planes as shown. BCC and FCC have rows of atoms closely spaced as in (*a*), making movement easier than in (*b*). HCP type metals are less ductile since many slip planes as in (*b*) are spaced closer (Neely/*Metallurgy* 2 Ed., © 1984 John Wiley & Sons, Inc.).

Figure 19
When stress is applied to a metal grain above its elastic range, slip occurs and the grain flattens and becomes elongated. Slip planes are often rotated when stress is applied (Neely/*Metallurgy* 2 Ed., © 1984 John Wiley & Sons, Inc.).

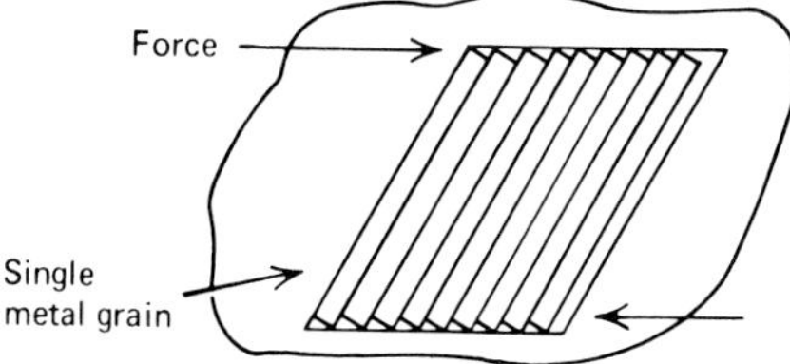

Figure 20
Slip planes as they appear in a micrograph resulting from the cold working of low carbon steel (Metals Handbook, Vol. 7, 8th ed., American Society for Metals, 1972, p. 135. With permission).

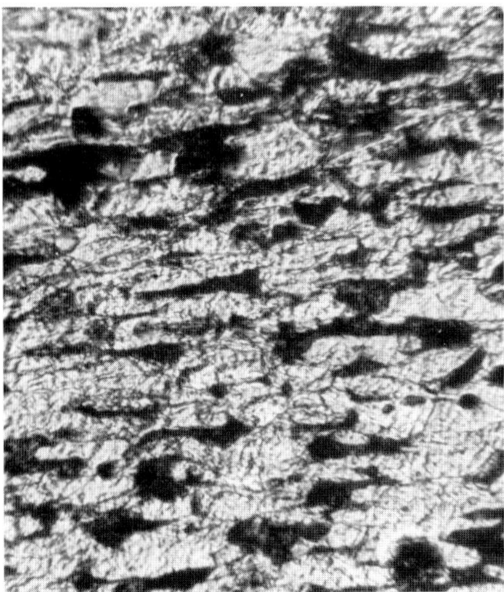

Figure 21
Micrograph of flattened grains in drawn wire showing partial recrystallization of the ferrite (white) grains.

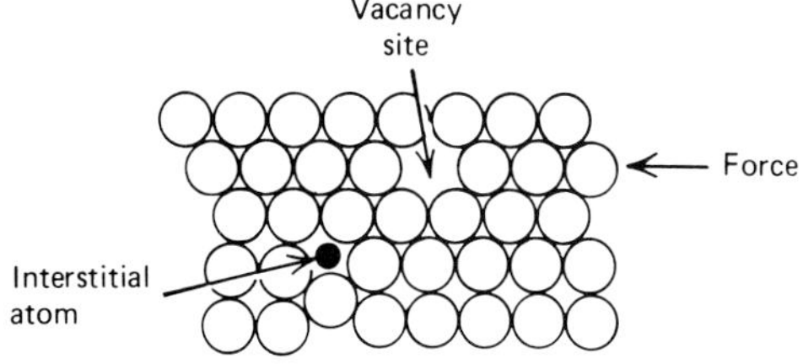

Figure 22
Rows of atoms move, one atom at a time, to fill in a vacancy or space created by dislocated atoms. Interstitial atoms jam up the slip planes of atoms, making the metal that contains them harder and less ductile (Neely/*Metallurgy* 2 Ed., © 1984 John Wiley & Sons, Inc.).

grain and flatten it (Figure 21). Each atom shifts, one at a time, and the process of slip is progressive, not quite like cards sliding over each other. It is believed that imperfections in metals, such as vacancies (spaces where there are no atoms), dislocations or distortions of the lattices, and impurity atoms are responsible for the phenomenon of slip that allows for permanent deformation of metals without rupture (Figure 22). When a vacancy site exists, atoms move one at a time, moving the vacancy site or dislocation along the lattice until there is no more room to move. When that happens in all of the grains, **work hardening** results, so that the material is no longer plastic and it becomes more brittle. If the load increases beyond the ultimate strength of the material when it is no longer plastic, the result is failure as a brittle fracture.

Work hardening or strain hardening is a problem to be dealt with in manufacturing where cold working processes are involved. After a certain amount of cold working, a metal must be subjected to a process called **annealing** to restore its grain structure to a softer, more plastic condition. This is called **recrystallization.** Annealing and recrystallization will be further discussed in Chapter 4, "Heat Treatment of Metals."

Grain Size

It can be seen why large grain metals have a greater capacity for slip, to be deformed, and to have more plasticity than fine grain metals. A fine grain steel, for example, is stronger than an equivalent coarse grain steel; that is, it can withstand greater stresses without being permanently deformed. Since each grain has a random orientation (direction of dendrite growth), slip in each grain ends at the grain boundary; small grains have a lesser capacity for slip than large grains do. For this reason, grain size is an important principle to consider when describing the nature of metals. Steel manufacturers provide mechanical properties charts and tables for each carbon, alloy, or tool steel which give, among other specifications, the grain size in terms of standard grain sizes (Figure 23). Ordinary machine steel may have a grain size of about 6 to 8. When a specimen of steel is fractured, the broken cross section often reveals the grains without magnification. These can be

Figure 23
Standard grain size numbers. The grain size per square inch at 100× (Courtesy of Bethlehem Steel Corporation).

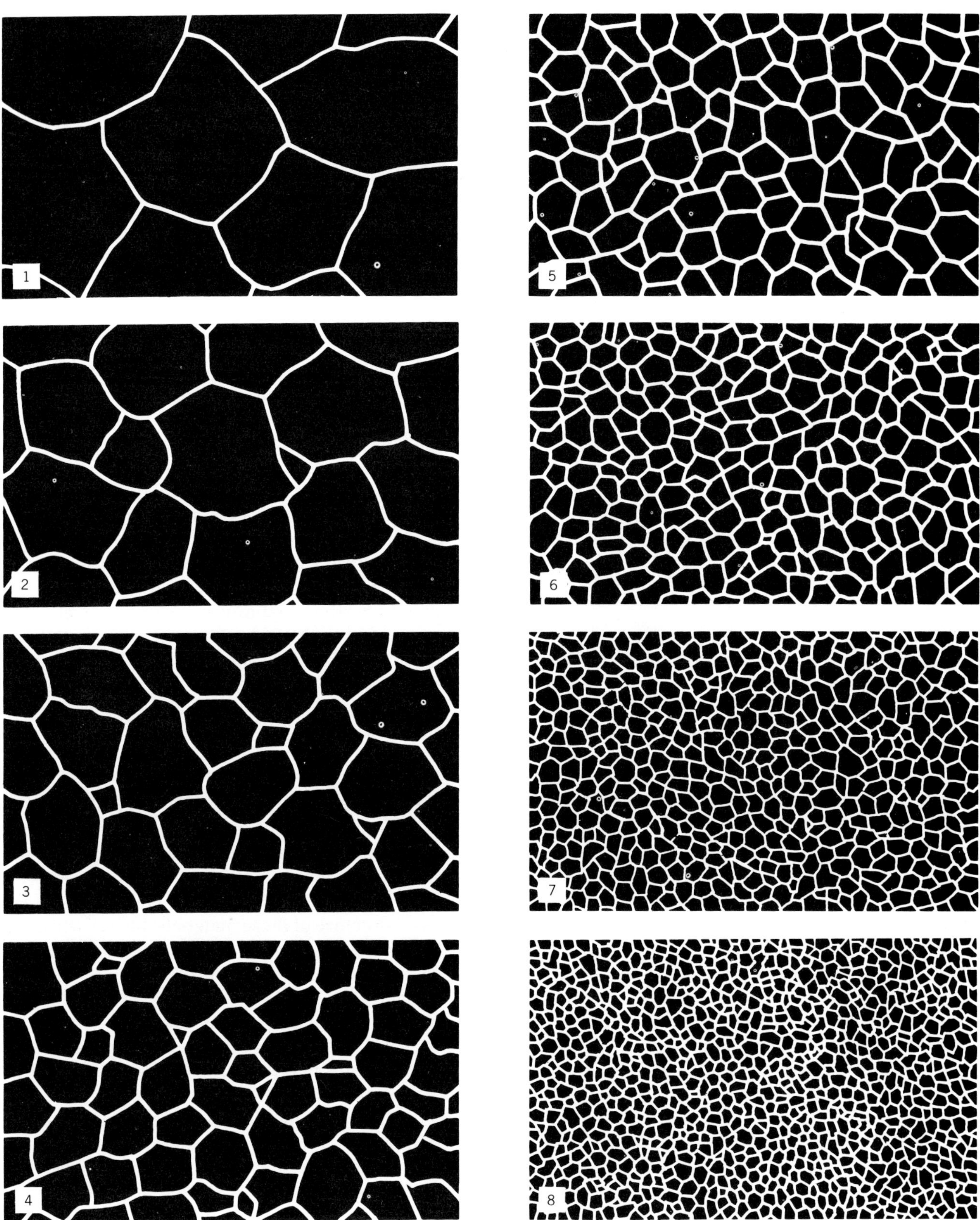

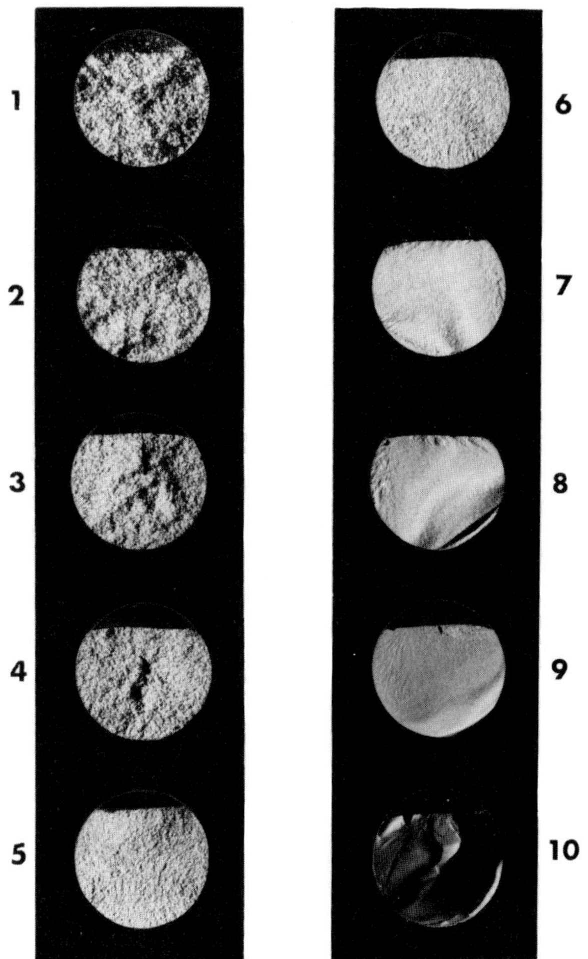

Figure 24
Shepherd grain size fracture standards. The fracture grain size test specimen can be compared visually with this series of 10 standards. Number 1 is the coarsest and number 10 is the finest fracture grain size (Photo courtesy of Republic Steel Corporation).

compared to the **Shepherd grain size fracture standards;** number 1 is coarsest and number 10 is finest (Figure 24).

Hardening Metals

Metals may also be hardened by locking up the slip planes with atoms of other elements or compounds, by alloying, or by trapping very small atoms such as carbon in the interstices (spaces or voids) of the iron lattice. When carbon steel is hardened, it is quenched (rapidly cooled) from a high temperature of about 1500° F (816° C) to room temperature. A plain carbon tool steel having about 1 percent carbon can contain all of the carbon in the FCC phase of iron at 1500° F (816° C) in the interstices of the lattice because there is more open space than in BCC. There is room for only a very tiny percentage of carbon in the BCC structure. Therefore, when the steel is suddenly quenched to room temperature, there is no time for the carbon atoms to leave the interstices and form a compound of iron carbide (Fe_3C) which is what happens when steel is slowly cooled. The result of the sudden cooling is that the trapped carbon atoms in the BCC structure are under a tremendous stress that causes the crystal to elongate to form a **body-centered tetragonal** unit crystal formation (Figure 25). This structure is called **martensite** and it is extremely hard and brittle. Hardening and heat treating metals is also covered in Chapter 4, "Heat Treatment of Metals."

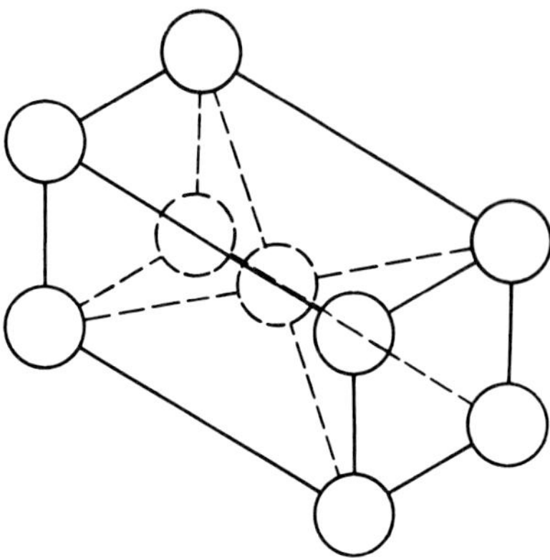

Figure 25
Body-centered tetragonal structure. This is the distorted form of the unit cell (Neely/*Metallurgy* 2 Ed., © 1984 John Wiley & Sons, Inc.).

PLASTICS, ELASTOMERS, AND CERAMICS

Of all the nonmetallic materials used in manufacturing, plastics, elastomers, and ceramics are probably the most important to understand. No attempt will be made here to go into great detail about the chemical, molecular, or crystal structure of these materials; but we will cover what is necessary to gain an understanding of their behavior for manufacturing purposes.

Crystalline Structure of Plastics

Soluble substances can be divided into two classes, the **crystalloids** and the **colloids.** The crystalloids, such as table salt or sugar dissolved in water, easily pass through a semipermeable membrane, but colloids are thick, viscous substances, like table jelly or gelatin, which will not pass through a semipermeable membrane. This colloidal behavior is seen in the synthetic plastics and rubber and is caused by extremely large molecules.

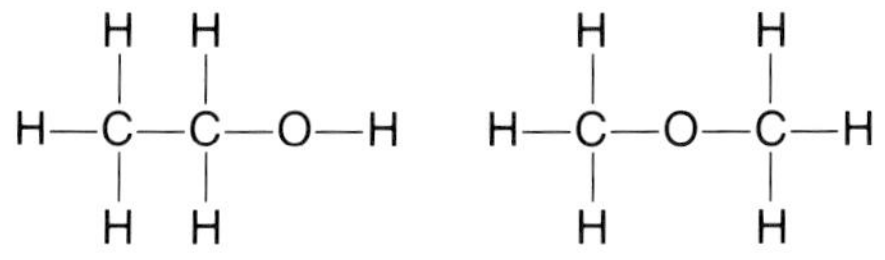

Figure 26
An example of structural isomerism.

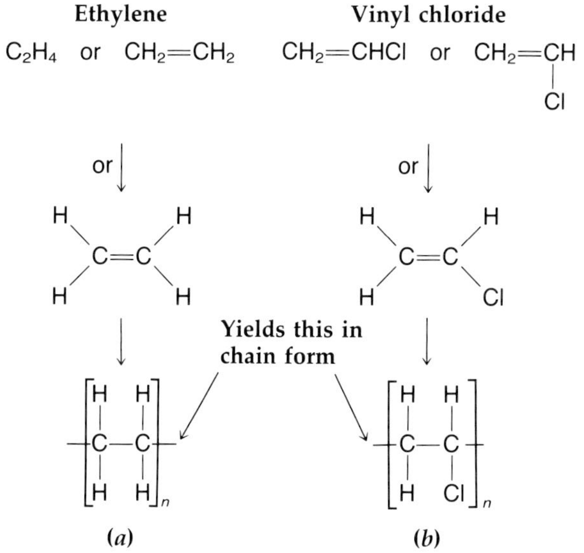

Figure 27
A monomer unit can become a branch source unit if it has one or more hydrogen atoms replaced by other elements. The subscript *n* indicates the indefinite length of the polymer.

These macromolecules do not form crystals in the same way as in metals, but many plastics do have crystalline forms. Some polymers such as polyethylene exhibit definite crystalline formations. Whereas others such as polystyrene with their long chains are amorphous; that is, under a microscope they might look like a bowl of spaghetti with no apparent order.

In organic chemistry and in the study of structural polymers, the phenomenon of **isomerism** is of some importance. Some compounds having the same number and kind of atoms can join to form different structures having completely different properties. This is roughly analogous to various crystalline phases in metals (allotropism). Many plastics are isomers. A simple example of structural isomerism is the way in which the atoms C_2H_6O can be arranged as ethyl alcohol and dimethyl ether (Figure 26).

Polymers

Natural organic materials such as leather, wood, cotton, and natural rubber are typically **polymers,** made up of long-chain molecules. Early discoveries in organic chemistry that produced substitutes for organic materials were not widely accepted and were considered to be poor replacements. For example, in 1866 John Wesley Hyatt first produced a synthetic material he called **celluloid** by combining camphor with nitrocellulose (gun cotton) that was made with slightly less nitric acid than usual. This new product was used as a substitute for the ivory that was obtained from the tusks of animals and was quite expensive. Soon products such as combs, handles, and billiard balls were made of this new synthetic material. Celluloid had two bad qualities; it had a tendency to burn rapidly when ignited and it deteriorated (discolored and cracked) with age. Movie film was first made of celluloid (cellulose nitrate) but later a nonflammable cellulose acetate was developed for movie film. Since this early discovery, a wide range of plastics, synthetic rubbers, textile fibers, adhesives, and paints have been developed by chemists. These are considered by some to be among our greatest scientific achievements in modern times.

These substances, often called high polymers, have molecular weights that are often hundreds of times greater than ordinary materials. These giant molecules are made up of small molecules called monomers (Figure 27). These are the building blocks from which the polymer chain is built (Figure 28). **Polymerization** is a chemical process in which many small molecules are linked together to form one large molecule. Celluloid was not the first polymerized plastic since the basic component of cellulose nitrate is cellulose, which in cotton or wood is already a high polymer. The first high polymer to be made from small molecules of raw materials was the strong, hard plastic called **bakelite** which was developed by the Belgian chemist Leo Baekeland in 1908. He was able to join the small molecules of phenol and formaldehyde to produce large

Figure 28
These are the molecule forms (structural formulas) that serve as monomers to produce polyethylene and polyvinyl chloride (PVC) as shown here. Thus, giant chain molecules may be formed from monomers as shown in Figure 29.

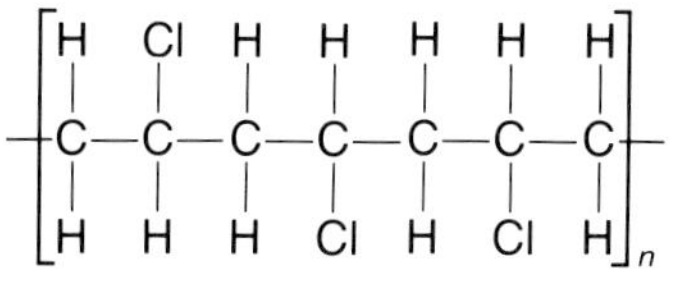

CH_2O

Formaldehyde molecule

C_6H_5OH

Phenol molecule

Bakelite polymer

Plus H_2O

Figure 29
The basic chemical structure of bakelite as it is produced from formaldehyde and phenol monomers.

molecules. Its basic structure is shown in Figure 29. There are three basic types of polymerization: **addition, copolymerization,** and **condensation.**

Addition polymerization takes place when similar monomers join to form a chain (Figure 30). This is also called linear polymerization. Copolymerization takes place when two or more different kinds of monomers are combined. Condensation polymerization is a process of combining long-chain molecules to form more complicated chains. When the reaction takes place there is a residue, usually water. See Figure 29. These chains when coiled exhibit plasticity, but some, when they are cross-linked, lose plasticity and become elastic (Figure 31). Rubber is an example. Natural rubber latex, when combined with sulfur and then heated, changes the nature of crude rubber. This process of cross-linking rubber latex, a natural polymeric substance (Figure 32), is called **vulcanization.**

Natural rubber consists of long chains that must be broken down to short chains by kneading or mastica-

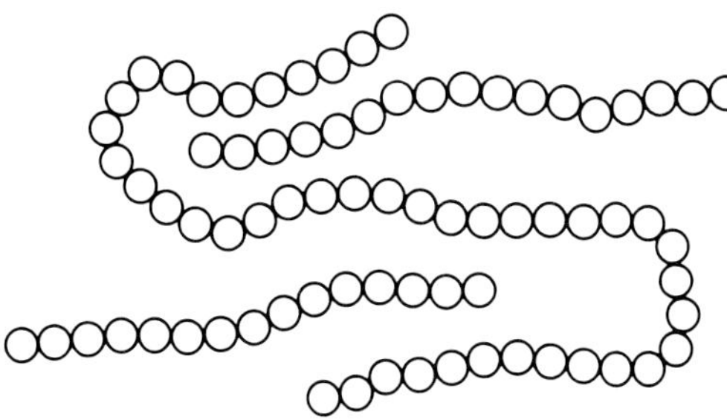

Figure 30
The spheres in this long-chain molecule represent atoms or monomers. Some chains are rigid, as found in heat-resisting materials, whereas others are elastic as in polyethylene, nylon, and cured rubber.

Figure 31
Cross-linking is only one of several ways that polymers can be strengthened.

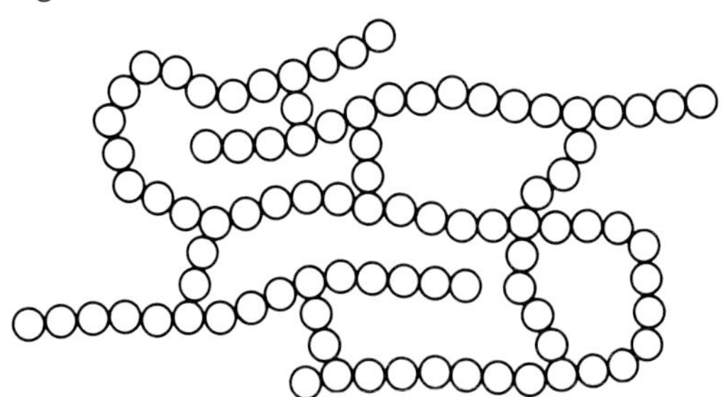

Figure 32
(*a*) Natural rubber long-chain molecules; wire model, unstretched. (*b*) Vulcanized rubber, cross-linked.

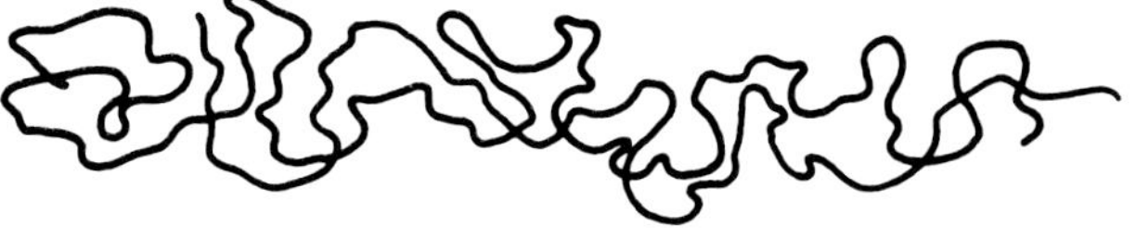

(*a*)

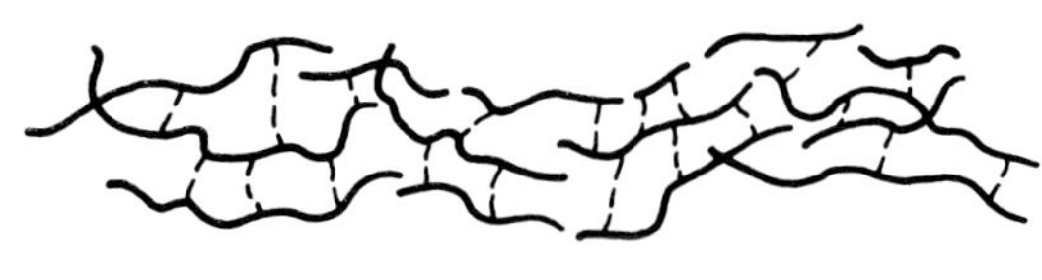

(*b*)

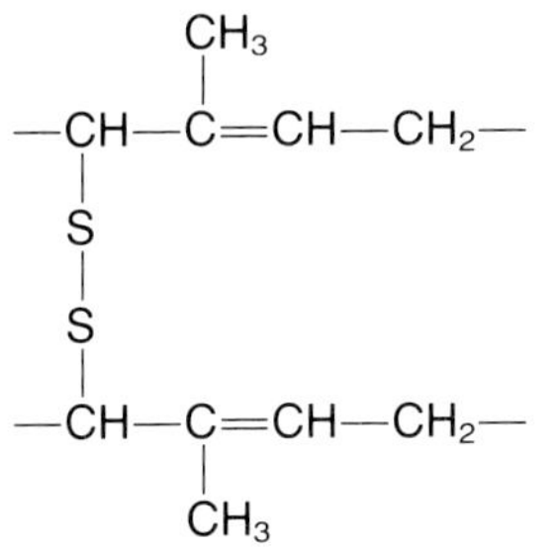

$$
\begin{array}{l}
\qquad\quad CH_3 \\
\qquad\quad | \\
-CH-C{=}CH-CH_2- \\
\;\; | \\
\;\; S \\
\;\; | \\
\;\; S \\
\;\; | \\
-CH-C{=}CH-CH_2- \\
\qquad\quad | \\
\qquad\quad CH_3
\end{array}
$$

Figure 33
General formula for vulcanized rubber.

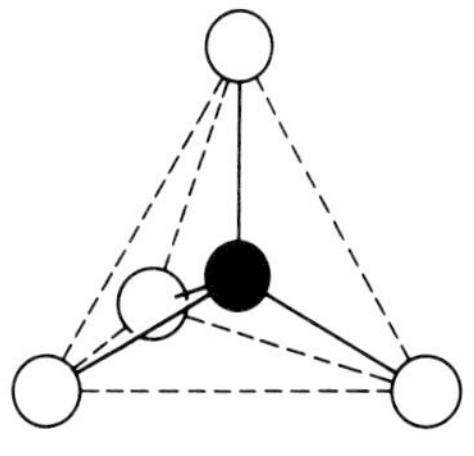

Figure 35
Unit structure of silicon dioxide (SiO_4). The darkened circle is silicon and the plain circles represent oxygen atoms (Neely/*Metallurgy* 2 Ed., © 1984 John Wiley & Sons, Inc.).

tion between hot rolls. Chemically, natural rubber is a polyisoprene, a hydrocarbon that has the empirical formula C_5H_8. A general formula of vulcanized rubber is shown in Figure 33.

Rubber can be made hard, soft, tacky, or resistant to oxidizers by using various additives. Accelerators are used to speed up the process of vulcanization which is a relatively slow process.

A great variety of synthetic rubbers have been developed. Many of these have properties different from those of natural rubber and are superior for some specific applications. Most of these synthetic rubbery materials are derived from petrochemicals (petroleum distillates) such as acetylene, butadiene, isoprene, and chloroprene (Figure 34). These chemicals and their derivation will be further discussed in Chapter 5.

Ceramics

The term *ceramics* includes a wide diversity of products. Clay bricks, drain pipes, insulators, spark plugs, synthetic diamonds, abrasives, glass, building stone, and concrete are among the many ceramic materials we use today.

The basic ingredients of clay are alumina and silica, that is, hydrated silicate of alumina ($Al_2O_3 \cdot 2SiO_2 \cdot 2H_2O$). Most clays contain impurities such as lime, iron oxide, magnesium, and potash. These impurities impart colors to the natural clay. Ferric oxide produces the red color seen in many clays. Shale and slate are two forms of clays that contain many impurities. Kaolin is a white clay in its pure state. It is used for the manufacture of chinaware, porcelain, firebrick, paper, rubber, and paint pigment.

Silicate structures give fired clays their hardness and stability (Figure 35). When linked together in a chain, the unit structures become Si_4O_7 (Figure 36). When these structures are combined with alumina, they become clay, a very plastic substance when wet, and rigid when dried. Clay is composed of very small crystalline, flat or platelike grains that tend to be stacked like packs

Figure 34
Rubbers and elastomers.

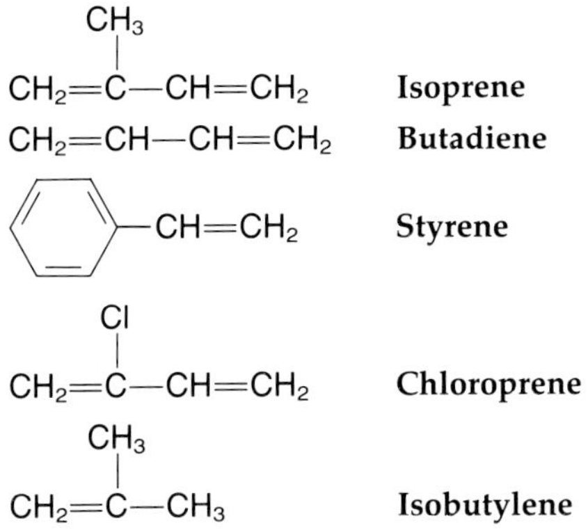

Formula	Name
$CH_2{=}C(CH_3){-}CH{=}CH_2$	**Isoprene**
$CH_2{=}CH{-}CH{=}CH_2$	**Butadiene**
$C_6H_5{-}CH{=}CH_2$	**Styrene**
$CH_2{=}C(Cl){-}CH{=}CH_2$	**Chloroprene**
$CH_2{=}C(CH_3){-}CH_3$	**Isobutylene**

Figure 36
Silicate units can link through any of the oxygen atoms as shown. This is one way in which a chain of silica tetrahedra can be joined. The dark circles represent silicon atoms and the plain circles are oxygen atoms (Neely/*Metallurgy* 2 Ed., © 1984 John Wiley & Sons, Inc.).

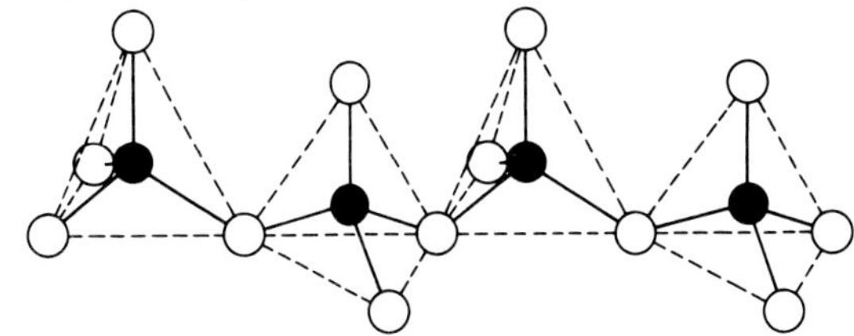

Figure 37
Platelike grains are like decks of cards. (*a*) These are wet and plastic. (*b*) These are dried, hard, and brittle.

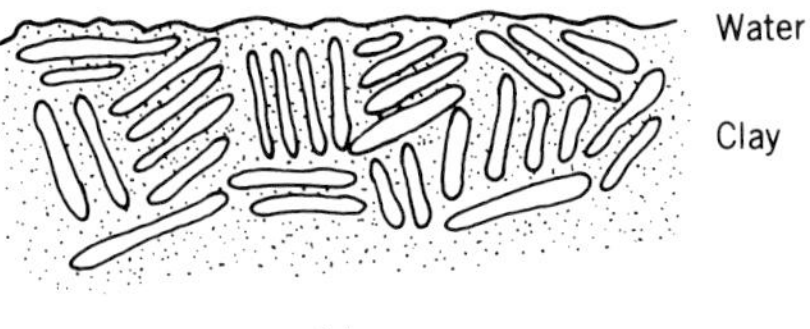

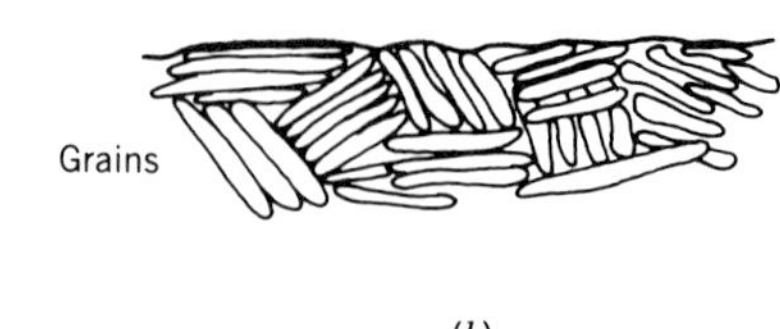

of playing cards, each pack oriented in a different direction. Each crystal plane is separated and surrounded by water which acts as a lubricant and a bond, allowing the grains to slide easily over one another (Figure 37). When the water is dried out as in sun-dried bricks, the lubrication is gone and the material becomes rigid but with low tensile and compressive strength. Clay in this state can easily be wetted and brought back to its plastic state. However, if it is fired, it loses that ability and it becomes somewhat stronger and quite brittle. The plates of hydrated silicate of alumina when fired combine to form new crystalline phases or glasses that permeate the entire structure, bonding the new crystals together into one fused mass.

Glass

The basic ingredient of glass is quartz (silica) SiO_2. When quartz is cooled, it forms a crystalline structure that is brittle and semitransparent. Glass is slow cooled to keep it from crystallizing and it contains other substances such as lime and sodium. The molecular structure of glass is amorphous, like a liquid; in fact, glass is considered to be a solid liquid. The raw material is silica sand which is quite plentiful, making glass a relatively inexpensive material.

Portland Cement

Concrete that is used to make roads, sidewalks, dams, buildings, and other permanent structures is composed

Figure 38
The hydration of portland cement. (*a*) Schematic of the formation of silica gel in the presence of water, 1 day. (*b*) Schematic showing fibers beginning to grow, 7 days. (*c*) Electron micrograph showing the growth of fibers at 7 days (3000 ×). (*d*) Schematic showing the final development of the strong fibers, 28 days. (*e*) Electron micrograph of completely hydrated portland cement at 28 days (5000 ×) (c and e are SEM photographs by Hugh Love, Portland Cement Association, Skokie, IL).

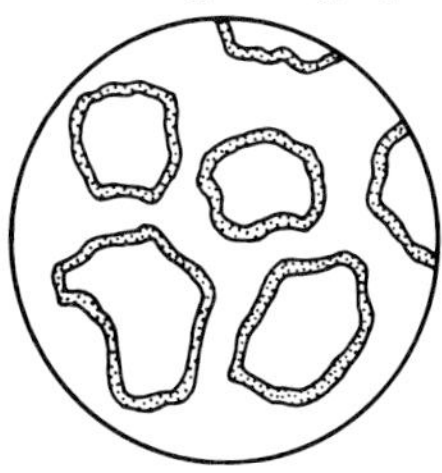

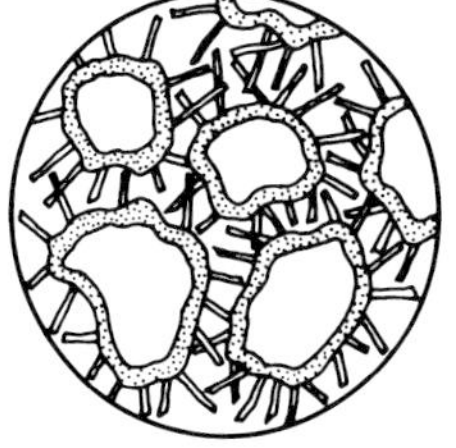

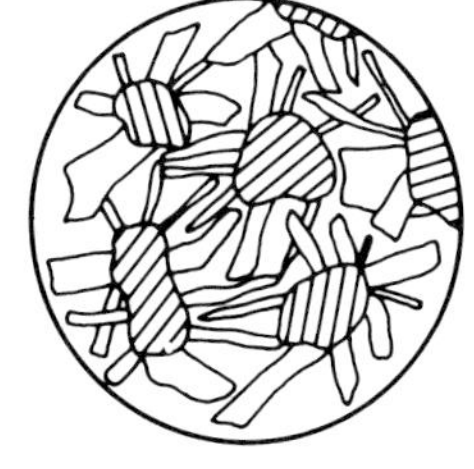

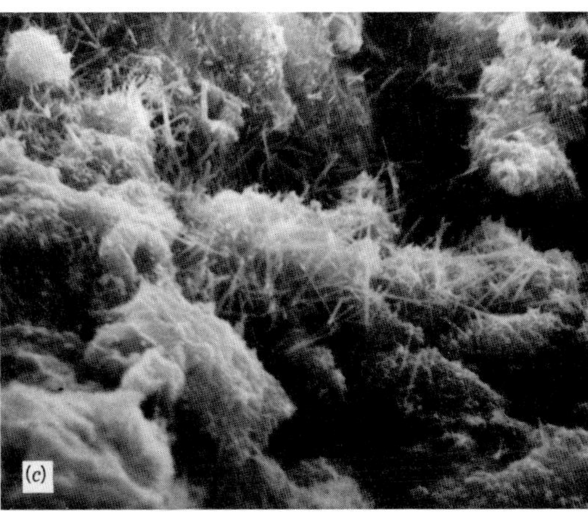

of portland cement, sand, and graded gravel. The cement when combined with water hardens to a rocklike consistency, binding the sand and gravel into one homogenous mass. This hardening takes place without the presence of air or drying out. In fact, it must be kept from drying out during the hardening process. This process is called **hydration.** Portland cement is basically a calcium silicate mixture predominantly containing tricalcium and dicalcium silicates. These materials are basically derived from shales, clays, and limestone with small amounts of iron oxide and dolomite. The alumina content of clay has an effect on the long-term strength of concrete. Other than impurities, the important ingredients of portland cement could be written as ($CaO \cdot SiO_2 \cdot Al_2O_3$). The action of hydration and hardening is that of a silica gel with fibers growing out of cement particles that join with adjacent particles to form a solid interlocked mass. The production and uses of portland cement and concrete will be covered in later chapters.

In Figure 38*a* the schematic shows the tiny cement particles to which mixing water has been added as they would appear magnified many thousands of times. A silica gel begins to form on their outer surfaces. Soon whiskerlike fibers begin to grow from the particles, and after 7 days they are interconnected, as can be seen in Figure 38*b*. Also the electron micrograph in Figure 38*c* reveals this condition at 7 days in hydrated portland cement. After 28 days this cement and the concrete made with it is about as hard as it will ever get. In moist conditions, as found in dams, the concrete may get slightly harder and stronger over the years. Figure 38*d* shows how these fibers develop into coarser and stronger bars or plates that interlock to form a solid rocklike mass. The electron micrograph, Figure 38*e*, is of hydrated portland cement at 28 days.

REVIEW QUESTIONS

1. Not considering the many subatomic particles, all atoms are made up of three basic components. Name them.
2. Briefly define *valence*.
3. Atoms with positive or negative valence are called ions. What causes this condition?
4. Name four kinds of bonding arrangements of atoms.
5. What does the type of bond in metals have to do with their elastic, plastic, thermal, and electrical behavior?
6. What is a space lattice in solid materials?
7. What are unit cells in metals?
8. When a metal changes from one unit cell structure to another at certain temperatures, it is said to be ______.
9. When a metal begins to solidify, many grains begin to grow. These are called dendrites. At what location do they stop growing?
10. Define *stress* and *strain*.
11. What is work hardening?
12. All other conditions being the same, which metal would be more plastic and ductile, a fine grained steel or a coarse grained steel? Which would be stronger (highest yield strength)? 150 TO POLY
13. When small molecules are linked together to form a large molecule, it is called ______.
14. A gummy substance such as natural rubber latex can be made into useful products by the process of cross-linking. In rubber this process is called ______.
15. The first high polymer was produced by joining the small molecules of phenol and formaldehyde. What is this well-known plastic called?
16. How can clay be plastic when wet, brittle when dry, and hard and strong when fired?
17. What is the basic ingredient of glass?
18. Portland cement hardens and bonds together in the presence of water. What is this process called and how does it progress to form a solid mass?

CHAPTER 2 EXTRACTION AND REFINEMENT OF COMMON METALS

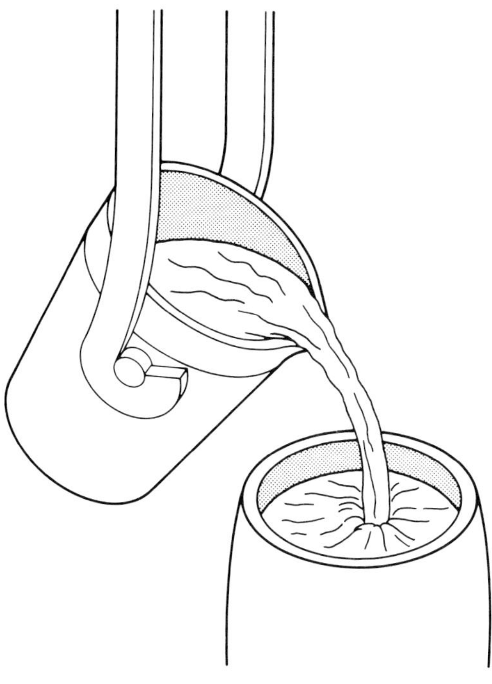

Metallic ores and other important minerals are found in virtually every continent; however, there is an unequal distribution of many of these minerals. For example, tin (Sn) is a metal that is abundant only in Africa, Bolivia, and Southeast Asia, and is found in a few other locations that are not major producers. Chromium ore is extensively mined in South Africa and in the Soviet Union; there are only minor sources of this strategic metal in the United States, so most of our chromium is supplied from these two areas.

Our modern civilization is probably more dependent on strategic metals than on oil. These metals include cobalt, chromium, manganese, platinum, columbium, strontium, titanium, and tantalum. Iron and aluminum are relatively plentiful, but even these materials are affected by the increasing cost of energy.

Mining originated in prehistoric times. Some metals were found in small quantities in their "native" or metallic state. Native copper, silver, and gold were found in streambeds and in outcroppings. These were shaped by hammering to make jewelry and crude tools. The need for larger quantities of metals brought about underground and surface mining methods that have been considerably improved in modern times. Processing and smelting ores to extract the metal has been historically a simple and limited process for metals such as iron, copper, silver, tin, lead, and zinc. Only a few pounds of iron could be produced in one day in primitive forges and furnaces. In modern times the daily output of one of our iron smelting furnaces is measured in tons. Other metals were not extracted and refined in commercial quantities until relatively recently. Some metals such as titanium, which are often called space-age metals, have been manufactured in quantity only since World War II. Iron bearing metals such as pig iron, cast iron, and steel are called **ferrous** and all other metals are called **nonferrous.**

OBJECTIVES

This chapter should enable you to:

1. Describe methods of mining and processing metallic ores.
2. Give an account of the development and methods used in the iron- and steelmaking processes.
3. Explain some of the principles and methods used in smelting several nonferrous metals.
4. Show the construction and principles of operation for several types of furnaces and refining vessels.

Figure 1
Mining machine uses whirling cutting teeth to rip coal from underground seams. Coal is conveyed backward to shuttle cars. Water sprays entrap the coal dust (American Iron and Steel Institute).

ORES AND MINING

An ore deposit is any geological deposit of nonrenewable material having a sufficiently high percentage of the needed element or metal that can be recovered at a profit. Most elements are thinly distributed over the earth's crust. Aluminum and iron are found in clays (alumino-silicates) in substantial quantities, but it would not be economical to process that material for the metal. Ores that occur in narrow veins are usually mined by underground methods (Figure 1). This is done by excavating vertical or sloping shafts and horizontal openings or rooms where the ore is removed, broken up, and hauled to the surface in mine cars or bucket and cable devices. When ores are found in large deposits, a surface mining technique, called open-pit mining (Figure 2), is used. Ore bodies are almost always covered by soil or other waste material, called overburden, which must be removed prior to mining the ore. In a copper mine near Bingham, Utah (Figure 3), an entire mountain has been excavated and removed for its copper ore. In some surface mining operations the top soil is removed, the ore is removed, and the soil is returned to its former place. This method helps to avoid excessive damage to the ecology of the region.

Figure 2
In an open pit mine on Minnesota's Mesabi Range, a power shovel dumps a 13-ton dipperful of iron ore into a truck. Although this is a natural ore, it must be beneficiated or improved in some way before it is shipped to blast furnaces (American Iron and Steel Institute).

Figure 3
An entire mountain has been removed from this gigantic open pit copper mine near Salt Lake City, Utah.

Iron Ore

Before the large deposits of rich hematite and magnetite ores in the Lake Superior Mesabi Range were depleted, as they are now, the 60 to 70 percent iron in those ores was quite profitable to extract. There are still vast deposits of iron ore in these regions, but the iron is bound up in a very hard rock formation that is difficult to mine and process and that has a relatively low percentage of iron. This material, called taconite, is now the major source of iron ore in that region. The most common natural iron ores are given in Table 1.

Mining Aluminum Ore

Aluminum is the third most abundant element on the earth, after oxygen and silicon; however, only a few deposits of two aluminum-bearing minerals are considered to be profitable to mine. These are gibbsite and boehmite, which are composed of aluminum oxide (alumina Al_2O_3) and water mixed with sand or clay and other minerals which must be removed. This ore is called bauxite, named for a French village, Les Baux, where the first deposit was found in 1821. Bauxite is usually mined by the open-pit method and is found on every continent except Antarctica.

Ore Treatments

The cost of shipping iron ores from the mine to the smelter can be greatly reduced if the unwanted rock and other impurities are removed prior to shipment. This is done with several processes called beneficiation. These processes include crushing, screening, tumbling, floatation, and magnetic separation. The refined ore is enriched to being over 60 percent iron by these methods and it is often formed into pellets before shipment. In the case of taconite, about two tons of unwanted minerals are removed for each ton of pellets shipped (Figure 4) and the tailings are returned to the mining site. Growing vetches, grasses, and trees on some of these barren landscapes has been a project of biologists and foresters.

Figure 4
Taconite ore powder, now free of much impurity and rich in iron, is mixed with coal dust and a binder, rolled into small balls in this drum pelletizer, and then baked to hardness (American Iron and Steel Institute).

Table 1
Some Common Ores of Industrial Metals

Metal	Ore	Color	Chemical Formula
Aluminum	Bauxite	Gray-white	Al_2O_3 (40 to 60%)
Cadmium	Sphalerite (zinc blende)	Colorless to brown to black	$ZnS + 0.1$ to 0.2% Cd
	Smithsonite	White	$Zn_4Si_2O_7OH \cdot H_2O$
Chromium	Chromite	Green	$FeO \cdot Cr_2O_3$
Cobalt	Cobaltite	Gray to silver-white	CoAsS
	Linnaeite (Cobalt pyrite)		Co_3S_4
	Smalt-ite	Bluish white or gray	$CoAs_2$
	Erythrite	Transparent, crimson, or peach-red	$Co_3(AsO_4)_2 \cdot 8H_2O$
Columbium (Niobium)	Columbite (Niobite)	Black	$(FeMn)(CbTa)_2O_6$
Copper	Chalcopyrite	Brass yellow	$CuFeS_2$
	Chalcocite	Lead gray	Cu_2S
	Cuprite	Red	Cu_2O
	Azurite	Blue	$2CuCo_3 \cdot Cu(OH)_2$
	Bornite	Red-brown	$FeS \cdot 2Cu_2S \cdot As_2S_5$
	Malachite	Green	$CuCo_3 \cdot Cu(OH)_2$
	Native copper	Reddish	Cu
Gold	Native (alluvial or in quartz)	Yellow	Au
Indium	Sphalerite	White to yellow	ZnS + low % of In
Iron	Magnetite	Black	Fe_3O_4
	Hematite	Reddish brown	Fe_2O_3
	Limonite	Brown to yellow	$2Fe_2O_3 \cdot 3H_2O$
	Goethite	Brown	$Fe_2O_3 \cdot H_2O$
	Siderite	Blue	$FeCO_3$
	Taconite	Flintlike	Fe_3O_3

Lead	Galena	Black	PbS
Magnesium	Sea water		$Mg(OH)_2$
Manganese	Pyrolusite	Black	MnO_2
	Manganite	Steel-gray to black	$Mn_2O_3 \cdot H_2O$
	Rhodochrosite	Rose-red	$MnCO_3$
Molybdenum	Molybdenite	Blue	MoS_2
Nickel	Niccolite	Pale copper-red	$NiAs$
	Pentlandite	Bronze-yellow	$(FeNi)_9S_8$
	Garnierite	Bright apple-green	$(MgNi)_3Si_2O_5(OH)_4$
Silver	Argentite	Black	Ag_2S
	Native	White	Ag
Tin	Cassiterite	Brown or black	SnO_2
Titanium	Anatase,	White	
	Brookite, and Rutile	Reddish-brown	TiO_2
Tungsten	Wolframite	Brownish to grayish black	$(FeMn)WO_4$
	Scheelite	Various	$CaWO_4$
	Ferberite		$FeWO_4$
	Huebnerite		$MnWO_4$
Uranium	Uraninite	Black	UO_2
	Pitchblende	Brown to black	U_3O_8
Vanadium	Vanadinite	Brown, yellow, red	$Pb_5(VO_4)_3Cl$
	Carnotite	Yellow	$K_2(UO_2)_2(VO_4)_2 3H_2O$
Zinc	Franklinite	Black	$ZnFe_2O_4$
	Willemite	Varying	Zn_2SiO_4
	Zincite	Red to orange-yellow	ZnO
Zirconium	Zircon (zirconium silicate sand)	Brown or gray square prisms or transparent	$ZrSiO_4$

Direct Reduction of Iron

In ancient times, the only known method of separating iron from its ore was by the process of direct reduction. This was done in a charcoal forge or furnace. Unlike copper ores, which yielded molten copper in these furnaces, iron would not melt at temperatures below 2799° F (1537° C) and the highest temperature that could be reached in those primitive smelters appears to have been about 2192° F (1200° C). Iron ore subjected to that temperature yields not a molten puddle of metal, but a spongy mass mixed with iron oxide and iron silicate with perhaps other impurities that together made up the slag. The iron worker removed this mass of sponge from the furnace, reheated it in a forge, and then literally squeezed the slag out of it by hammering it on an anvil. The hammering produced a wrought iron with very little carbon interspersed with a few stringers of slag that had not been eliminated. This wrought iron was not much stronger than copper and was a poor substitute for bronze because it could be hardened somewhat only by hammering, as could copper and bronze. However, with the discovery of steel (combining iron with a small amount of carbon and heat treating it), it became a far superior metal. This method of smelting iron and steel has long since been abandoned and replaced by using the blast furnace and steel-making furnaces.

Sponge Iron and Powder

Since the 1920s a new method of direct reduction of ores has been used to produce sponge iron and iron powder, by-passing the blast furnace stage. Many of these "direct-reduced" ores have a lower percentage of impurities than molten iron from a blast furnace. Pellets are formed in this process that contain as much as 95 percent iron and can be economically shipped from remote locations to be remelted in steel-making furnaces. Iron powder is also produced by direct reduction processes; much of this is used in the powder metallurgy industry. See Chapter 10.

Wrought Iron

The wrought (worked) iron of ancient times had some advantages over steel. It had less tendency to corrode (rust) and it had a fibrous quality (from the stringers of slag) that gave it a certain toughness. Although the ancient process of making wrought iron is obsolete, a modern method, called the Aston process, is used to produce a wrought iron in large quantities. Molten pig iron and steel are poured into an open hearth furnace with a prepared slag that removes most of the carbon. The slag cools the molten metal to a pasty mass that is later squeezed in a hydraulic press to remove most of the slag. Wrought iron is used for iron work such as railings for stairways and for the production of pipe and other products subject to deterioration by rusting.

PIG IRON

Until about 1350 A.D., all of the iron made in the known Western world was made by the direct reduction process. It was the development of the shaft or blast furnace at that time (although it might have been developed earlier in China) that made possible the high production rate of iron that was needed for the industrial revolution. The blast furnace (Figure 5) made possible a molten, castable iron product that contains about 4 percent carbon and some impurities. The high carbon content caused it to be somewhat brittle and it has little usefulness for tools and cutting instruments or other products. When refined in cupola furnaces, it is an excellent metal for some machinery parts, cooking utensils, and other cast iron products. However, the greatest usefulness of the blast furnace is in an intermediate step in the steelmaking process (Figure 6).

The operation of a modern blast furnace can be seen in Figure 7. The three raw materials needed to make pig iron are iron ore, coke, and limestone. They are put into the furnace at intervals, making the process continuous. About two tons of ore, one ton of coke, and a half ton of limestone are required to produce one ton of iron.

Figure 5
One of four blast furnaces at Bethlehem Steel Corporation's Sparrows Point, MD plant can process more than 3000 tons of molten iron per day (Courtesy of Bethlehem Steel Corporation).

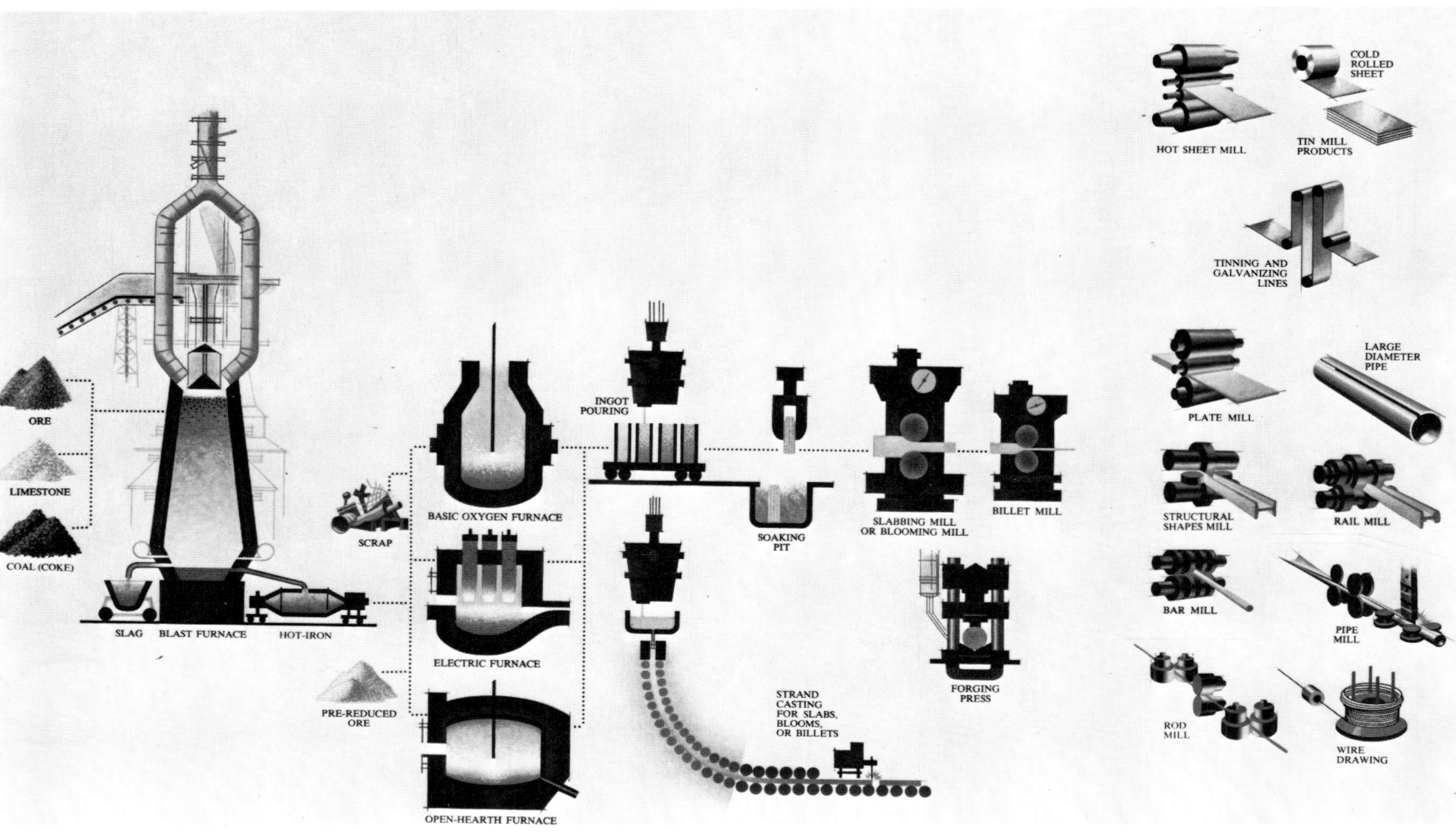

Figure 6
Flow chart of the iron and steel production process, showing the route from ore to iron to finished steel (Courtesy of Bethlehem Steel Corporation).

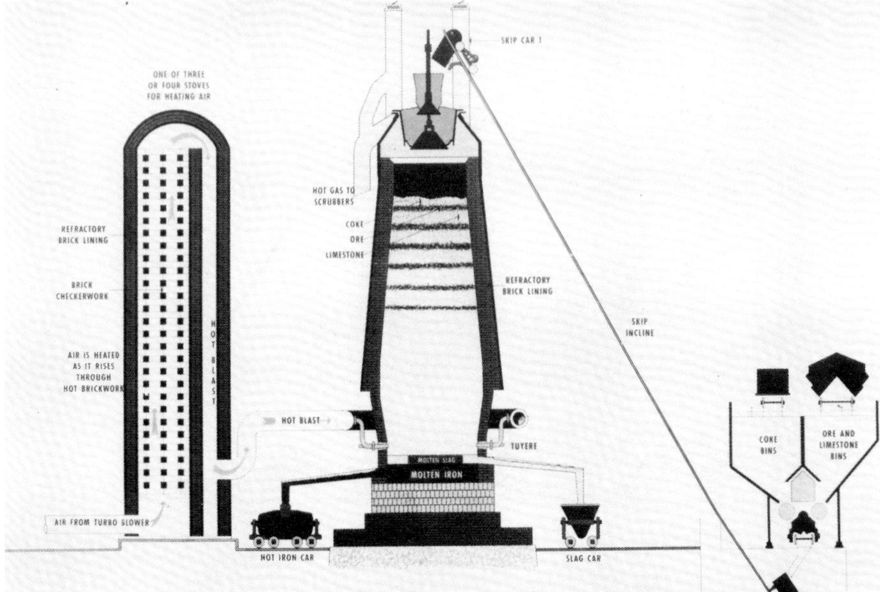

Figure 7
Cross section of a blast furnace. Iron ore is converted into pig iron by means of a series of chemical reactions (Courtesy of Bethlehem Steel Corporation).

Coke is a residue left after soft coal is heated in the absence of air. The gases are driven off and the remaining hard, brittle, and porous material is the resultant coke which contains 85 to 90 percent carbon with some impurities. Coke is produced in coke ovens (Figures 8 and 9) where the volatile elements that are driven off the coal are collected. Many useful products are made from these gases: fuel gas, ammonia, sulfur, oils, and coal tars. From the coal tars come many important products, such as plastics, dyes, synthetic rubbers, perfumes, sulfa drugs, and aspirin.

The operation of the blast furnace is called **reduction.** This is a process in which oxygen is removed from a compound, in this case, iron ore. The oxygen is combined with carbon to form carbon dioxide (CO_2), and the metallic iron is released from its oxide state, leaving molten iron with about 4 percent carbon. The solid materials coke, limestone, and iron ore are dumped

Figure 8
An 850,000 ton per year battery of 80 slot-type, side by side coke ovens (Courtesy of Bethlehem Steel Corporation).

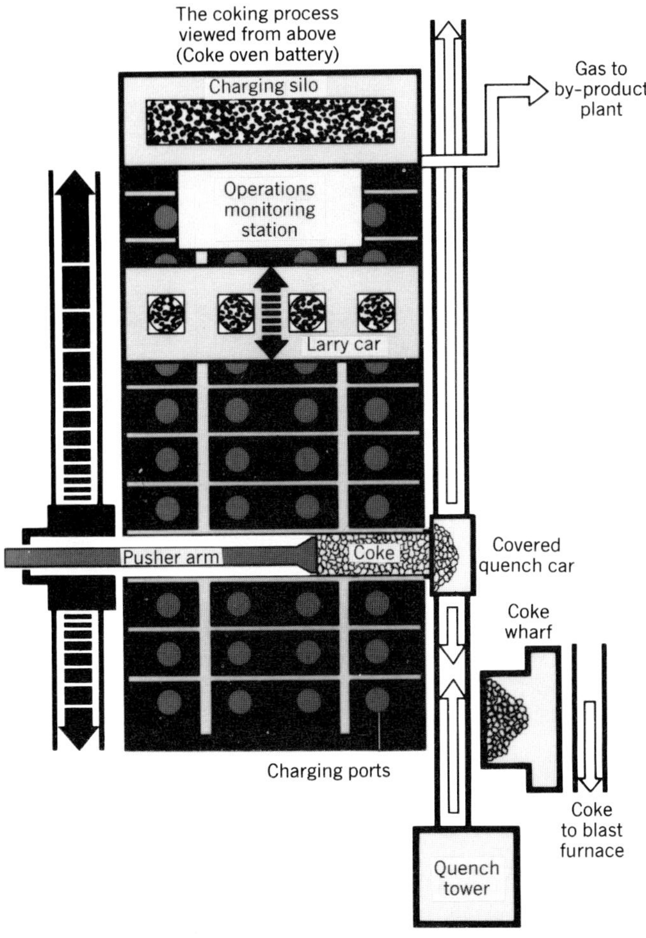

Figure 9
The operation of a coke oven battery at Bethlehem Steel Corporation's Sparrows Point, MD, plant is similar to the drawing. The larry car (or charging machine) is loaded with coal from a charging silo, then charges the coal into the individual ovens through ports in the oven roofs. After baking for approximately 18 hours and reaching a final temperature of at least 1800° F (982° C), coke is pushed from the ovens by the pushing machine at left. Hot coke is then moved by way of a quench car to a quench tower where it is flooded with approximately 16,000 gal of water to cool it before being unloaded at a coke wharf to await further processing (Courtesy of Bethlehem Steel Corporation).

into a hopper at the top of the furnace at regular intervals and heated air is blasted in at the bottom. The air is heated in stoves containing refractory brick that is heated by exhaust gases from the furnace. The stoves alternate between blasting hot air to the furnace to being in a heating cycle. The fuel burns near the bottom of the blast furnace and the rising heat meets the descending charge of coke and ore, producing very high temperatures of about 3000° F (1649° C). As the molten iron is released, it settles to the bottom and a slag forms on top of it. The slag is drawn off at intervals and hauled away in slag cars to a dumping area. This slag is ground up and used as an aggregate for concrete products and asphaltic concrete roads. The molten iron is topped (drawn off) at intervals and collected in an insulated transfer car (Figure 10). Then it is taken to a steelmaking furnace and charged into it. Some of the iron is used elsewhere and is remelted and refined in foundries for casting purposes. For shipping, it is cast in pigs (Figure 11). In earlier times the pigs were made in open sand molds consisting of a groove or trough with many small molds on each side, reminding one of a sow and pigs, hence the name *pig iron.*

STEELMAKING

Pure iron is a soft, easily bent metal having a limited usefulness. Steel is iron combined or alloyed with other metals or nonmetals such as carbon. Plain carbon steel actually contains a small amount of the metal manganese as a scavenger to remove unwanted oxygen and to control sulfur. Sulfur is difficult to remove from steel, and iron sulfide (the form it naturally takes in iron) is deleterious to steel, but if manganese is present, it combines more readily than iron with the sulfur, forming manganese sulfide which is not harmful to the steel. For this reason no steel is made without manganese. It is the element carbon that causes steel to be such a useful material since it can be hardened by heat treatments. Only a small amount of carbon is needed to

Figure 10
The blast furnace is tapped after slag removal, and the molten iron flows into ladles or hot-metal cars. Most iron is used in steelmaking (American Iron and Steel Institute).

Figure 11
Automatic pig casting machine. Some blast furnace iron is poured to solidify in molds. Iron foundries are major users of cast pigs (American Iron and Steel Institute).

make steel: .20 to .30 percent for low carbon steel, .40 to .60 percent for medium carbon steel, and .60 to over 1 percent for high carbon steel. Steel can contain up to 2 percent carbon, but over that amount it is considered to be cast iron, in which excess carbon forms graphite. Steel can be given many different and useful properties by alloying the iron with other metals such as chromium, nickel, tungsten, molybdenum, vanadium, and titanium and with nonmetals such as boron and silicon.

Furnaces

The principal purpose of the steelmaking furnace is to remove most of the carbon from pig iron. Then a measured amount of carbon is added to the molten steel to give it the desired properties. Steel scrap is also utilized and other elements are added to improve the properties of the steel. A now obsolete furnace, the Bessemer converter, made possible the high tonnage production of steel in the mid-nineteenth century. Great steel ships, railroads, bridges, and large buildings were built as a result of this new source of steel being available. However, many of the impurities remained in the steel and, since a blast of air was used to burn out the carbon, the nitrogen in the atmosphere also became an impurity that weakened the steel.

The three modern processes and furnaces used to produce steel are **basic oxygen, open hearth,** and **electric.** These furnaces are not used interchangeably. Each system requires different energy sources and different sources of raw materials. The kind of facility is therefore chosen for economic reasons and the availability of raw materials and energy sources.

The basic oxygen furnace (BOF) (Figures 12 and 13) is designed to produce a high quality steel in a relatively short time, about one 200 to 300 ton heat (a single batch) per hour as compared to the 100 to 375 tons per heat in 5 to 8 hours of the open hearth process. The principal raw material used in the basic oxygen furnace is molten pig iron from the blast furnace. Steel scrap is charged into the furnace and lime is used as a fluxing agent that reacts with the impurities and forms a slag on top of the molten metal. The water cooled lance is lowered to within a few feet of the charge to blow a stream of oxygen at more than 100 pounds per square inch (PSI) on the surface of the molten bath. The oxidation of the carbon and impurities causes a violent rolling agitation which brings all the metal into contact with the oxygen stream. The ladle is first tipped to remove the slag and then rotated to pour the molten steel into a ladle. Carbon, usually in the form of anthracite coal pellets, is added in carefully measured amounts and other elements are also added to produce the desired quality in the steel. A schematic cross section of a basic oxygen furnace is shown in Figure 14.

Open hearth furnaces get their name from a shallow steelmaking area called a **hearth** that is exposed or open to a blast of flames that alternately sweeps across the

Figure 12
Basic oxygen furnace being charged with molten pig iron (American Iron and Steel Institute).

Figure 13
Charging basic oxygen furnace with scrap (American Iron and Steel Institute).

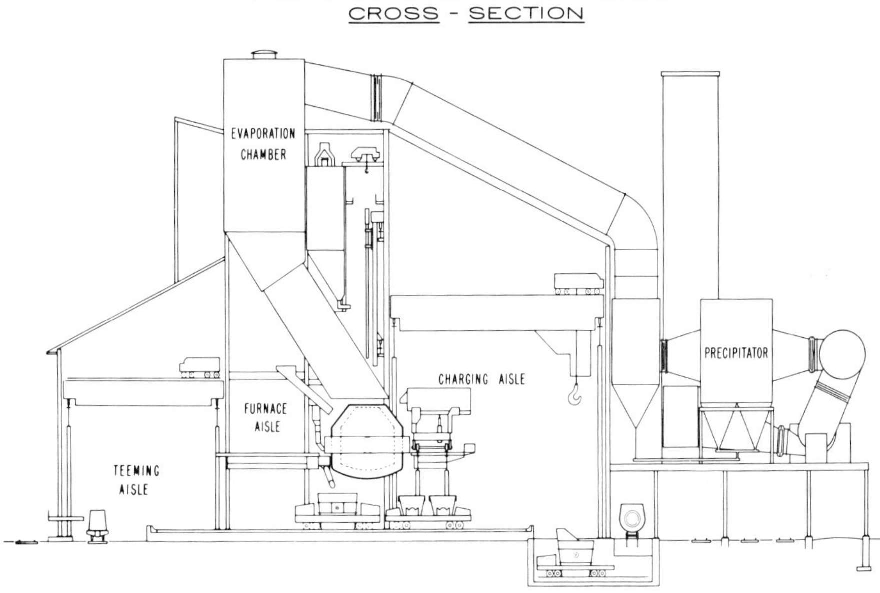

Figure 14
Schematic of a basic oxygen furnace showing the facilities needed to charge scrap and molten iron into the vessel and receive the steel after the oxygen-blowing process is complete (American Iron and Steel Institute).

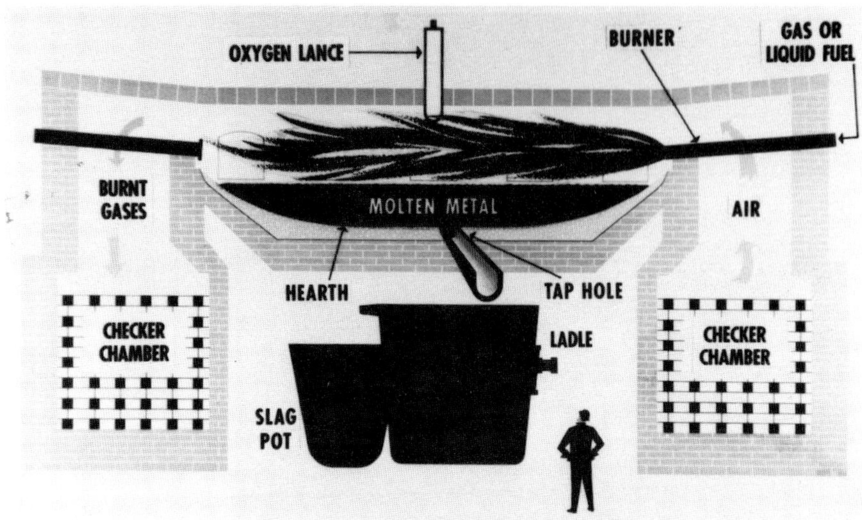

Figure 15
Simplified cutaway diagram of a typical open-hearth furnace. Oxygen may be injected through one or more lances. In some cases it is introduced through the burners to improve combustion (Courtesy of Bethlehem Steel Corporation).

hearth from one end for a period of time and then from the other end of the furnace (Figure 15). The furnace is charged (loaded) from a door in the side facing the charging floor. Theoretically, an open hearth furnace can operate with blast furnace iron alone or steel scrap alone, but most operations use both in about equal proportions, varying with the price of scrap and the availability of molten iron. Scrap is placed in the furnace with a charging machine which usually serves a line or series of open hearth furnaces in a single building (Figure 16). Other elements are added to improve the steel. Mill scale forms on the surface of hot steel and falls off as a black scale; it is actually one form of iron oxide (Fe_3O_4), called magnetite as an ore. In steel mills it is used in steelmaking furnaces to reduce carbon content. Aluminum ferrosilicon is added if the steel is to be killed. A **killed steel** is one that has been sufficiently deoxidized to prevent gas evolution in the ingot mold, making a uniform steel. If the steel is not killed, it is called **rimmed steel** because the gas pockets and

Figure 16
Closeup of charging machine arm thrusting a box of baled scrap through an open-hearth furnace door (American Iron and Steel Institute).

Figure 17
Tapping an open-hearth furnace. Molten steel fills the ladle; the slag spills over into the slag pot (Courtesy of Bethlehem Steel Corporation).

Figure 18
Molten steel from a basic oxygen furnace is teemed into ingot molds (Courtesy of Bethlehem Steel Corporation).

holes from free oxygen form in the center of the ingot and the rim near the surface of the ingot is free of defects. Most of these flaws, however, are removed by later rolling processes (discussed in Chapter 8). When the contents of the heat or melt are acceptable and the temperature is right, the furnace is tapped and the molten metal is poured into a ladle (Figure 17). As with the BOF, carbon is added to the contents of the ladle to adjust the carbon content. Aluminum pellets are added to the ladle to deoxidize the molten metal and produce a finer grain size.

The ladle with its load of molten steel is picked up by a traveling crane and brought over a series of heavy cast iron ingot molds. The steel is teemed (poured) into the molds by means of an opening in the bottom of the ladle (Figures 18 and 19).

Electric furnaces lined with basic refractory (Figures 20 to 23) produce a high quality steel from selected scrap, lime, and mill scale. Most alloy and tool steels are made in electric furnaces, but for these high-quality steels selected scrap must be used. Tin cans and automobile scrap contain such contaminants as tin and copper that spoil the steel for these purposes. Electric furnaces are most competitive where low-cost electricity is available and where very little coal or iron ore is found. The charge is melted by the arcing between carbon electrodes and metal scrap. In some cases, the electrodes are 2 feet in diameter and 24 feet long.

Figure 19
Once steel is poured into an ingot mold and allowed to solidify, the mold is stripped off, exposing a glowing hot ingot (Courtesy of Bethlehem Steel Corporation).

Figure 20
Scrap is being charged into an electric furnace that produces 185-ton heat of steel every $2\frac{1}{2}$ hours (Courtesy of Bethlehem Steel Corporation).

Advanced Melting and Refining

Standard steelmaking furnaces are not capable of removing all impurities and unwanted gases. When higher grade specialty steels are needed, further processing is necessary. There are two types of these refining vessels, those that receive molten metal from conventional steel furnaces and those that remelt to refine it (Figures 24*a* to 24*h*).

Vacuum Stream Degassing Figure 24*a*. Ladles of molten steel from conventional furnaces are taken to vacuum chambers where the molten steel is poured into a pony ladle. There the molten stream of steel is controlled and broken up into droplets when exposed to the vacuum within the chamber. At this point large quantities of undesirable gases escape from the steel and are drawn off by the vacuum pump.

Vacuum Ladle Degassing Figure 24*b*. Of the several kinds of ladle degassing facilities in current use, one of the most important is shown here. The furnace ladle is placed beneath a heated vacuum vessel that has a descending suction nozzle. The entire vessel is lowered into the ladle of molten steel which is forced through the suction nozzle into the heated vacuum chamber. Gases are thus removed in the vacuum chamber which is then raised so that the molten steel is returned by gravity into the ladle. The process is repeated several times.

Figure 21
In a rare view, the interior of an electric steelmaking furnace resembles an amphitheater. The furnace interior is lined with refractory brick and special water-cooled panels to withstand an interior temperature of 3000° F (1649° C). Steel scrap is dumped into the furnace through the roof and is melted electrically and then refined into high-quality steel (Courtesy of Bethlehem Steel Corporation).

Figure 22
The entire electric furnace is tilted during a tapping operation in which molten steel flows into a waiting ladle (Courtesy of Bethlehem Steel Corporation).

Figure 23
A cross section of an electric furnace (Courtesy of Bethlehem Steel Corporation).

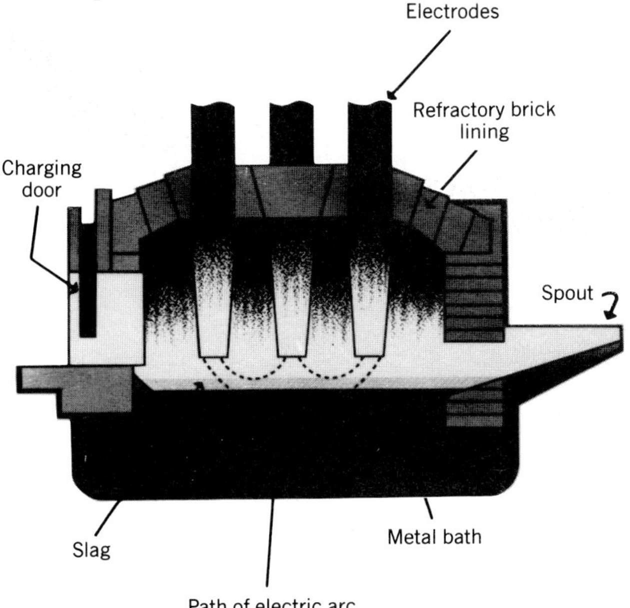

Argon–Oxygen Decarburization Figure 24*c*. Argon, an inert gas, is used in this refining vessel to dilute the carbon–oxygen atmosphere in the melt. The argon and oxygen gases have separate tuyeres so they can enter the bath of molten steel simultaneously. In the case of stainless steel which must have a very low carbon content, this process increases the affinity of carbon for oxygen, producing harmless CO_2. This also minimizes the oxidation of chromium and a less expensive form of chromium can be used. The argon, being inert, also stirs the molten steel, promoting equilibrium between the molten metal and the slag.

Vacuum Oxygen Decarburization Figure 24*d*. The vessel, full of molten steel, is placed in an induction stirrer which may be mounted or permanently installed on a transfer car. Then a separate roof is placed over the ladle. The roof contains an oxygen lance and is equipped for vacuum degassing. The metal is degassed (decarburized) while being stirred. Then a second roof is placed over the ladle through which additives and alloys are dropped from a hopper into the melt.

Electron Beam Melting Figure 24*e*. This purification process involves a cascading action of molten steel carried on in a vacuum. The drawing shows steel melted in an induction heated crucible inside a vacuum chamber. This crucible tilts, pouring partially degassed steel into an induction heated ladle in an adjoining vacuum chamber. The electron beam hits the steel imparting thermal energy to keep it molten as it flows over a water-cooled copper hearth. The steel is continuously cast into ingots.

Vacuum Induction Melting Figure 24*f*. In this process, molten steel, or more commonly steel scrap, is charged into an induction furnace situated in a vacuum chamber. Undesirable gases are removed by a vacuum pump after which the molten refined steel is poured into a trough by tilting the furnace. The steel is then conveyed to a holding ladle in an adjoining vacuum chamber and ingot molds are filled from this ladle.

Consumable Electrode Melting Figure 24*g*. Sometimes called the vacuum arc process or furnace, this vessel remelts steels produced by other methods and allows it to solidify into ingots, usually in a cylindrical shape. A solid steel bar is slowly lowered on a control rod into a vertical water-cooled mold. The electric current forms an arc which melts the bar much like the electrode melts at the arc in electric arc welding. The gaseous impurities are drawn off by the vacuum in the chamber as the molten steel progressively fills the mold. The consumable electrode melting process is also used

Figure 24
Remelting processes that are used mostly to produce highly sophisticated alloy and specialty steels for the world's new technologies. (*a*) Vacuum stream degassing. (*b*) Vacuum ladle degassing. (*c*) Argon–oxygen decarburization. (*d*) Vacuum oxygen decarburization. (*e*) Electron beam melting. (*f*) Vacuum induction melting. (*g*) Consumable electrode melting. (*h*) Electroslag melting. (American Iron and Steel Institute).

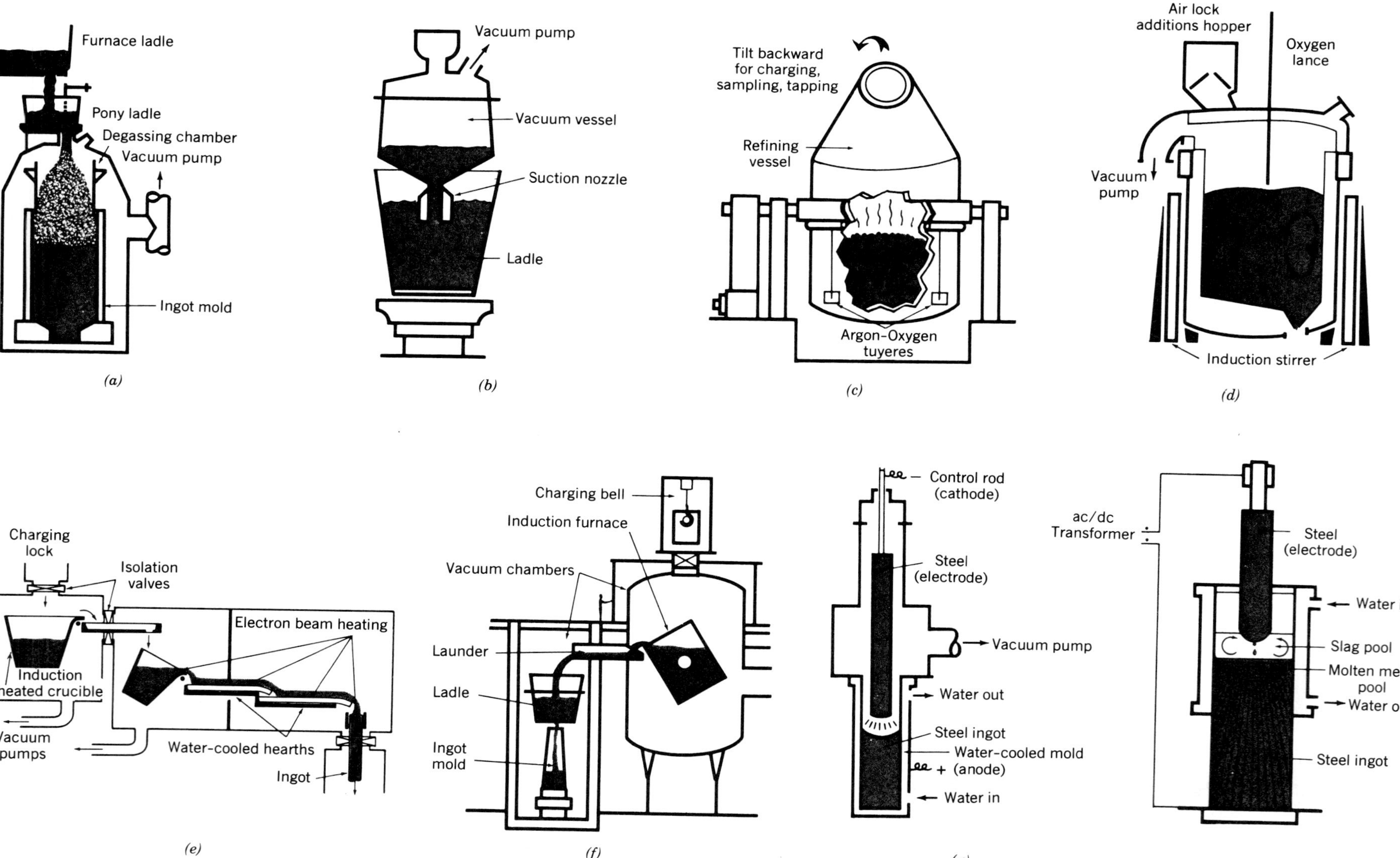

in refining some nonferrous metals such as titanium and zirconium. See the flow sheet of zirconium production in this chapter.

Electroslag Melting Figure 24*h*. There is a strong similarity between the electroslag process and the consumable electrode process, but there are some further technological refinements. As in the consumable electrode process, the electrode is suspended in a mold and connected to an electrical power source. Instead of a vacuum, a powdered slag, which quickly melts, is used to refine the steel. The molten droplets from the steel electrode drop through the liquid slag and collect in a molten pool beneath the slag and quickly solidify to form the ingot. Impurities are removed by floatation in the slag and a thin layer of slag also solidifies between the ingot and mold walls resulting in a smooth surface free of defects.

ALUMINUM EXTRACTION

The bauxite ore from which aluminum is obtained contains many unwanted impurities such as iron oxide, silicon, titania, and water. The ore is first crushed to a powder and mixed in a solution of strong caustic soda (sodium hydroxide, NaOH) and dumped into large tanks or digesters where more caustic soda or soda ash and lime are added to make up for losses.

The alumina (aluminum oxide, Al_2O_3) is dissolved in the caustic soda, forming a sodium aluminate solution leaving the impurities which are insoluble. A series of separation processes that involve filtration and precipitation separates the alumina from unwanted solids called red mud. This aluminum hydrate is then heated to 1800° F (982° C) in kilns to drive off the chemically attached water. The resulting alumina is a fine white powder that is about half aluminum and half oxygen bonded so tightly that neither heat nor chemical action alone can separate them.

Smelting Aluminum

An abundant supply of low cost electricity is required to competitively produce aluminum. For this reason, aluminum smelters are usually located in areas where hydroelectric generating plants are established. One great advantage of using hydroelectric power for smelters is that low cost electricity can be purchased in off-peak hours and when excess power is available in high water run-off periods.

Smelting is the process by which alumina is reduced; that is, separated into its two constituents, aluminum and oxygen (Figure 25). In a modern smelter, the alu-

Figure 25
The process of making aluminum (Courtesy of Aluminum Company of America).

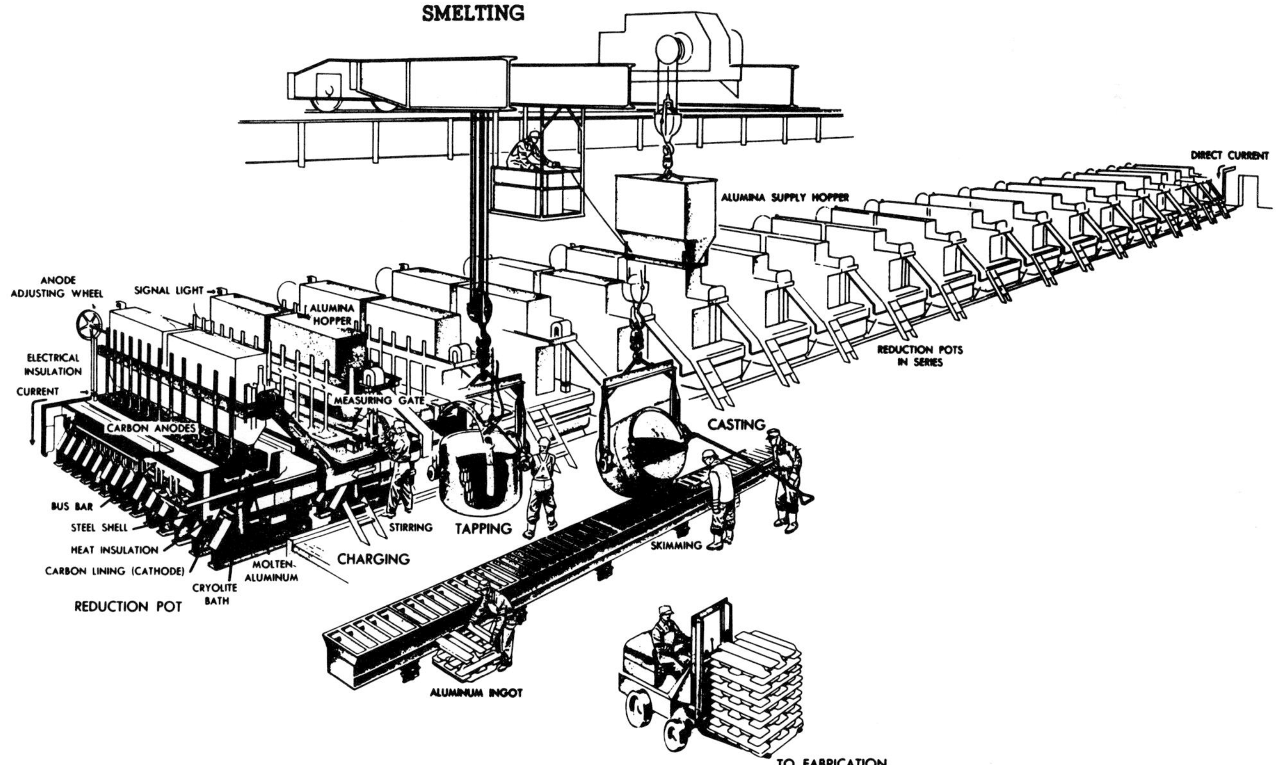

mina is dissolved in a bath of molten cryolite in electrolytic furnaces called pots. Cryolite (sodium aluminum fluoride, Na_3AlF_6) is a material, mined in Greenland, that allows the passage of an electric current, whereas alumina does not conduct electricity. The smelting process is continuous; low voltage—high amperage current is passed through the cryolite bath from carbon anodes suspended in the melt to a carbon lining in the bottom of the pot which serves as a cathode. This powerful electric current separates the dissolved alumina into pure aluminum and oxygen.

The molten aluminum collects at the bottom of the pot and is periodically siphoned off while the oxygen combines with the carbon anode and is released as carbon dioxide. A small portion of aluminum fluoride is added from time to time to compensate for small fluoride losses. About two pounds of alumina will yield one pound of aluminum and about one-half pound of carbon anode is consumed for every pound of aluminum produced.

COPPER SMELTING

The two main classes of copper ores are oxides and sulfides. The oxidized ores can simply be heated with carbon to release the copper metal. However, most available ores are sulfides and perhaps half the world's copper deposits are in the form of chalcopyrite ore ($CuFeS_2$). These sulfide ores require more complex treatment in the melting process.

The ore is first crushed and then ground to a powder in ball mills. The powder is enriched by a floatation process that removes much of the unwanted material. Silica is added and the mix is melted in a furnace (usually a reverberatory type) producing two liquid layers: a lower layer of copper matte (cuprous sulfide plus sulfides and oxides) and an upper layer of silicate slag which is drawn off leaving cuprous and ferrous sulfides called matte. Air under pressure is blown through the molten copper matte in a converter. This operation is one of oxidation–reduction similar to the iron producing blast furnace. The iron is first removed, and in a second blow the sulfur is removed and becomes sulfur dioxide. The molten copper is poured into ladles and at this stage it is about 98 to 99 percent pure and is cast into ingots. When further refinement is necessary, the copper anodes are placed in electrolytic cells that produce 99.95 percent pure copper. The impurities collect at the bottom of the tanks in a sludge from which other metals such as gold, silver, and other rare and precious metals are later removed.

Nickel is often associated with copper and iron ores as a nickel sulfide concentrate. The smelting process for nickel ores is similar to that of copper in which the ore is roasted, smelted in an electric furnace to produce a matte. It is then placed in a converter and further refined before casting the nickel–copper anodes. The nickel is then separated from other metals by the electrolytic refining process.

MAGNESIUM EXTRACTION

There are several important ores of magnesium, but the most important source is in seawater which contains about 0.13 percent dissolved magnesium. Thermal reduction is used to extract magnesium from ores such as roasted dolomite ($MgO \cdot CaO$). The electrolytic method is used to extract magnesium from seawater to which calcium hydroxide ($Ca(OH)_2$) has been added. This causes the magnesium in the water to precipitate as magnesium hydroxide ($Mg(OH)_2$) which is converted to magnesium chloride ($MgCl_2$) by adding hydrochloric acid (HCl). This product is dried and electrolyzed to produce chlorine gas and magnesium metal.

LEAD AND ZINC SMELTING

The most common form of lead ore is the sulfide (PbS), known as galena. It is often associated with zinc and frequently with silver. The ore is ground into small pieces and the impurities are separated by floatation. The lead ore is smelted in reverberatory furnaces or in water-jacketed blast furnaces to produce molten lead which is drawn off at the bottom of the furnace.

SPACE-AGE METALS

The metals columbium, titanium, hafnium, zirconium, and tantalum alloys are used in the aerospace industry, in jet aircraft, rockets and missiles, and for nuclear reactors. Columbium (niobium) is a relatively lightweight metal that can withstand high temperatures and can be used for the skin and structural members of aerospace equipment and missiles. Columbium, a superconductor of electricity, presents a possible way to store and transfer large quantities of energy in the future.

Titanium is extensively used for jet engine components (rotors, fins, and compressor parts) and other aerospace parts because it is as strong as steel and 45 percent lighter. Titanium–aluminum–tin alloys possess high strength at high temperatures and, since titanium forms an alloy with many metals, it can be made to have properties that make it useful for various applications.

Hafnium and zirconium are always found together and their properties are so similar that there has been no commercial reason to separate them except one.

Zirconium and hafnium have opposite properties in relation to neutron flow in a nuclear reactor, so they have to be separated for such use. Hafnium has the property of stopping the flow of neutrons by absorbing them, thus stopping the reaction taking place in the reactor core. In contrast, zirconium freely allows neutrons to flow through itself, allowing the fission process to occur. These metals make possible the precise control of nuclear reactors. Zirconium and its alloys were originally developed to encapsulate uranium in nuclear reactors, but because of its superior corrosion resistance, it is being used in the chemical industry and for surgical implants. Pure zirconium is a reactive metal that will burn in air with a brilliant white light. It has long been used as the light source in photo flash bulbs.

Tantalum often occurs as the mineral columbite–tantalite. The metal tantalum is very malleable and ductile and has a high melting point 5425° F (2996° C). It is used as a replacement for platinum in chemical, surgical, and dental instruments and equipment. Tantalum is also used in vacuum furnaces and to make electrolytic capacitors.

Tantalum and columbium are probably more closely related in more ways than any other two metals in the periodic table. They occur together in the same ores. Both are highly immune to attack by strong acids and both are more ductile (easily formed) than any other refractory (high temperature) metal.

EXTRACTION

Columbite or tantalite ore is usually separated from other minerals by washing and magnetic separation at the mine. Ores processed for use as alloys of steel are normally converted to ferro–columbium–tantalum by reduction in electric furnaces.

The production of pure columbium or tantalum requires a more elaborate treatment in which the ore is ground and fused with caustic soda. The impurities are washed away and the fused product is dissolved in acids to form salts which are separated and purified by crystallization. Both columbium and tantalum are made by powder metallurgy in which the metal is prepared as a powder by crushing and washing with acids. The powders are compacted by a hydraulic press to produce bars of about one square in. in cross section and about 30 in. long. These are sintered in a vacuum chamber, clamped on each end to a water-cooled terminal, and subjected to the flow of a heavy electric current, raising it to a fusing temperature. After sintering, the bar is cold-worked by rolling or forging.

Titanium ores are converted to titanium tetrachloride and reduced with magnesium or sodium in an inert gas atmosphere. The product of this chemical reaction is a spongy mass of metal that must be melted to form an ingot before it can be further processed. Since titanium is a reactive metal, it will react with oxygen in the air and burn while in the molten state. Therefore, melting must be carried out in a vacuum or in an inert gas atmosphere. The titanium sponge is crushed and then compacted into briquettes in a hydraulic press. These briquettes are welded together in a weld tank to form an electrode which is melted in a water-cooled copper crucible in a vacuum-arc furnace. The resulting ingot is processed in rolling mills in much the same way as steel products are formed.

Zirconium is processed in a similar manner as titanium; however, hafnium and zirconium are separated by precipitation and calcining, which produces zirconium oxide (ZrO_2) and hafnium oxide (HfO_2). The product is chlorinated and then reduced with magnesium, and finally the magnesium is distilled from the zirconium metal, leaving a porous spongelike material. The sponge is broken up, crushed, blended, and then pressed into compact bars. These are welded together to form an electrode (Figure 26). An ingot is formed in a vac-

Figure 26

Pressed bars of zirconium sponge welded together to form an electrode (Courtesy of Teledyne Wah Chang Albany, Albany, Oregon).

Figure 27
Flow chart of zirconium production (Courtesy of Teledyne Wah Chang Albany, Albany, Oregon).

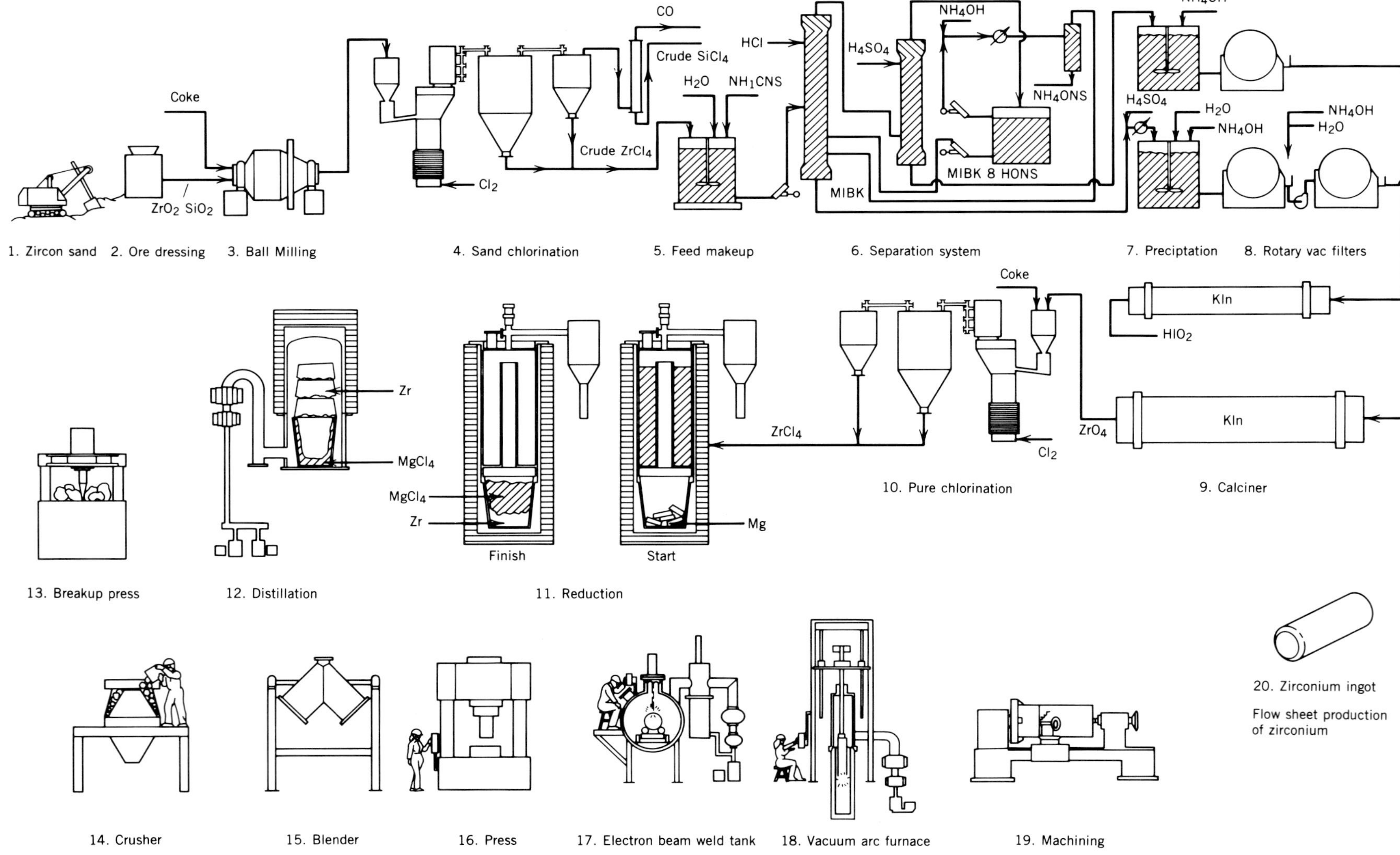

Figure 28
Zirconium ingots ready to be processed into useful products (Courtesy of Teledyne Wah Chang Albany, Albany, Oregon).

uum arc furnace by the comsumable arc melting process (Figure 27). The ingot must be machined on the surface to remove scale and impurities. The ingots (Figure 28) are then fabricated into products by forging, hot or cold rolling, or extrusion.

Zirconium is also a reactive metal that can burn in the presence of oxygen or in the air. This property and others can be greatly altered in those reactive metals by alloying them with other metals. For example, columbium–titanium alloys are produced for superconducting wire and other products. Titanium is alloyed with aluminum and molybdenum to give it a high tensile strength of about 170,000 PSI.

REVIEW QUESTIONS

1. Are titanium and zirconium ferrous or nonferrous metals? Explain.
2. A large proportion of common clay, which is available almost everywhere on earth, is aluminum. Why is bauxite used for the extraction of aluminum instead?
3. Iron ore is usually processed at the mine site to upgrade it and remove unwanted impurities. Why is this done at the site?
4. What is the name of the furnace that first made possible the smelting of iron in a molten state? What product is made in this furnace?
5. What three raw materials are used to make pig iron?
6. Why does all steel contain the element manganese in small quantities?
7. Why is cast iron a poor material to use for making cutting tools such as knives?
8. How much carbon can steel contain?
9. Name the three modern types of steelmaking furnaces.
10. What energy source is a necessity for smelting aluminum?
11. The final stage of refinement of 99.95 percent copper is called ____________________.
12. Space-age metals such as titanium are extracted by complex processes. What is the major energy source for these processes: coal, coke, petroleum oil, or electricity?

CASE PROBLEM

Case 1
Cast Iron versus Steel

You are required to choose a metal for manufacturing carpenter's claw hammers. Nonferrous metals have been ruled out because they are all too costly or have insufficient strength; this leaves cast (pig) iron and steel. Making the hammer heads of cast iron is the least expensive method since the metal is cheaper and the hammer head can be readily cast in sand or reusable permanent molds. Thus, their sale price would be much lower than the price of hammer heads made of steel forgings or castings. What other consideration should be taken in this choice? Would you choose cast iron or steel to make these hammer heads?

CHAPTER 3 METALLURGICAL SCIENCE

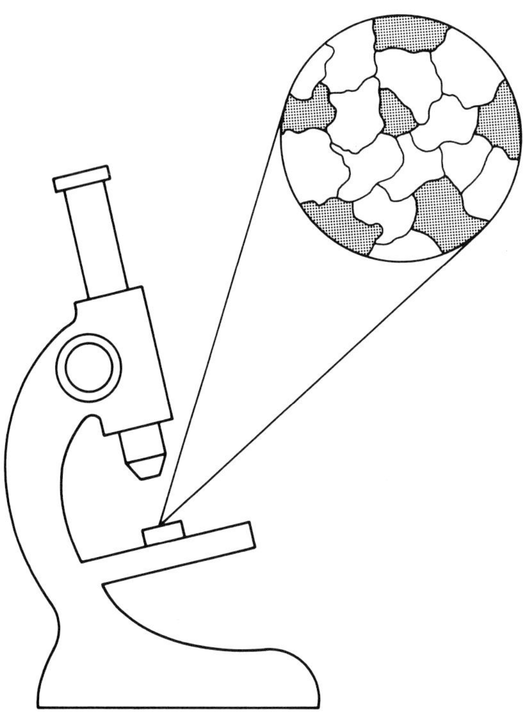

Extractive or **process metallurgy** is the technology of extracting metals from their ores and refining them for further processing. This includes mining procedures, furnace operations, and ingot casting, methods covered in Chapter 2. **Physical metallurgy** is the science or study of the properties, structure, behavior, and composition of metals and how to adapt metals to make useful products. In the past, discoveries about the behavior of metals were fortuitous accidents in which the underlying principles and causes were not understood. Perhaps some accomplishments in modern physical metallurgy are still made by trial and error, but for the most part metallurgy is a science in which new metal alloys are planned and developed to meet exacting specifications, such as those for jet engine impeller blades. Scientific equipment such as light and electron microscopes is used to view and analyze the crystal structure of metals, and testing machines are used to pull, push, bend, twist, and impact metal specimens to determine their properties. The behavior of metals is no longer shrouded in mystery and folklore as it was in ancient times; much is known today about metals even at the atomic level. The atomic and crystalline structure of metals was dealt with in Chapter 1. This chapter will be confined to the study of physical metallurgy and the behavior of metals under various conditions.

OBJECTIVES

This chapter should enable you to:

1. Understand the behavior of metals in terms of their mechanical properties.
2. Describe the method of testing metals to determine their various properties.
3. Make some material selections on the basis of mechanical and physical properties.
4. Explain how the microstructure of metals can be seen and photographed.
5. Show how mechanical parts, steel structures, and pipelines can be tested for flaws without damaging them for use.

MECHANICAL AND PHYSICAL PROPERTIES OF METALS

In the field of manufacturing in which metals are cast, shaped, or formed, the mechanical properties of the metal selected for processing must be known. Some of these properties are the relationship between stress and strain, elasticity, strength, ductility, malleability,

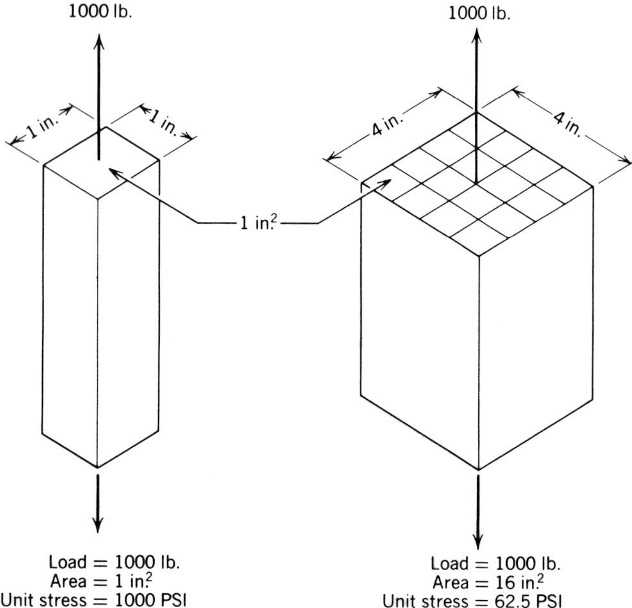

Figure 1
Unit stress (White, Neely, Kibbe, Meyer, *Machine Tools and Machining Practices*, Vol. II © 1977 John Wiley & Sons, Inc.).

hardness, brittleness, creep, and toughness. Various tests and testing machines are used to determine the mechanical properties. These tests are usually destructive; that is, the specimen or part is destroyed by testing.

Engineers and scientists need to have an understanding and a means of measurement of the physical properties of metals. Some of these properties are the melting point, coefficient of thermal expansion, electrical and thermal conductivity, specific gravity, magnetic susceptibility, and reflectivity. Tests for determining the physical properties of metals are often nondestructive. Nondestructive testing is discussed later in this chapter.

Mechanical Properties

An elastic material is one that can be deformed and then return to its original shape when the load is removed. **Elasticity** involves the relationship between stress and strain. Although these terms are often used interchangeably, they have separate and distinct meanings in the science of metallurgy. **Unit stress** is determined by the load applied to a metal and is defined as the resistance of a material to external elements such as force, load, or weight measured in pounds per square inch (PSI) (Figure 1). **Unit strain** is the movement or stretching of a metal when a stress is applied. Unit strain is the amount of deformation (tension or compression) in a unit length (original or gage). Elasticity is expressed by *Hooke's Law* as the degree to which an elastic body bends or stretches out of shape (strain) which is in direct proportion to the force (stress) acting upon it (Figure 2). Within the elastic range, the metal will return to its original shape without permanent deformation when the applied load is released. Metals are elastic only to a certain stress level, above which they become plastic and permanently deform. Below this **elastic limit,** also called the proportional limit, stress and strain are proportionate; that is, for every unit of loading there is always the same amount of strain or movement. This can be seen on the stress–strain diagram as a straight line, called the elastic range (Figure 3).

Plasticity is the property that allows metals to be permanently deformed beyond the elastic limit without failure by rupture or splitting. Clay, when wet, is an example of a plastic substance that can be easily formed. However, when metals reach a certain limit of strain,

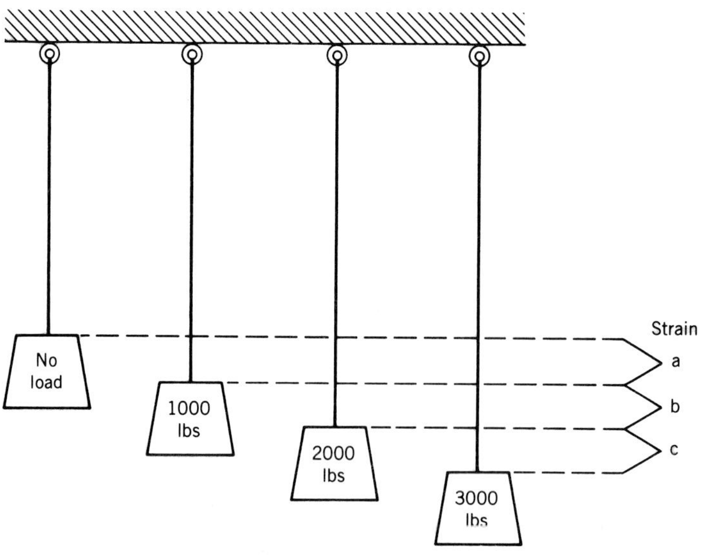

Figure 2
Diagram of the elastic properties of metals as explained in Hooke's Law. Four weights, of which one represents zero load, are suspended by identical metal wires. The three loads toward the right of the diagram are heavier in the equal progression. The resultant movements *a*, *b*, and *c*, called strain, are also equal.

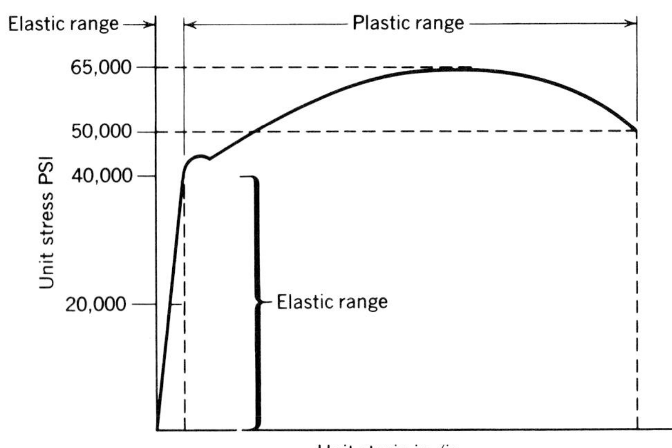

Figure 3
Stress–strain diagram for a ductile steel (White, Neely, Kibbe, Meyer, *Machine Tools and Machining Practices*, Vol. II, © 1977 John Wiley & Sons, Inc.).

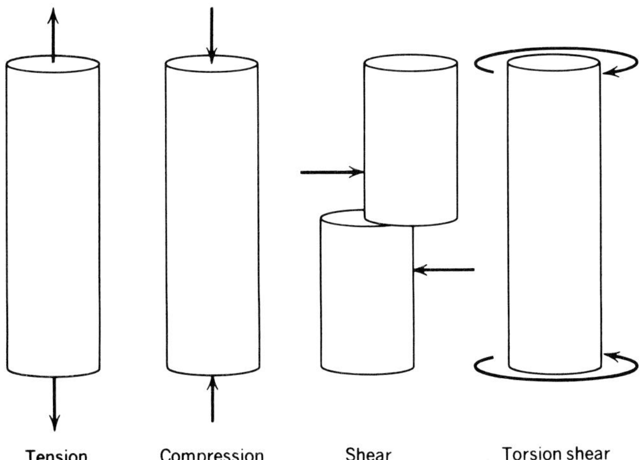

Figure 4
The four types of stresses (White, Neely, Kibbe, Meyer, *Machine Tools and Machining Practices*, Vol. II, © 1977 John Wiley & Sons, Inc.).

Figure 5
Digital readout of torque valves on a torque testing machine (Photograph courtesy of Tinius Olsen Testing Machine Co., Inc., Willow Grove, PA).

they are no longer plastic but become brittle and suddenly fail. This is called the ultimate strength (see Figure 3).

Strength of metals is their ability to resist deformation or changing shape when external forces are applied. There are four types of stresses involved in the strength of materials: **tensile, compressive,** and **shear** (Figure 4). **Torsion** shear is found in rotating parts such as axles and machinery shafting (Figure 5). Some metals such as steel have nearly equal compressive and tensile strengths, but cast iron has a relatively low tensile strength compared to its higher compressive strength. Shear strength is lower than tensile strength in virtually all metals (Table 1). The tensile

Table 1
Material Strength

Material	Modulus of Elasticity (PSI)	Allowable Working Unit Stress			
		Tension	Compression	Shear	Extreme Fiber in Bending
Cast iron	15,000,000	3,000	15,000	3,000	
Wrought iron	25,000,000	12,000	12,000	9,000	12,000
Steel, structural	29,000,000	20,000	20,000	13,000	20,000
Tungsten carbide	50,000,000				

Material	Elastic Limit (PSI)		Ultimate Strength (PSI)		
	Tension	Compression	Tension	Compression	Shear
Cast iron	6,000	20,000	20,000	80,000	20,000
Wrought iron	25,000	25,000	50,000	50,000	40,000
Steel, structural	36,000	36,000	65,000	65,000	50,000
Tungsten carbide	80,000	120,000	100,000	400,000	70,000

SOURCE John E. Neely, *Practical Metallurgy and Materials of Industry*, Second Edition, Wiley, New York, © 1984.

Figure 6
A specimen with a strain gage attached to it is placed in a tensile testing machine to be pulled. Data will be plotted on an *X–Y* graph, shown on the right (Photograph courtesy of Tinius Olsen Testing Machine Co., Inc., Willow Grove, PA).

Figure 7
Tensile testing machines are also equipped to make compressive tests; here a concrete sample is being tested (Photograph courtesy of Tinius Olsen Testing Machine Co., Inc., Willow Grove, PA).

strength of a specimen is the total amount of stress loading it can withstand before breaking apart. To determine tensile strength, metal specimens are "pulled" on a machine called a tensile tester (Figures 6 and 7) until they break. A continuous record of stress load and strain movement is often recorded (Figure 8) and the information is placed on a graph called a stress–strain diagram. Standard specimens are usually used with a cross-sectional area that can be easily calculated (Figure 9). For example, the most common C-1 standard has a diameter of .505 in., making the area .2 in.^{2} A specific gage length is used to record elongation, usually 2 in. $\pm$ .005 in. Tensile strength in pounds per square inch is determined by dividing the maximum load (in pounds) by the original cross section (in square inches) before testing. This is called unit stress. **Unit stress** equals the load divided by the area.

$$\text{unit stress} = \frac{\text{load}}{\text{area}}$$

Figure 8
Tensile testing data system. A printout of several tests made on a tensile tester (Photograph courtesy of Tinius Olsen Testing Machine Co., Inc., Willow Grove, PA).

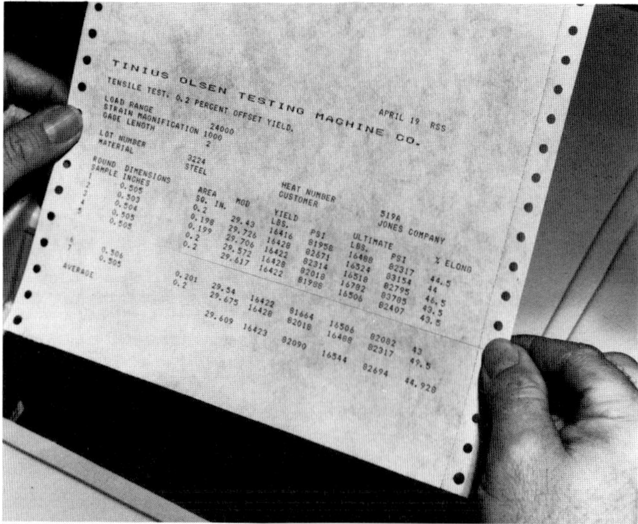

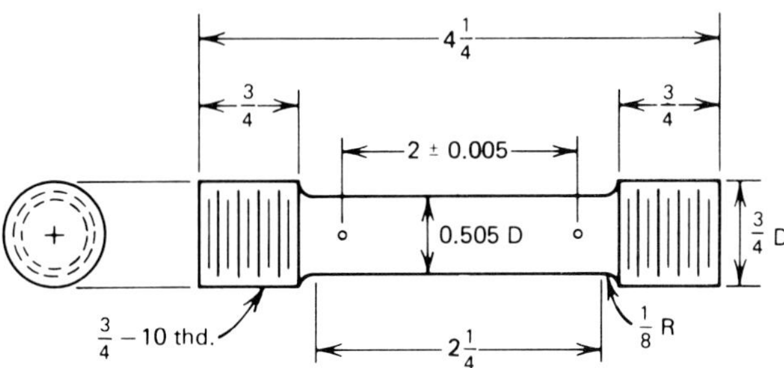

Figure 9
This standard test specimen has a cross-sectional area of .002 in. and a gage length of 2 in. Round numbers make calculations for percent reduction of area and percent elongation easier (Neely/*Metallurgy*, 2 Ed., © 1984 John Wiley & Sons, Inc.).

The tensile test not only reveals the yield point, ultimate strength, and strain, it also determines the relative **ductility** of a metal specimen. The amount (percent) of elongation and reduction of area at the point of failure is an indication of ductility (which is the property that allows a metal to deform permanently when in tension). Any metal that can be drawn into wire is ductile. Soft steel, aluminum, gold, silver, and nickel are ductile metals.

Malleability is the ability of metals to be deformed permanently when loaded in compression. Metals that can be rolled, hammered, or pressed into flat pieces or sheets are malleable. Ductile metals are usually also malleable but there are some exceptions. Lead is very malleable, but not very ductile and cannot be drawn into wire very easily, but it can be extruded in wire or bar form. Gold, tin, silver, iron, and copper are malleable metals.

Hardness is generally defined as the resistance to penetration or indentation. The greater the hardness, the smaller the indentation with the same penetrator tip and pressure. However, the property of hardness is related to the elastic and plastic properties of metals and certain hardness tests are based on rebound or elastic hardness. Also, certain scratch tests are used to determine abrasive hardness or resistance to abrasion. Hardness is also related to tensile strength so it can be assumed that a hard steel is also a strong one and one resistant to wear.

The hardness test that uses pressure and a penetrator or indenter is widely used for industrial applications because of its simplicity and ease of operation. The *Rockwell* (Figures 10 and 11) and *Brinell* (Figures 12 and

Figure 10
Rockwell-type hardness testing machine (Copyright © 1985 King Tester Corporation).

Figure 11
Testing a part on a Rockwell-type hardness tester (Copyright © 1985 King Tester Corporation).

Figure 12
A large Brinell hardness tester with a separate readout panel (Photograph courtesy of Tinius Olsen Testing Machine Co., Inc., Willow Grove, PA).

Figure 14
Brittle fracture of a shaft.

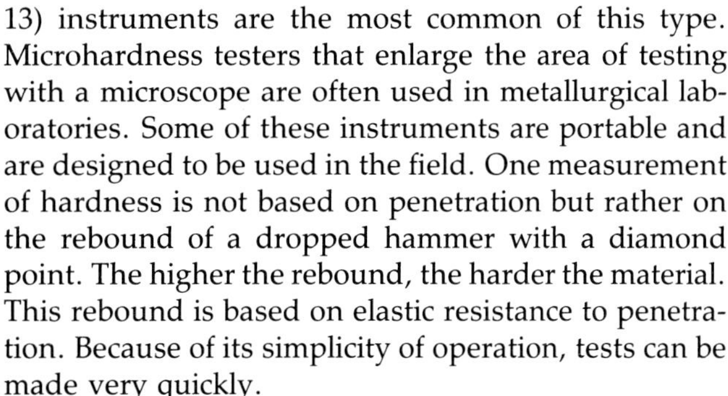

13) instruments are the most common of this type. Microhardness testers that enlarge the area of testing with a microscope are often used in metallurgical laboratories. Some of these instruments are portable and are designed to be used in the field. One measurement of hardness is not based on penetration but rather on the rebound of a dropped hammer with a diamond point. The higher the rebound, the harder the material. This rebound is based on elastic resistance to penetration. Because of its simplicity of operation, tests can be made very quickly.

A material that will not deform plastically under a load is said to be **brittle** (Figure 14). A brittle break or failure in metals can be identified by a crystalline surface, often with a tiny rock candy appearance. Sometimes the individual crystals are so small that the surface has a somewhat smooth appearance to the eyes, but under magnification the crystals can be seen. Of course, metals and some other materials may exhibit other properties such as ductility or malleability along with brittleness. When a metal specimen is pulled apart in a tensile tester, a considerable amount of deformation can be seen in a ductile metal before a brittle failure takes place. However, very hard steel and cast iron show little or no ductility when pulled apart, so we say those metals are brittle, meaning they are completely brittle. A normally ductile metal can exhibit complete brittleness under some circumstances. A sharp notch

Figure 13
The Brinell microscope is used for measuring indentation diameters for simple Brinell hardness testers. Some Brinell testers, such as the one in Figure 12, are fully automatic and do not require a microscope (Copyright © 1985 King Tester Corporation).

that concentrates the load in a small area or at one point can reduce plasticity (Figure 15). Also low temperatures can have an embrittling effect on some steels. The behavior of metals at low temperatures is discussed later in this chapter.

Notch toughness, also called impact strength, is the ability of a metal to resist rupture from impact loading when there is a notch or stress raiser present. The property of toughness is not necessarily related to ductility, hardness, or tensile strength. However, as a rule, a brittle metal with low tensile strength such as cast iron will fail under low shock loads and hardened, tempered tool steels show a high impact strength and coarse-grained metals have lower shock resistance than fine-grained metals. A notch or groove will lower the shock resistance of a metal. Test specimens are made with given dimensions and a specific notch shape (Figure 16).

Notch toughness is measured on a device called an Izod–Charpy testing machine in which either of the two styles of testing can be used (Figure 17). A weighted, swinging arm or pendulum strikes a specimen that is

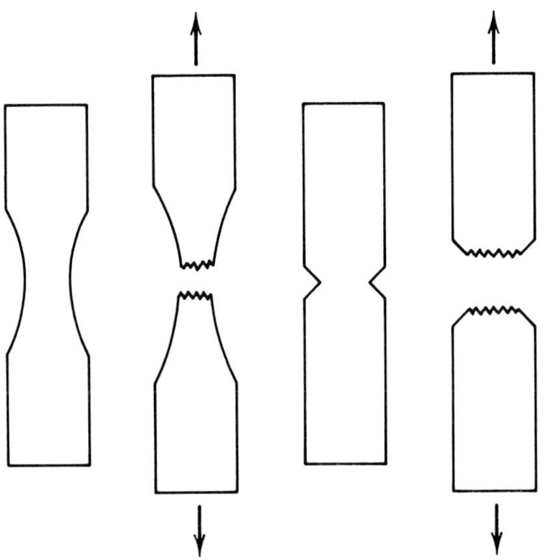

Figure 15
Notching and its effect on plasticity. The specimen (left) when pulled becomes deformed before breaking, showing ductility. The notched specimen of the same material (right) exhibits complete brittleness.

Figure 16
Test specimen specifications for Charpy and Izod test (Photograph courtesy of Tinius Olsen Testing Machine Co., Inc., Willow Grove, PA).

Izod test specimens

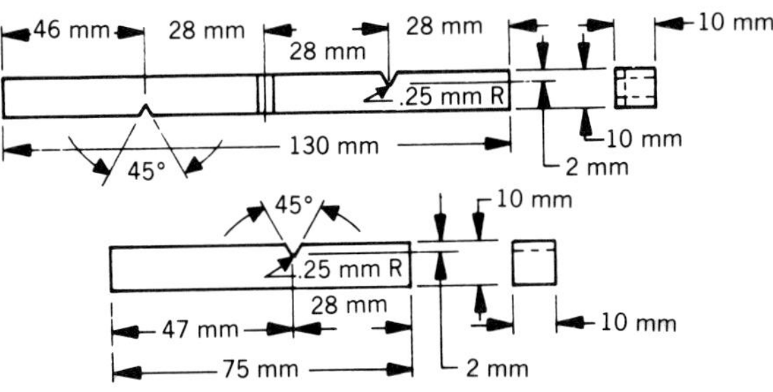

Charpy test specimen

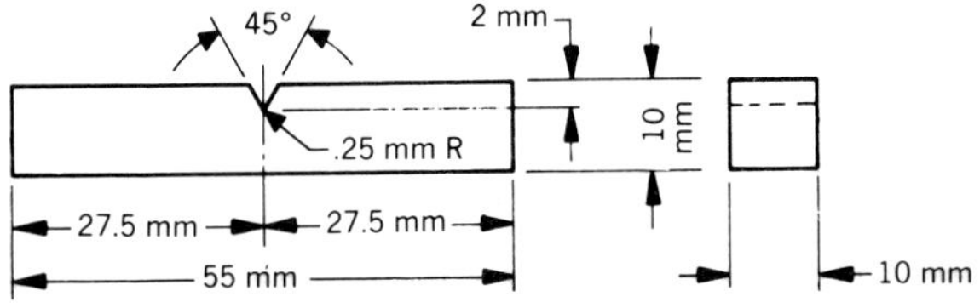

Tension impact specimen

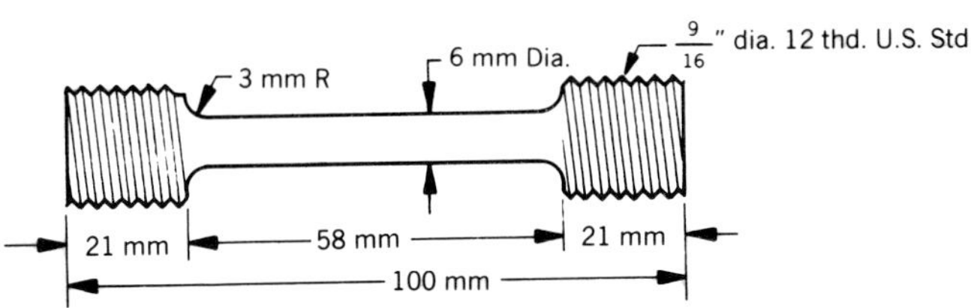

Figure 17
Izod–Charpy testing machine (Photograph courtesy of Tinius Olsen Testing Machine Co., Inc., Willow Grove, PA).

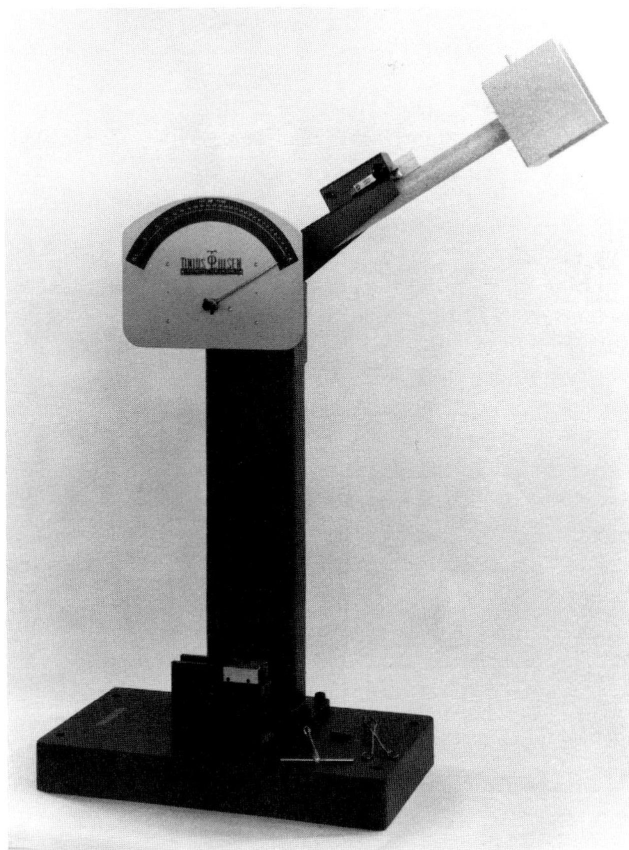

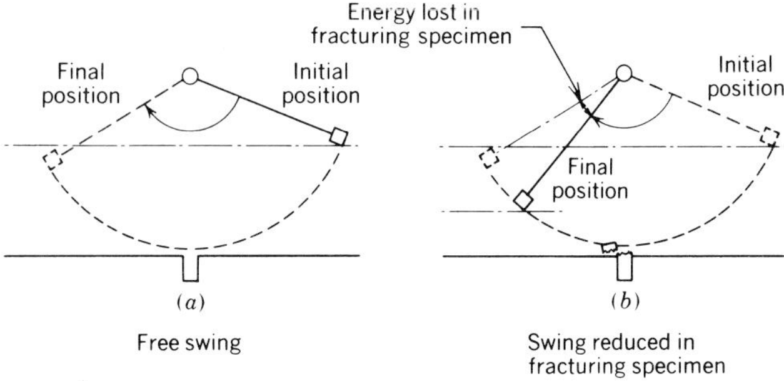

Figure 18

The method by which (Izod) impact values are determined (White, Neely, Kibbe, Meyer, *Machine Tools and Machining Practices*, Vol. II, © 1977 John Wiley & Sons, Inc.).

Figure 19

Mounting of the Izod specimen (Photograph courtesy of Tinius Olsen Testing Machine Co., Inc., Willow Grove, PA).

Figure 20

Mouting of the Charpy specimen (Photograph courtesy of Tinius Olsen Testing Machine Co., Inc., Willow Grove, PA).

clamped in a vise, breaks it, and continues swinging. It will not swing as high as the starting position because of the energy absorbed in breaking the specimen (Figure 18). The difference in starting and ending height is measured in foot–pounds. The pendulum will not swing as high for tough metals as for brittle metals. The difference between the two styles of testing, Izod and Charpy, is in the mounting of test specimens (Figures 19 and 20). The Izod specimen is mounted horizontally and the Charpy specimen is mounted vertically.

Metals at low temperatures are tested on Izod–Charpy machines to determine the effect of temperature drop on various metals. Metals that remain ductile at low temperatures show a slow steady decrease in ductility as the temperature drops. Metals that become brittle at low temperatures show a temperature range in which ductility and, more important, toughness drop rapidly. This **transition zone** can be seen in Figure 21. When the notch-bar specimens show half brittle and half ductile failures, the transition temperature has been reached. Metal for parts that are designed for low-temperature service must be selected so their operating temperature is well above their transition zone of embrittlement.

Figure 21

Appearance of Charpy V-notch fractures that were obtained in a series of tempered martensite of hardness 30Rc (Illustration courtesy of LTV Steel Company).

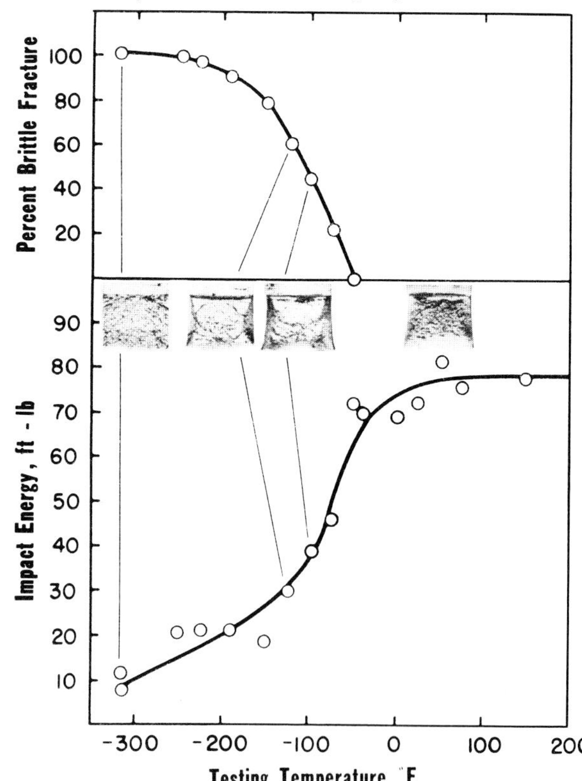

Table 2
Creep Strengths for Several Alloys

Alloy	70° F Tensile Strength (PSI)	Creep Strength (PSI) 800° F—Stress for 1 percent Elongation per 10,000 hr	1200° F—Stress for 1 percent Elongation per 100,000 hr	1500° F—Stress to Failure
.20 percent carbon steel	62,000	35,100	200	1,500
.50 percent molybdenum .08 percent to 20 percent carbon steel	64,000	39,000	500	2,600
1.00 percent chromium .60 percent molybdenum .20 percent C steel	75,000	40,000	1,500	3,500
304 stainless steel 19 percent chromium 9 percent nickel	85,000	28,000	7,000	15,000

SOURCE John E. Neely, *Practical Metallurgy and Materials of Industry*, Second Edition, Wiley, New York, © 1984.

This transition brittle zone of metals has caused some spectacular failures in the past, before it was known that some low carbon steels would fail at temperatures lower than 0° F (−18° C). Some of these failures were in pipelines, storage tanks, bridges, and ships breaking in half at sea. These failures are almost instantaneous, with the split moving along pipelines at 600 feet per minute (FPM). Plain carbon steels having coarse grain have higher transition temperatures than fine-grained and alloy steels. Certain alloying elements in steel, particularly nickel, tend to lower the transition temperature. Stainless steel containing nickel is used to contain liquefied gases at temperatures below −150° F (−101° C). Aluminum and monel are also used at cryogenic temperatures without embrittlement. High strength, low alloy steels (HSLA) that contain such elements as columbium, cerium, molybdenum, and silicon were developed for Arctic pipelines.

The **modulus of elasticity** is actually an expression of the stiffness of a metal. It is also called *Young's Modulus*. If, within the elastic range, the unit stress is divided by the corresponding unit strain at any given point, the result will be the modulus of elasticity for that material; thus,

$$\text{modulus of elasticity} = \frac{\text{unit stress}}{\text{unit strain}}$$

The modulus of elasticity for some common metals is given in Table 1. Carbon, alloy, and hardened or soft steels all have about the same modulus, which varies between 29 million and 30 million PSI. This means a steel piece of a given size and shape will twist or bend about the same amount with a given load within the elastic range. Therefore, by replacing a mild steel part with an alloy part, its tensile strength may be increased but not its stiffness and it will still deflect the same amount if it is the same size. However, if it is exchanged for a metal having a higher modulus, such as tungsten carbide, which is 50 million PSI, then it would not bend or deflect so easily. Surprisingly, cast iron will deflect to some degree within its elastic range, showing a stiffness that is about half that of steel.

Creep is a continuing plastic flow at a stress below the yield strength of a metal. For the most part, creep is a high-temperature phenomenon that increases as the temperature increases. Table 2 gives some stress loads at various temperatures required to cause elongation by creep. Some metals have been designed for high temperature service such as for aircraft jet turbine engine blades that show no signs of creep at temperatures that would melt many metals. Creep tests are made at elevated temperatures under a certain stress in creep-testing machines (Figure 22) over a period of many hours.

Metal fatigue can occur when metal parts are subjected to repeated loading and unloading, particularly in cyclic reversals of stress, such as is seen in a rotating shaft having transverse or one-side loading. Fatigue failures can occur at stresses far below the yield strength of a material with no sign of plastic deformation. When

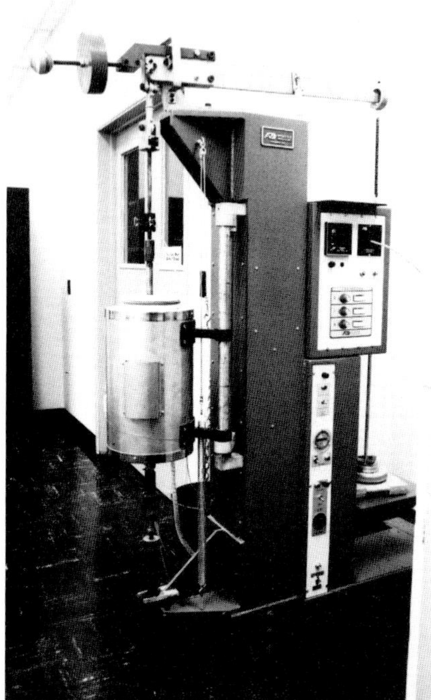

Figure 22
Creep testing machine in Oregon Bureau of Mines testing laboratory.

Figure 23
Fatigue on a shaft.

Figure 24
Fatigue of Ti-6A1-4v, 28,000 cycles at 8000 lb. (10 Hz). Transmission electron microscope replica (5400 ×) (Oregon Bureau of Mines).

machine parts that are subject to cyclic loading are designed, fatigue strength may be more important than yield or ultimate strength. Fatigue is not time dependent and can be initiated by any number of factors such as machining tool marks or welding undercuts and localized stress caused by a welding bead. Fatigue can be accelerated by a corrosive atmosphere and by higher frequency stress reversals. A typical fatigue break pattern of a rotating shaft can be seen in Figure 23. Fatigue break patterns on a shaft usually can be seen in three stages: the first is a slow progression of striations that has a smooth surface, the second can be seen as coarse striations (like on a clam shell) (Figure 24) that progresses more rapidly, and the third is the brittle, crystalline section that suddenly fails when the shaft strength finally becomes less than the applied stress. Metals are tested for their fatigue strength (endurance test) on a fatigue testing machine that rotates a specimen under various loads and conditions. Fatigue life can be considerably increased by surface hardening or by carburizing or shot peening.

Physical Properties of Metals

The **thermal and electrical conductivity** of metals are related. If a metal such as silver or copper is a good conductor of electricity, it is also a good conductor of heat. Pure metals are better conductors than their alloys. This is the reason that copper or aluminum used for electrical wiring must be uncontaminated. Electrical copper grades are 99.97 percent pure. Resistance to the flow of electricity can be altered by several factors.

1. It increases as the temperature increases.
2. It increases as a result of cold-working metal.
3. It increases with the amount of impurities and alloying elements present.

In the case of electrical elements, resistance is an advantage in which the electric current produces and liberates heat.

Metals may look alike but have very different physical properties. For example, one can scarcely tell the difference between zinc, nickel, stainless steel, and silver by appearance, but if you held a bar of equal length of each in your hand and heated the other ends, you would quickly drop the silver bar but could hold the

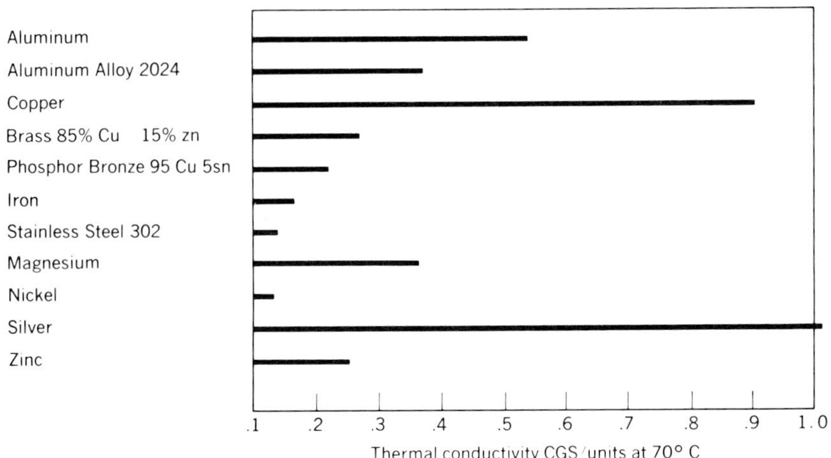

Figure 25
Comparison of thermal conductivity (White, Neely, Kibbe, Meyer, *Machine Tools and Machining Practices*, Vol. II, © 1977 John Wiley & Sons, Inc.).

others for several minutes before feeling any heat. Figure 25 is a comparison of thermal conductivity for some common metals.

Most metals expand when heated and contract when cooled. This change in dimension is called the **coefficient of thermal expansion.** Each metal expands and contracts at a different rate when heated or cooled. This physical property of metals can cause difficulties in construction and manufacturing. Engineers plan use of expansion joints in paved highways and bridges to avoid buckling. When a metal device is to be subjected to repeated heating and cooling, dissimilar metals that expand at different rates can cause abnormal stresses and subsequent damage when rigidly fastened together. If a machinist measures a finished dimension while the part is still hot from machining, the part may be under the tolerance limit when it is cool and it will have to be scrapped. For example, brass expands at the rate of 0.0001 in. per degree Fahrenheit per unit length (or diameter). A machined brass sleeve, three inches in diameter, has a tolerance of plus .001 in. or minus .000 in. The temperature rises 100° F (38° C) above room temperature. When it cools back to room temperature, it has shrunk undersize (.00001 in. × 100 × 3 = .003 in.) and must be scrapped. The coefficient of thermal expansion for some common metals can be seen in Figure 26.

Some metals have a high degree of **reflectivity** and are used as light reflectors and in insulation as heat deflectors. Aluminum, nickel, chromium, tin, and bronze are among the metals having high reflectivity, whereas lead has virtually none; yet reflectivity is one of the characteristics that distinguishes metals from other substances.

Ferromagnetism is characterized by an attraction between certain metals that have an ability to retain a residual magnetic force. The ferromagnetic metals are iron, cobalt, and nickel. Of these, iron is the most commercially important metal used extensively in electrical machinery. Soft iron does not retain much residual magnetism when removed from a magnetic field. This quality makes it useful for electromagnets and electric

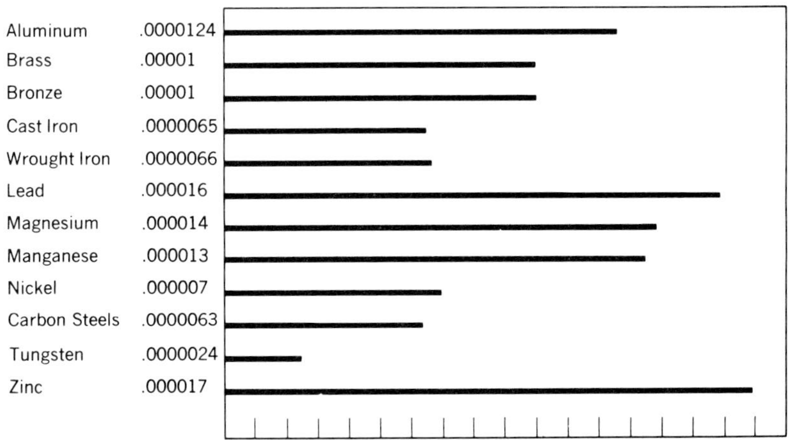

Figure 26
Coefficient of thermal expansion per degree Fahrenheit per unit length (White, Neely, Kibbe, Meyer, *Machine Tools and Machining Practices*, Vol. II, © 1977 John Wiley & Sons, Inc.).

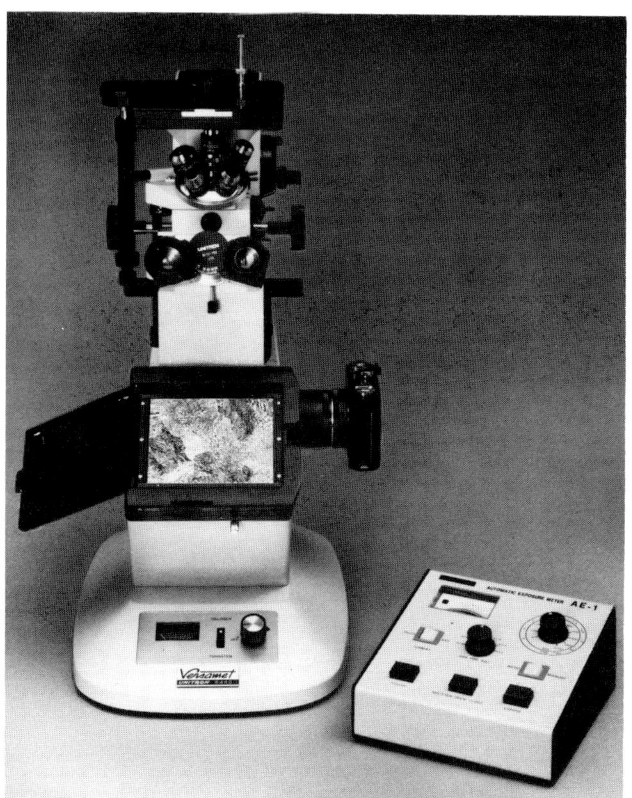

motors in which the field is turned on and off or reversed repeatedly. Powerful permanent magnets are made by alloying metals such as cobalt, nickel, iron, and aluminum.

METALLURGICAL MICROSCOPY

Metal surfaces are examined by means of metallurgical microscopes and associated techniques of surface preparation. Structures in metals not readily visible to the naked eye may be seen at high-power magnification from 10 times the diameter (10×) to 3000× for light

Figure 27
The Versamet metallographic microscope has the advantage of having a display screen, so more than one person may observe the magnified specimen. Also photographs may be taken of the enlarged specimen (*Versamet* is a trademark and a copyright of Unitron Inc.).

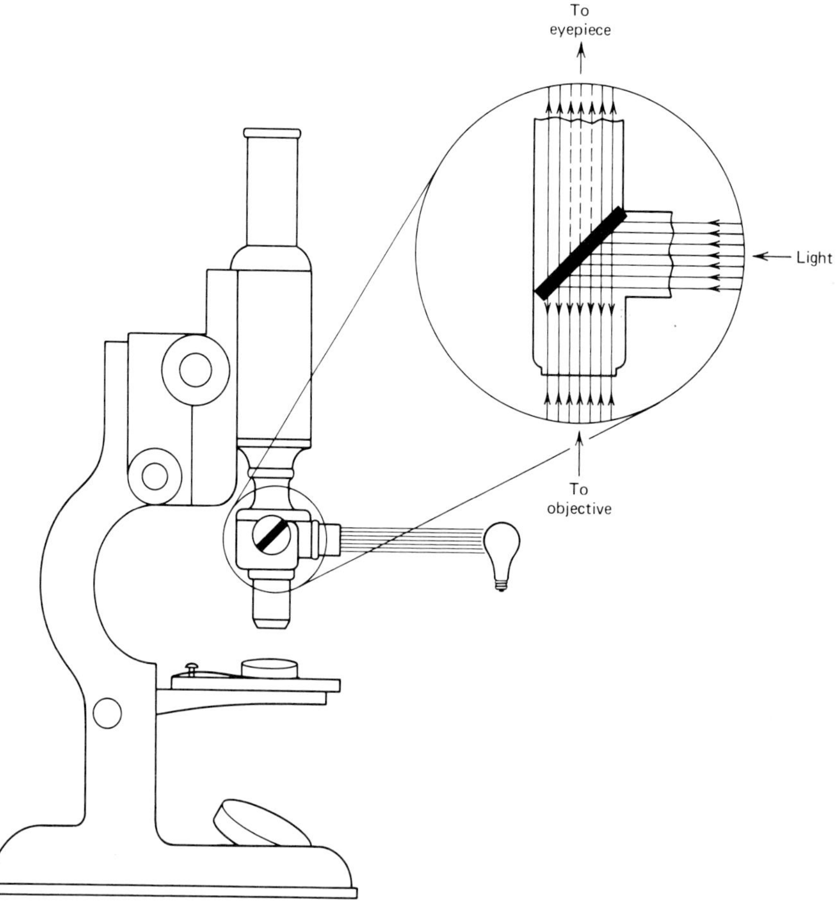

Figure 28
Illumination in a metallurgical microscope (Neely/*Metallurgy* 2 Ed., © 1984 John Wiley & Sons, Inc.).

Figure 29
BMEC inverted stage microscope. The specimen is placed upside down on the stage on the top of the microscope. A 35mm camera can also be attached to this microscope (BMEC is a trademark and a copyright of Unitron Inc.).

microscopes and many more thousands of diameters with the electron microscope. Techniques of photomicrography in which a camera is attached to the microscope make possible photographic illustrations of specimens (Figure 27). Metallurgical microscopes, unlike those used to see through a specimen, use reflected light from an internal source (Figure 28). Inverted stage microscopes are widely used to study metals (Figure 29).

Specimens to be analyzed are cut from a metal part with a metallurgical saw (Figure 30). The specimen is mounted in plastic in a press (Figures 31 and 32) and

Figure 31
Metallurgical press. After the specimen is placed in the press, granulated thermoplastic is poured in. It is then heated under pressure, forming a solid cylindrical shape with the specimen embedded on one side (Beuhler, Ltd.).

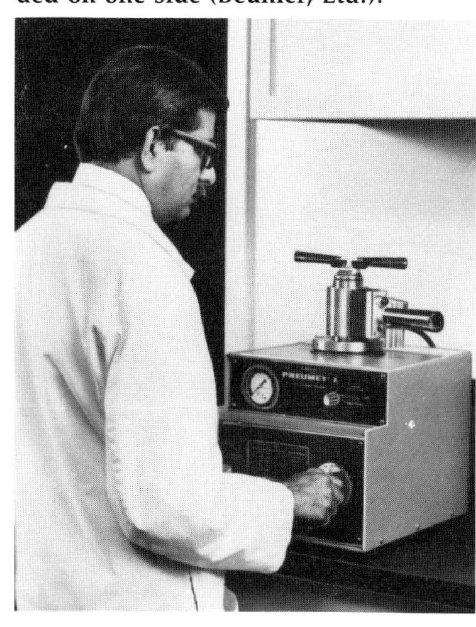

Figure 30
Metallurgical saw. The specimen is cut off with a thin abrasive blade to which coolant is applied. This keeps the specimen from becoming overheated causing damage to the metallurgical structure (Beuhler, Ltd.).

Figure 32
Encapsulated specimen.

Figure 33
Rough grinding the specimen (Beuhler, Ltd.).

Figure 35
Polishing table (Beuhler, Ltd.).

rough ground (Figure 33). Next, the encapsulated specimen is ground on successively finer grit abrasive surfaces (Figure 34) after which it is polished on a rotating cloth-covered disk (Figure 35) or by other techniques such as electropolishing or vibratory polishing. When the specimen has a mirror finish, it is etched in a reagent which is most commonly a dilute acid, varying with the type of metal being prepared. Etching causes surface characteristics of the metal to be visible; some of these are grain boundaries, grains, impurities, and microstructures in the grains such as pearlite, bainite, ferrite, iron carbide (cementite), and martensite (Figure 36).

Ferrite is more or less pure iron. Pearlite is a relatively soft form of alternate layers of cementite and ferrite. Bainite is harder than pearlite and is actually a tough form of martensite. Acicular martensite is an extremely hard form of steel. The microscopic study of metals can help us to readily determine the condition of the specimen. Of course, much more information about metals is gained by metallurgists by studying microstructures in metals with powerful microscopes.

Figure 34
Finish grinding the specimen on HandiMet® (Beuhler, Ltd.).

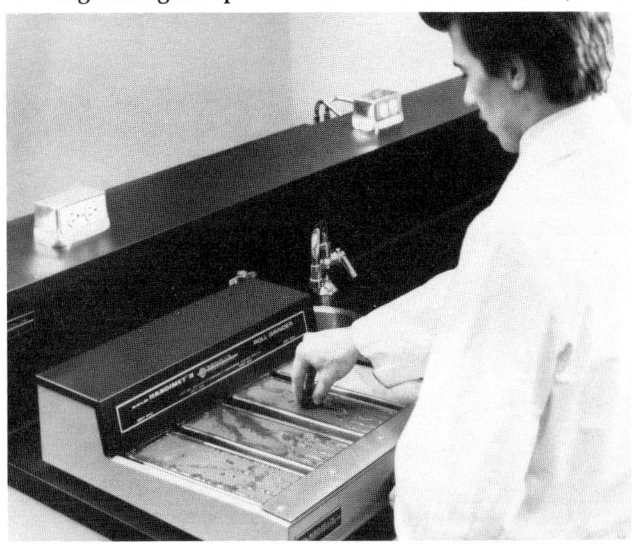

NONDESTRUCTIVE TESTING

As the name implies, nondestructive testing in no way impairs the part for further use. It does not measure mechanical or physical properties, but instead identifies defects such as voids, inclusions, or cracks that will later lead to failure of the part. Also, improper handling of materials that cause changes in the microstructure can lead to failure. For example, a nick or dent in a highly stressed part such as a spring can cause early failure. At present, even these aberrations can be detected on a production basis. Nondestructive methods can also verify hardness or heat treatment in many cases. Since testing for defects in the early stages of the manufacturing process can substantially reduce rejects, nondestructive testing is closely related to quality control. (See Chapter 20.) It has been assumed in what has been called the damage-tolerant philosophy in the study of fracture mechanics that all metals have microscopic flaws such as discontinuities and microcracks on the surface that will lead to ultimate failure. With

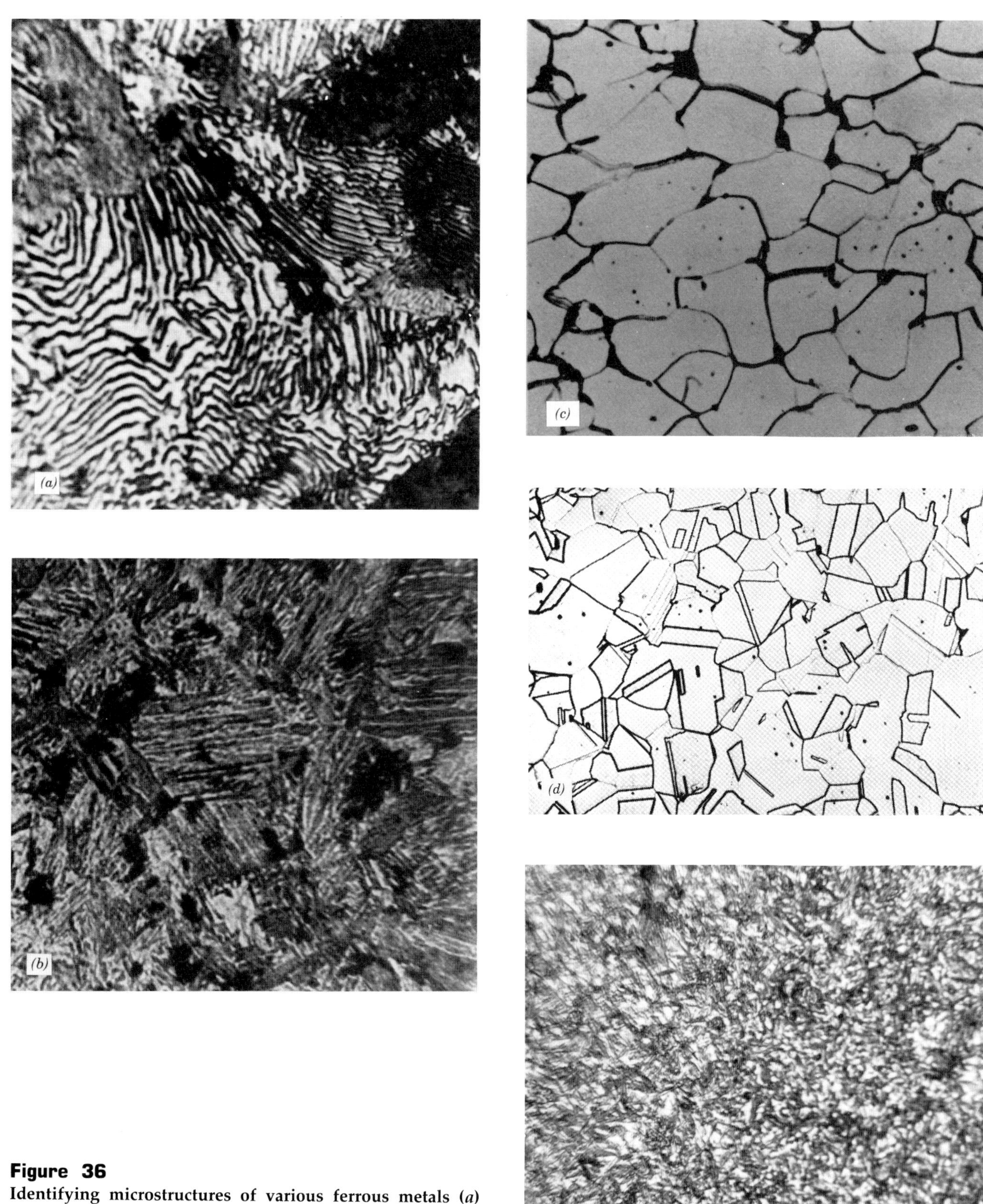

Figure 36
Identifying microstructures of various ferrous metals (*a*) pearlite, (*b*) bainite, (*c*) ferrite (annealed low carbon steel), (*d*) austenite, and (*e*) martensite. (*c*) and (*d*) are from *Metals Handbook,* Vol. 7, 8th ed., American Society for Metals, 1972, pages 9 and 86. With permission.).

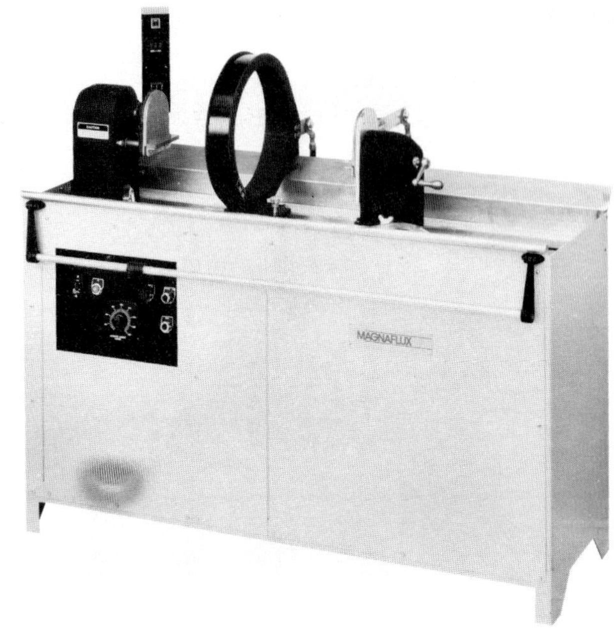

Figure 37
Magnetic-particle testing machine (Courtesy of Magnaflux Corporation, Chicago).

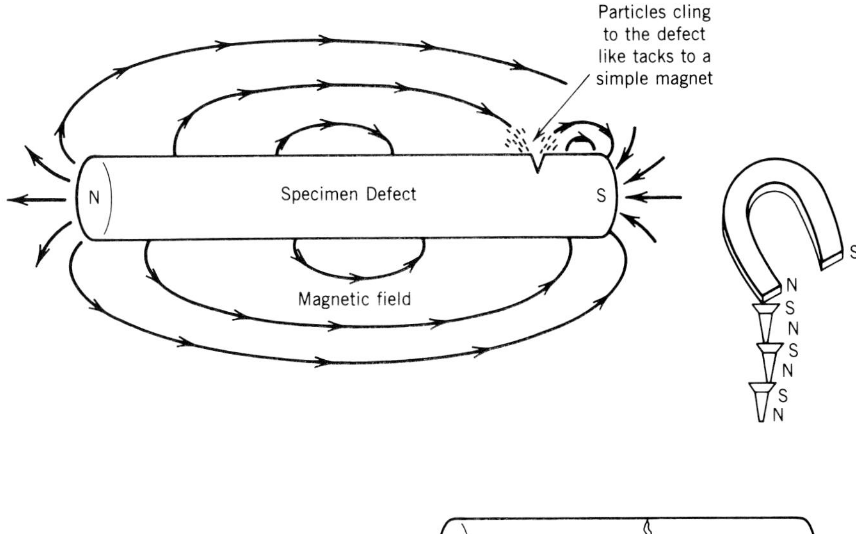

Figure 38
By inducing a magnetic field within the part to be tested and applying a coating of magnetic particles, surface cracks are made visible; in effect, the cracks form new magnetic poles. Particles cling to the defect as tacks would to a simple magnet (Courtesy of Magnaflux Corporation, Chicago).

inspection and testing methods, a safe service life or stress-life curve can be determined. Along with a safety factor, the design is known as a safe-life design. With the increasing use of composites and HSLA steels in automobiles and other equipment, nondestructive testing will be needed even more in manufacturing and maintenance operations.

The most common types of nondestructive testing are:

1. Magnetic-particle inspection
2. Fluorescent penetrant inspection
3. Dye penetrants
4. Ultrasonic inspection
5. Radiography (X ray and gamma ray)
6. Eddy current inspection

Magnetic-particle inspection methods are used to detect surface flaws in ferromagnetic materials such as iron and steel (Figures 37 to 40). One method starts with sprinkling dry iron powder on a magnetized part. A more widely used method starts with pouring a liquid in which the iron powder is suspended over the

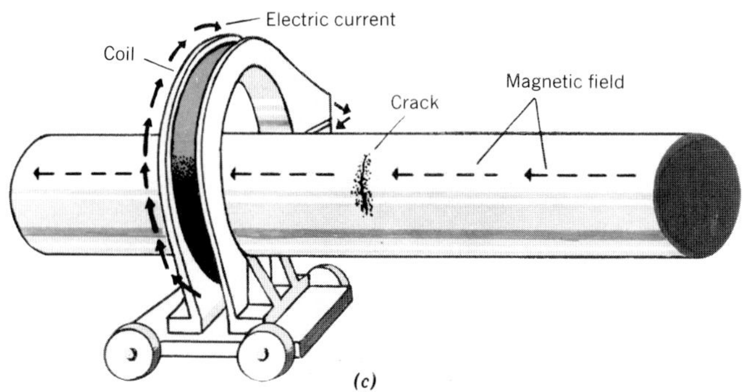

Figure 39
Longitudinal method of magnetization (Courtesy of Magnaflux Corporation, Chicago).

Figure 41
With Zyglo–Pentrex, this vanadium alloy stainless forged hip-ball joint used in medical prosthesis is revealed to have a crack, detected in manufacture. This could result in severe injury to patient (Courtesy of Magnaflux Corporation, Chicago).

Figure 40
Circular method of magnetization (Courtesy of Magnaflux Corporation, Chicago).

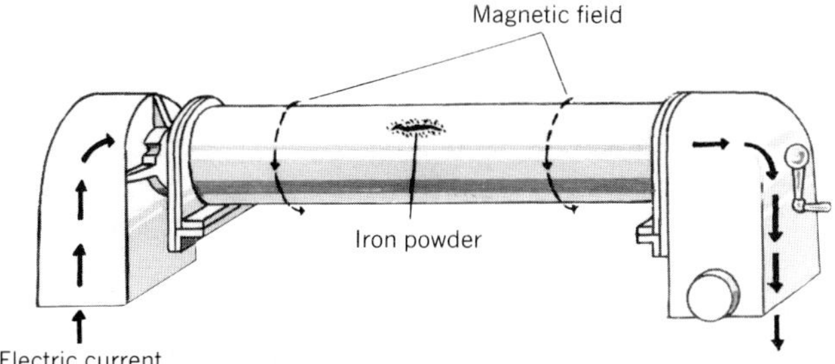

magnetized workpiece; this is then viewed with a fluorescent light. Surface cracks only a few millionths of an inch wide can be seen. This method of testing is also used in automated manufacturing systems with advanced laser scanning techniques to sort out questionable parts from good ones.

Fluorescent penetrants are available in sealed pressure spray cans for application for visual examination in any material such as metals, ceramics, and plastics. A fluorescent light is also used with this process to detect cracks. Testing systems range from huge automated systems to portable test kits for field inspection. *Zyglo* is the Magnaflux copyrighted name for this method (Figure 41). Dye penetrants also depend on visual inspection but do not use fluorescent light. Essentially, a dye is caused to seep into cracks and then is wiped off, and a developer is applied. A bright color then indicates the location of cracks (Figures 42 and 43). This method, called Spotcheck®, is also useful for field testing.

Figure 42
Spraying Spotcheck developer on a gear (Courtesy of Magnaflux Corporation, Chicago).

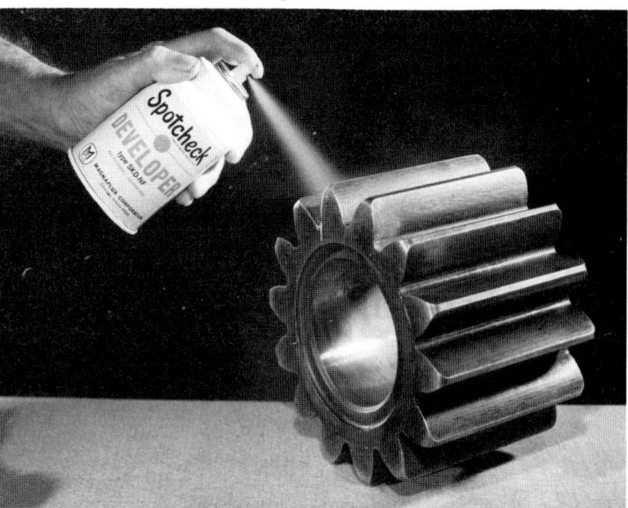

Figure 43
Defect is now visible on the gear (Courtesy of Magnaflux Corporation, Chicago).

Figure 44
How ultrasonic sound waves are used to locate flaws in material (White, Neely, Kibbe, Meyer, *Machine Tools and Machining Practices*, Vol. I, © 1977 John Wiley & Sons, Inc.)

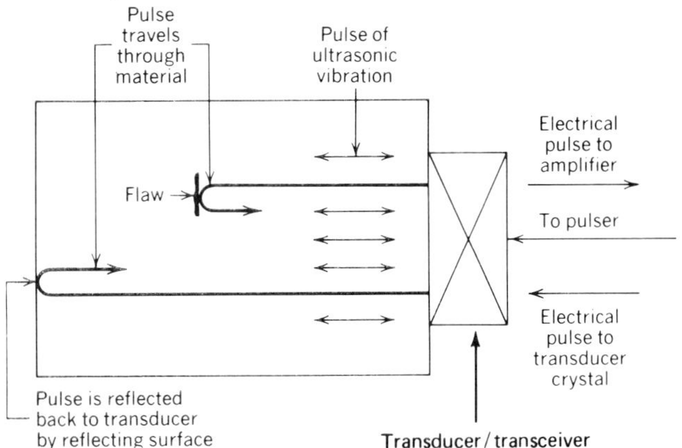

Ultrasonic inspection makes use of high-frequency sound waves that are sent through a test piece by means of a transducer which is usually a piezoelectric crystal that converts electrical energy to mechanical energy (Figures 44 and 45). A transceiver that has the opposite function of converting mechanical energy (vibrations) to electrical pulses picks up the ultrasound waves and displays them on an oscilloscope (Figure 46) as signal pulses called pips. In the pulse–echo system, part of the sound wave is reflected back from the other side of the material to the transducer/transceiver and is seen as a second pip on the oscilloscope. If there is a flaw, a smaller pip will be seen between the two, indicating the distance from the surface to the flaw. The through-

Figure 45
Through-ultrasonic transmission (White, Neely, Kibbe, Meyer, *Machine Tools and Machining Practices*, Vol. I, © 1977 John Wiley & Sons, Inc.)

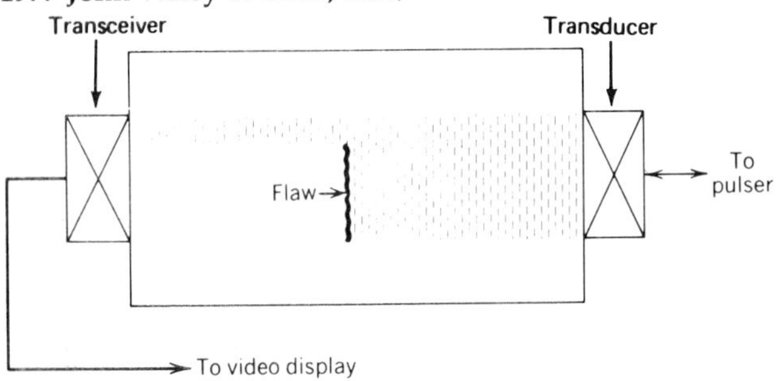

Figure 46
Ultrasonic pulse–echo system (White, Neely, Kibbe, Meyer, *Machine Tools and Machining Practices*, Vol. I, © 1977 John Wiley & Sons, Inc.)

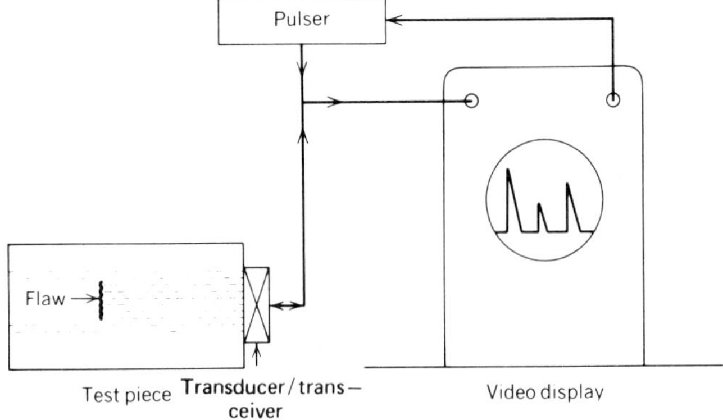

Figure 47
Portable ultrasonic machine instruments can be used for routine manual inspection under field conditions. This is very useful for structural weld testing and corrosion surveys (Courtesy of Magnaflux Corporation, Chicago).

transmission method uses a transducer on one side and a transceiver on the other. Ultrasonics is one of the best ways to check for cracks in pipelines and for structural weld testing (Figure 47). This method of inspection is very adaptable to automated manufacturing systems and is widely used for inspection for internal defects, cracks, laminations, variation in thickness, and weld bonds.

Radiography, X-ray, and gamma ray inspection have been widely used on pipelines and manufactured parts to detect flaws, especially internal defects such as porosity (Figures 48 and 49). There are several disadvantages of this system. Since X-ray and gamma radiation are both hazardous, technicians using this process are required to be trained. Radiography is also expensive and time consuming and there is a considerable time delay until the film is developed. The test is also two dimensional, possibly missing flaws viewed at another angle. However, a new process in which a video image and tape that is made in real time (instantly) has been developed. In the past, attempts at real-time imaging of X rays had poor resolution and were not practical.

Eddy-current inspection systems can test only electrically conducting materials. When a test piece is placed in or near a coil in which alternating current produces a magnetic field, an eddy current is induced in the test piece. This current causes a change in impedance in the magnetic field that is read on a meter or oscilloscope

Figure 48
An experienced technician is preparing to make a gamma ray inspection of a section of pipe (Courtesy of Magnaflux Corporation, Chicago).

Figure 49
X-ray method of testing. A specialist makes a test of pilot-run parts to evaluate manufacturing techniques and procedures (Courtesy of Magnaflux Corporation, Chicago).

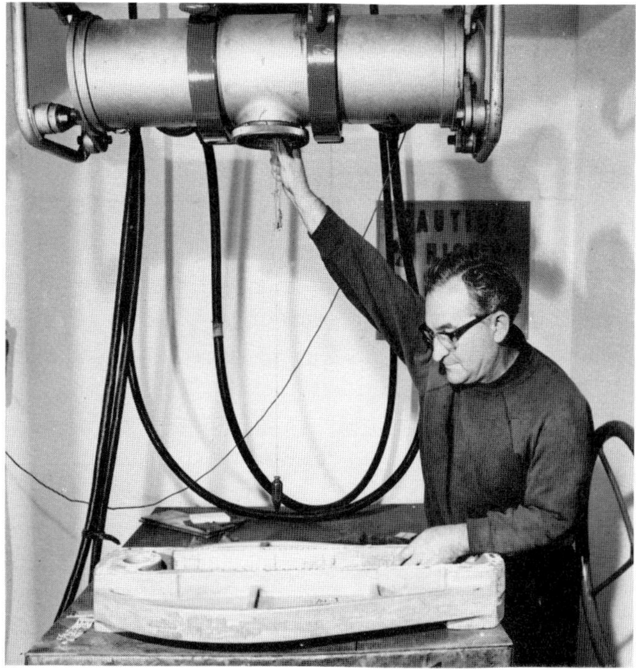

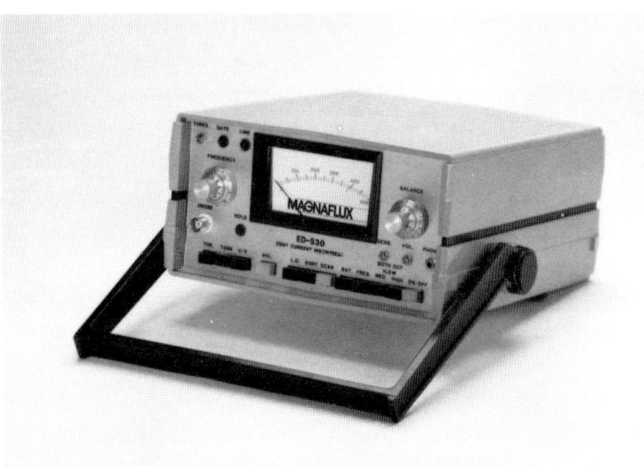

Figure 50
Eddy current testing instrument for nondestructive comparison of magnetic materials. It is used to separate materials according to their metallurgical properties, such as alloy variations, heat treat condition, hardness differences, internal stress, and tensile strength (Courtesy of Magnaflux Corporation, Chicago).

(Figure 50). Variations such as seams, thicknesses, and other physical differences can quickly be detected with this method. Eddy-current testing is quite adaptable to automated systems. In fact, computerized data analysis systems for eddy-current testing have been developed in which analog data are converted to digital printout or magnetic tape storage.

REVIEW QUESTIONS

1. What is the difference between elasticity and plasticity? Give a definition for each.
2. How can tensile strength of a specific metal be determined?
3. What property does a metal possess when it can be easily drawn into a wire, but cannot be rolled into a sheet without splitting?
4. Define hardness as it is measured on a Rockwell or Brinell testing instrument.
5. When a material breaks suddenly with no sign of deformation, it is said to be ______.
6. Some very ductile metals, such as certain low carbon steels, can change their properties at low temperatures at their particular transition zone. How can this information help an engineer design machinery that will be operating in subzero temperatures?
7. Steel has a modulus of elasticity of 29 million PSI and tungsten carbide is 50 million PSI. If a steel and a tungsten carbide bar of the same dimensions were each loaded on one end and supported on the other as cantilever beams, which would bend the most?
8. When a metal is loaded well within the elastic range at a temperature of 800° F (427° C) and it slowly deforms permanently, the phenomenon is called ______.
9. Welds, machining tool marks, and stress reversals, such as those in rotating parts, can cause a progressive failure even on a lightly loaded part such as a shaft. There is no sign of ductility, only a smooth surface having striations like a clam shell. How would you identify this kind of failure? What is it called?
10. Which is the better conductor of heat or electricity, a pure metal or an alloy?
11. By what means is it possible to see the internal structure of metals at high magnification? How is this done?
12. How can machinery parts and pipeline welds be tested for flaws without destroying them? Name several processes.

CASE PROBLEMS

Case 1
Case of the Twisted Shaft

The shafting shown in Figure 51 was taken from a mill transfer chain that was frequently overloaded, causing the shaft (made of low carbon steel) to fail by torsion shear. It has obviously been loaded in torsion beyond its elastic limit into the plastic range and has taken a permanent set. Since the sprocket and bearing bores are not easily changed in order to use a larger diameter shaft, what kind of replacement shaft would solve the problem? What different mechanical properties must it possess so it will not again fail by torsion shear?

Figure 51

Case 2
Broken Shaft

A hammer mill that was located outside a building was driven by an 800 hp electric motor and its main shaft was 8 in. in diameter. It had operated for years; the only maintenance needed was to lubricate bearings and replace hammers and grates. One winter morning the temperature dropped to −10° F and the machine operator turned the switch to start the machine which was unloaded and free-running. The main shaft instantly broke in two in a brittle-type fracture. What could have been the problem and how could it be avoided in the future?

Case 3
Piston Clearance

An aluminum piston 4 in. in diameter moves in the cylinder of a cast iron engine block and it must have clearance diameter of .003 in. when hot. The temperature rise is from 70 to 160° F. What should the clearance between the bore diameter and the piston be if the linear coefficient of expansion per unit length (inch) per degree F is .00001244 for aluminum and .00000655 for cast iron when the temperature is at 70° F?

CHAPTER 4 HEAT TREATMENTS OF METALS

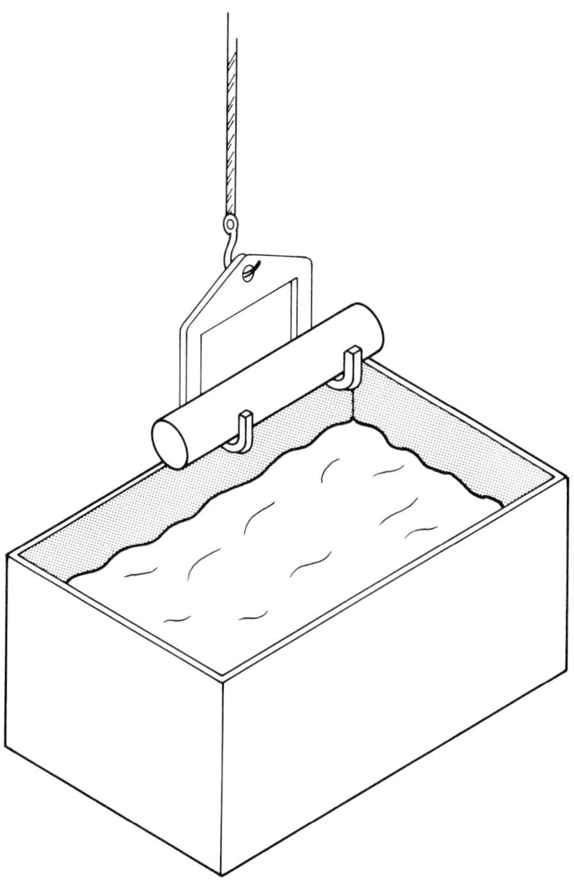

From ancient times, the immense value of metals to the progress of civilization could be attributed to their ability to be easily shaped plastically at ambient temperatures and even more easily at high temperatures. Second only to their elastic and plastic behavior is the ability of carbon-bearing iron, called steel, to become very hard when heated and suddenly cooled as in a quenching medium such as water. Other heat treatments such as temper drawing, annealing, and stress relieving also have a profound effect upon metals. The high quality of swords made in Damascus has been legendary since Alexander the Great was said to have cut the Gordian knot with one and the Saracens wielded them against the Crusaders. The process of making those swords was a well-kept secret that was never discovered and has since been lost. It is now believed that the unique forging process and, most importantly, the heat-treating process used produced this unusually hard and tough grade of carbon steel.

Today, there are many kinds of alloy and tool steels, each of which often require special heat treatments in the manufacturing process to give them the qualities necessary to fit the required specifications. In addition, many nonferrous metals such as aluminum and titanium are heat treated to increase their strength and durability. The manufacture of machinery, automobiles, modern aircraft and space vehicles, and consumer products would not be possible without the technology of heat treatments of metals.

OBJECTIVES

This chapter should enable you to:

1. Use an iron–carbon diagram to explain the basic principles involved in heat treating steel.
2. Use an I-T diagram and cooling curves to explain the various microstructures found in carbon steel as a result of cooling rates.
3. Describe how steel is hardened and tempered.
4. Explain several methods of techniques to surface harden steel and the significance of selective surface hardening in manufacturing.

5. State the principles involved in precipitation hardening.
6. List the various types of annealing and stress relief and explain their purposes and uses for industrial applications.
7. Explain the factors involved in recrystallizing low-carbon work-hardened steel.

THE IRON–CARBON DIAGRAM

In order to understand the basic principles involved in heat treating iron and steel, a person should have an understanding of the iron–carbon diagram (Figure 1). In Chapter 1, Figure 11, the cooling curve diagram shows the phase changes in iron from liquid to solid delta iron at 2800° F (1538° C) and then to gamma iron or austenite as it cools to about 1600° F (869° C). The iron then undergoes transformation to alpha iron or ferrite as it cools further. Since that diagram represents

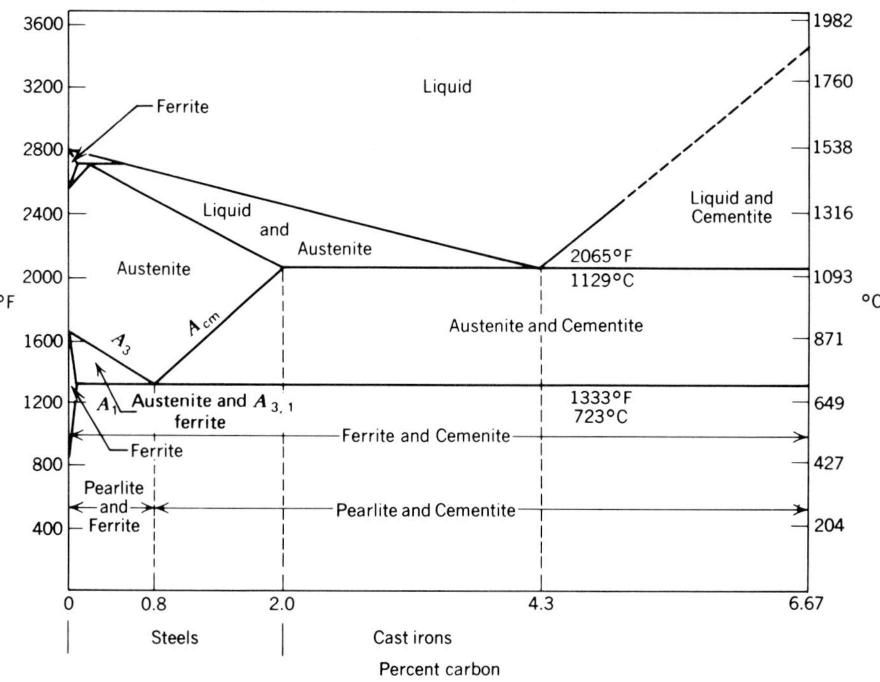

Figure 1
Simplified iron–carbon diagram showing temperatures in equalibrium or very slow cooling (White, Neely, Kibbe, Meyer, *Machine Tools and Machining Practices*, Vol. II, © 1977 John Wiley & Sons, Inc.).

Figure 2
The iron–carbon diagram showing the steel portion (White, Neely, Kibbe, Meyer, *Machine Tools and Machining Practices*, Vol. II, © 1977 John Wiley & Sons, Inc.).

pure iron, the same changes in solid iron can be seen at the left hand vertical line on the iron–carbon diagram. The percentage of carbon in the iron increases toward the right side of the diagram, ending at 6.67 percent carbon. The diagram ends there because 6.67 percent C by weight is the amount of carbon in the compound, iron carbide (Fe_3C), and no other useful information would be given by including more carbon. Most of the diagram is devoted to cast irons because 2 percent carbon and above produces cast iron; 2 percent carbon and under produces steel.

The area within the dotted lines (Figure 2) is the steel portion of the iron–carbon diagram and is the area most important to heat treaters. **Eutectic** is a word meaning the lowest melting point. On either side of the eutectic point of the cast iron portion of the diagram, the metal solidifies slowly over a temperature range from the liquidus to the solidus line in which the metal is somewhat "mushy," but at the eutectic point, it solidifies instantly like a pure metal. The word **eutectoid** in the steel portion is similar but not the same as *eutectic*. Since the metal is now in a solid state, the word does not refer to solidification but rather to the lowest transformation temperature from austenite to ferrite. Again, on either side of the eutectoid the transformation is gradual over a temperature range, but at the eutectoid, it is almost instantaneous. As noted in Chapter 1, up to 2 percent carbon can be contained in austenite as a solution. We can see this in Figure 1 in the area above the A_3 and A_{cm} lines.

When austenite containing .8 percent carbon (eutectoid) is slowly cooled, there is just sufficient carbon present to produce a 100 percent pearlitic microstructure (Figure 3). Pearlite consists of alternating plates of an iron–carbon compound called cementite (Fe_3C) and relatively pure iron, called ferrite (Fe). Less carbon than eutectoid, when slowly cooled, produces whole grains of pearlite and grains of ferrite in proportions depending on the amount of carbon present. When these hypoeutectoid steels cool from a point on the A_3 line to the A_1 line, they undergo a transformation over a range of temperature drop in which the carbon in the austenite forms the pearlite grains of eutectoid steel, drawing the carbon from the prior austenite grains and leaving ferrite at those grain boundaries (Figure 4). Full regular grains of ferrite will form, depending on the cooling rate and carbon content (Figure 5). When hy-

Figure 3
SAE 1090 steel slowly cooled, showing very coarse pearlite (500×).

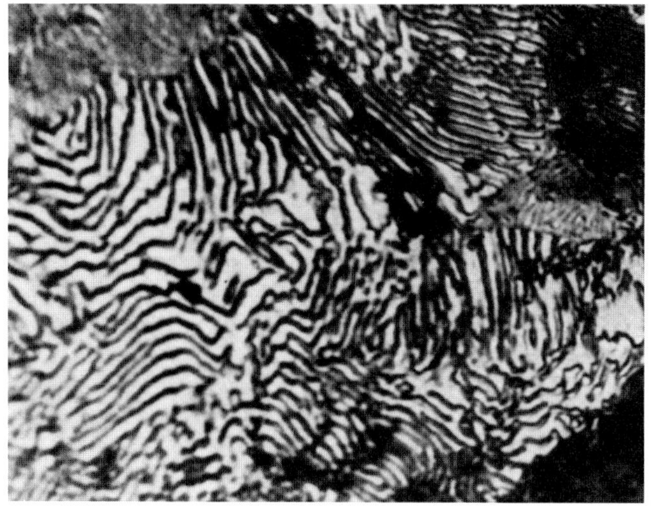

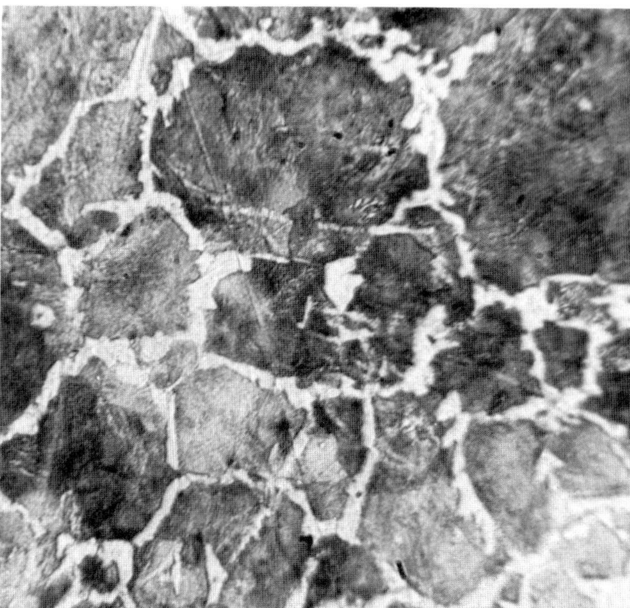

Figure 4
Ferrite at the grain boundaries in SAE 1040 hypoeutectoid steel (500×).

Figure 5
Low-carbon steel with full ferrite grains (500×).

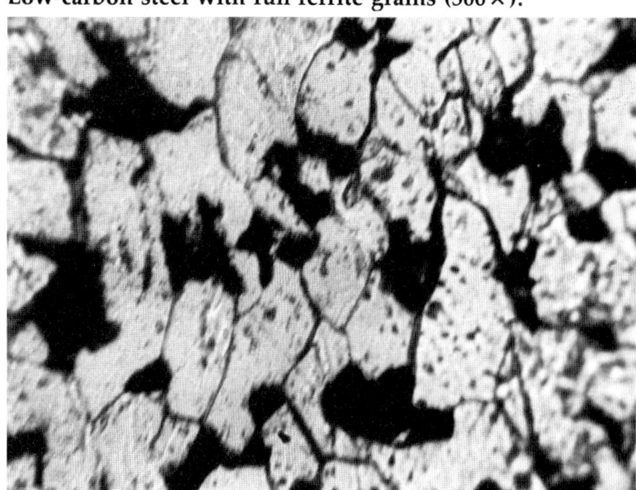

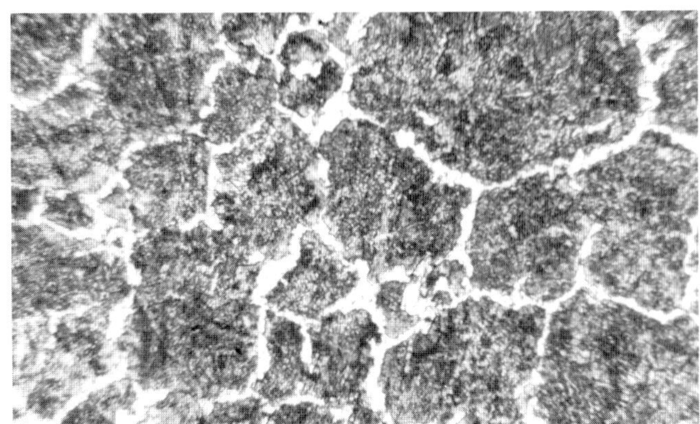

Figure 6
SAE 1095 hypereutectoid steel showing cementite at the grain boundaries. Massive cementite and ferrite both etch white (500×).

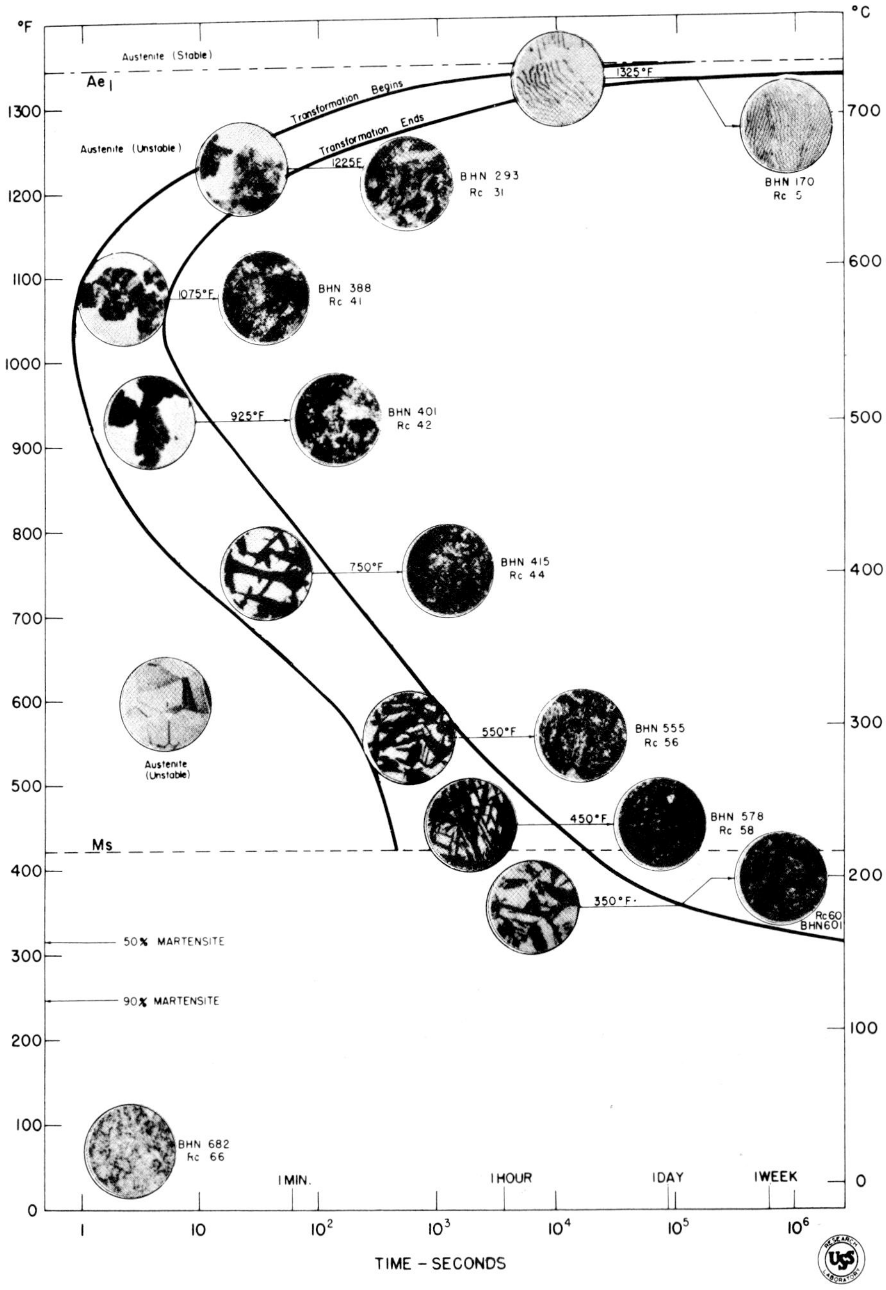

Figure 7
I–T diagram of .89 percent carbon steel (Copyright United States Steel).

pereutectoid steel (steel containing more carbon than eutectoid) is slowly cooled from a point on the A_{cm} line, the excess carbon over .80 percent begins to form cementite at the prior austenite grain boundaries since the A_{cm} line is the point of solubility of cementite. Within the grains the eutectoid composition is formed and at the $A_{3,1}$ line, the grains become 100 percent pearlite with a network of cementite at the grain boundaries (Figure 6).

Now microstructures other than those shown in the study of slow-cooled plain carbon steel in the iron–carbon diagram will form if the cooling rates are more rapid. These other microstructures are best explained in an isothermal–transformation (I–T) diagram (Figure 7). Sometimes these diagrams are referred to as a TTT or 3T (time–temperature–transformation) diagram. It can be seen on the I–T diagram that when austenite is cooled to a point below its stable temperature (A_1 on the iron–carbon diagram), it will begin to transform into any of a number of transformation products, depending on the cooling rate. If it is held at a constant temperature (isothermal) for a given length of time, it will transform into one particular kind of microstructure. If, on the other hand, it is so suddenly cooled that the cooling curve No. 1, as shown in Figure 8, does not cut into the "nose" of the I–T diagram, no transformation can take place that would form microstructures that are relatively soft, such as pearlite or bainite, but instead would cool the austenite to an area below the M_s line where martensite is formed. Therefore, the formation of martensite, if cooled rapidly enough, is temperature and not time dependent as are the other microstructures. Martensite is very hard and brittle and must be tempered to be useful (see "Tempering Carbon Steel" later in this chapter). The .90 percent C steel I–T diagram (Figure 8) shows cooling curve No. 2 with a very slow cooling rate as when steel is annealed to soften it. This produces a soft coarse pearlitic microstructure in this carbon steel that is easily machined. Cooling curve No. 3 represents a steel that is air-cooled or normalized in which mixed microstructures, such as fine pearlite and possibly upper bainite, are formed.

Hardening Plain Carbon Steel

Plain carbon steel contains no alloying elements other than carbon and small percentages of those elements such as manganese that are necessary in steel manufacture. It is used for knives, files, and fine cutting tools such as wood chisels since it will hold a keen edge. Carbon steel is much easier to heat treat than the alloy and tool steels are.

The process of hardening steel is carried out in two operations. The first step is to heat the steel to a temperature that is slightly above the A_3 and $A_{3,1}$ lines on

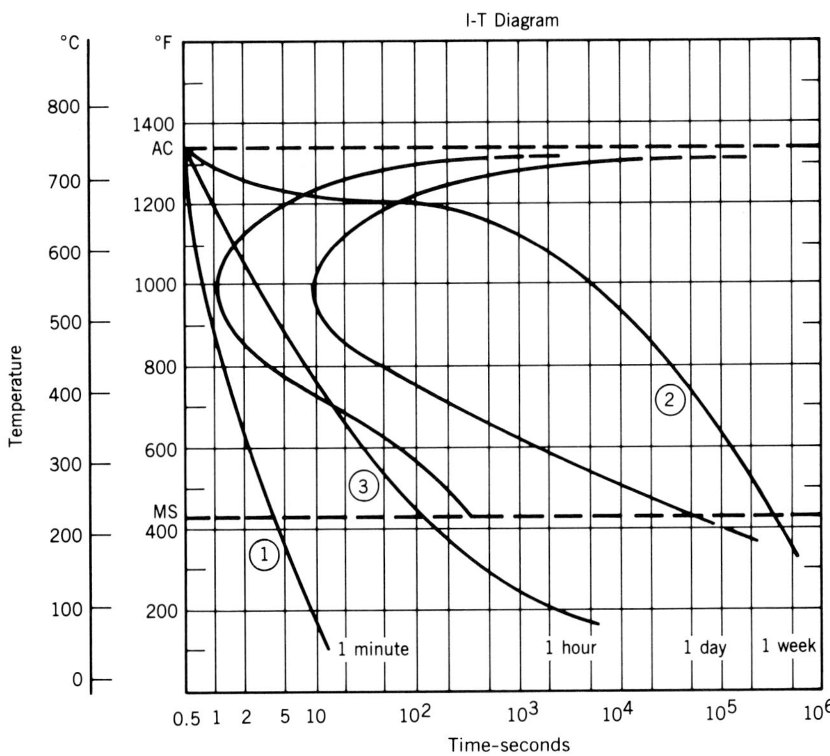

Figure 8
Three cooling curves on an I–T diagram of .90 percent carbon steel.

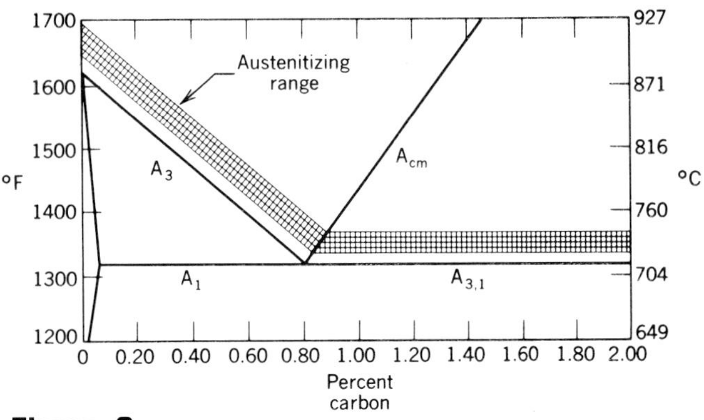

Figure 9
Diagram showing hardening temperature of steels (austenitizing range) (White, Neely, Kibbe, Meyer, *Machine Tools and Machining Practices*, Vol. II © 1977 John Wiley & Sons, Inc.).

the iron–carbon diagram (Figure 9). This operation is called austenitizing by metallurgists. The austenitized steel (face-centered cubic crystal structure) contains all the carbon in the interstices (spaces or voids in the lattice structure). The second step is to cool the red-hot metal so quickly that it has no chance to transform into softer microstructures, but still holds the carbon in solution in the austenite. This operation is called quenching (Figure 10). **Quenching media,** such as brine,

Figure 10
Temperature gradients and other major factors affecting the quenching of a gear. The gear was quenched edgewise in a quiescent (not agitated) volatile liquid (From *Metals Handbook*, Vol. 2, 8th ed., American Society for Metals, 1964, p. 16. With permission).

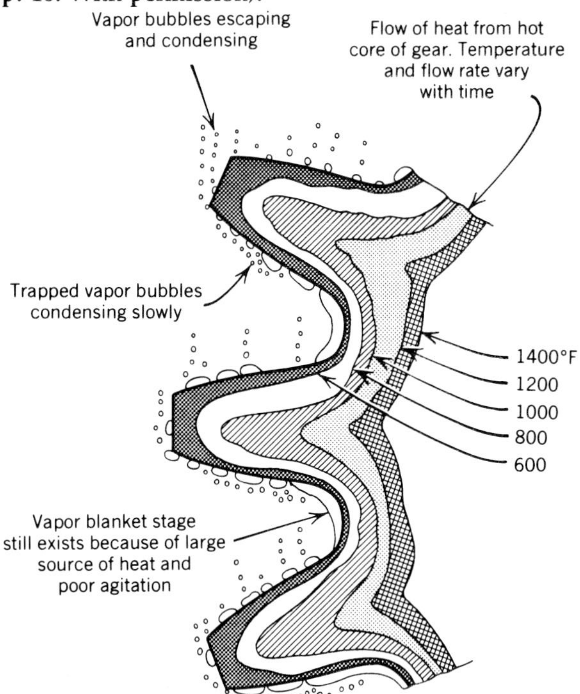

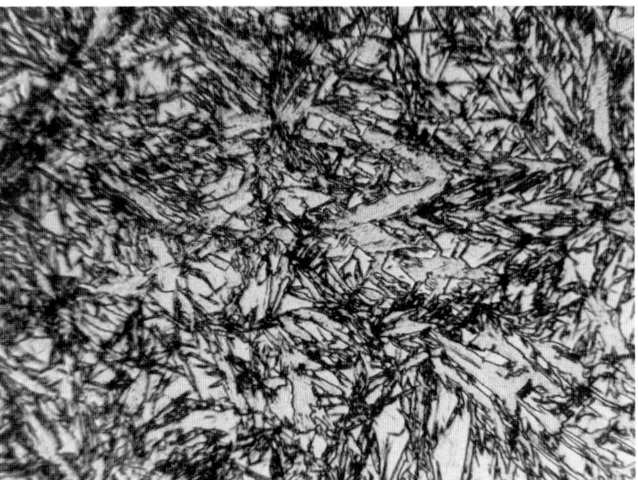

Figure 11
The microstructure of martensite (500×).

tap water, fused salts, oil, and air, all have different cooling rates. Slower cooling is necessary for tool steels and rapid rates are needed for plain carbon steel. Rapid quenching can produce cracking in thicker sections and therefore is normally used on small or thin sections with low mass and for plain carbon steels.

The austenitized steel, when quickly cooled to a point below the M_s temperature, attempts to transform to its natural crystal structure of body-centered cubic which can contain almost no carbon in its interstices. The result is a distorted, elongated cubic structure that is very hard and is called martensite (Figure 11).

TEMPERING (DRAWING) PLAIN CARBON STEEL

Martensite is very brittle until it is tempered or drawn, as some heat treaters call it. A tool that is hardened and not tempered would break in pieces when first used. Tempering is simply a process of reheating the hardened steel to a much lower temperature than that used for hardening. In this process the steel loses some of its hardness, depending on the temperature, and gains toughness, losing some brittleness. Hardening and tempering temperatures are given in Table 1. In Table 2 it can be seen that the higher the tempering temperature is, the softer the metal is. Oxide colors that form on the clean surfaces of steel in this temperature range show heat treaters the approximate temperature of the metal. This color method of tempering was used by blacksmiths to determine the temperature before plunging the part into a water tank to stop the heating action. It is still used to some extent in small shops, but more exact methods are used in the manufacturing of heat-treated steel parts.

Table 1
Temperatures and Colors for Heating and Tempering Steel

	Colors	Degrees Fahrenheit	Process	
Heat Colors	White	2500		
		2400	High Speed Steel Hardening (2250–2400° F)	
	Yellow to white	2300		
		2200		
		2100		
	Yellow	2000		
		1900		
	Orange to red	1800	Alloy Steel Hardening (1450–1950° F)	
		1700		
		1600		
	Light cherry red	—		
		1500		
	Cherry red	1400	Carbon Steel Hardening (1350–1550° F)	
		1300		
	Dark red	—		
		1200		
		1100		
	Very dark red	1000		
		900		
	Black red in dull light or darkness	800		High Speed Steel Tempering (350–1100° F)
		700	Carbon Steel Tempering (300–1050° F)	
Temper Colors	Pale blue (590° F)	600		
	Violet (545° F)	—		
	Purple (525° F)	500		
	Yellowish brown (490° F)	—		
	Straw (465° F)	400		
	Light straw (425° F)	—		
		300		
		200		
		100		
		0		

SOURCE Pacific Quality Steels, "Stock List and Reference Book," No. 85, Pacific Machinery and Tool Steel Company, 1981.

Table 2
Temper Color Chart

Degrees C°	Degrees F°	Oxide Color	Suggested Uses for Carbon Tool Steels	
220	425	Light straw	Steel-cutting tools, files, and paper cutters	Harder
240	462	Dark straw	Punches, dies	
258	490	Gold	Shear blades, hammer faces, center punches, and cold chisels	
260	500	Purple	Axes, wood-cutting tools, and striking faces of tools	
282	540	Violet	Springs, screwdrivers	
304	580	Pale blue	Springs	
327	620	Steel gray	Cannot be used for cutting tools	Softer

SOURCE White, Neely, Kibbe, Meyer, *Machine Tools and Machining Practices,* Volume I, Wiley, New York, © 1977.

Figure 13
Large heat treating furnace (Bureau of Mines).

Figure 12
Small electric furnace.

Figure 14
Drawing of one type of salt bath furnace. Electrodes immersed in the salt produce the heating effect.

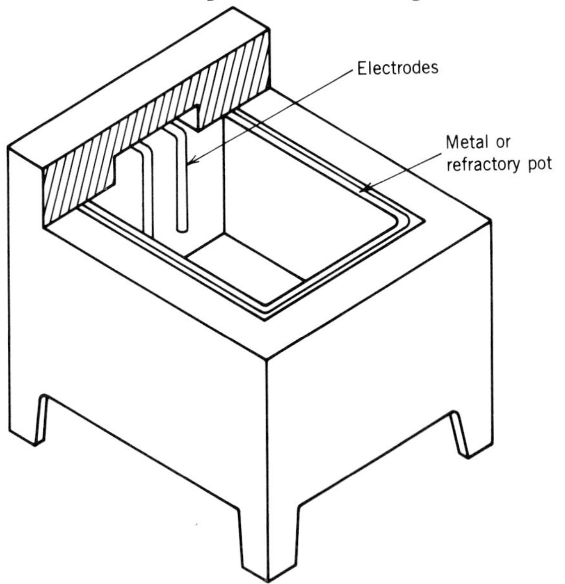

HEAT TREATING EQUIPMENT

Almost any method can be used that will heat steel sufficiently to get it red hot for the purpose of hardening it. Blacksmiths used forges and charcoal or coke. A heating torch is sometimes used, but the most reliable method is to use a temperature-controlled furnace or a molten salt or metal bath (Figures 12 to 14). These high temperature salt baths used for heating metals to the hardening temperature are sometimes called pot furnaces. Metal baskets are used to lower small parts into the furnace. In manufacturing, the heat treating process is often done in a continuous furnace that has a heat resistant belt slowly moving parts through the furnace. The parts are correctly heated for quenching

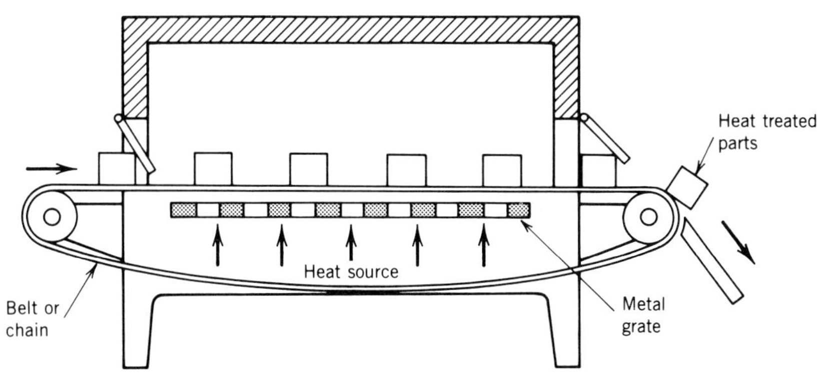

Figure 15
Continuous heat-treating furnace.

Figure 16
Thermcouple sensor. The twisted wires of a thermocouple produce a feeble electric current when heated. Furnace temperatures are controlled from this sensor.

as they emerge from the furnace (Figure 15). All heat treating furnaces are built of a refractory (heat resistant material), usually firebrick, and are controlled by an electronic relay system that turns the furnace off and on. The temperature is sensed by a bimetal thermocouple that is placed directly in the furnace (Figure 16).

SURFACE HARDENING PROCESSES

Flame Hardening and Induction Hardening

These processes in manufacturing are also usually continuous (Figures 17 and 18). Surface hardening of metal parts for greater wear resistance has been common

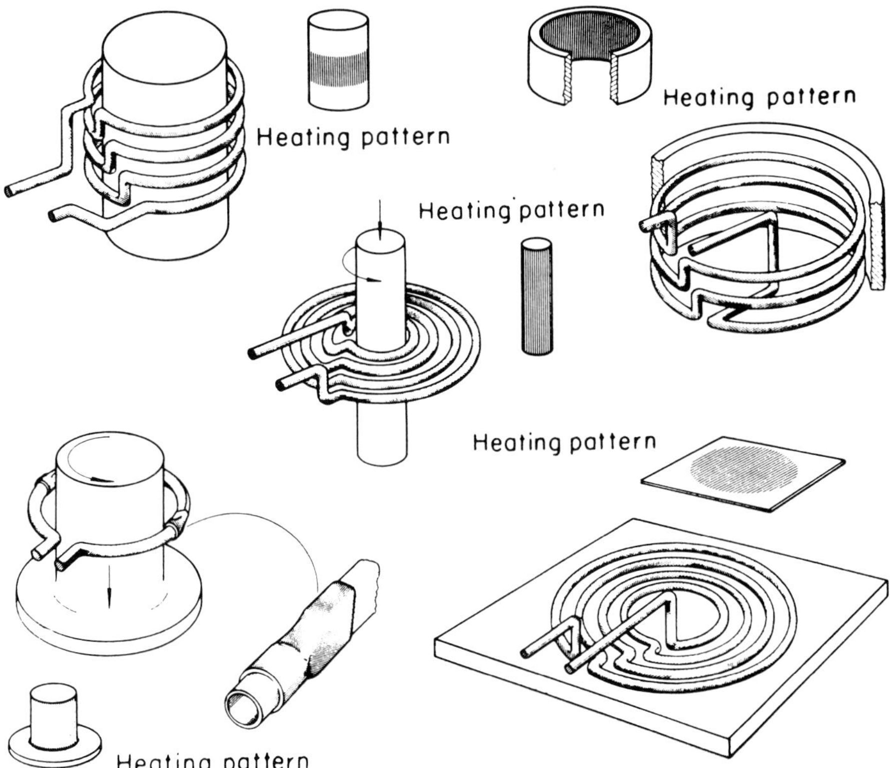

Figure 17
Induction hardening. Typical work coils for high-frequency units (*Metals Handbook*, Vol. 2, 8th ed., American Society for Metals, 1964, p. 172. With permission).

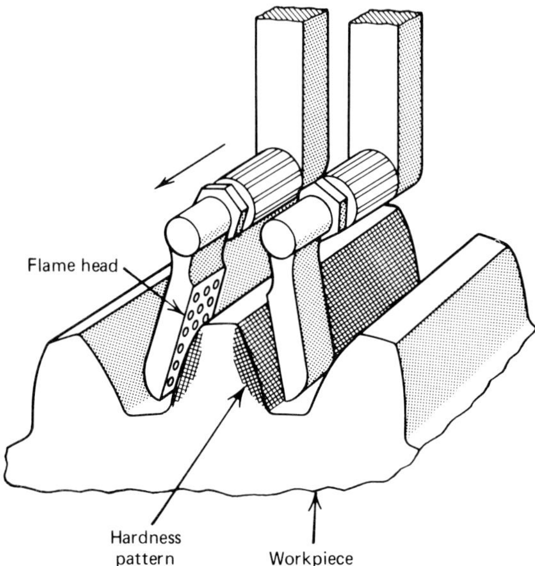

Figure 18
Hardness pattern developed in sprocket teeth when standard flame tips were used for heating. When space permits this method to be used, hardening one tooth at a time results in low distortion (*Metals Handbook*, Vol. 2, 8th ed, American Society for Metals, 1964, p. 200. With permission).

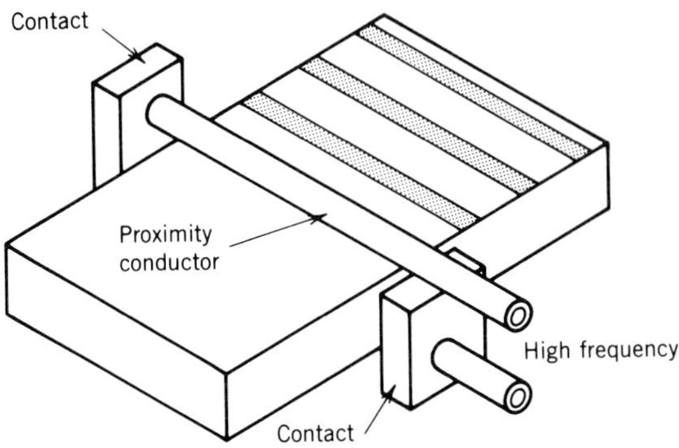

Figure 19
Selective high frequency resistance heating on a flat surface. A water-cooled "proximity conductor" is placed close to the surface to be heated through which a high frequency current is applied.

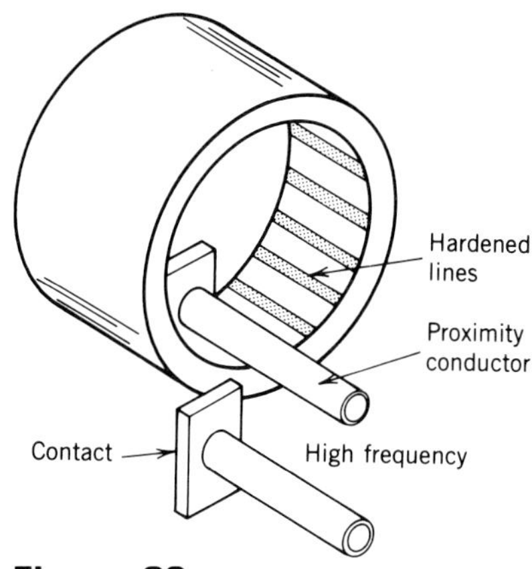

Figure 20
Internal surfaces are selectively hardened by high-frequency heating.

practice in the manufacture of automotive components, industrial machinery, and hardware for many years. Surface hardening creates a hard case around the relatively soft core of the steel part. Steels that contain sufficient carbon and also some cast iron can be quickly heated on the surface. When the steel contains very little carbon, the surface chemistry may be altered by diffusing carbonaceous or nitrogenaceous substances into the surface followed by heating and quenching in most cases. Some of these processes are carburizing, nitriding, carbon–nitriding, and cyaniding. Low carbon steel can be carburized and surface hardened to a depth of about .003 in. by heating it with a torch to about 1700° F (927° C) and rolling it in a carbon compound such as Kasenit® followed by reheating and water quenching. In order to harden to $\frac{1}{16}$ in. deep, the part must be packed in the carburizing compound and held at that temperature for about eight hours. Nitriding produces a harder case with a lower temperature and less distortion. Other methods produce a more uniform, harder case than carburizing in a shorter time. Some of the disadvantages to these methods of surface hardening are that the entire part must often be heated and quenched, altering its entire chemical structure. Rising energy costs and the need for increased production efficiency have brought about the development of new methods.

Selective Surface Hardening

Although induction hardening has been used for many years to harden small parts or ways on machine tools, newer processes make use of this hardening process in a selective manner in which wear surfaces are hardened only in stripes, moving progressively along the surface (Figures 19 and 20). This is done on flat surfaces, inside cylinders, and for bearing races on shafts. In these quick-heating processes the heated area is self-quenched by the adjacent cold metal, resulting in a shallow hardened area in the form of a line (stripe) or spiral. Electron beam equipment is also capable of producing selective hardened areas, but it is usually done in a vacuum. Laser systems can operate in ambient conditions for selective heat treating (Figure 21). Their primary advantage is accessibility to selective areas on a metal part where other systems cannot be easily ap-

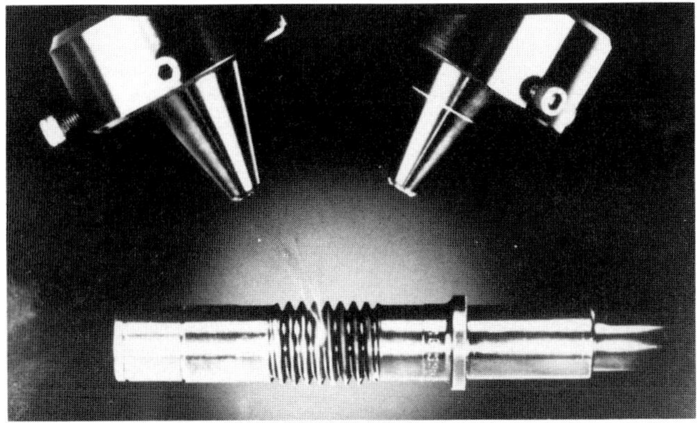

Figure 21
Laser hardening a worm gear. This is an example of selective heat treating. Instead of the entire part being hardened, which could cause distortion, just the contact faces of the gear are hardened (Coherent, Inc., Palo Alto, CA).

plied. A longer part life is claimed for laser-hardened parts. One disadvantage to laser hardening is that a nonreflective coating on the part to be hardened is sometimes needed. High-frequency resistance heating is an alternative method that can produce a hardened line or stripe in one "shot" without the traversing operation. The cycle time is also shorter and is quite adaptable to conventional automation. Tempering is not normally used in surface-hardening processes.

Precipitation Hardening

Some metal alloys, both ferrous and nonferrous, can also be hardened by solution heat treatment and aging processes called precipitation hardening. Some stainless steels that cannot be hardened by ordinary heat treatment have an alloying element or compound that begins to precipitate as small particles that lock up the slip planes after the metal has been heated to put the elements in solution and then quenched. The period of time these particles takes to precipitate (often two or three days) is called aging. The 2000 series of aluminum can be solution heat treated and aged to produce a light, tough and strong metal that is used extensively in aircraft manufacture.

Aluminum containing about 4 percent copper is heated to about 1000° F (538° C) for a period of time to bring the copper into solution (Figure 22). When the aluminum is quenched in water to room temperature, the solubility of the copper is drastically decreased and it forms a compound of copper aluminide which slowly disperses along grain boundaries and slip planes to harden the alloy (Table 3). Artificial aging is often practiced to speed up the process but it is not recommended in some cases, since it can decrease corrosion resistance. Some copper, nickel, magnesium, and titanium alloys, and those of some other metals can be hardened by the process of precipitation heat treatments.

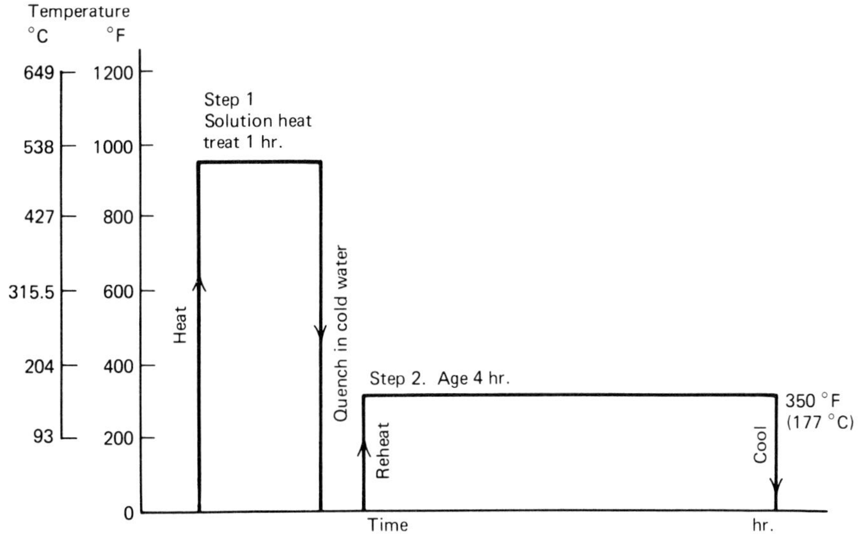

Figure 22
Process of solution heat treating and artificial aging of 2014-T6 aluminum alloy, $\frac{1}{4}$ in. thick (Neely/*Metallurgy*, 2 Ed. © 1984 John Wiley & Sons, Inc.).

Table 3
Solution Heat Treatment with Aging Times and Temperatures for Some Commercial Hardenable Aluminum Alloys

Designation	Soaking Temperature (° F)	Soaking Time[a] for Various Thicknesses (min)			
		Up to .032 in.	Over .032 to .125 in.	Over .125 to .025 in.	Over .250 in.
2014-T6	925 to 950	20	20	30	60
2017	925 to 950	20	20	30	60
2117	890 to 950	20	20	30	60
2024	910 to 930	30	30	40	60
6061-T6	960 to 1010	20	30	40	60
7075	860 to 960	25	30	40	60

Designation	Aging Temperature (° F)	Aging Time (hr)
2014-T6	345 to 355	2 to 4
	355 to 375	½ to 1
2017	Room temperature	96
2117	Room temperature	96
2024	Room temperature	96
6061-T6	315 to 325	50 to 100
	345 to 355	8 to 10
7075	245 to 255	24
	315 to 325	1 to 2

SOURCE John E. Neely, *Practical Metallurgy and Materials of Industry,* Second Edition, Wiley, New York, copyright © 1984.
[a]Soaking time begins after the part has reached temperature.

ANNEALING, RECRYSTALLIZATION, AND STRESS RELIEF

The term *anneal* is often used when referring to any one of several processes: full anneal, normalizing, spheroidize anneal, process anneal, and stress relief. When tool steel that has been hardened by heat treatment must be softened for the purpose of machining or subsequent heat treatment, the full anneal is used. This process consists of heating the tool steel to a temperature about 50° F (28° C) above the A_3 or $A_{3,1}$ lines (Figure 23). The part is then allowed to remain in the furnace while it cools slowly to room temperature. Alternately, the part could be placed in a noncombustible insulating material such as ashes. The cooling rate is dependent on the type of tool steel being annealed. Several hours of cooling is usually required for plain carbon tool steel; other kinds often take much longer. Full anneal is necessary when hardened structures such as iron carbides, other metallic carbides, and martensite are present. These are found in the higher carbon steels (about .40 to 2 percent) and some alloy and tool steels.

Lower carbon steels, between .30 and .60 percent carbon, that have been hardened, can be normalized in order to soften them sufficiently for machining. The normalizing process does not completely soften these steels but leaves them just hard enough for efficient machining. A metal that is too soft will not machine well because it is too "gummy," leaving a rough torn surface finish, whereas a somewhat harder metal will have a much smoother surface. Normalizing is also used to refine castings and forgings since it recrystallizes the coarse, irregular grains to produce a fine-grained

Figure 23
Temperature ranges used for heat treating carbon steel (White, Neely, Kibbe, Meyer, *Machine Tools and Machining Practices,* Vol. I, © 1977 John Wiley & Sons, Inc.).

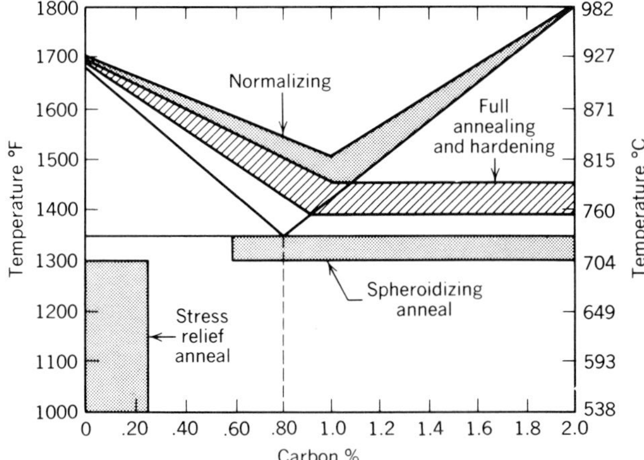

metal that is stronger. It is used to prepare tool steel parts for hardening heat treatments by removing machining and other stresses. Normalizing is usually carried out in a furnace by slowly raising the temperature to about 100° F (56° C) above the A_3 or just above the A_{cm} lines. Then the part is removed from the furnace and allowed to cool slowly in still air.

Both the full anneal and the normalizing process require the temperature to be raised above the transformation lines on the diagram in Figure 1 in order to dissolve the carbides. Stress-relief anneal and spheroidizing anneal in steel may be carried out at a temperature lower than the A_1 line represents. When medium to high carbon steels must be annealed to their softest possible condition, a spheroidize anneal is often used. Full anneal for some of these high carbon steels will not soften them sufficiently for some cold working and machining operations. For example, commercially available drill rod contains almost 1 percent carbon and is usually spheroidized. In that state it is softer and more ductile than low carbon cold-rolled steel. Spheroidizing is carried out at a temperature just under 1300° F (704° C) in a furnace for about four hours. Carbides slowly form into globules or tiny spheres in the microstructure instead of the harder lamellar pearlite (Figure 24).

Stress relief anneal is closely related to prior cold working of metals. In this process it is not carbides that are softened, but the ferrite grains that have been hardened by cold working. Therefore, this process is confined to low carbon steels and many nonferrous metals. In industrial processes such as wire drawing, cold-worked metal that must be softened so further cold working can be done is subjected to process anneal. This is substantially the same thing as stress relief anneal except it is often carried out in closed containers with an inert gas to prevent oxidation of the surface. It is often called bright annealing. Cold-worked grains become flattened, signifying a loss of ductility. When the metal has been raised to its recrystallization temperature, the flattened grains reform into a soft whole grain structure (Figures 25 and 26).

Figure 24
Micrograph of spheroidized carbon steel (500×).

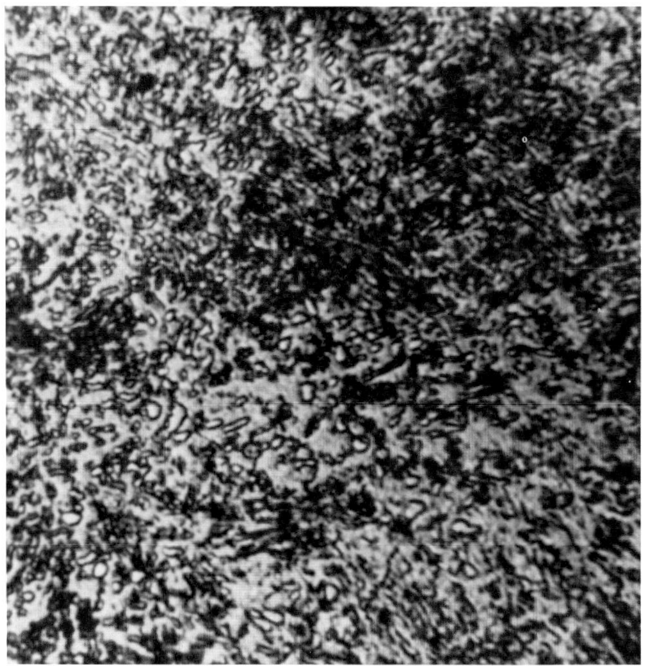

Figure 25
Microstructure of .10 percent carbon steel, cold rolled. Both ferrite and pearlite grains have been flattened by plastic deformation (*Metals Handbook*, Vol. 7, 8th ed., American Society for Metals, 1972, p. 10. With permission).

Figure 26
The same .10 percent carbon steel as in Figure 25 but annealed at 1025°F (552°C), the recrystallization temperature of ferrite. Ferrite grains are mostly reformed to their original state, but the pearlite grains are still distorted because the transformation temperature was not reached (1000×) (*Metals Handbook*, Vol. 7, 8th ed., American Society for Metals, 1972, p. 10, With permission).

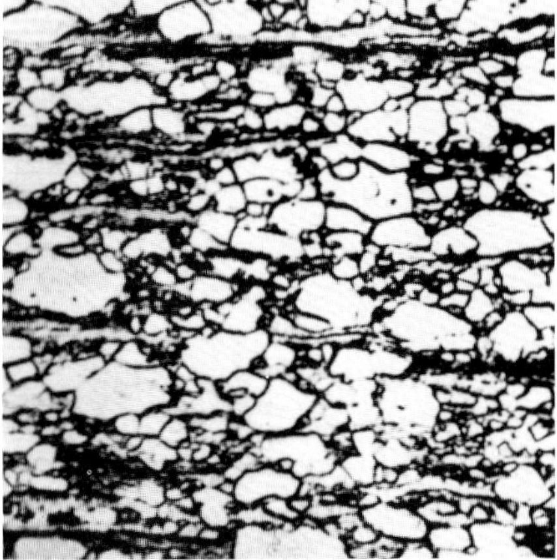

Table 4
Recrystallization Temperatures of Some Metals

Metal	Recrystallization Temperature °F
99.999% aluminum	175
Aluminum bronze	660
Beryllium copper	900
Cartridge brass	660
99.999% copper	250
Lead	25
99.999% magnesium	150
Magnesium alloys	350
Monel	100
99.999% nickel	700
Low carbon steel	1000
Tin	25
Zinc	50

SOURCE John E. Neely, *Practical Metallurgy and Materials of Industry*, Second Edition, Wiley, New York, copyright © 1984.

Recrystallization takes place when cold worked metals are heated to their specific recrystallization temperatures (Table 4). The stored energy from cold working is then released, allowing the distorted grains to reform by migration of grain boundaries and joining of adjacent grains. This process is called grain growth. The following factors affect recrystallization:

1. A minimum amount of deformation is necessary for recrystallization to occur.
2. The larger the original grain size, the greater the amount of cold deformation is required to give an equal amount of recrystallization with the same time and temperature.
3. Increasing the time of anneal decreases the temperature necessary for recrystallization.
4. The recrystallized grain size depends mostly on the degree of deformation and, to some extent, on the annealing temperature.
5. Continued heating, after recrystallization (reformed grains) is complete, increases the grain size.
6. The higher the cold working temperature, the greater will be the amount of cold work required to give equivalent deformation.

Stress relief is often needed for castings and weldments. Large welded structures such as tanks are sometimes stress relieved by covering them on the outside with thermal insulation blankets and heating them on the inside with propane burners to the recrystallization temperature.

Thermal stress relief is preferred for most manufacturing processes. However, vibratory stress relief (VSR) is often used for cast or welded structures that are too large to fit into a heat-treat furnace. Contraction from welds on nonuniform cooling in castings or stresses removed or induced by machining can cause dimensional instability. In order to be able to use VSR effectively, three principles must be understood:

1. Residual stresses can cause metal components to be more rigid than stress-free ones.
2. Effective loading of a structure by means of resonance requires close control of a vibrator's frequency.
3. Proper instrumentation is required to display the pertinent VSR data.

In order to relieve stresses, the entire part must be made to resonate and induced vibration must be very near to its natural resonant frequency which is generally in the 50-Hz to 100-Hz range. The dynamic loading causes plastic flow where any concentration of high-level residual stress is present, such as in welds. Flexure at these places combined with the residual stress is sufficient to momentarily exceed the yield point. Electrically-powered portable vibrators are attached to the part to be stress relieved with clamps and, in addition, elastic isolation cushions support the part at strategic locations. It seems that there is no upper limit on size for this stress-relieving technique; however, there have been some serious difficulties in its use in the past. Welds are sometimes shot peened to relieve residual weld stresses. The difference from other methods is that shot peening induces surface compressive stresses to counteract the tensile stresses in the weld.

REVIEW QUESTIONS

1. To what temperature should hypoeutectoid (less than eutectoid) plain carbon steel be heated for hardening purposes as seen on the iron-carbon diagram?
2. If a .80 percent plain carbon steel is austenitized (heated to a red heat) and quenched in water to room temperature so that the cooling curve does not cut into the "nose" of the I-T diagram, what will the resultant microstructure be?
3. On the right side of the iron carbon diagram, what kind of metal is found when the carbon content is over 2 percent?

4. State the difference between the eutectic and eutectoid points.
5. What happens to a quench-hardened tool or machine part that was not tempered when it is subjected to shock loads or bending stress?
6. What does the process of temper drawing do for hardened carbon steel?
7. Name two kinds of equipment that are used for heating steel to a precise hardening temperature.
8. How can a steel that contains insufficient carbon to be hardened be surface hardened?
9. By what means can a small area or stripe be selectively hardened on a large steel part?
10. How can some nonferrous metal alloys such as aluminum or monel be hardened?
11. What annealing method should be used for softening a hardened high carbon tool steel?
12. A low-carbon steel part has been partly formed toward its final shape and now is in a work hardened condition. Any further cold working will cause it to tear. At this point, heating to a red heat would distort it too much to recover its shape with further work. How can this part be restored to a soft condition so it can be cold worked to its final shape?

CASE PROBLEMS

Case 1
The Case of the Badly Worn Pin

The soft steel pin in Figure 27 was removed by a millwright in a manufacturing plant from a hydraulic cylinder clevis at the end of a cylinder rod. It was originally a straight cylindrical pin, but it has worn almost completely through before a mill shutdown allowed it to be removed. The holes in the clevis, which is made of soft cast steel, were also elongated accordingly. The clevis was periodically lubricated but the extreme forces still caused excessive wear. Considering what you have learned in this chapter, what could be a possible solution to this problem of excessive wear?

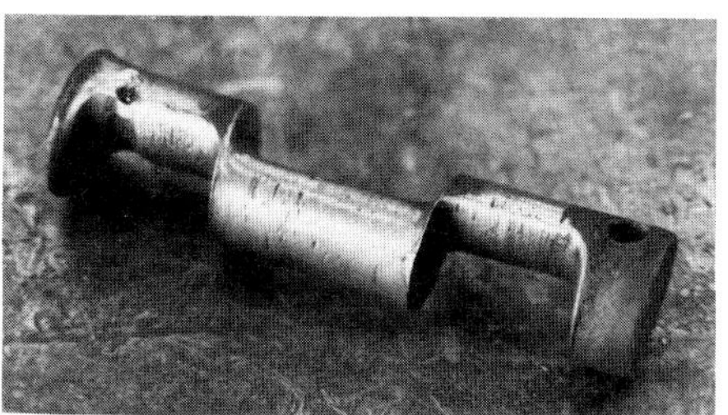

Figure 27

Case 2
Strength versus Weight Problem

A structure for an aircraft part (made of 1000 series aluminum) has a certain weight limitation. After all weight reduction by engineering design had been done, the part was still too heavy. No metal other than aluminum is permissible in this case. How can the part weight be further reduced and still have sufficient strength?

CHAPTER 5 EXTRACTION AND REFINEMENT OF COMMON NONMETALLIC MATERIALS

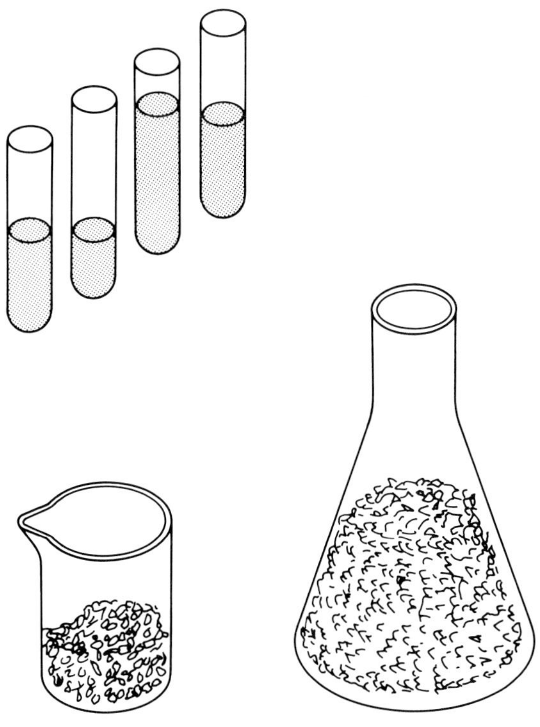

Stone, clay, and wood are among the oldest engineering materials used by humans to construct buildings and aqueducts, to make containers to store foodstuffs and contain liquids, and to make tools and weapons. The Romans discovered a way to make fluid stone that would solidify into a hard mass. It was called pozzolan cement, because it was first made from the volcanic ash at Pozzuoli, Italy. After this ash was combined with slaked lime it would begin to harden. Modern portland cement, from which concrete foundations, sidewalks, and dams are made, is similar in some ways to that ancient Roman cement.

The making of glass by fusing silica sand with another substance dates back to at least 3000 B.C. It was used for making jewelry, containers, small window panes, and other artifacts; its use for plate glass windows, however, has appeared only relatively recently because of the difficulties of producing large, flat sheets with optical quality.

Although wood has been displaced by other materials to some extent in recent times in the manufacture of furniture and in building construction, it continues to be important as a structural material. Wood has the advantage of being a renewable resource and, if forestry programs of replanting harvested trees are maintained, there will always be a source that will not be depleted.

Chemicals from which the commercially important plastic resins are derived are themselves derived from natural products that are quite plentiful at the present time. Those raw materials include coal, water, limestone, sulfur, salt, and petroleum. An Arizona desert wildflower called the popweed (known scientifically as *lesquerella*) may some day partially replace petroleum as a source for plastics. Combining popweed oil with a compound derived from castor oil (from the castor bean) produces a polyester that is made entirely from natural sources. Popweed oil can also be polymerized by itself into a soft rubbery material. Jojoba (*Simmondsia californica*), a desert shrub native to the Southwest, produces a waxy oil which is being tested for use as a substitute for the sperm whale oil presently used in some industrial lubricants. Meadowfoam, grown commercially in the Willamette Valley in Oregon, produces a waxlike oil from its seed. It is also a substitute for sperm whale oil, and is presently used as an oil in cosmetics. Such renewable resources will be needed in the future as petroleum oil becomes more scarce.

Petrochemicals, that is, petroleum products such as plastics and elastomers, have only recently appeared on the scene and have rapidly displaced the use of wood, metal, and glass in some areas. The great advantage of plastics is that a special material can be created for a specific purpose that will fit the requirements better than any natural material. Thus, surgical implants, quiet long-lasting gears, and nondeteriorating rubbers are some items that are produced by chemical synthesis of raw materials. All plastics are classified as either thermosetting or thermoplastic, each group having distinctly different qualities under conditions of repeated heating and cooling.

All of these materials will be discussed in detail in this chapter. However parts manufactured from these materials will be discussed in Chapter 15, "Plastics and Composites Processing."

OBJECTIVES

This chapter should enable you to:

1. Identify the raw materials derived from petrochemicals that are used to produce many synthetic materials.
2. Explain the properties of some thermomelt and thermoset plastics that are used to manufacture products.
3. State the uses for certain additives and finishes for plastic materials.
4. Relate the means by which elastomers are derived.
5. Name the uses and explain the significance of asphalts and lubricants in the manufacturing industry.
6. Demonstrate how ceramic materials are produced.
7. Assess the significance of wood materials as a manufacturing material.

PETROCHEMICALS

Chemicals made from crude oil, natural gas, and liquefied natural gas are the basic raw materials for many of the synthetic fibers, sealants, paints, synthetic rubbers, plastics, and other products. These products are typically polymers which are made from small molecules called monomers. These are the building blocks of these synthetic materials and they derive their names from the Greek *mono* (one), *poly* (many), and *meros* (parts). Polymer chains are composed of monomer cells (raw materials) that are combined to produce these useful materials (Table 1).

Table 1
Products from Petrochemicals

Products	Petrochemical Monomer (Raw Materials)
Antifreeze fluid, Mylar	Ethylene glycol
Butyl rubber	Isobutene
Epoxy resins, polycarbonates (Lexan®) acetone, phenol, bisphenol-A	Isopropylbenzine
Ethylene glycol antifreeze, Mylar (Dacron®) ethyl alcohol	Ethylene
Mylar, poly-para-xylene (Parylene®)	Xylenes
Neoprene rubber	Chloroprene
Nylon synthesis, neoprene, and polybutadiene rubber (source)	Butadiene
Orlon: chloroprene and neoprene rubber	Acetylene
Phenol, phenolic resins, styrene, and polystyrene	Benzene
Phenolic plastics	Toluene
Phenolic resins, epoxy resins	Phenol
Polypropylene, isopropylbenzine, isopropyl alcohol	Propylene
Polystyrene	Styrene
Polystyrene (nonsoluble)	Divinylbenzene
Production of acetylene	Methane
Synthetic rubber	Isoprene

SOURCE John E. Neely, *Practical Metallurgy and Materials of Industry*, Second Edition, Wiley, New York, 1984.

By the process called catalytic cracking, heat is used to induce a chemical reaction in the crude oil with the help of catalysts, usually composed of refractory oxides of aluminum, silicon, and magnesium. High temperature boiling components of petroleum are broken down into gasoline and oils. The remaining gaseous material is made up of compounds having from one to four atoms. Among these are ethylene, propylene, and the butylenes. Natural gas, from which the liquefied gases butane and propane are derived, is also a source of propylene, ethylene, hydrogen, and methane obtained by means of the process of thermal cracking. Acetylene gas, made from the hydration of calcium carbide, is also a source of plastic and rubbery materials. Most acetylene gas is now produced from methane rather than from calcium carbide. Chloroprene and polychloroprene (neoprene) are derived from acetylene.

Nitrogen fertilizer made from natural gas and pesticides made from other petrochemicals are both vital to food production. About 60 percent of the fibers used for clothing are synthetic and are mostly based on petrochemicals; exceptions are rayon, acetate, and acetic anhydride. Many of today's medicines are derivatives of petrochemicals; examples are antihistamines, antibiotics, sulfa, penicillin, and aspirin. Automobiles use plastic gears and body parts; our homes have synthetic materials in such things as carpets, siding, heating and cooling systems, furnishings, and housewares. The petrochemical materials and derivatives covered in this chapter will be the plastics, rubbers, adhesives, lubricants, and asphalts.

PLASTICS (THERMOSETTING)

All plastics are formable when heated, hence, the prefix *thermo*. The difference between the two types of plastic is that the thermoplastics, after once being hardened in the heating process, can be reheated and formed into new shapes a number of times without much change in properties. Thus, thermoplastic scrap can be reheated and reused. However thermosetting types take on a permanent set after they are heated (and sometimes placed under pressure) after which they cannot be remelted or reformed. Additional heating will only destroy them by breaking the permanent cross-links in the polymer chain. Some types of thermosets do not require heat to harden, but are cross-linked by combining liquids in definite proportions. The mixture then hardens in a preset time. Some examples are epoxy, silicone, polyurethane, and polyester. Thermoset scrap cannot be reprocessed; however, the ever-increasing amount of automobile thermoset plastic scrap is being considered as a recyclable material and a low-cost natural resource. According to research done at Lehigh University's research center, this plastic scrap can be ground and used as a filler or aggregate in acrylic polymer concrete to produce medium-strength products. These could compete with portland cement and could perhaps be used as construction materials for some applications where light weight and low density are needed.

As a general rule, the thermosets must be combined with fillers and reinforcing agents to give them the mechanical properties needed for molded parts. For example, most of the epoxies are quite brittle but, when combined with glass fibers, they become strong and tough materials.

Plastics have several general characteristics in common. They are among the lightest in weight of the engineering materials. They are all relatively good heat insulating materials since they have a relatively low thermal conductivity, and nearly all plastics are good electrical insulators. However, a recently developed polyacetylene plastic is a good conductor of electricity and shows great promise in production of plastic batteries, possibly solar panels, and plastic wire for motors and generators. This family of plastics can be considered as "synthetic metals." For the most part, plastics can be obtained in a wide range of colors, either transparent or opaque; a few exceptions, such as Bakelite, have a natural color that is often acceptable for a product. Plastic items can usually be produced in one operation such as casting, molding, or extrusion. Although many plastics are not nearly as strong as metals, some have been developed that are almost as strong as some metals, and sometimes they exceed metals in other properties such as toughness and resistance to abrasion.

Alkyds

The Alkyds are polyester derivatives, produced by the reaction of an alcohol and an acid. The name *alkyd* comes from the words *ALCohol* and *acID*. Their most important use is in paints and lacquers. Also molded automobile ignition parts and light switches are made from this thermoset because they do not break down at higher temperatures and are not affected by moisture, conditions found around automobile engines.

Allyl Plastics

The allyl plastics, produced from an alcohol, are very hard, scratch resistant, clear plastics used for optical glass—prisms and lenses in eyeglasses (Figure 1), radar domes, and window glass. Although a very costly thermoset, it is far superior to the clear acrylic plastic that is also used for transparent shapes. It resists most solvents such as gasoline, benzene, and acetone, and has a high tensile and compressive strength.

Aminos

Urea and melamine belong to the family of amino plastics that have an unlimited range of colorability. The

Figure 1
Safety glasses with plastic lens.

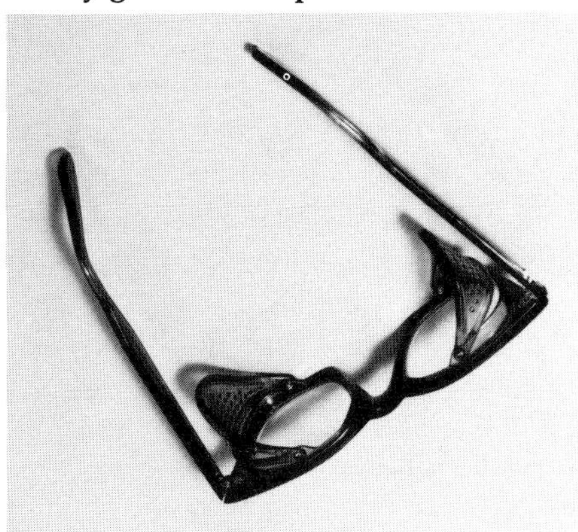

urea–formaldehyde plastics are produced by reacting urea with formaldehyde in the presence of a catalyst and mixing in a filler. These are used for knobs, handles, and household fixtures. Urea–formaldehyde plastics are used as adhesives for bonding plywood, foundry sand cores, and other materials. Melamines are similar but more complex plastics of the amino family and are harder, more shock, heat, and water resistant than the urea types, and have good resistance to acid and alkali attack. Because of their water resistance, they are used to bond exterior plywood. They are used for electrical parts such as circuit breakers and terminal blocks, and because of their ability to resist high temperatures and their chemical stability, the melamines are used for dinnerware and abuse-resistant furniture surfaces.

Epoxies

These plastic materials are quite brittle (although a few types are tough and elastic), and have good electrical properties and chemical resistance. They cure at normal temperatures and pressures. Though they have good bonding properties as adhesives, their shock load strength is poor. Epoxies can be cast in molds and are used in paints and for surface coatings for metals.

Phenolics

The phenolic–formaldehyde plastics are low-cost molding compounds that possess high strength, high temperature resistance, good dimensional stability, and good electrical properties. The most common type, Bakelite, is made from phenol and formaldehyde with a filler material such as wood flour. Bakelite is used for electrical switches and automobile distributor caps. Canvas, linen, or paper is sometimes used as laminations to reinforce the plastic. Examples of these phenolic laminates are Micarta® and Formica®. The major drawback of the phenolics is the lack of colorability. They are commonly brown or black.

Polyesters

Polyesters belong to a class of plastics related chemically to the alkyd type and contain two different types: "saturated" and "unsaturated." The saturated types are best known as film and fibers such as Dacron® and Mylar®. The unsaturated types are used in molding and casting. There are also many subtypes in each group. Polyester can be cast in molds without heat and pressure, but a disadvantage is that it has considerable shrinkage when hardening. This plastic is widely used with glass fiber and other reinforcements to provide high strength for such applications as boats, tanks, and automobile bodies. Polyesters can be either thermosetting or thermoplastic depending on formulation.

Polyurethanes

These plastic materials may be either thermosetting or thermoplastic. Since these are rubbery plastics, they are often made in a foamed structure for mattresses and insulation and are a substitute for foamed rubber. The thermosetting types can be cast as tough, wear-resistant elastomeric materials and they can be rigid or flexible depending on formulation.

Both urethane elastomers and rigid products are produced by the casting, injection molding, and coating process. Since they have good chemical and weather resistance along with high resilience and elasticity, they are used in a number of products such as oil seals (Figure 2) that can last up to eight times longer than rubber. Some of these seals are graphite impregnated to make them self-lubricating. Urethanes also have load-dampening ability and very good abrasion resistance, making them ideal materials for industrial solid truck tires (Figure 3). They are also used for caster wheels (Figure 4), gears, drive belts, floor coverings, and roofing materials. Metal objects such as furniture are sometimes covered with a tough, resilient skin of polyurethane which imparts both aesthetic form and a colorful, useful surface on the part, while the steel core provides the needed high strength.

A synthetic leather produced by DuPont, called Corfam®, is a polyester-reinforced urethane. It can be made porous or sealed with silicone. It was developed for shoe-upper parts, but is used in many other applications such as to make industrial gaskets and shaft seals.

Silicones

Silicone plastics are a type of synthetic rubber that is one of the highest temperature resistant plastics; they may be used at temperatures up to 500° F (260° C) (Figure 5). Because of their excellent electrical properties, they are widely used in the electrical industry for en-

Figure 2
New urethane compounds for hydraulic seals will last eight times longer than rubber and provide leak-free sealing up to 10,000 PSI (Disogrin Industries).

Figure 3
Urethane industrial solid truck tires (Maine Rubber International).

Figure 4
These solid polyurethane caster wheels are available in many different sizes, shapes, and colors (The Hamilton Caster & Mfg. Co.).

capsulation and components. Silicones are also used as sealants and caulking since they do not lose their rubbery qualities or crack with age and they have good bonding characteristics.

Molds can be made with room-temperature vulcanizing silicone rubbers that require a separate catalyst as a hardener that is mixed in at the time of use. Silicone products are also used as lubricants. Since they do not tend to break down in hostile environments, the silicones have been very successful for use as prosthetic devices, such as artificial heart valves, and for reconstructive surgery near the surface of the body.

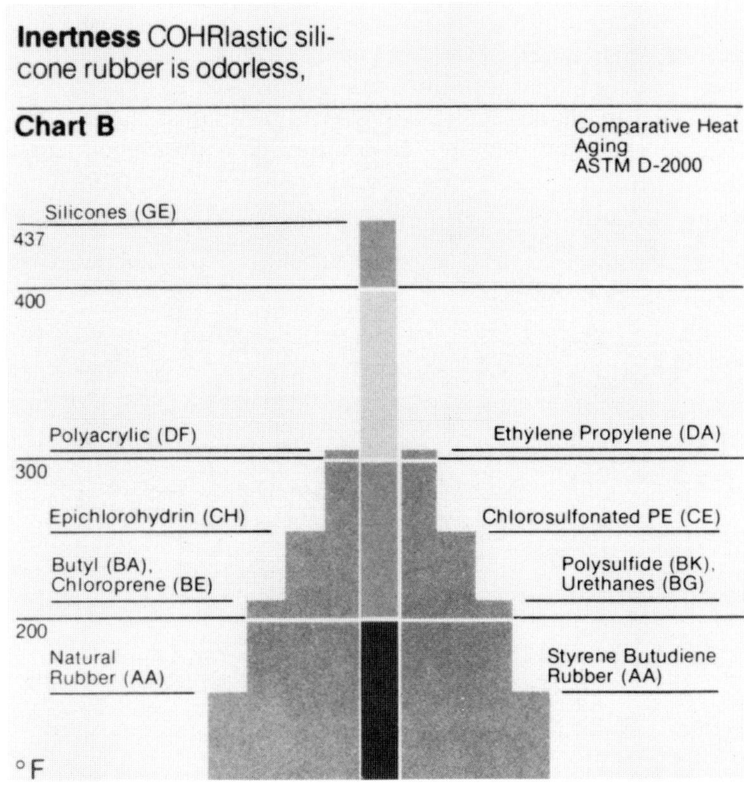

Figure 5
Comparative heat aging graph of silicones (CHR Industries, Inc.).

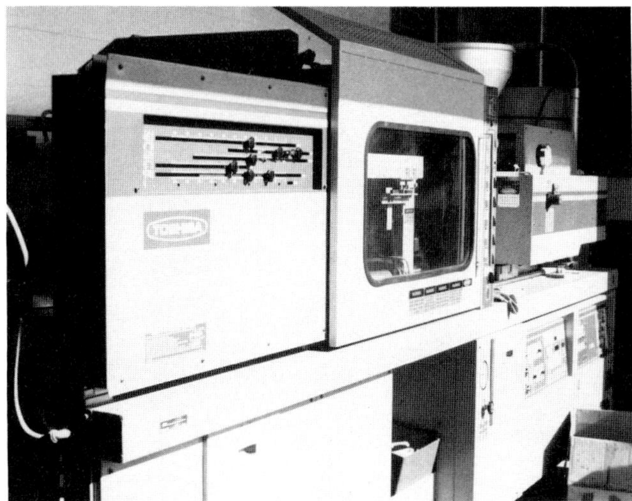

Figure 6
Modern injection molding machine with microprocesser that automatically controls cycling times (Hergert's Industries, Inc.).

Figure 7
The hopper on the injection molding machine must be frequently filled while the machine is in operation (Hergert's Industries, Inc.).

PLASTICS (THERMOPLASTIC)

Thermoplastic materials have the cost-saving advantage over thermosets in that they are injection molded (Figures 6 and 7) and therefore have faster cycling times than compression or transfer molded thermosets (Figures 8 to 10). When thermosets can be injection molded, deflashing and other secondary operations are often unnecessary. Thermoplastics are generally less brittle than most thermosets and can also be bonded by the process of ultrasonic welding. Also, since they retain some flexibility, thermoplastics can be designed with snap-fit parts, eliminating screws and other fasteners. Many thermoplastics can be used as a "living hinge" because they can flex thousands of times without breaking. Although thermosets have been the workhorse of the electrical industry applications because of their good electrical properties and dimensional stability, thermoplastics are rapidly taking over in this application. Electrical and electronic standards are changing, highlighted by the trend toward miniaturized parts that must endure ever higher operating temperatures. Tiny, more complex parts are more easily

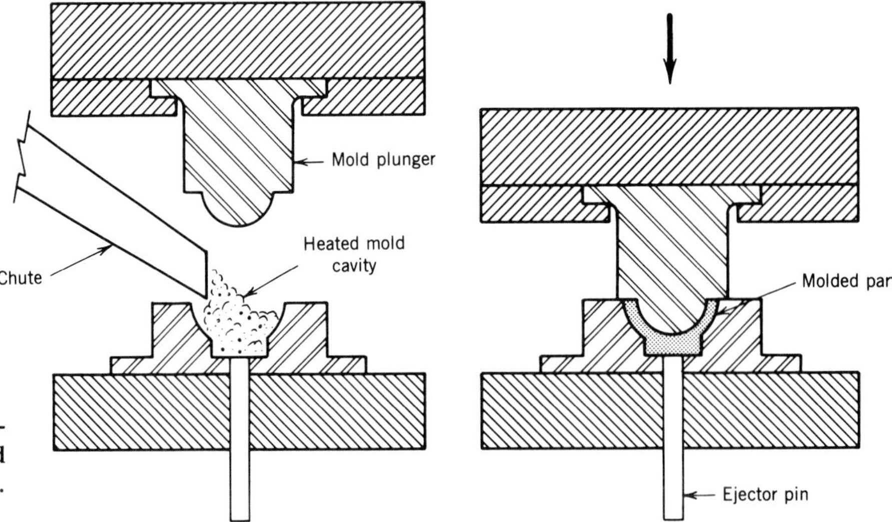

Figure 8
Compression molding. Granulated plastic is placed in a heated mold (left) and compressed (right) until object is formed. It is then removed with an ejector pin.

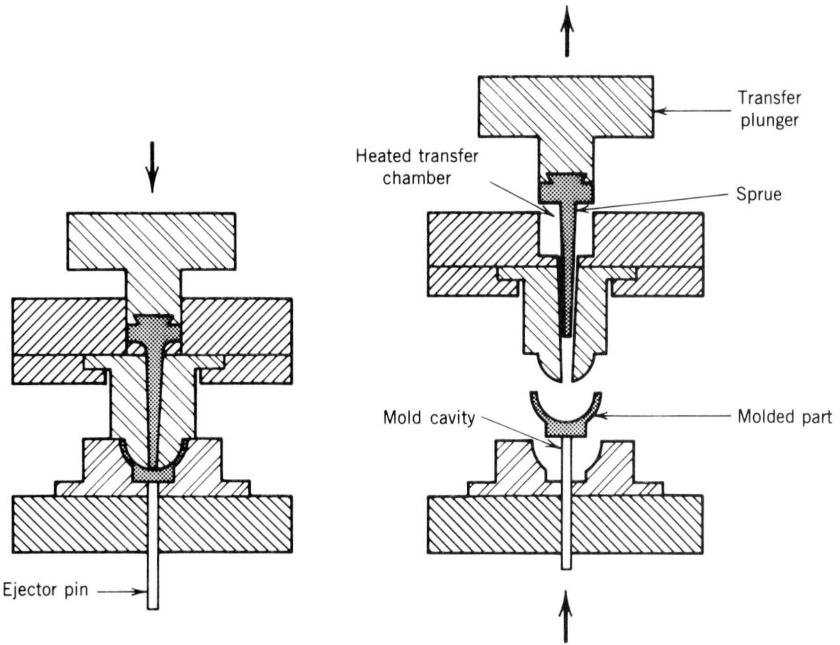

Figure 9
Transfer molding. Plastic is heated in an upper chamber and forced into the mold (left) by a second plunger. The object is formed and ejected (right).

produced by injection molding techniques, so new thermoplastics have been and are being developed to fulfill the requirements for new products.

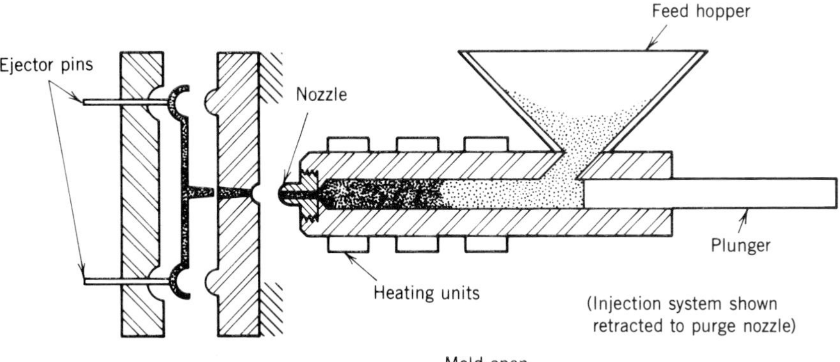

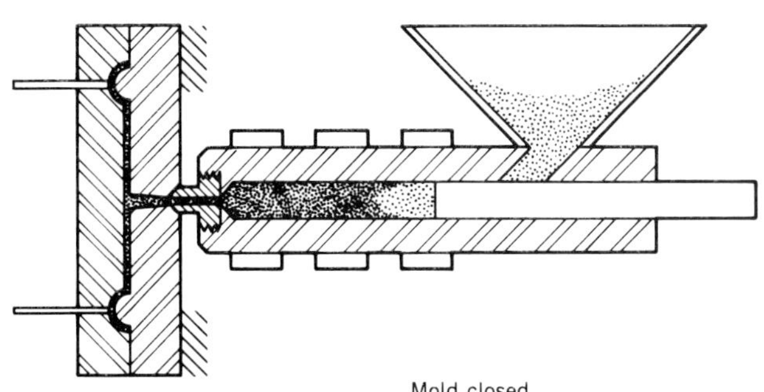

Figure 10
Principle of injection molding. Granulated thermoplastic is moved forward by the ram and passes through a series of heaters that melt the plastic. It is then injected with pressure into a closed mold where the molten plastic quickly solidifies. The mold is then opened and the part with the sprue is ejected. This cycle is then automatically repeated.

Fillers are often added to thermoplastics to gain properties such as dimensional stability, high-temperature resistance, heat or electrical conductivity, or magnetic properties. Plasticizers are added to improve their flexibility and stabilizers are added to prevent deterioration by light, heat, and oxidation. Coloring and fluorescent additives give these thermoplastics more customer appeal. Surface finishes, such as paint or electroplated metals, are also applied to plastics to enhance their appearance.

Acrylonitrile–Butadiene–Styrene (ABS) Copolymer

These very tough, very strong thermoplastics have very high tensile strengths and have an unusual combination of high rigidity and high impact strength even at very low temperatures. These copolymers can be obtained in different combinations of heat resistance, flexibility, and toughness. Their major disadvantages are flammability and solubility in many solvents.

The first of the styrene plastics was introduced in 1938 and was called polystyrene. It has a very low strength, but is one of the least expensive plastics, and for this reason is used for disposable products such as knives, forks, spoons, and water glasses. Styrene–acrylonitrile (SAN) is a styrene copolymer that is stronger and has a better chemical resistance than polystyrene. However, ABS is, by far, the most outstanding plastic making a wide range of specific materials available. Since it has high impact resistance, it is used in luggage, safety helmets, and such automobile parts as grilles. It can be plated in large volumes and can be blended with polyvinyl chloride (PVC) to become flame resistant.

Acetals

Acetal thermoplastics, also called engineering resins because of their high strength, have excellent engineering properties, very high tensile strength, stiffness, good dimensional stability, low moisture absorption, and resistance to creep under load. There are two types of acetals, homopolymer and copolymer. They are very similar in most respects, but the homopolymer is stiffer, stronger, and more impact resistant. It also has a self-lubricating quality that makes it useful for bearings and gears. The DuPont acetal Delrin® is a homopolymer that is used for such products as gears, butane cigarette lighters (because butane lighter fluid will not affect it), and car door handles (Figure 11).

Mineral-filled acetal copolymer resins have potential applications in such products as stereo cassette cases, spools for video cassettes, toys, and zippers. In the plumbing industry, faucet cartridges, shower mixing valves, and shower heads molded of acetal homopolymer are replacing those made of brass and zinc.

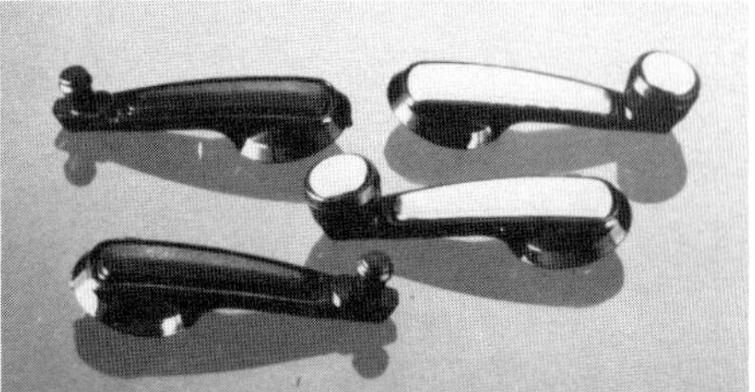

Figure 11
Delrin® in interior automotive door handles is not only lighter in weight than die-cast metals, it also eliminates the need for any finishing operations (Engineering Polymers Division, DuPont Company).

Acrylics

Widely used for their light transmission properties and optical clarity, these plastics have excellent resistance to outdoor weathering and dimensional stability; that is, they tend to keep their shape over a wide range of temperatures. However, they will soften at about 200° F (93° C). The acrylics are not very scratch resistant and have a low shock resistance; therefore, they are not as satisfactory for making eyeglasses as are the allyl thermosets. Acrylics can be made in brilliant transparent colors and are used for the manufacture of automobile tail lights, jewelry, and novelty items. Lucite® and Plexiglas® are acrylics that are made in the form of tubes, sheets, and rods. Acrylic parts are easily machined from these standard shapes.

Cellulosics

In their natural form, cellulosics do not melt, but with proper chemical treatment they can be molded in the same standard equipment as other thermoplastics. Cellulosic plastics are among the toughest of all plastics. There are five basic groups: nitrate, acetate, propionate, butyrate, and ethyl cellulose. Cellulose nitrate is the toughest of these but has the drawback of being very flammable. Celluloid is one product made from this material. Cellulose acetate is nonflammable, but like the nitrate, embrittles with age. The propionate and butyrate types do not have that problem and have better chemical resistance. The ethyl type possesses a very high shock resistance, so it is used for plastic products that may be handled and accidentally dropped, such as flashlight cases and toys. Cellulose plastics are also made into thin sheets and are sold under the name *cellophane*. Cellulosics tend to creep or cold flow when under load.

Fluoroplastics

The fluorocarbon family includes polytetrafluoroethylene (TFE), chlorotrifluoroethylene (CTFE), fluorinated ethylene propylene (FEP), and polyvinylidene fluoride

(PVF_2). TFE has the greatest chemical resistance of all plastics. However, TFE is difficult to mold. FEP is more easily molded but still possesses the properties of TFE, yet it is more expensive than the other types to produce. CTFE is similar to the other fluoroplastics, but has a much lower permeability to water and gases and is the least permeable of all plastics. PVF_2 has a greater stiffness than the other types and is used for rigid pipe and in load-bearing applications. Also it is the easiest type to extrude.

Fluorocarbons have the lowest coefficient of friction of any plastic and they have a very high chemical resistance; none of the common chemicals will react with them at all. In fact, adhesives for the most part will not adhere to them, although some adhesives were developed to bond fluoroplastics to metals. Fluoroplastics are useful for nonstick cooking surfaces, such as baking pans and frying pans that need not be greased. Paints made of these materials have exceptional resistance to weather and sunlight. Fluoroplastics are used for labware, valve seats, O rings, and bearings. TFE fibers are sometimes used as fillers in acetal resins to improve thermal and frictional qualities.

Inomers

Inomers are plastics that have both ionic and covalent bonds. They are clear plastics with high optical clarity and have excellent toughness. They have high chemical resistance and are easily formed. Films are hard to tear, but once a tear begins, propagation is easy.

Nylon

Nylon is the name given by DuPont to a family of polyamides. It was first used for the production of fibers for such products as webbing, parachutes, and stockings. The most widely used nylons are Type 6, Type 6/6, Type 6/10, Type 11, and Type 12. All nylons have the properties of toughness, fatigue resistance, low friction, and inertness to aromatic hydrocarbons. Nylons 6 and 6/6 possess some incredible qualities when combined with other materials such as glass fibers. They can replace metals in many applications, such as automobile speedometer gears (Figure 12), bicycle wheels, and carburetors (Figure 13). Nylon 6/10 is more flexible than nylons 6 or 6/6 and has less tendency to absorb moisture (which can alter the physical dimensions of a plastic part) and thus can be used in water mixing valves. Nylons 11 and 12 are commonly used in film and tubing.

Figure 12
Heat-stablized Zytel® (nylon) has been used for speedometer take-off, driven, and governor gears since 1954. In this difficult environment they have performed as reliably as the metals they have replaced, and they weigh much less and are much less expensive (Engineering Polymers Division, DuPont Company).

Figure 13
This lawnmower carburetor molded of Minlon® 10B (a reinforced nylon) is a first among four-cycle engines (Engineering Polymers Division, DuPont Company).

Polycarbonates

Polycarbonate resins can be used for many of the same products that can be made of nylon since they have similar properties. They are extremely strong and are very difficult to break. For this reason they are used for unbreakable bottles and as a window glass replacement where crime and vandalism are problems. The General Electric polycarbonate Lexan® is an excellent glazing product, which is made with a special surface coating that resists ultraviolet light which causes yellowing and impact degradation in polycarbonates. This nonyellowing coated product is used for architectural glazing and for solar energy systems. When polycarbonate sheets are laminated and fused with proprietary layers to a thickness of .75 in., they outperform bullet resistant glass against medium power weapons. Polycarbonate foams are other important products that are used for thermal insulation and sound-dampening suitable for automobile and business machine use.

Polyester

These thermoplastic polyester molding compounds are listed chemically as polybutylene–terephthalate (PBT) and polytetra–methylene terephthalate (PTM). Compared with the thermosets, thermoplastic polyesters possess better molding characteristics and more rapid cycle times because of their rapid crystallization rate. Glass fiber is added to improve heat resistance and strength. Applications include automotive electrical parts, gears, housings, and pump impellers.

Polyimide

This engineering resin can exist both as a thermoset and as a thermoplastic; but unlike other thermoplastics it does not melt. Polyimides are difficult to injection mold, requiring high molding temperatures 660° F (349° C), a long cycle time, and high injection pressures. These resins can withstand 500° F (260° C) continuously and 900° F (482° C) for a short time without breaking down. Polyimide resists most chemicals, except for strong alkaline solutions and strong inorganic acids, and it has poor weathering resistance. Polyimides have almost no creep even at high temperatures and, when certain fillers such as glass or graphite fibers are used, the thermal expansion is close to that of metals. They are used for bearing sleeves, seals, and thrust washers.

Another engineering resin, polyamide–imide, can also resist high temperatures of 500° F (260° C) for long periods. Because of its toughness, high strength (30,000 PSI at room temperature), and heat resistance, it has many applications in the automotive industry and in making cookware and electrical equipment.

Polyolefins

The polyolefins include polyallomer, polypropylene, and polyethylene. Polyethylene is a very strong, flexible resin that has an unusually high flex life, which makes it an excellent material for "living hinges." Polyethylene is used for squeeze bottles, ice cube trays, film for boil-in packaging, and vending machine food containers. High density polyethylene is used to make pressure pipe (Figure 14). Since it can be made either flexible or stiff, it is used for a great variety of housewares and toys. Polyethylene terephthalate (PET) is a more recently developed plastic resin that is being increasingly used for beverage bottles. Since beverage bottles usually contain no reinforcing fibers, the thermoplastic resin is highly recyclable and millions of pounds per year of PET bottles and scrap are recovered and reprocessed. A polycarbonate polyethylene terephthalate blend (PC/PET) was developed and is valued for its high impact strength at ambient and low temperatures as well as its resistance to chemical attack.

Figure 14
Construction workers on street in New York compare section of 48-in. diameter high density polyethylene pressure pipe, the largest size previously available, with a section of newly available record 63-in. diameter "MLSO-ppd/305" high density polyethylene pressure pipe (M. L. Sheldon Plastics Corp., New York, N.Y.).

Polypropylene is used for electrical insulation because of its outstanding electrical properties. Polyallomer possesses many of the properties of polyethylene and polypropylene, but it can be processed more easily.

Polysulfone

This engineering resin has a very low to high constant use temperature of 400 to 500° F (204 to 260° C) and good impact strength. It is resistant to salt solutions, acids, alkalies, and most solvents and can be processed by extrusion, injection molding, and compression molding. These resins can also be ultrasonically welded and machined. Because of their good electrical insulating properties, they are appropriate for use in printed circuit boards, switches, lamp housings, and electrical insulation.

Polyvinyl Chloride

There are two basic types of polyvinyl chloride (PVC), rigid and flexible. Rigid PVC is used in pipes and pipe fittings and other construction materials. Flexible PVC has replaced rubber in many applications, such as in making gaskets, vacuum tubing, garden hoses, shower curtains, floor mats, and toys. PVC can also be foamed and coated with other solutions, can be fabricated by most processes, and may be joined by heat sealing and solvent welding. Plastic products manufacture will be covered in Chapter 14.

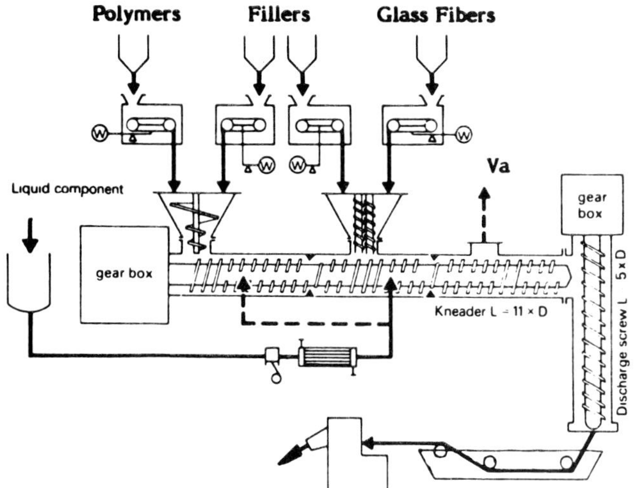

Figure 15
Flow chart showing continuous compounding of engineering plastics (Buss-Condux, Inc.).

ADDITIVES AND FINISHES

Various additives, pigments, and fillers are combined with polymer resins by mixing and kneading machinery. It is now possible to compound these materials continuously as a manufacturing process (Figure 15).

Additives

Since the first plastic materials were synthesized, the list of uses and applications for products has grown significantly. In many cases plastic products are replacing metal products and competing with other products made from natural materials such as wood and leather. This widespread and diversified use of plastics is partly possible because of additives that alter and strengthen the plastic resins to make them more easily processed and more useful. Some of the additives used are as follows.

Antioxidants These materials impart ultraviolet stability for such resins as ABS, polyethylene, polypropylene, and polystyrene, and they impart melt–flow retention, making them easier to mold.

Blowing agents When plastics must be processed at low temperatures, as required in sponge plastic and rubber applications, these additives allow production speeds to be increased and create voids to give porosity or low density.

Colorants Pigments and dyes for plastics give them their brilliant colors and are important for giving them sales appeal. Some pigments are designed to provide a solid, opaque color, whereas others tint the clear resin to impart a translucent coloration. Luminous colors are provided for some products, such as bicycle reflectors, to make them easily seen. A new luminous-green pigment was developed which is said to be nontoxic and safe for toys and glows in the dark for about eight hours after being exposed to a light source.

Fillers These materials can reduce the cost of plastic products by reducing resin usage. They lower the part weight and still retain, or even increase, physical properties such as tensile and compressive strength. Common fillers are wood flour, quartz, limestone, clay, and metal powders.

Plasticizers Some plastics such as vinyls are normally hard, brittle materials, but by adding a plasticizer they can be made soft and flexible.

Reinforcements Resins tend to have rather low physical strengths, but by adding reinforcements, their strengths can be markedly increased. Some reinforcing materials for plastics are glass fiber, mica, asbestos, jute, sisal, graphite, carbon, and ceramic. Containers such as large tanks are formed by winding filaments of various materials such as glass, boron, and graphite with a plastic binder such as epoxy over a form that rotates while the filament moves in circumferential, longitudinal, and helical patterns, or a combination of these to produce a very high-strength product.

Stabilizers Some plastic resins such as styrene and vinyl are subject to degradation or breakdown in the presence of oxygen or heat. PVC compounds are stabilized by barium–cadmium–zinc compounds and phenols can be added to styrene to stop degradation.

Plastic Finishes

Almost every type of molded plastic part can be decorated to bring about wear, scratch, and chemical resistance, as well as for aesthetic appeal. The molding process can impart a decorative surface with raised or depressed designs in geometric patterns such as basket weave, pebble, or leather texture. Texturing allows the manufacturer to mask or hide flaws, flow lines, sink marks, pinholes, and swirls, thus lowering the rate of part rejection. Other finishes are made with lacquers, enamels, and decorative overlays. Plastics can be metallized on the surface by the process of vacuum metalizing. This is done by vaporizing metal filaments onto the plastic surface and condensing them in a vac-

Figure 16
Bright metallic surface on this plastic toy appears to have been applied by the process of sputtering or by condensation.

uum chamber. Gold, silver, nickel, chromium, and pure aluminum may be deposited by this method. Another widely used method of depositing metals on the surface of plastic parts is called sputtering (Figure 16). This method ionizes the metal in a vacuum chamber by using an electrical discharge between two electrodes. This forms a metallic cloud inside the chamber that fogs the plastic parts with a thin layer of metal. Sputtering techniques are widely used to plate copper and other metals on printed circuit boards (PCBs) for the electronics industry. Glass–epoxy laminates are usually used for PCBs and the copper-plated circuits printed on them are easily soldered to the electronic components.

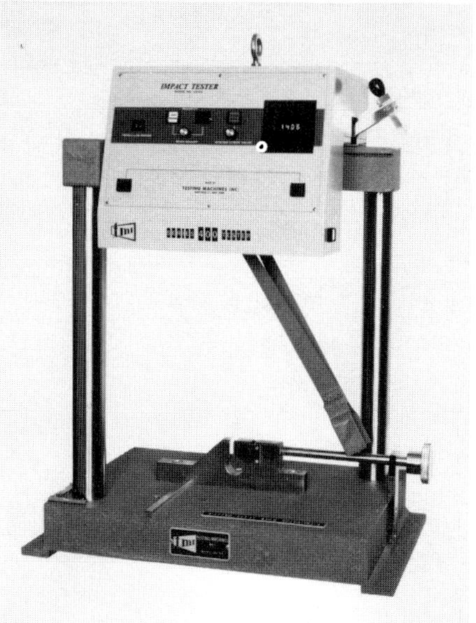

Figure 17
Izod–Charpy impact tester used for determining impact strength of plastics, ceramics, and light metals. This device uses a sensing system with a rotary transducer and a digital display to replace the analog dial. Its range is between 0 and 1 and 0 and 45 ft lb (Testing Machines Inc.).

Engineering Properties of Plastics

When a plastic material must be selected from the many available types for the manufacture of a new product, the choice must be based on the various engineering properties of the plastic. For example, if a gear were made of a brittle material such as acrylic with no reinforcing material, it would shatter when a load was applied, but using a high-performance plastic such as one of the acetals and nylons would make excellent gears for many purposes. However, if a less expensive, disposable plastic product were needed, it would be noncompetitive and wasteful to select an expensive engineering plastic, and a low-cost material such as polystyrene should be selected. Some products must be made highly resistant to impact; some of these are automobile windshields, doors, bumpers, telephones, and toys. The engineer should probably choose the plastic for these items from among the high-impact polymers such as the polycarbonates, polyesters, and the newer polyphenylene oxide-based polymers and polyetherimides. Plastic resin manufacturers make the properties of their plastics available to engineers and designers. See Table 2 for an example of a properties chart.

Standard tests are used to assure the engineer that the selected plastic material will meet the requirements of the product. These tests were set up by the American Society for Testing and Materials (ASTM). The many tests for plastics can be divided into several general

Figure 18
Slip and friction tester. This device can make several tests, tensile tests, measurements of the coefficient of plastic surfaces, and peel tests for plastic laminates (Testing Machines Inc.).

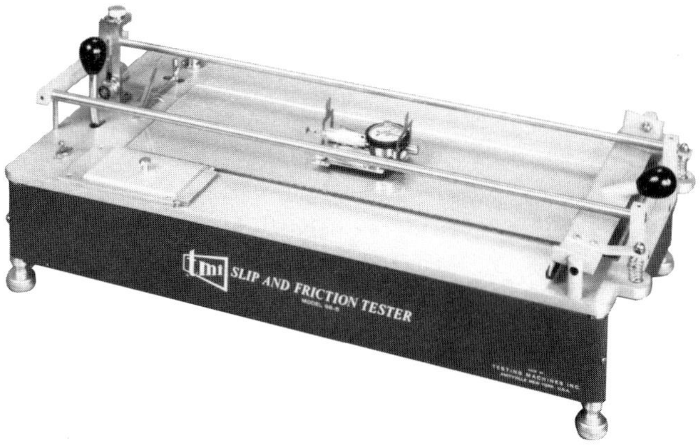

Table 2
Properties Chart of Polyamide-Imide[a]

	Dixon TL Series Polyamide-Imide			
	TL-01[b]	TL-02[c]	TL-03[d]	ASTM Test Method
I. Mechanical Properties				
Tensile strength (psi)	19,600	26,900	27,300	D-638
Elongation %	6	12	4.1	D-638
Flexural strength (psi)	26,400	30,700	40,500	D-790
Compressive strength (psi)	21,000	32,000	—	D-695
Impact strength (ft–lb/in.)	1.1	2.5	1.6	D-256
II. Thermal Properties				
Heat distortion temperature (° F at 264 psi)	525	525	550	D-648
UL temperature rating (° F)	410	410	410	
Coefficient of thermal expansion (linear—in./in./° F)	2.4×10^{-5}	2.0×10^{-5}	1.1×10^{-5}	D-696
Thermal conductivity (BTU/hr/ft²/° F/in.)	2.5	1.7	—	UL94
Flammability rating	V-O	V-O	V-O	
Limiting oxygen index %	42	42	42	
III. Electrical Properties				
Dielectric constant, 1MHz	—	3.5 to 4.0	—	
Dissipation factor, 1MHz	—	.001 to .009	—	D-150
Dielectric strength (V/mil)	—	600	—	
IV. Physical Properties				
Specific gravity	1.45	1.4	1.6	
Density (lb/in.³)	.05	.05	.06	D-792
Hardness	M109	M119	—	D-785
Water absorption %	.22	.28	.18	D-570
Radiation resistance	Resistant	Resistant	Resistant	
V. Friction and Wear Ratings				
Coefficient of friction	Good	Fair	—	
Wear life	Excellent	Fair	—	
VI. Chemical Resistance	Excellent	Excellent	Excellent	

[a]All statements, information and data given herein are believed to be accurate and reliable but are presented without guaranty, warranty, or responsibility of any kind, expressed or implied.

[b]Dixon TL-01 is a self lubricating grade with excellent continuous use temperature.

[c]Dixon TL-02 is an unfilled material for structural applications.

[d]Dixon TL-03 is a fiber glass filled structural material with high stiffness and low thermal expansion.

SOURCE *Dixon Data: Flow Control Products Unit*, Dixon Industries Corporation, Bristol, Rhode Island, © 1984.

areas. The most common are mechanical properties, thermal properties, electrical properties, physical properties, and special characteristics. Mechanical properties include tensile strength, hardness, compressive strength, impact strength, and flexural strength (Figures 17 and 18). Thermal properties include Underwriters Laboratories (UL) ratings, heat distortion, coefficient of thermal expansion, thermal conductivity, and flammability rating. Electrical properties are dielectric constant, dielectric strength, and dissipation factor. Physical properties are specific gravity, density, hardness, water absorption, and radiation resistance. Special characteristics include weatherability, optical clarity, chemical resistance, and colorability.

Plastics designers selecting plastics for food packaging must consider the government regulations set forth by the National Institute for Occupational Safety and Health (NIOSH). Also toxicity in the manufacture of plastics comes under the recommendation of this organization to the regulatory agency, Occupational Safety and Health Administration (OSHA).

RUBBER, ELASTOMERS, AND ADHESIVES

The term *rubber* was originally used only for the natural product that is obtained from a thick, milky fluid (latex) that oozes from certain plants when they are cut. Most latex comes from the Para' rubber tree (*Hevea brasiliensis*) that grows in South America, Southeast Asia, and Sri Lanka. Natural rubber in the form of cured latex is a sticky, gummy substance that has a limited usefulness for rubber products. In 1839 Charles Goodyear discovered that this latex could be greatly strengthened and its elastic properties were improved when he added a small amount of sulfur to it and heated the mixture. The process is called vulcanizing. By varying the sulfur content and adding certain pigments such as carbon black, fillers, and softeners, rubber can be made to any hardness from very soft to very hard, such as ebonite. When rubber products need to be strengthened such as in automobile tires, textile cords or fabrics are used. The term *elastomer* has been generally accepted as a broad term for those materials, synthetic or natural, possessing rubbery qualities, such as high resilience, extensibility, and elastic properties.

Artificial Elastomers

Synthetic rubbers were first created in the early 1930s and were subsequently developed because of the uncertainty of the supply of natural rubber. This first rubbery synthetic was derived from acetylene gas and was a long chain molecule called polychloroprene, better known as neoprene. Natural rubber is somewhat more resilient than neoprene, but it tends to deteriorate in sunlight (ultraviolet) and swells when in contact with petroleum oil, neoprene has better resistance to these factors. However, chemical additives such as antioxidants, plasticizers, and stabilizers are now added to natural rubber and other elastomers to overcome these difficulties. One synthetic rubber, polyisoprene, seems to have almost the same properties as natural rubber, and some others such as the rubbery plastics have certain specific advantages over natural rubber.

The latex is compounded with additives and vulcanizers in a mixing machine to form a homogenous mass. The mix is then placed in a mill with chilled rolls (to prevent premature vulcanization). The rubber compounds are made into sheet form on calender rolls (Figure 19). The manufacture and vulcanizing of rubber products is covered in Chapter 15.

Adhesives

Adhesives can be separated into two groups: those that bond two surfaces together as a kind of fastener, and those that adhere to a crack or joint for the purpose of making a fluid or air-tight seal. Sealants are very similar to adhesives in that they must adhere to the sealing surface even though they usually do not have the high bonding strength of adhesives.

Prior to the development of synthetic adhesives, bonding agents were glues made from the hides and hooves of animals or parts of fish. Pastes were made of tree gums, starch, or casein. All of these products were easily loosened by the action of moisture. Since there is little molecular attraction between these materials and the parts to be bonded, they are not very strong adhesives. At the present time, almost any material can be bonded to another if the right adhesive is used. Most adhesives used today are plastics, either thermoplastic or thermoset. Sealants are a type of ad-

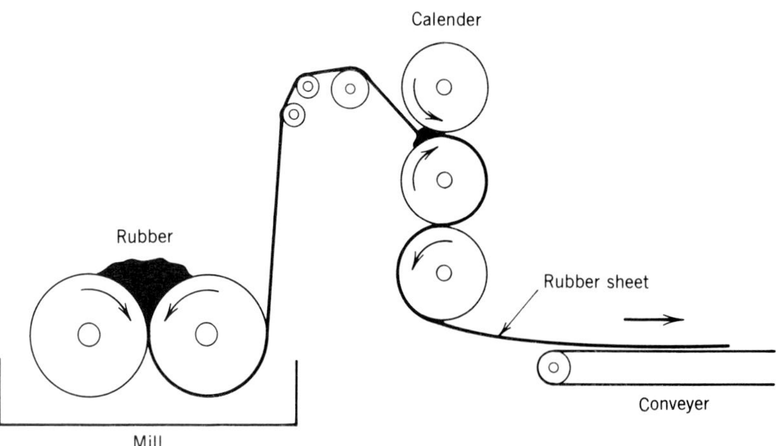

Figure 19
Rubber calender rolls.

hesive that bonds to a material but has the function of sealing rather than fastening two parts together. Some types of sealants can remain soft and flexible for years, maintaining a good seal.

Because of their flexibility, structures bonded with adhesives resist damage from vibration better than riveted or bolted assemblies. A riveted joint must resist high stress at each rivet hole and cracking can result, but adhesive bonding is made over the entire surface, reducing stress concentration. Also connectors, such as rivets or bolts, can be a cause of corrosion, allowing galvanic currents to flow which allows corrosion to take place. In contrast, adhesives prevent galvanic current flow because they normally have a high electrical resistance.

Nitrile rubber epoxy resins are used for bonding honeycomb structures for aircraft (Figure 20). Two-part epoxy kits are available for general use in which a resin and a catalyst are measured out and mixed together. When mixed, this cement hardens in minutes or hours depending on the formation. Epoxy resins with fiber glass cloth are used on the hulls of ships and boats to give them a strong leak-proof surface. Automobile body repairs are often made with fiber glass and resin.

The cellulosic plastics are used in model cement and some household cement. Polyvinyl acetate is a white glue used in woodworking. Polyethylene is often used as a hot-melt adhesive. Polysulfides, polyurethanes, and silicones are used as sealants for caulking of buildings and boat hulls. Phenolic resins have been used for years as a bonding material for abrasive grinding wheels. Most adhesives will harden in the presence of air (aerobic conditions) but some types such as the cyanoacrylates will harden only when they are not exposed to oxygen in the atmosphere (anaerobic conditions) and only when spread out as a thin film between two surfaces. Cyanoacrylates, popularly known as super glue, will instantly bond smooth flat surfaces with a high bonding strength but will not bond uneven or porous surfaces. Table 3 lists some of the kinds of adhesives and their uses.

As with other plastics and rubbers, additives are used in adhesives to impart needed properties such as resilience and toughness. When bonding agents are required to conduct electricity, conducting additives such as carbon black, graphite, or metal flakes or powders are used. Certain destructive tests are used to determine joint strength and efficiency for various adhesives and bonding materials. Among these are shear, tensile, compression, impact, and peel tests. Cleavage (a variation of the peel test), tack (adhesion), creep, and fatigue tests are also very common (Figure 21). There are several specifications applicable to adhesive bonding: two of them are the Military, Federal Adhesive Specifications and those of the American Society for Testing and Materials (ASTM).

Figure 20
Bonded honeycomb aluminum structures for aircraft wing parts.

Figure 21
The glueability tester tests the strength of a bond formed between adhesive and material being tested under specific, carefully controlled circumstances (Testing Machines Inc.).

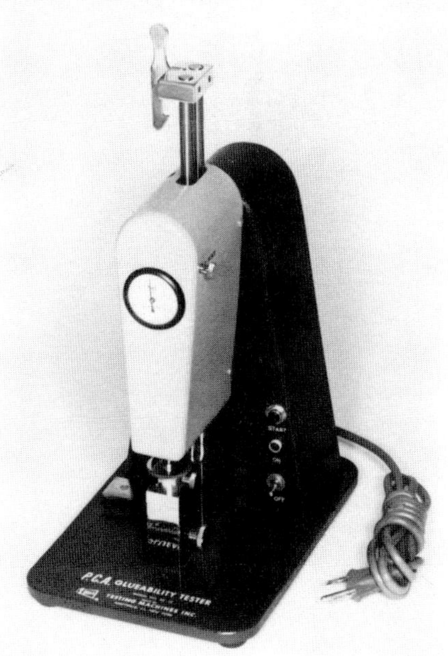

Table 3
Adhesives and Their Uses

Name	Advantages	Disadvantages	Uses
Anaerobic cyanoacrylates (popularly called super glue)	High strengths (2000 psi) in a few minutes	High cost, thicknesses greater than .002 in. cure slowly or not at all, does not cure in presence of air	Production speed in bonding small parts especially metal to metal and metal to plastics
Acrylic	Clear with good optical properties, ultraviolet stability	Moisture sensitive, high cost	Bonding plastics
Cellulose esters and ethers	Very soluble	Moisture sensitive	Bonding of organic substances such as paper, wood, and some fabrics
Epoxies: modified, nitrile, nylon, phenolic, polyamide, vinyl	Low temperature curing, adheres by contact pressure, good optical qualities for transmitting light, high strength, long shelf life	Cost, low peel strength in some types	Bonding microelectron parts, aircraft structural bonding, bonding cutting tools, and abrasives
Phenolics (Bakelite)	Heat resistance, low cost	Hard, brittle	Abrasive wheels, electrical parts
Polyimides	Resistance to high temperature (600° F or 315.5° C)	Cost, high cure temperature (500 to 700° F or 260 to 371° C)	Aircraft honeycomb sandwich assemblies
Polyester (anaerobic)	Hardens without the presence of air	Limited adhesion, cost	Locking threaded joint
Polysulfide	Bonds well to various surfaces found in construction	Low strength	Caulking sealant

Polyurethane (rubbery adhesive)	Strong at very low temperatures, good abrasion resistance	Low resistance to high temperatures, sensitive to moisture	Flexible and rigid foam protective coatings, vibration damping mounts, coatings for metal rollers and wheels
Rubber adhesives: butadienes, butyl, natural (reclaim), neoprene, nitrile	High strength bonds with neoprene (4000 psi), oil resistance (nitrile)	Low tensile and shear strength, poor solvent resistance	Adhesives for brake linings, structural metal applications, pressure sensitive tapes, household adhesives
Silicones	High resistance to moisture, can resist some temperature extremes	Low strength	Castable rubbers, vulcanized rubbers, release agents, water repellents, adhesion promoters, varnishes and solvent type adhesives, contact cements
Urea-formaldehyde resins	Low cost, cures at room temperatures	Sensitive to moisture	Wood glues for furniture and plywood
Vinyl: polyvinyl-butyralphenolic vinyl plastisols, polyvinyl acetate and acetals, polyvinyl chloride	High room temperature peel, ability to bond to oily steel surfaces (vinyl plastisols)	Sensitive to moisture	Bonds to steel (brake linings), safety glass laminate, household glue, water and drain pipe (PVC), household wrap

SOURCE John E. Neely, *Practical Metallurgy and Materials of Industry*, Second Edition, Wiley, New York, © 1984.

PETROLEUM PRODUCTS (LUBRICANTS AND ASPHALTS)

In 1859 Colonel E. L. Drake struck oil in the little Pennsylvania town of Titusville. Its early impact on the economy was to replace whale oil as fuel for lamps. This, no doubt, helped to save the whales for posterity but, as everyone today knows, petroleum now has many more uses—it fuels our industries and our transportation systems and provides the raw materials for many of the products we use. Without the lubricants, oils, and greases all of our industrial machinery would stop running and transportation would come to a halt.

As it comes from the oil wells, petroleum crude is composed of a large variety of hydrocarbon molecules in many chemical combinations. By the process of distillation and condensation, these constituents are separated into several classes. These are gases, gasoline, distillate fuel, and lubricating oil, with the residual heavy end products being used in the production of heavy lubricating-oil stocks, bunker fuels, and asphalt. Chemically, the carbon and hydrogen molecules in petroleum exist in many combinations giving various properties to oils from different sources of crude. In addition, all petroleum oils contain small amounts of sulfur and nitrogen which affect the stability and corrosiveness of the oil. The hydrocarbon molecules found in petroleum can be grouped into three classes—paraffins, naphthalenes, and aromatics. Of these, the paraffins and naphthalenes have the greatest stability and are the principal constituents of lubricating oils that are used in automobile crankcases and in other machinery. The aromatic oils have good solvency characteristics and are used for certain chemical formulations and for ink oils.

Greases

Greases are actually lubricating oils held by absorption and capillary action in the fibrous or granular structure of a thickening agent (clay or lamp black). The oil used for greases that are intended to withstand high pressures at relatively slow speeds is usually one of higher viscosity than one intended for normal or higher speeds. The most important quality of greases is their adhesiveness to metallic surfaces and their "stay-put" property. Greases are extensively used to pack wheel bearings and steering linkages in automobiles and for lubricating manufacturing machinery.

Asphalts

Asphalt is a black, sticky substance that is also separated from crude petroleum. However, some asphalts occur naturally in pits or lakes that are residues of crude oil in which the lighter fractions (more volatile parts) have evaporated over a period of many thousands of years. Asphalts can generally be divided into two categories, paving asphalts (asphalt cements) and liquid asphalt materials. Thermoplastic asphalt cements are solid but can be heated to a liquid condition for use as roofing tar and for paving. Liquid asphalts are not heated but are diluted with solvents to provide the proper consistency. The liquid asphalts harden over a period of time as the lighter fractions evaporate, but the molten (heated) asphalt hardens when it cools.

Asphalts are used in the manufacture of waterproof papers and textiles and for waterproof linings for irrigation canals and reservoirs. However, plastics are beginning to replace asphalt products in many of these applications. Asphaltic concrete is used extensively for road and highway construction and for parking lots and driveways.

CERAMIC MATERIALS

Ceramic materials include clay, silica, glass, natural stone, and portland cement (with which concrete is made). Ceramics are held together by ionic and covalent bonds which are very strong but rigid and much less ductile than the metallic bond, making them subject to brittle failure. Because they are good thermal and electrical insulators and can withstand high temperatures, ceramic materials are extensively used in the electrical industry and for furnace linings.

Clays

Clays vary considerably according to where they are mined. Ordinary clays containing impurities such as iron oxide (which gives them their reddish color) are used to manufacture brick for building construction and firebrick for furnaces. Kaolin, a purer white clay composed mostly of alumina and silica, is used in the manufacture of earthenware, fine china, porcelain, paper products, and firebrick. Fire clay has less than 10 percent impurities and is used for the manufacture of clay products and for joining firebrick in furnace construction.

Porcelain enamels are composed of quartz, feldspar, borax, soot ash, and other substances. These are used to coat iron, steel, or aluminum for wear and decorative purposes for products such as stoves, refrigerators, pots, and pans.

Some recently developed kinds of ceramic materials are known as fine ceramics or high technology ceramics. These include insulation materials for electronic substrates, transition-metal oxides (semiconductors), heat resistant materials for engines, cutting tools, and corrosion resistant materials. Fine ceramics, along with plastics technology, automation, and robotics appear to be among the rapidly growing areas of high technology.

Glass

Glass is a supercooled rigid liquid. It is in a metastable state; that is, it can go to a stable lower energy state of crystallization only by passing through an intermediate higher energy state. In effect, glass has become too cold to freeze from the liquid. Common types of glasses are based on silicon dioxide (SiO_2) which occurs abundantly in nature as quartz and cristobalite and as a part of many of the silicate minerals. Natural volcanic glass, called obsidian, was used by primitive peoples for arrowheads and spear points. Quartz and silica sand are common raw materials for the production of commercial glasses and silica. Additives can alter the glass, giving it special properties. For example, the common soda–lime glass has a lowered softening temperature and is used for windows, containers, and lamp bulbs. Borosilicate glass, with a low thermal expansion, does not tend to crack when unevenly heated; consequently, it is used for laboratory glassware, large telescope mirrors, and household cookware such as Pyrex. Lead–alkali–silicate glass has an even lower softening point and is known as flint glass for optical purposes and crystal glasses for tableware. For a high refractive index, the lead oxide (PbO) content may be as much as 65 percent.

Glass has excellent colorability and is a versatile material that has excellent chemical resistance, with the exception of hydrofluoric acid which is used for the purpose of etching glass objects. A considerable amount of glass plate is used in modern building construction and glass fiber is used extensively as a reinforcement material in plastic resins and in the communications industry.

Portland Cement

The invention of portland cement is generally credited to Joseph Aspdin, an English mason who obtained a patent for it in 1824 and named his product portland cement because the concrete made of it resembled a natural limestone on the Isle of Portland. Materials used in the manufacture of portland cement are lime, silica, alumina, and iron. Each of these can come from any of a number of source materials. For example, alumina comes from clay, shale, slag, aluminum ore refuse, and a number of other source materials.

Selected raw materials are pulverized and proportioned to produce the desired chemical composition. The prepared mix is fed into the upper end of a sloping rotating kiln. A powdered coal, fuel oil, or gas is burned at the lower end of the kiln, producing temperatures of 2600 to 3000° F (1427 to 1649° C). The resulting portland cement clinker is cooled and then pulverized and a small amount of gypsum is added to regulate the setting time of the cement. This finely ground product, which will pass through a sieve with 40,000 openings in a square inch, is the finished portland cement, ready to be shipped. Typical steps in the manufacture of portland cement are illustrated in the flow chart (Figure 22). Each manufacturer of portland cement uses a trade or brand name under which the product is sold.

Concrete

Concrete is commonly seen in sidewalks, streets, dams, and building foundations. It is a mixture of portland cement, sand, and gravel plus water and, sometimes, additives. These ingredients (sand and gravel are called aggregates) are mixed together in specific proportions to produce a fluid mass called concrete. It is then poured or pumped into molds where it is left to harden. The forms or molds may be removed in a few days. Ordinary concrete takes about 28 days to completely cure, providing it is kept wet, at which time it is nearly as hard and strong as it will ever be.

Specialty Concretes

Among the many admixtures used to improve portland cement concrete is flyash, a finely divided residue resulting from the combustion of coal. It has a pozzolanic effect in that it acts like a cement in the presence of lime. Its main value lies in its ability to reduce the percentage of portland cement needed in concrete, thus reducing the cost. It also imparts a higher chemical resistance than ordinary concrete. Like flyash, blast furnace slag can also reduce the percentage of cement needed in concrete; however, when granulated and activated by sodium hydroxide, it can completely replace portland cement in some cases. Silica fume, a powdery byproduct of electric arc furnace production of ferrosilicon alloys and metallic silicon, also behaves like natural pozzolan and it gains strength earlier in the curing cycle than flyash and slag concretes do.

Polymer Concretes A growing technology involves the introduction of polymers into concrete. Latex modified portland cement concretes have high strengths, reduced porosity, and water resistance. Polymer–cement concrete has better abrasion, higher strength, and impact resistance than conventional concrete, making it useful for industrial floor applications. Polymer–portland-cement concrete is mixed with either a monomer or a polymer in a liquid, powdery, or dispersed phase that is allowed to cure along with the cement or is polymerized in place.

Polymer concrete is a composite material in which an aggregate mixture is combined with a monomer that is polymerized in place; no portland cement is used. Monomers used in polymer concrete include polyester–styrene resins, furan, vinyl esters, and epoxy. Also sulfur–concrete and sulfur impregnated concrete are included in a broad definition of polymer concretes.

(1)

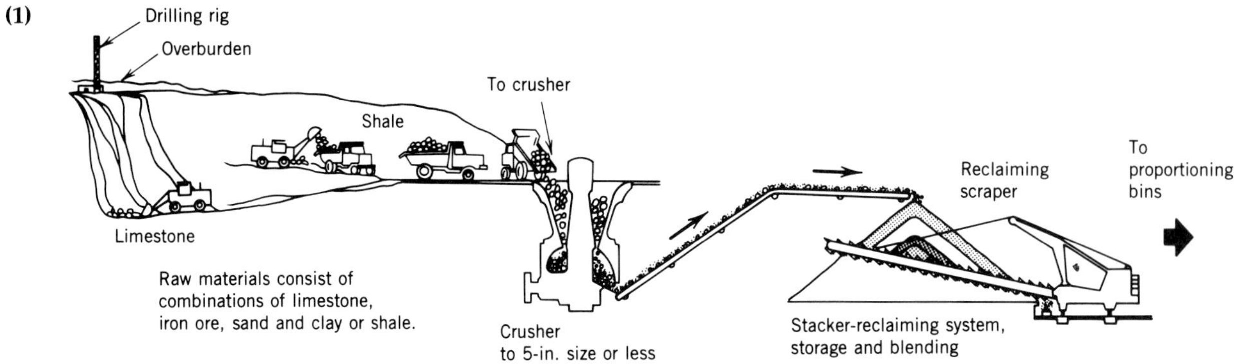
Drilling rig
Overburden
To crusher
Shale
Limestone
Reclaiming scraper
To proportioning bins
Raw materials consist of combinations of limestone, iron ore, sand and clay or shale.
Crusher to 5-in. size or less
Stacker-reclaiming system, storage and blending

(2)

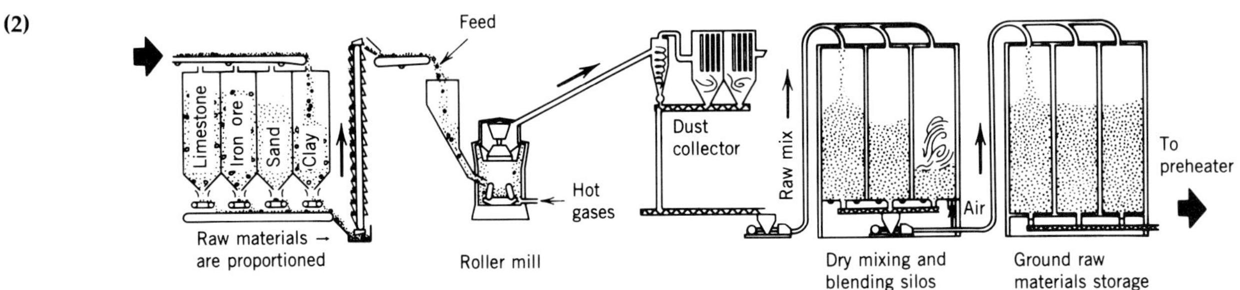
Feed
Limestone
Iron ore
Sand
Clay
Dust collector
Raw mix
Hot gases
Air
To preheater
Raw materials are proportioned
Roller mill
Dry mixing and blending silos
Ground raw materials storage

(3)

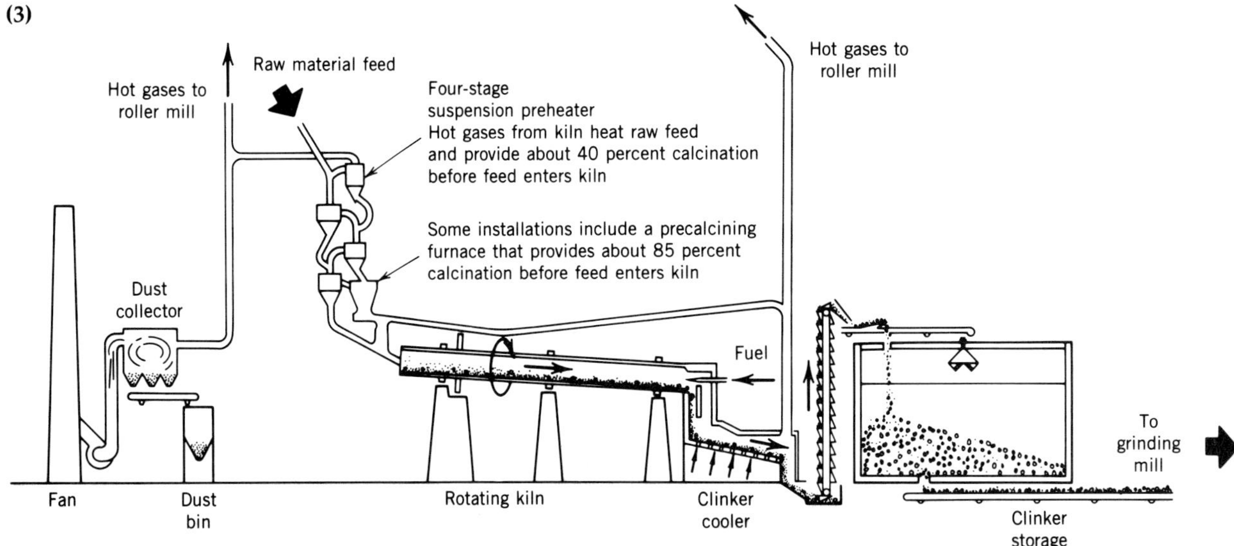
Hot gases to roller mill
Raw material feed
Four-stage suspension preheater
Hot gases from kiln heat raw feed and provide about 40 percent calcination before feed enters kiln
Hot gases to roller mill
Some installations include a precalcining furnace that provides about 85 percent calcination before feed enters kiln
Dust collector
Fuel
To grinding mill
Fan
Dust bin
Rotating kiln
Clinker cooler
Clinker storage

(4)

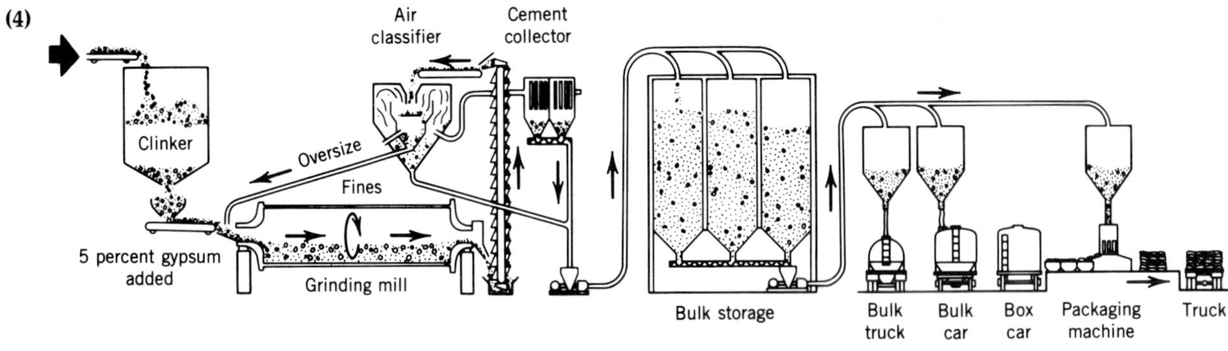
Air classifier
Cement collector
Clinker
Oversize
Fines
5 percent gypsum added
Grinding mill
Bulk storage
Bulk truck
Bulk car
Box car
Packaging machine
Truck

Figure 22
Flow chart showing the steps in the manufacture of portland cement by the dry process using a preheater. (1) Quarry and blending of raw materials. (2) Proportioning and fine grinding of raw materials. (3) Kiln system. Preheating, burning, cooling, and clinker storage. (4) Finish grinding and shipping (Oregon Portland Cement Company).

Polymer concrete materials are excellent for patching because of their rapid curing time (15 to 60 min). A major disadvantage is that they cannot withstand higher temperatures.

Fiber Reinforced Concrete Steel reinforcing bars have long been in use in portland cement concrete. Concrete mixtures using fibers act as secondary reinforcement in concrete and can greatly increase its tensile strength. Small fibers can prevent crack growth in concrete structures. Typical fibers include asbestos, cotton, rayon, glass, steel, acrylic, nylon, polyester, polyethylene, and polypropylene.

WOOD PRODUCTS

Forest industries, in which wood products such as lumber, plywood, masonite, and particleboard are manufactured, supply a large portion of the building materials for homes and offices and for furniture manufacture. Paper products produced from wood pulp are extensively used in packaging and in printing. Wood is composed of cellulose, lignin, and small amounts of inorganic materials which make up the ash when wood is burned.

The engineering properties of wood vary a great deal in different species of trees and even within one species because of imperfections such as knots and checks. Moisture markedly decreases the mechanical properties of wood. Because of this and the shrinkage of "green" wood when it dries, construction-grade lumber and other wood products are either air or kiln dried before they are finished (planed or sanded) to size. Wood may be classified into two broad categories, **hardwood** and **softwood**.

Hardwoods

Deciduous trees shed their leaves each year and are classed as hardwoods. The definition generally refers to the difficulty of sawing rather than its hardness as the term is used in engineering. In fact, balsa, although it is very light in weight, soft, and porous, is classed as a hardwood because it comes from a deciduous tree.

Oak trees are found in many varieties in the United States, but they can be generally divided into two classes, white oak and red oak. White oak is more resistant to decay and finer grained than red oak, and it is more difficult to work. Red oak is used extensively in furniture manufacture, flooring, and plywood. Oaks, sugar and black maples, and birch growth in the northern forests from Minnesota east and south to Virginia. Sweet gum, ash, and tupelo are found in Texas and some other southern states. Hickory, oak, black walnut, and basswood grow in the prairie states. Oregon maples are not as hard or strong as sugar and black maples. Walnut is found in most eastern and northern states and in some far-western states.

Softwoods

Coniferous trees have cones and needles instead of flat leaves and are sometimes called evergreen trees because they do not lose their needles in the fall but keep them throughout the year. The coniferous trees are classed as softwoods even though some varieties have a very tough, hard wood. For example, the sitka spruce was once used for structural frames in aircraft because it was considered to have the highest strength-to-weight ratio of any wood. Because of its resonant qualities, it is also used for musical instruments.

The pines are the most widely used softwoods. Western white pine, sugar, ponderosa, and lodgepole pine are found in the western and northwestern parts of the United States. Southern yellow pine is grown in the southeastern part of the United States.

Douglas firs and white fir are grown in the far West and are widely used in building construction. Firs are somewhat harder and stronger than pines. Spruce, hemlock, redwood, and cedar are generally found in the western United States and are used in building construction, for boats, mine timbers, and various wood products.

REVIEW QUESTIONS

1. From what raw materials are most plastics derived?
2. What is the source of natural rubber?
3. What polymer derived from acetylene gas is called polychloroprene (neoprene)?
4. From what basic source of raw material do we

get most of our pesticides, fertilizers, clothing fibers, and many medicines such as antibiotics?

5. What is the major difference between thermosetting and thermoplastic materials?
6. Which type of material, thermosetting or thermoplastic, can be easily injected by pressure into a mold?
7. If you needed a plastic material for a product that had to resist a variety of chemicals and the effects of sunlight, must have a nonstick surface, and withstand high temperatures such as would be found in a baking oven, what would be your choice: (*a*) silicone plastic, (*b*) polyurethane plastic, (*c*) Bakelite, or (*d*) fluorocarbon plastic?
8. Name two items in which glass reinforced nylon 6/6 can replace metal where considerable loading is involved.
9. Some plastics are colored or tinted with pigments whereas others have external surface finishes made with lacquers, enamels, or overlays. How can a metal film be applied to a plastic product?
10. In what process is natural latex combined with sulfur and heated to produce the flexible substance we call rubber?
11. Modern aircraft are made with less rivets and more adhesive bonded structures than in the past. Name two advantages in using bonded structures over riveted ones.
12. Of the three classes of hydrocarbon oils, the aromatics are used for solvents and ink oils. What is the principal use of the other two types, paraffins and naphthalenes?
13. The tarry substance used for roofing and highway paving is a petroleum derivative. What is it called?
14. The use of clay for bricks, porcelains, and electrical insulation is well known. New uses for special ceramics, called fine ceramics or high technology ceramics, are being developed. Name two applications for these special ceramic materials.
15. Traditional construction materials include brick, stone, metal, and glass. Also plastic materials are being used for many products in the construction industry. Besides these, what are the two major material types used in construction?

CASE PROBLEMS

Case 1
Plastic Color Problem

A small plastic product made of Bakelite is black in color and has had a good sales record for years. However, sales have begun to drop off significantly. It was found that the competition is using bright colors in their product and they are taking over the market. In what way can this company improve its product, add color, and increase productivity to lower the price?

Case 2
Air Cylinder Stick–Slip Problem

A small air cylinder was manufactured to move a machine part. The piston seal was made of neoprene rubber which had a tendency to take a set or stick to the cylinder wall when it was not operated for a given time. This caused the mechanism to stick–slip, that is, to hesitate and then jump forward too fast. Rubber O rings, cups, and other types of rubber seals did not solve the stick–slip problem which was obviously a seal problem. From what synthetic material would you choose to make this seal in order to eliminate this sticking problem?

CHAPTER 6 SELECTION AND APPLICATION OF MATERIALS

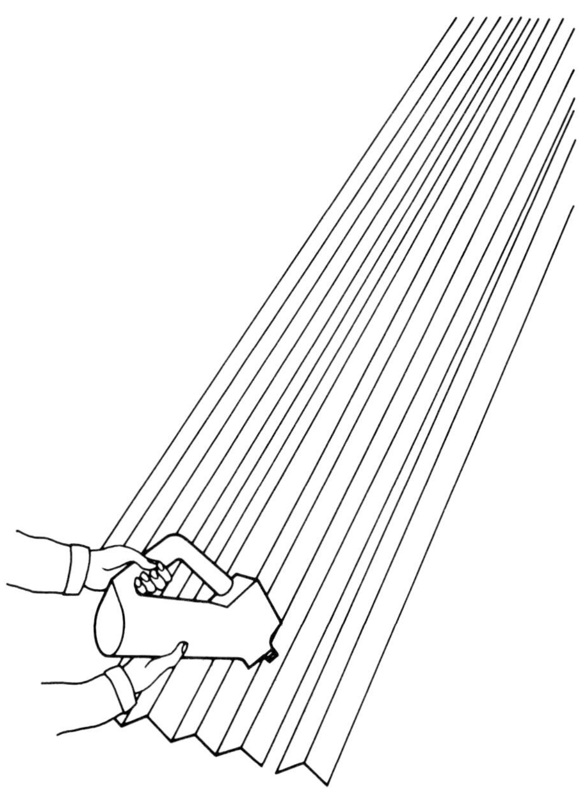

When only carbon steel, cast iron, wrought iron, and a few nonferrous metals were commonly used, identification was a relatively simple process. Spark testing on a grinder was a sufficient means of separating the three ferrous metals and, for the most part, the nonferrous metals could be identified by their color. Of course, nonferrous metals such as gold, silver, and copper began to be alloyed (combined) as early as the time of Archimedes. As the story goes, after he made the discovery of the natural law of buoyancy, he immediately applied it to solving the riddle of whether a golden crown contained pure gold or was cheapened with an alloying element having lesser value.

In our modern world there are literally thousands of alloys, both ferrous and nonferrous. Selection and identification of metals is therefore a far more complex process today, requiring classification systems and sophisticated testing equipment. Even more important than metals identification is the need for methods and equipment for testing and analyzing other materials of industry, especially those that present radiation hazards and contain toxic materials.

With so many new engineering materials being developed in the areas of plastics, composites, metals, and ceramics, it can be difficult to make choices for manufacturing materials. Although a new product may have more sales appeal when it is made of plastic, it might not have sufficient strength or heat resistance. Other products may be too costly to fabricate from metal but can be made from a selected plastic at a high production rate by the injection molding process and still have satisfactory engineering properties. Some of these factors in material selection will be covered in this chapter.

OBJECTIVES

This chapter should enable you to:

1. Identify many metals and alloys by their numerical classification systems.
2. Describe some of the characteristics of many commonly used metals and alloys.
3. Name some methods used to identify and analyze various metal alloys and materials.

4. State some of the methods and factors involved in selecting materials for manufacturing purposes.

CLASSIFICATION SYSTEMS FOR METALS

The most common numerical system used to classify carbon and low alloy steels in the United States was developed by the Society of Automotive Engineers (SAE) and the American Iron and Steel Institute (AISI) for steels used in manufacturing. The American Society for Testing Materials (ASTM) developed specifications for carbon and alloy structural steels. Tool steels are classified under their own system which covers seven major types. Stainless steels are classified under the chromium and chromium–nickel types and cast irons are identified by ASTM numbers. Nonferrous metals are also classified under specific systems for each metal or alloy. Color coding of steels has long been used as a means of identification, but there is no universal code; each manufacturer, distributor, or local manufacturing plant has its own code. It is a very useful system on a local basis as long as the color coding is not cut off, leaving the remainder of the stock unbranded.

The SAE classification system for alloy steels uses a four- or five-digit number (Table 1). The first number on the left designates the type of steel. For example, plain carbon steel is denoted by the number 1, 2 is a nickel steel, 3 is a nickel–chromium steel, and so forth. The second digit indicates the approximate percentage of the predominant alloying element. The third and fourth digits, represented by X, always represent the amount of carbon in hundredths of a percent. For plain carbon steel it is normally between .08 and 1.70 percent. SAE1040 is plain carbon steel containing .40 percent carbon. SAE4140 is a chromium–molybdenum alloy steel containing .40 percent carbon and about 1 percent of the major alloy, molybdenum. The AISI system is basically the same as the SAE system with the addition of certain capital letter prefixes. The two systems are usually combined. The AISI prefixes are:

B—Acid Bessemer

C—Basic open-hearth carbon steel

CB—Either acid Bessemer or basic open-hearth carbon steel at the option of the manufacturer

D—Acid open-hearth carbon steel

E—Electric furnace alloy steel

Table 1
SAE–AISI Numerical Designation of Alloy Steels*

Carbon steels	
Plain carbon	10xx
Free-cutting, resulfurized	11xx
Manganese steels	13xx
Nickel steels	
.50% nickel	20xx
1.50% nickel	21xx
3.50% nickel	23xx
5.00% nickel	25xx
Nickel–chromium steels	
1.25% nickel, .65% chromium	31xx
1.75% nickel, 1.00% chromium	32xx
3.50% nickel, 1.57% chromium	33xx
3.00% nickel, .80% chromium	34xx
Corrosion and heat-resisting steels	303xx
Molybdenum steels	
Chromium	41xx
Chromium–nickel	43xx
Nickel	46xx and 48xx
Chromium steels	
Low-chromium	50xx
Medium-chromium	51xx
High-chromium	52xx
Chromium–vanadium steels	6xxx
Tungsten steels	7xxx
Triple alloy steels	8xxx
Silicon–manganese steels	9xxx
Leaded steels	11Lxx (example)

SOURCE John E. Neely, *Practical Metallurgy and Materials of Industry*, Second Edition, Wiley, New York, © 1984.

* x represents percent of carbon in hundredths.

Tool Steels

Tool steels are mostly high carbon or high alloy in carbon steel. However, a few types do have low carbon or alloy to give them special properties. As the name implies, these special steels were designed for tools, dies, molds, machinery parts, and for certain specific applications. They all require heat treatments such as hardening and temper drawing to prepare them for use. Symbols of one or more than one letter have been assigned to each of the seven major groups of tool steels. Each group is designated by its special purpose or use by the quenching medium used. The classification is as follows:

1. Water-hardening tool steels
   W—high carbon steels
2. Shock-resisting tool steels
   S—medium carbon, low alloy
3. Cold-work tool steels
   O—Oil-hardening types
   A—Medium alloy air-hardening types
   D—High carbon, high chromium types

Table 2
Types of Commonly Used Tool Steels

Type of Steel	Examples
Water hardening:	
Straight carbon tool steel	W1, W2, W4
Manganese, chromium, tungsten:	
Oil-hardening tool steel	O1, O2, O6
Chromium (5.0%):	
Air-hardening die steel	A2, A5, A10
Silicon, manganese, molybdenum:	
Punch steel	S1, S5
High-speed tool steel	M2, M3, M30 T1, T5, T15

SOURCE John E. Neely, *Practical Metallurgy and Materials of Industry*, Second Edition, Wiley, New York, © 1984.

4. Hot-work tool steels
   H—H10 to H19, inclusive, chromium base types
   H20 to H39, inclusive, tungsten base types
   H40 to H59, inclusive, molybdenum base types
5. High-speed tool steels
   T—Tungsten base types
   M—Molybdenum base types
6. Special purpose tool steels
   L—Low alloy types
   F—Carbon tungsten types
7. Mold steels
   P—P1 to P19, inclusive, low carbon types
   P20 to P39, inclusive, other types

Some examples of commonly used tool steels and their numbers are given in Table 2.

STRUCTURAL STEELS

Standard ASTM specifications are used to designate carbon and alloy structural steels. These steels are produced in steel mills as standard shapes (Figure 1) and are used in the construction of buildings, bridges, pressure vessels, and for other structural purposes. Engineers who design steel structures and welders are more likely to use the ASTM specifications for steels than designers of machinery, heat treaters, and machinists who would probably need the SAE–AISI system to refer to the steels they would be using. These products are usually in the form of bar stock, strip, plate, and round bars and are used extensively in manufacturing steel products. Some bar stock is alloy steel such as AISI 4140 which is used to produce tough heat-treated parts such as gears and machine parts. AISI 8620 is a carburizing grade used where an extremely tough

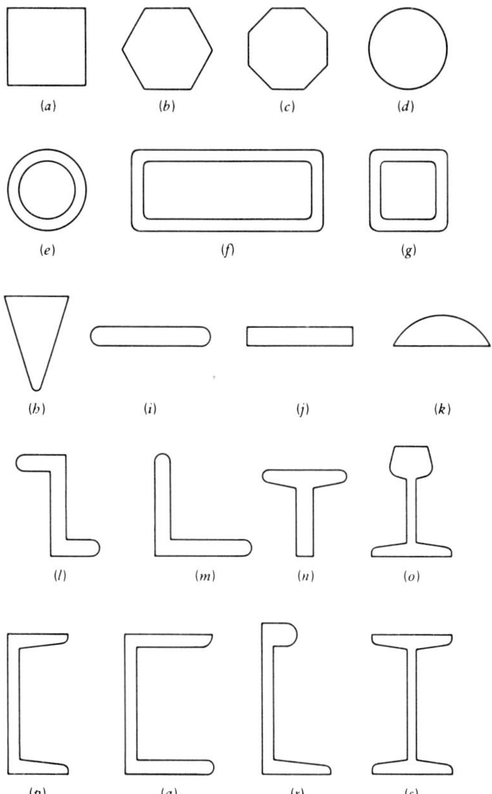

Figure 1
Steel shapes used in manufacturing. (*a*) Square HR or CR. (*b*) Hexagonal. (*c*) Octagon. (*d*) Round. (*e*) Tubing and pipe (round). (*F*) HREW (hot rolled electric welded) rectangular steel tubing. (*g*) HREW square steel tubing. (*h*) Wedge. (*i*) HR flat bar (found edge spring steel flats). (*j*) Flat bar (CR and HR). (*k*) Half round. (*l*) Zee. (*m*) Angle. (*n*) Tee. (*o*) Rail. (*p*) Channel. (*q*) Car and ship channel. (*r*) Bulb angle. (*s*) Beams—I, H, and wide flange (Neely/*Metallurgy* 2 Ed., © 1984 John Wiley & Sons, Inc.).

material with a high surface hardness is required. However, much of the steel used in manufacturing is of mild steel (low carbon) AISI 1020. Low carbon steel comes from the mill with either a cold finish or hot rolled surface. Cold finish (CF) is simply a broad term for cold rolled (CR) or cold drawn (CD). These cold-worked steels are tougher and stronger than are hot-rolled (HR) steels that have the same carbon content.

Cold-drawn steels have a relatively smooth finish without any scale. Hot-rolled steels are covered with a black mill scale that is formed when the steel is at rolling or forging heat by the action of oxygen. This scale, which is chemically Fe_3O_4, must be removed before the steel can be subjected to cold-finishing operations. Cold-finished steels are more expensive than hot-rolled steel, so they are not normally used when less accurate dimensions are acceptable and where heating or welding is to be done. Steel shafting is available as low carbon or resulfurized cold-drawn steel and as alloy steel ground and polished (G and P) which has

Table 3
Classification of Stainless Steel

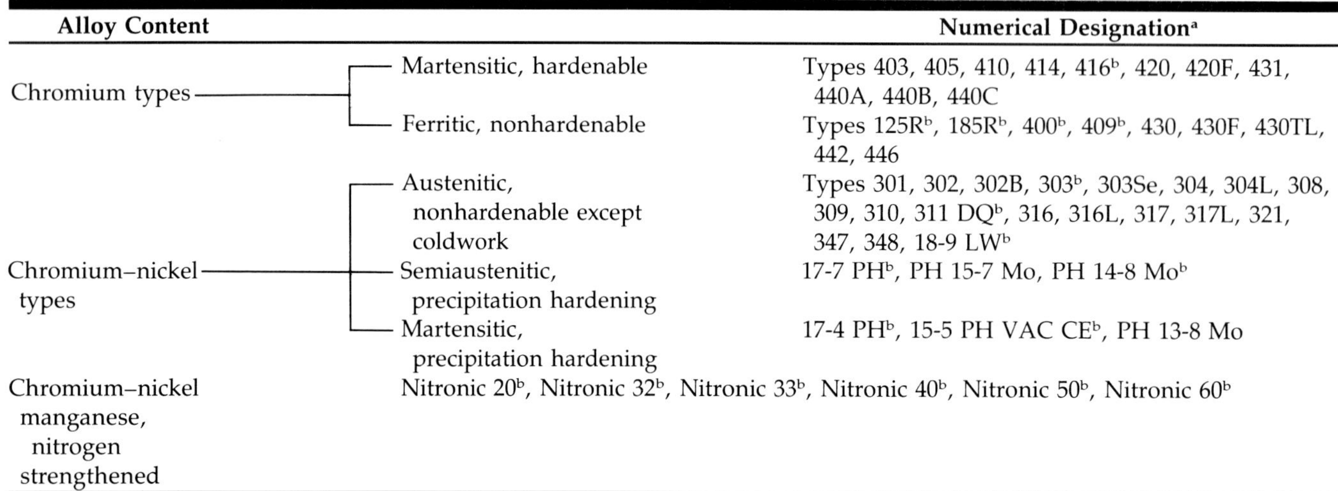

Alloy Content		Numerical Designation[a]
Chromium types	Martensitic, hardenable	Types 403, 405, 410, 414, 416[b], 420, 420F, 431, 440A, 440B, 440C
	Ferritic, nonhardenable	Types 125R[b], 185R[b], 400[b], 409[b], 430, 430F, 430TL, 442, 446
Chromium–nickel types	Austenitic, nonhardenable except coldwork	Types 301, 302, 302B, 303[b], 303Se, 304, 304L, 308, 309, 310, 311 DQ[b], 316, 316L, 317, 317L, 321, 347, 348, 18-9 LW[b]
	Semiaustenitic, precipitation hardening	17-7 PH[b], PH 15-7 Mo, PH 14-8 Mo[b]
	Martensitic, precipitation hardening	17-4 PH[b], 15-5 PH VAC CE[b], PH 13-8 Mo
Chromium–nickel manganese, nitrogen strengthened	Nitronic 20[b], Nitronic 32[b], Nitronic 33[b], Nitronic 40[b], Nitronic 50[b], Nitronic 60[b]	

SOURCE Armco Inc., Specialty Steels Division, Copyright 1984, Armco Inc., Middletown, Ohio.

[a]Nitronic, 17-4 PH, 15-5 Ph, 17-7 Ph, Ph 15-7 Mo, 21-6-9 are registered trademarks of Armco Inc. Middletown, Ohio. PH 13-8 Mo, 18 SR, 18-9 LW, 18-2 Mn, 18-3 Mn, 22-13-5, and 12 SR are trademarks of Armco Inc.

[b]Special Armco stainless steels.

a brighter surface finish than cold-drawn steel and is held to closer dimensional tolerances. It is also a much higher strength steel and is much more expensive than cold-drawn, low carbon steel. Cold-finished steel strip is produced in rolling mills for use in punch press work for blanking and drawing operations. It is also provided in a variety of surface finishes and colors that are not damaged by the manufacturing process. Tool steels and alloy steels are usually referred to by a trade name, but they usually are also identified by number.

HSLA Steels

Low carbon steels such as the ASTM-A36 have been in use since the 1920s, but a relatively new series of construction steels called high strength/low alloy (HSLA) have greatly increased in importance. They were originally developed primarily for automotive weight reduction because a thinner sheet, plate, or structural shape or a smaller diameter bar will almost always do the work of a thicker, heavier, plain carbon steel. These steels are appropriate for many other applications where high strength, formability, and weldability are needed. Some of these HSLA steels in plate, bar, and structural shapes include A709 and A737 for bridges; A441, A572, and A633 for low temperature use; and A242 and A588 for weather resistant construction steels. Full ASTM specifications may be found in materials and welding data handbooks.

STAINLESS STEELS

Stainless steels, which are alloys, are resistant to the corrosive action of the atmosphere and most reagents because of the relatively large proportion of chromium present in them. Chromium in quantities of less than 11 to 12 percent in steel will inhibit but not eliminate corrosion, but in those containing more than 12 percent corrosion is completely eliminated under normal circumstances. However, many grades of stainless steel have a limited resistance to chlorine and sulfur ions. Therefore, they should not be used in the presence of sea water and hydrochloric or sulfuric acids. Some stainless steels were developed to resist these corrosives as well as most others.

There are two major series of stainless steel, the chromium (400 series) and the chromium–nickel (300 series). In the 400 series there are two types, the hardenable (by quenching and tempering) and the nonhardenable (Table 3). The 300 series includes the austenitic types that contain 8 percent or more nickel which gives it higher resistance to corrosion and high temperatures, but this series cannot be hardened by quenching. Since it has such high ductility, it can be work hardened until it is almost as hard as the martensitic type. The ferritic type in the 400 series contains almost no carbon and is actually a chromium–iron alloy. It is the least expensive of the stainless steels and is used for pots and pans and building trim. The martensitic types can be hardened by heating and quenching in oil or air and are then tempered. Included in the chromium–nickel types are the precipitation hardening stainless steels that harden over a period of time after solution heat treatment. Stainless steels are more difficult to machine than mild steels and they tend to work harden quickly with dull tools, but with proper tooling and cutting fluids they lend themselves to good production rates, especially the free machining types (Table 4).

Table 4
Stainless Steels and Their Uses

Armco Stainless Steel	Composition (%) Cr	Ni	C Maximum	Other Significant Elements	Forms[a] Bar	Rod and Wire	Sheet	Strip	Billets	Pipe and Tubing	MAJOR CHARACTERISTICS	METALLURGICAL STRUCTURE
Nitronic 20	22.00 to 24.00	7.00 to 9.00	0.28 to 0.38	Mn 1.5–3.5 N 0.28–0.40	▼	▼			▼		Designed for elevated temperature use where high mechanical strength and resistance to oxidation and sulfidation are important.	Austenitic, age-hardenable by heat treatment
Nitronic 32	16.50 to 19.00	0.50 to 2.50	0.15	N 0.20–0.45 Mn 11.00–14.00	▼	▼			▼		Twice the yield strength of Type 304 and comparable corrosion resistance, nonmagnetic.	Austenitic, nonhardenable by heat treatment
Nitronic 33	17.00 to 19.00	2.25 to 3.75	0.08 maximum	N 0.20–0.40 Mn 11.5–14.5	▼	▼	▼	▼	▼	▼	Twice the yield strength of Type 304, comparable corrosion resistance and better stress corrosion cracking resistance, nonmagnetic.	Austenitic, nonhardenable by heat treatment
Nitronic 40	19.00 to 21.50	5.50 to 7.50	0.08	N 0.15–0.40 Mn 8.00–10.00	▼	▼	▼	▼	▼		High strength with good resistance to oxidation and toughness at sub-zero temperatures.	Austenitic, nonhardenable by heat treatment
Nitronic 50	20.5 to 23.5	11.5 to 13.5	0.06	N 0.20–0.40 Mn 4.00–6.00 Mo 1.50–3.00	▼	▼			▼		Approximately twice the yield strength of Type 316 and more corrosion resistant. Nonmagnetic.	Austenitic, nonhardenable by heat treatment
Nitronic 80	16.00 to 18.00	8.00 to 9.00	0.10	Mn 7.00–9.00 Si 3.50–4.50 N 0.08–0.18	▼	▼			▼		Excellent wear and galling characteristics, twice the yield strength of Type 304 with comparable corrosion resistance.	Austenitic, nonhardenable by heat treatment
311 DQ	17.25	4.50	0.04	Mn 2.50 N 0.15 Cu 2.4			▼	▼			Improved drawability and strength over Types 301 and 304.	Austenitic, nonhardenable by heat treatment
Type 301	16.00 to 18.00	6.00 to 8.00	0.15	—		▼	▼	▼	▼	▼	High work-hardening rate combines cold-worked high strength with good ductility.	Austenitic, nonhardenable by heat treatment
Type 302	17.00 to 19.00	8.00 to 10.00	0.15	—	▼	▼	▼	▼	▼	▼	Basic, general purpose type with good corrosion resistance and mechanical properties.	Austenitic, nonhardenable by heat treatment

(continued)

Table 4 *Continued*
Stainless Steels and Their Uses

Armco Stainless Steel	Composition (%) Cr	Ni	C Maximum	Other Significant Elements	Forms[a] Bar	Rod and Wire	Sheet	Strip	Billets	Pipe and Tubing	MAJOR CHARACTERISTICS	METALLURGICAL STRUCTURE
Type 303	17.00 to 19.00	8.00 to 10.00	0.15	S 0.15 min	▼	▼			▼		Machinability superior to that of most regular Type 303 grades.	Austenitic, nonhardenable by heat treatment
Type 304	18.00 to 20.00	8.00 to 10.00	0.08	—	▼	▼	▼	▼	▼	▼	Low carbon variation of Type 302, reduces carbide precipitation during welding.	Austenitic, nonhardenable by heat treatment
Type 304L	18.00 to 20.00	8.00 to 10.00	0.03	—	▼	▼	▼	▼	▼	▼	Extra-low carbon content minimizes harmful carbide precipitation due to welding.	Austenitic, nonhardenable by heat treatment
18-9 LW	17.00 to 19.00	8.00 to 10.00	0.10	Cu 3.00–4.00	▼	▼	▼	▼	▼		Special Armco type with low work-hardening, excellent cold-heading characteristics.	Austenitic, nonhardenable by heat treatment
Type 308	19.00 to 21.00	10.00 to 12.00	0.08	—	▼	▼			▼		High Cr and Ni produce good heat and corrosion resistance, used widely for welding rod.	Austenitic, nonhardenable by heat treatment
Type 308L	19.00 to 21.00	10.00 to 12.00	0.03	—	▼	▼			▼		Extra-low carbon variation of Type 308 eliminates harmful carbide precipitation during welding.	Austenitic, nonhardenable by heat treatment
Type 309	22.00 to 24.00	12.00 to 15.00	0.20	—	▼	▼			▼		High strength and resistance to scaling at high temperatures.	Austenitic, nonhardenable by heat treatment
Type 309S	22.00 to 24.00	12.00 to 15.00	0.08	—	▼	▼			▼		Similar to Type 309 but carbon lowered to minimize carbide precipitation and improve weldability. Welding wire.	Austenitic, nonhardenable by heat treatment
Type 310	24.00 to 26.00	19.00 to 22.00	0.25	—	▼	▼			▼		Higher alloy content improves basic characteristics of Type 309.	Austenitic, nonhardenable by heat treatment
Type 310S	24.00 to 26.00	19.00 to 22.00	0.08	—	▼	▼			▼		Similar to Type 310 but carbon lowered to minimize carbide precipitation and improve weldability. Welding wire.	Austenitic, nonhardenable by heat treatment

Type 316	16.00 to 18.00	10.00 to 14.00	0.08	Mo 2.00–3.00	▼	▼	▼	▼	▼	▼	Mo improves general corrosion and pitting resistance and high temperature strength over that of Type 302.	Austenitic, nonhardenable by heat treatment
Type 316L	16.00 to 18.00	10.00 to 14.00	0.03	Mo 2.00–3.00	▼	▼	▼	▼	▼	▼	Extra-low carbon version of Type 316, minimizes harmful carbide precipitation due to welding.	Austenitic, nonhardenable by heat treatment
Type 317	18.00 to 20.00	11.00 to 15.00	0.08	Mo 3.00–4.00	▼	▼			▼	▼	Higher alloy content improves basic advantages of Type 316.	Austenitic, nonhardenable by heat treatment
Type 317L	18.00 to 20.00	11.00 to 15.00	0.03	Mo 3.00–4.00	▼	▼			▼	▼	Extra-low carbon version of Type 317 for welded structures.	Austenitic, nonhardenable by heat treatment
Type 321	17.00 to 19.00	9.00 to 12.00	0.08	Ti × C, min	▼	▼	▼	▼	▼	▼	Stabilized to permit use in 800 to 1500° F range with reduced carbide precipitation.	Austenitic, nonhardenable by heat treatment
Type 347	17.00 to 19.00	9.00 to 13.00	0.08	Cb + Ta 10 × C, min	▼	▼			▼		Characteristics similar to Type 321, stabilized by Cb and Ta.	Austenitic, nonhardenable by heat treatment
Type 348	17.00 to 19.00	9.00 to 13.00	0.08	Ta 0.10 max Co 0.20 max Cb + Ta 10 × C, min	▼	▼			▼		Similar to Type 347 but Ta reduced for atomic energy applications.	Austenitic, nonhardenable by heat treatment
22-4-9	20.00 to 23.00	3.00 to 5.00	0.60	Mn 7.00–10.00 N 0.30–0.50	▼	▼			▼		Good hot hardness at elevated temperatures.	Austenitic, age-hardenable by heat treatment
17-4 PH	15.00 to 17.50	3.00 to 5.00	0.07	Cu 3.00–5.00 Cb + Ta 0.15–0.45	▼	▼	▼	▼	▼		Combines excellent corrosion resistance, high strength and hardness, low temperature hardening, and good fabricating characteristcs.	Martensitic, precipitation hardened
15-5 PH	14.00 to 15.50	3.50 to 5.50	0.07	Cu 2.50–4.50 Cb + Ta 0.15–0.45	▼	▼	▼	▼	▼		Similar in properties and characteristics to Armco 17-4 PH but has superior transverse ductility and toughness.	Martensitic, precipitation hardened
PH 13-8 Mo	12.25 to 13.25	7.50 to 8.50	0.05	Mo 2.00–2.50 Al 0.90–1.35	▼	▼			▼		Excellent transverse properties, has superior strength and toughness.	Martensitic, precipitation hardened

(continued)

17-7 PH	16.00 to 18.00	6.50 to 7.75	0.09	Al 0.75–1.25	▼	▼	▼	▼	▼	▼	Special ultra-high strength with good formability, excellent fabricating characteristics.	Semiaustenitic, precipitation hardened
PH 15-7 Mo	14.00 to 16.00	6.50 to 7.75	0.09	Mo 2.00–3.00 Al 0.75–1.50	▼	▼	▼	▼	▼		Special type similar to Armco 17-7 PH but with higher strength.	Semiaustenitic, precipitation hardened
400	12.00 to 13.00	—	0.05	Al 0.50 max			▼	▼		▼	Corrosion resistance comparable to Type 410 at lower cost. No Ti for better surface.	Ferritic, nonhardenable
Type 403	11.50 to 13.00	—	0.15	Si 0.50 max	▼	▼			▼		Special high-quality variation of Type 410 for highly stressed parts.	Martensitic, hardenable
409	10.00 to 12.00	—	0.06	Ti 0.50	▼	▼	▼	▼	▼		Lowest cost of all stainless sheet and strip, provides economical corrosion and oxidation resistance for applications where surface appearance is not important.	Ferritic, nonhardenable
Type 410	11.50 to 13.50	—	0.15	—	▼	▼	▼	▼	▼	▼	Low cost general purpose stainless steel. Wide use where corrosion is not severe.	Martensitic, hardenable
Type 416	12.00 to 14.00	—	0.15	S 0.15 min	▼	▼			▼		Has much better machinability than most regular Type 416 grades.	Martensitic, hardenable
Type 430	16.00 to 18.00	—	0.12	—	▼	▼			▼		Most popular of chromium types. Combines good corrosion and heat resistance and mechanical properties.	Ferritic, nonhardenable
Type 430F	16.00 to 18.00	—	0.12	S 0.15 min	▼	▼			▼		Free machining version of Type 430.	Ferritic, nonhardenable
18 SR	18.00	0.50	0.06	Si 1.0 Al 2.0			▼	▼			Excellent resistance to high-temperature scaling.	Ferritic, nonhardenable
12 SR	12.00	—	0.02	Si 0.50 Al 1.2 Ti 0.30 Cb 0.60			▼	▼			More oxidation resistant and more creep resistant than Type 409 stainless steel.	Ferritic, nonhardenable

SOURCE Armco Inc., Specialty Steels Division, Copyright 1984, Armco Inc., Middletown, Ohio.

[a]Some Armco stainless steel grades may be subject to heat lot order accumulations. Plate items are available from several producers.

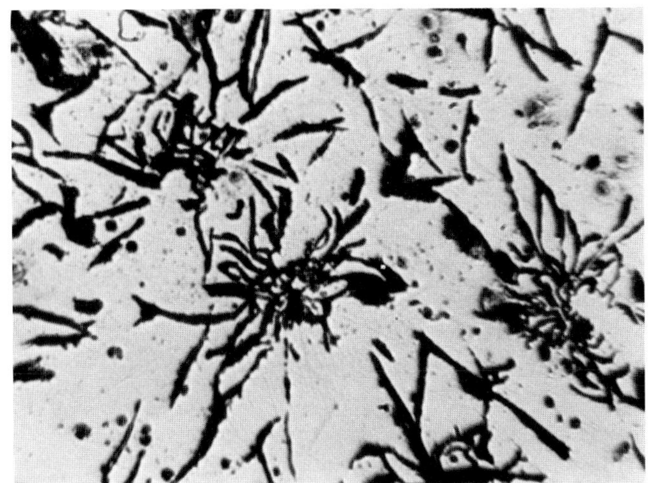

Figure 2
Gray cast iron showing graphite flakes, unetched (500 ×).

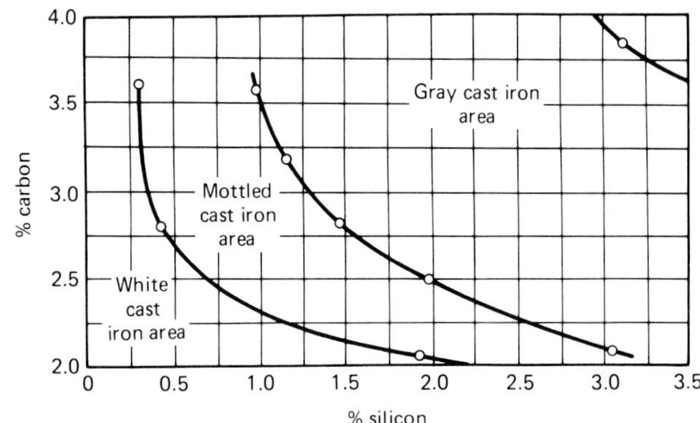

Figure 3
Composition limits for white, gray, and mottled cast irons (Neely/*Metallurgy* 2 Ed., © 1984 John Wiley & Sons, Inc.).

CAST IRONS

Cast iron is essentially an alloy of iron, carbon, and silicon. Theoretically, it can contain from 2 to 6.67 percent carbon, but commercial cast iron rarely has more than 4 percent carbon. Since austenite can contain only 2 percent carbon in solution, any more than that amount will precipitate out to form flakes of graphite (Figure 2) when cast iron is slowly cooled, or cementite plus some ferrite if it is rapidly cooled. Also, graphitization is increased as the silicon content is increased (Figure 3). The iron–graphite equilibrium diagram (Figure 4) shows the various microstructures found in cast iron.

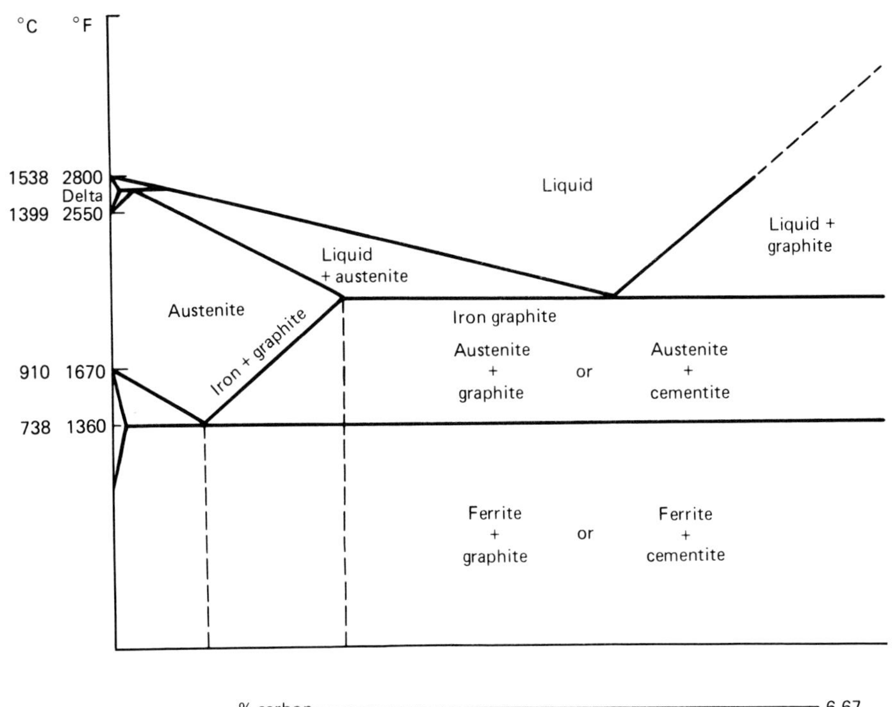

Figure 4
Iron–graphite equilibrium diagram (Neely/*Metallurgy* 2 Ed., © 1984 John Wiley & Sons, Inc.).

Table 5
Classes of Gray Iron

ASTM Number	Minimum Tensile Strength (psi)
20	20,000
25	25,000
30	30,000
35	35,000
40	40,000
45	45,000
50	50,000
60	60,000

SOURCE John E. Neely, *Practical Metallurgy and Materials of Industry*, Second Edition, Wiley, New York, © 1984.

Gray Cast Iron

Gray cast iron is widely used to make castings and machinery parts since it is easily machined. Machine bases, tables, and slideways are made of gray cast iron because it tends to remain dimensionally stable after an aging period and it makes an excellent bearing and sliding surface because of the graphite content. Cast iron has a lower melting point than steel and possesses good fluidity so it is easily cast into sand molds, even those having intricate designs. When low cost but serviceable machine tool housings and frames are needed, the choice is almost always gray cast iron or one of its alloys because of its ability to deaden vibration and minimize tool chatter. Gray cast irons are often alloyed with elements such as nickel, chromium, vanadium, or copper to toughen and strengthen them. Unalloyed gray cast iron has two major drawbacks as an engineering material; it is quite brittle and it has a relatively low tensile strength so it is not used where high stresses or impact loads are found. Cast irons may be ferritic or pearlitic in microstructure which affects their strength. Gray cast irons are classified according to their tensile strength by ASTM numbers (Table 5). They can be welded but the high carbon content and their tendency to form extremely hard, brittle microstructures next to the weld make good welds difficult without preheating the base metal.

White Cast Iron

Unlike gray cast iron, white cast iron is virtually nonmachinable because it is so hard. It contains massive cementite, which is the hardest substance in iron and steel. For this reason, it is mostly used for its abrasion resistance in such applications as wear plates in machinery. There are two ways white cast iron can be formed: by lowering the iron's silicon content and by rapid cooling, in which case it is called chilled cast iron. White cast iron cannot be welded.

Malleable Cast Iron

A cast iron that possesses all of the positive qualities of gray cast iron but has increased ductility, tensile strength, and toughness is called malleable cast iron. It is produced from white cast iron by a prolonged heat treating process lasting for about 30 hours. This process essentially graphitizes the white cast iron to form tiny spheres of graphite instead of flakes as in gray cast iron (Figure 5). The iron matrix surrounds the graphite in this form, making it a more homogenous metal, whereas graphite flakes (which are very soft and weak) cut across the iron matrix in every direction causing it to be weak and brittle.

Nodular Cast Iron

Essentially the same purpose is achieved in nodular cast iron as in malleable cast iron; graphite spheres are formed, but without the extensive heat treatment. Nodular cast iron is known by several names: nodular iron, ductile iron, and spheroidal graphite iron. Like malleable cast iron, it has toughness, good castability, machinability, good wear resistance, weldability, low melting point, and hardenability. The graphite is formed into spheres in the ladle by adding certain elements such as magnesium and cerium to the melt just prior to casting. A vigorous, often spectacular, mixing reaction takes place that forms the graphite into tiny balls dispersed throughout the mix. Malleable and nodular cast iron castings are often used for machine parts where there is a higher tensile stress and moderate impact loading. These castings can be a less expensive alternative to steel castings or steel weldments.

NONFERROUS METALS

Nonferrous metals vary considerably in density (absolute weight per unit volume), color, melting points, and mechanical properties. Some, often called noble metals, such as gold and platinum, have a high resistance to corrosion. Many of the so-called space-age metals such as titanium and columbium were only relatively recently extracted from their ores in commercial quantities and even aluminum, which we see everywhere today, was first extracted from ores in quantities only about 100 years ago. Most nonferrous metals and alloys are classified in a numerical system peculiar to that metal.

Aluminum

Aluminum is white in color or white–gray if the surface is oxidized. Unlike the oxidation or rust of iron, which continues to form, the thin oxidized surface of aluminum prevents further corrosion. Aluminum weighs only 168.5 lb/ft^3, compared to 487 lb/ft^3 for steel, and pure aluminum melts at 1220° F (660° C).

Figure 5
Micrograph of malleable cast iron (100 ×) (*Metals Handbook,* Vol. 7, 8th ed., American Society for Metals, 1972, p. 95. With permission).

Aluminum is classed as either wrought or cast. Wrought aluminum is cold worked and strain hardened by rolling, drawing, extruding, forming, or hot working by forging. Molten aluminum is cast by gravity into sand and permanent molds or injected at high pressure into dies. Cast aluminum is generally less hard and strong than wrought products, yet some types can be hardened by heat treatments.

Pure aluminum resists corrosion better than aluminum alloys so, where corrosion is a problem, an alloy should not be used. However, rolled sheet stock of aluminum alloy is sometimes sandwiched between two very thin sheets of pure aluminum (Figures 6 and 7). Thus, the advantage of the high strength alloy and the corrosion resistant pure aluminum are combined in one sheet. These clad aluminum sheets are used extensively in the aircraft industry.

There are several classification systems of specifi-

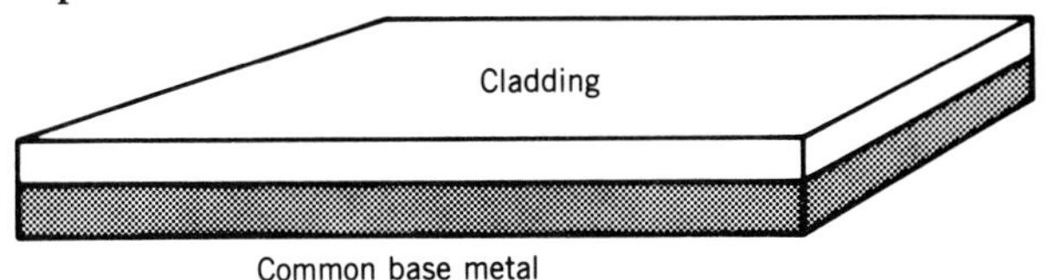

Figure 6
Metal cladding. By cladding we use less of the scarce and expensive metals (Bureau of Mines).

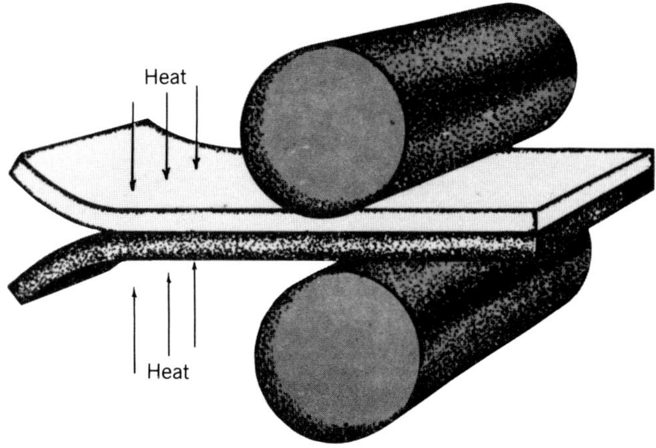

Figure 7
Roll bonding is one method of cladding metals. But if the heated surfaces oxidize, they will not bond (Bureau of Mines).

Table 6
Aluminum and Aluminum Alloys

Code Number	Major Alloying Element
1xxx	None
2xxx	Copper
3xxx	Manganese
4xxx	Silicon
5xxx	Magnesium
6xxx	Magnesium and silicon
7xxx	Zinc
8xxx	Other elements
9xxx	Unused (not yet assigned)

SOURCE John E. Neely, *Practical Metallurgy and Materials of Industry*, Second Edition, Wiley, New York, © 1984.

cations used to identify aluminum alloys: federal, military, the Society of Automotive Engineers (SAE), and the American Society for Testing and Materials (ASTM). The system most used by manufacturers is the one adopted by the Aluminum Association in 1954, which is shown in Table 6.

The first digit in Table 6 represents the alloy type. The second digit, here represented by the letter *X*, indicates any modifications made to the original alloy, and the last two digits identify either the specific alloy or impurity. The first entry in the series represents nearly pure aluminum. An 1120 aluminum contains no alloying element and has .20 percent pure aluminum above 99 percent. This type cannot be hardened by heat treatment. In the 2000, 6000, and 7000 series alloying elements make it possible to harden these alloys by solution heat treatment and aging. The alloys that cannot be heat treated are usually designated in the 1000, 3000, 4000, or 5000 series. However, all of these alloys are work hardenable and therefore have an additional designation for temper. In this case, temper denotes hardness from cold working.

Temper Designation Temper designations for those alloys that are not heat treatable sometimes follow the four-digit alloy series number.

—F As fabricated. No special control over strain hardening or temper designation.

—O Annealed, recrystallized wrought products only. Softest temper.

—H Strain hardened, wrought products only. Strength is increased by work hardening.

The symbol —H is always followed by two or more digits. The first digit, which can be 1, 2, or 3, denotes the final degree of strain (work) hardening.

—H1 Strain hardened only.

—H2 Strain hardened and partially annealed.

—H3 Strain hardened and annealed.

The second digit denotes the final degree of strain hardening.

2 $\frac{1}{4}$ hard

4 $\frac{1}{2}$ hard

6 $\frac{3}{4}$ hard

8 full hard

9 extra hard

For example, 3003—H16 is an aluminum–manganese alloy, strain hardened to $\frac{3}{4}$ hard temper. A third digit is sometimes used to indicate a variation of a two-digit —H temper.

Heat Treatment Classifications For those aluminum alloys that can be hardened by **solution heat treatment** and **precipitation hardening** or aging, the symbol —T follows the four-digit number. Numerals 2 to 10 follow this letter to indicate the sequence of treatment.

—T2 Annealed (cast products only)

—T3 Solution heat treated and cold worked

—T4 Solution heat treated but naturally aged

—T5 Artificially aged only

—T6 Solution heat treated and artificially aged

—T7 Solution heat treated and stabilized

—T8 Solution heat treated, cold worked, and artificially aged

—T9 Solution heat treated, artificially aged, and cold worked

—T10 Artificially aged and then cold worked

For example, aluminum rivets may be made of 2024—T4 (an aluminum-copper alloy) which is solution heat treated, riveted into place (which is cold working), and then naturally aged.

Cast Aluminum Classification
Aluminum casting alloys are generally divided into three categories: sand casting alloys, permanent mold alloys, and die casting alloys. A classification system similar to that of wrought aluminum alloys is used (Table 7). For example, the general purpose sand casting aluminum–copper alloy 212 has an ultimate strength of about 23,000 PSI. The die casting alloy 356—T6 is used for transmission cases and truck parts and has a tensile strength about 38,000 PSI.

Table 7
Cast Aluminum Alloy Designations

Code Number	Major Alloy Element
1xx.x	None, 99 percent aluminum
2xx.x	Copper
3xx.x	Silicon with Cu and/or Mg
4xx.x	Silicon
5xx.x	Magnesium
6xx.x	Zinc
7xx.x	Tin
8xx.x	Unused series
9xx.x	Other major alloys

SOURCE John E. Neely, *Practical Metallurgy and Materials of Industry*, Second Edition, Wiley, New York, © 1984.

Cadmium

Cadmium is used as an alloying element for low-temperature-melting metals such as solder, type casting metals, bearing metals, and storage batteries. It has a blue-white color and is commonly used as a protective plating on steel parts such as screws, bolts, and washers. Cadmium compounds are extremely toxic and can cause illness when inhaled. Welding, heating, or machining these plated parts can produce toxic fumes. These fumes can be avoided with proper ventilation.

Chromium

Chromium is a hard, slightly grayish metal that can take a brilliant polish and is often seen on automobile bumpers and trim as chromium plate on steel. It is very corrosion resistant and, when used as an alloying element in steel in quantities over 12 percent, it produces stainless steel; in lesser quantities, it produces a high-strength alloy steel. Chromium is too brittle to have any direct applications in its pure massive state.

Cobalt

This hard silver–white metal is widely used as a matrix or binder in which tungsten carbide particles are combined and formed into cutting tools by the powder metallurgy process. It is also used as an alloying element for high-speed and stellite cutting tools, for resistance wire, and for heat resistant alloys used for jet engine blades. Dental and surgical alloys contain large percentages of cobalt and chromium; for example, 62 percent Co and 29 percent Cr in alloy A. They are not attacked by body fluids and do not cause irritation of the tissues as many other metals do. Cobalt is resistant to ordinary corrosion in air. Some iron–nickel–cobalt–chromium alloys are called super-strength alloys because they are used where high strength and resistance to high temperature are required. Some of these alloys have the same expansion rate as glass and so are used for glass-to-metal joints and seals. Strong permanent magnets are made with cobalt and other metals. The familiar Alnico magnet material is an alloy of aluminum, nickel, cobalt, and iron. Cobalt and nickel are two of the few nonferrous metals that can be attracted to a magnet.

Columbium (Niobium) and Tantalum

Columbium, tantalum, and vanadium are all related chemically and they appear as a vertical group in the periodic table of the elements. However, columbium and tantalum are truly sister elements in more ways than one. Both occur in the same ores and have the same melting points. Unlike the other refractory (high-temperature) metals, they are both ductile and malleable at room temperature and therefore they can be easily formed. They are very immune to attack by strong acids with the exceptions of fuming sulfuric and hydrofluoric acids. The anodic oxide films they form are stable semiconductors and it is because of this property that they are used in capacitors and rectifiers. Columbium is used as a carbon stabilizer in stainless steels and as an alloying element for low temperature service steels such as those used in Arctic pipelines. Tantalum is used in electronics and for surgical implants. Tantalum carbide, along with tungsten carbide and titanium carbide, is widely used for cutting tools in machining operations.

Copper and Copper-Based Alloys

Copper is a soft, reddish metal which has a number of valuable properties as a pure metal and in various combinations with other metals. Its high thermal and electrical conductivity, which is second only to that of silver, is of major significance. Copper has a high formability, corrosion resistance, and medium strength. To American industry, the term *copper* refers to that element with less than .5 percent impurities or alloying elements. Copper-based alloys are those having more than 40 percent copper.

Tough-Pitch Copper This copper contains a carefully controlled amount of oxygen, between .02 and .05 percent. Electrolytic refined tough-pitch copper is the most widely used type for electrical conductors and for building trim, roofing, and gutters. It lends itself to high tonnage production after being cast into wire, bars, and billets for further fabrication. Tough-pitch copper can withstand the ravages of time and weather because it does not harden with age and develop season cracks.

Deoxidized Copper This type of copper differs from tough-pitch copper chiefly in its lower electrical and thermal conductivity. It also has a somewhat higher ductility and is more readily formed and for this reason it is the most commonly used copper for the manufacture of tubular products such as those used in domestic and industrial plumbing.

Oxygen-Free Copper This type is the purest of commercial coppers (99.92 percent minimum) and therefore is called a high-conductivity copper. It is used in electrical and electronic equipment, radiators, refrigeration coils, and distillers. Industrial copper is classified by a series of SAE numbers; for example, SAE No. 75 is a 99.90 percent deoxidized copper used for tubes.

Low Alloy Copper Copper is often alloyed with very small percentages of other metals from a fraction of 1 percent to approximately 2 percent for the purpose of imparting such qualities as corrosion resistance, higher operating temperature, and increasing tensile strength and machinability. These additives include arsenic, silver, chromium, cadmium, tellurium, selenium, and beryllium. Copper cannot be hardened by heat treatments, but with the addition of about 2 percent beryllium it can be sufficiently hardened by precipitation and aging so that it can be used for making springs, flexible bellows, and tools. Because of their nonsparking quality, beryllium–copper tools are used in explosion-hazardous environments such as mines, powder factories, and some chemical plants. Since beryllium is quite expensive, an alloy containing only .4 percent Be with 2.6 percent Co was developed that is useful for some purposes, but the straight beryllium–copper alloy develops a higher strength and hardness by heat treatment than the cobalt-bearing alloy.

Brass

Brass is essentially an alloy of copper and zinc. As the relative percentages of copper and zinc vary, there is a corresponding variation in properties such as color, strength, ductility, and machinability. Brass colors usually range from white to yellow, and alloy brasses range from red to yellow. Alloy brasses rarely contain more than 4 percent of the alloying elements, either singly or collectively. These elements are manganese, nickel, lead, aluminum, tin, and silicon. However, in the plain brasses containing only copper and zinc, the zinc content can be as much as 45 percent. Gilding metal used for jewelry contains 95 percent copper and 5 percent zinc, whereas Muntz metal, which is used for sheet stock and brazing rod, contains 60 percent copper and 40 percent zinc. Lead is sometimes added to brass to increase its machinability. In general, because of their ductility and malleability, brasses are easily cold worked by any of the commercial methods such as drawing, stamping, spinning, and cold rolling.

Bronze

Bronze is an alloy of copper and tin and has been used for thousands of years. The tin-bronze today is known commercially as phosphor bronze because a small percentage of phosphorus is generally added as a deoxidizing agent in the casting of these tin–copper alloys. Phosphor bronze can range from 1.25 to 10 percent in tin content. Bronze colors generally range from red to yellow. Some bronzes have good resistance to corrosion near sea water. There are many types of bronzes which derive their names from an alloying element other than tin. Two of these are the silicon–bronzes and the aluminum–bronzes which cover a wide range of metals. Phosphor–bronze can be cold worked and silicon–bronze can be hot and cold worked, but aluminum–bronzes are not generally used for severe cold-working operation. Most bronzes are easily cast into molds and they make good antifriction bearings. As with brasses, lead is often added to improve machinability. Both brasses and bronzes are designated by an SAE standard number system. For example, red brass casting metal is SAE standard No. 40, and SAE standard No. 64 is phosphor–bronze casting metal. These standard alloys used in the automotive industry can be found in reference books.

Die-Cast Metals

Die casting, which will be covered in Chapter 7, is a means of producing identical castings at a rapid rate by injecting molten, low melting point alloys into a mold with intense pressure. The mold then opens and the solidified part is ejected. Such small parts as carburetors, toys, and car door handles are mass produced by the process of die casting. Die-cast metals, also called pot metals, are classified into six major groups.

1. Tin-base alloys
2. Lead-base alloys
3. Zinc-base alloys
4. Aluminum-base alloys
5. Copper, bronze, or brass alloys
6. Magnesium-base alloys

The specific content of the alloying elements for any of the die-cast alloys may be found in materials reference books.

Indium

Indium is a very soft metal that can be scratched with the fingernail. It is silver-white in color, similar to platinum, and has a brilliant metallic luster. Indium adheres to other metals on contact and the molten metal clings to the surface of glass and wets it. Indium is used in bearing materials for wear reduction by diffusing it into metallic surfaces or as a coating of indium and graphite. Lead–indium alloys are used for solders and brazing materials. Indium in its pure form is not used as a manufacturing material; it is used only as it is alloyed with other metals.

Lead-Base Materials

Lead is a soft, heavy metal that is somewhat silvery in color when newly cut but gray when oxidized. It has been a very useful metal since ancient times when it was used for plumbing, utensils, and jewelry. Pure lead is so soft that it has limited usefulness, but when alloying constituents are added, its hardness is increased. Antimony in quantities up to 12 percent is added to harden and improve the mechanical properties of lead. Although lead has a high density, it has low tensile strength, and low ductility; therefore it cannot be easily drawn into wire. It does have the property of exceptionally high malleability which allows it to be rolled or compressed into thin sheets or foil. Lead shapes and wire cable sheathing can be easily extruded (pushed through a die). Lead has good corrosion resistance and is a good shielding material for gamma and X-ray radiation. But lead compounds are very toxic and adequate protection is necessary in handling and manufacturing lead products.

Babbitt A widely used antifriction lead alloy that is used for antifriction bearings is called babbitt. Babbitts can be lead, tin, or cadmium based. Lead-based babbitts contain up to 75 percent lead with antimony, tin, and some arsenic. Tin babbitts contain from 65 to 90 percent tin with antimony, lead, and some copper added. These are the higher grades and are more expensive than lead-based types. Cadmium-based babbitts resist higher temperatures than the other types. They contain from 1 to 15 percent nickel, a small amount of copper, and up to 2 percent silver.

Solders Tin and lead are combined in various proportions to form solders that are used to join metals at low temperatures. The most common general purpose solders contain about 50 percent of each metal. They are sometimes called 50–50 solders. The canning industry uses solders with only 2 to 3 percent tin. When there is a need to lower the melting point so the metal to be joined will not warp, cadmium or bismuth is added to the lead–tin alloy.

Terne Plate Terne metal is an alloy of tin and lead, the tin content being between 10 and 25 percent. Sheet steel coated with this alloy is called terne plate. This product has a high resistance to corrosion and is used for roofing and other architectural needs. Terne plate forms a good base for paints and in deep drawing of metal articles; the plating provides a good die lubricant. These coatings are applied to steel by the hot-dip process, or they are sprayed on or electrodeposited.

Magnesium

When pure, magnesium is a soft silver-white metal that resembles aluminum but is much higher in weight (108.6 lb/ft^3 compared to 168.5 lb/ft^3 for aluminum and 487 lb/ft^3 for steel). When alloyed with other metals such as aluminum, zinc, or zirconium, magnesium has quite high strength-to-weight ratios, making it a useful metal for some aircraft components. Magnesium is also used in cast iron foundries as an additive in the ladle to produce nodular iron. Household goods, typewriters, and portable tools are some of the many items made of this very light metal. Magnesium when finely divided will burn in air with a brilliant white light. Magnesium alloys are designated by several numerical systems: Military, AMS, SAE, ASTM, and Federal, all of which may be found in reference handbooks. Magnesium can be shaped and processed by practically all the methods used to manufacture other metal products. It can be cast by sand, permanent mold, or die-cast methods. It can be rolled, extruded, forged, and formed by bending, drawing, and other methods.

Manganese

This silver-white metal is seldom seen in its pure state since it is normally used as an alloying element for other metals. In steel production it is used as a deoxidizer and to control sulfur. All steel contains small percentages of manganese. Large percentages (12 percent or more) cause steel to become austenitic and no longer ferromagnetic. High manganese steel quickly work hardens and has excellent abrasion resistance as a result, so it is used extensively in making earth moving machinery, rock crushers, and conveyors. Manganese is also used in some stainless steels.

Molybdenum

As a pure metal, molybdenum is used in high temperature applications such as for filament supports in lamps and electron tubes. It is used as an alloying element in steel to promote deep hardening and to increase tensile strength and toughness.

Nickel and Nickel-Based Alloys

Nickel is a silvery white metal that is noted for its corrosion resistance. It is widely used for electroplating on other metals, especially steel, as a protection coating. It is also used as an alloying element in other metals and, like manganese, a large percentage (50 percent) of nickel in steel, or 8 percent Ni and 18 percent Cr as in austenitic stainless steel, causes it to become austenitic and lose its ferromagnetism.

Monel An alloy of 67 percent nickel and 28 percent copper plus some other impurity metals such as cobalt, manganese, and iron is called Monel. It is a tough alloy that is machineable, ductile, and corrosion resistant. Monel is widely used to make marine equipment such as pumps, valves, and fittings that are subjected to salt water, which is very corrosive to many metals.

Nichrome and Chromel These nickel–chromium–iron alloys are used as electric resistance elements in electric toasters and heaters. Inconel is similar to nichrome and chromel but is used to manufacture parts to be used in high temperature applications.

Nickel–Silver This alloy could be considered one of the copper alloys since it contains a high percentage of copper. A common nickel–silver contains about 55 percent copper, 18 percent nickel, and 27 percent zinc. The color is silver-white, making it a substitute for silver tableware and other such products. Because of the nickel content, it has good corrosion resistance.

Precious Metals

Gold and silver have long been used for coinage and jewelry because they tend to resist deterioration by corrosion and are relatively rare. Both are fairly evenly distributed over the earth's surface although not in sufficient amounts for economical recovery except in a few locations. The six metals of the platinum group are palladium (Pd), rhodium (Rh), iridium (Ir), osmium (Os), ruthenium (Ru), and platinum (Pt). Along with gold and silver these metals are generally called the precious metals group. Gold is used in dentistry and in the electrical and chemical industries. Silver has many uses in manufacturing and is well known in its use in photographic film. The specialty metals are all used for catalysts, as alloying elements, and in many industrial applications.

Tin

Tin is whiter in color than either silver or zinc. It has good corrosion resistance so it is used to plate steel. Vast quantities of tin plate are used in the food processing industry. Tin is one of the constituents of bronze and is used in babbitts and solders. Because of the rapidly increasing price of tin in recent times, other less expensive metals or materials have been developed to replace it in some products.

Titanium

This silver-gray metal weighs about half as much as steel and, when alloyed with other metals, is as strong as steel. Like stainless steel, it is a relatively difficult metal to machine, but machining is accomplished with rigid setups, sharp tools, and proper coolants. The greater application for titanium and its alloys is in making jet engines and jet aircraft frames. However, many promising civilian, military, naval, and aerospace uses for titanium are being considered because of its high corrosion resistance strength, and because it weighs less than steel. At the present time, its high cost precludes many of these proposed applications.

Tungsten

Tungsten is one of the heaviest of the metals and has the highest known melting point of any metal (6098° F or 3370° C). It is not resistant to oxidation at high temperatures and, when used as lamp filaments, it must be in a vacuum or inert gas atmosphere. Tungsten is used for electrical contacts and for welding electrodes. It is used as an alloying element in tool steels and in cutting-tool alloys. The carbides of tungsten are the most widely used and most valuable cutting tools. Tungsten carbide powder combined with cobalt powder is compressed into tool shapes and sintered in a furnace in a process called powder metallurgy.

Uranium and Thorium

Both thorium and uranium are radioactive metals and alpha emitters. Competent authorities should be consulted regarding the handling of these metals and their ores. Uranium is a silver-white metal in its pure state and it can be machined like other metals. If finely divided, uranium will burn in air. Uranium is an important fuel for nuclear reactors. Thorium, a dark gray metal, is also used in nuclear reactors. Thorium oxide with 1 percent cerium oxide is used for gaslight mantles.

Vanadium

Vanadium is a silver-white, very hard metal that oxidizes when exposed to air. It is used almost exclusively as an alloying element in steel to impart improved impact resistance (toughness) and better elastic properties. Vanadium contributes to abrasion resistance and high temperature hardness in cutting tools and it promotes finer grain in steels, giving them toughness. Usually not more than 1 percent vanadium is alloyed with steels.

Zinc

Zinc is a white metal that is valued for its corrosion protection of steel. The familiar galvanized steel is coated with zinc by any one of several processes: hot dip, electrogalvanizing, sherardizing (heating steel in zinc dust), metal spraying, or zinc-dust paint. When zinc and iron are galvanically (electrically) in contact under corrosive conditions, the zinc becomes anodic and serves as a sacrificial metal, corroding in preference to the steel. The greater consumption of zinc is for galvanizing steel. Zinc-based die-cast metals also consume a high percentage of zinc production, as do brass and rolled zinc to a lesser extent.

Zirconium

Zirconium and titanium are similar in appearance and physical properties and, like the other reactive metals, zirconium will burn in air or explode when finely divided. In fact, it was once used as an explosive primer and as a flashlight powder for use in photography because of the brilliant white light it gives off when it explodes. Zirconium alloys are still used in flash bulbs. Perhaps one of its most notable uses is in nuclear reactors. It has good corrosion resistance and stability, and therefore has been used for surgical implants. Like titanium it is a somewhat difficult metal to machine.

MATERIALS IDENTIFICATION

We soon learn to identify natural materials such as wood, leather, stone, glass, and earthy materials, such as clay, by sight and touch. Similarly, we can learn to identify many manufacturing materials by sight. However so many varieties of these materials have exactly the same appearance, that other means must be used for identification purposes. For example, in the past when a distinction had to be made between low carbon and high carbon steel that look alike on the surface, identification was made by means of spark testing (Figure 8). The blacksmith simply touched the two samples on a grinding wheel and observed the sparks; the high carbon steel produced a brilliant display of lines and sparks and the low carbon steel showed few carrier lines and a few sparks. This method is still sometimes used today to compare unknown steels with known samples in steel yards and shops although its accuracy is directly related to the operator's experience. Although it is still a useful tool, it is not a reliable method of identifying the alloy content of steel. Also, very few nonferrous metals produce a spark; exceptions are nickel, titanium, and zirconium. Nickel produces short carriers with no spark, similar to that of stainless steel, and titanium produces a brilliant white display of carrier lines, each having a spark at the end. The display of zirconium carrier and spark is identical to that of titanium (Figure 9).

Metals can often be identified by their reaction to certain acids and alkaline substances. A clean surface of a particular metal will become darkened or colored when a certain reagent is placed on it. For example, an aluminum sample can be quickly distinguished from a magnesium sample by placing a drop of a copper sulfate solution on each. The aluminum will not be affected but the magnesium will be blackened. Copper sulfate will turn ordinary steel a copper color, but it will not affect any type of stainless steel, nickel, Monel, or Inconel, all of which have a similar appearance. There are also many acid tests. Commercial spot testers are available in kits for the purpose of identifying metals. Magnetic testing is another means of identifying look-alike metals. Most ferrous metals, and those in which nickel or cobalt is the primary constituent, are attracted to a magnet, but some stainless steels, including the 300 series and the precipitation hardening types, 17-4 Ph and 15-5 Ph, may or may not be magnetic. Austenitic stainless steel (300 series) is not attracted to a mag-

Figure 8
Spark testing high carbon steel.

Figure 9
Spark testing titanium.

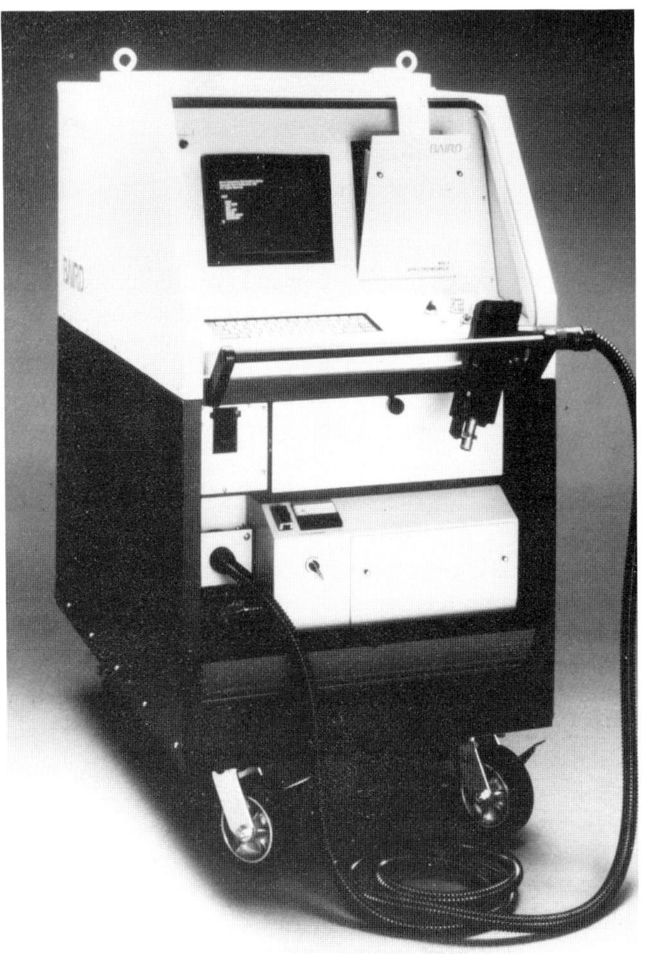

Figure 10
Spectromobile. A wheel-mounted spectrometer facilitates rapid, on-site identification and sorting of steel. It identifies and quantifies all key elements (alloying and residual) in metals and determines carbon content to within plus or minus .05 percent (Photograph courtesy of Spectrochemical Products Division, Baird Corporation).

net unless it is cold worked; then it is attracted to the extent to which it is cold worked. Of course, none of these methods will identify all of the elements or how much of each is in a sample of metal. But this can be done by spectrographic analysis in the laboratory. Speedier results are obtainable with a portable unit (Figure 10). This device identifies and quantifies key elements (alloying and residual) plus carbon in metals. Fuess spectroscopes work on the same principle as the spark test except that the reflected light is refracted through a prism. Since each element produces characteristic emission lines, it is possible to identify which elements are present. It is also possible to compare the results to standard spectra as an aid in identifying alloys. Mobile spectrometers work on the Fuess principle and are not able to sort alloys but are preset for a particular matching test. They are calibrated to a known standard and can quickly identify that material with a match signal.

Using an X-ray analyzer is probably one of the best methods of sorting and verifying metals (Figures 11 and 12). Its greatest drawback is that it costs more than other systems. Scrap metals are among the most dif-

Figure 11
In as little as 5 sec the X-site portable X-ray analyzer can identify metals such as this bar stock. The abbreviated alloy name appears on the LED display within a few seconds (Photo courtesy of Kevex Corporation © 1982).

Figure 12
Metal parts in a foundry are analyzed for content of various elements. This is done with an X-ray source that produces characteristic emissions from every element in the sample. A console with a microprocessor is connected to the hand-held unit that processes the data (Photo courtesy of Kevex Corporation © 1982).

ficult metals to identify and sort. High-integrity steel products must have carefully sorted scrap. All of the previously mentioned methods may be applied to this task of sorting scrap metals. However, it seems that the more accurate a method is the more the system costs.

Plastics Identification

Because there is such a great variety of plastics and more are being developed every day, identification of plastics is very difficult by ordinary means. Certain solvents will dissolve certain types of plastics, but this would not be a reliable method of testing because many types of plastics react to the same solvent. However, as discussed earlier, when heat is applied to a thermoplastic material, it will melt and solidify again when cool and it can be reused, but heat applied to a thermosetting material will cause it to break down instead of melting and it cannot be reused. Of course, in either case, this would be a destructive test for the plastic object involved.

Some plastics with chemical odors can be identified by holding a sample in a flame. An odor of phenol indicates a phenolic material, and an acrid odor with a yellow flame indicates cellulose acetate or vinyl acetate. An odor of burnt wood indicates ethyl cellulose. Melamine has a fishy odor when burned and casein smells like heated milk. Polystyrenes give off an odor of domestic heating and cooking gas, whereas polyethylene and polyolefins produce little or no odor, burn with a smoky flame, and melt like a candle while burning. **Note:** Cellulose nitrate is very flammable, especially in thin films or sheets. Precautions must be taken to protect hands, face, and eyes as well as to prevent the danger of fire hazard when conducting any of these tests.

FLUID ANALYSIS

Industrial process oils are used in a broad range of applications in manufacturing processes. An overview of these petroleum products is seen in the Telura® numbering systems of the Exxon Company, U.S.A. (Table 8) The applications for these oils are in Table 9.

The Telura line of process oils consists of 19 grades suitable for a broad range of applications. The line consists of six general types of products.

1. **Aromatic** Aromatic oils have good solvency characteristics. They are used to make proprietary chemical formulations and ink oils.
2. **Naphthenic** Naphthenic oils are characterized by low pour points and good solvency properties. They are used to make printing inks, shoe polish, rust-proofing compounds, and dust suppressants, and they are used in textile conditioning, leather tanning, and steam turbine flushing. The colors of the low-viscosity Telura® oils in this category are relatively light.
3. **Extracted Naphthenic** These oils have outstanding color and color stability and good low-temperature characteristics. They are used in proprietary chemical formulations, dust suppressants, and rust-proofing compounds. Of the extracted naphthenic products, Telura 415 and 417 meet the requirements of Federal Food and Drug Administration (FDA) Regulation 21 CFR 178.3620(c) for mineral oils used in products that may have incidental food contact.
4. **Paraffinic** Paraffinic oils are characterized by low aromatic content and light color. They are used in furniture polishes, ink oils, and proprietary chemical formulations.
5. **Extracted Paraffinic** These oils offer light colors, low volatilities, relatively low pour points, and improved color stability. They are used in waste disposal system fluidizing oils, textile oils, and autoclave oils. Telura 607, 612, 613, and 619 are the oils in this category that conform to the requirements of FDA Regulations 21 CFR 178.3620(c).

Table 8
Telura® Numbering Systems

The three-digit Telura grade number provides a key to the properties of the oil. The first digit designates the type of product, as follows:

1—Aromatic
2—Intermediate aromatic[a]
3—Naphthenic
4—Extracted naphthenic
5—Paraffinic
6—Extracted paraffinic
7—Special grades

The second digit is indicative of the approximate viscosity in centistokes (cSt) at 40° C, as follows:

0—<10	5—50 to 59
1—10 to 19	6—60 to 99
2—20 to 29	7—100 to 999
3—30 to 39	8—1000 to 1999
4—40 to 49	9—2000

The third digit is an arbitrary number used to discriminate between grades of similar viscosity.

SOURCE Reprinted with the permission of Exxon Corporation, "Telura® Industrial Process Oils," Exxon Corporation, 1984.

[a]Products in this category are not currently available.

Table 9
Applications for Telura Industrial Process Oils

	Telura																		
	126	171	309	323	343	401	407	415	417	515	521	607	612	613	619	662	668	671	797
Absorption oil in gas plants			●									●							
Autoclave oil																●			
Canning machinery lubricant								●											
Carbon paper manufacturing				●										●					
Chain oil		●																	
Coke oven absorption oil			●																
Compressor cylinder oil			●	●															
Concrete form oil				●							●								
Cordage oil				●					●		●								
Cutting oil base		●		●							●								
Drilling mud						●													
Dust suppressant		●	●	●							●								
Fiber finishing												●							
Foam depressant									●						●				
Furniture polish		●								●									
Glass mold lubricant							●	●											
Heat transfer oil for low temperatures						●	●												
Honing oil								●											
Impregnating oil								●											
Ink oil		●		●					●	●									●
Leather tanning				●	●						●								
Oil field chemicals	●					●													
Plywood waterproof coating				●															
Proprietary chemical formulations		●		●	●		●			●					●				
Quench oil											●				●				
Rust proofing			●	●					●		●								
Sealant component																●	●		
Shoe polish				●						●									
Slushing oil			●																
Spray oil base													●						
Steel roll oil base				●					●										
Textile conditioning oils				●			●								●				
Turbine flushing oil			●																
Waste disposal system fluidizing oil																		●	
Water treating chemical				●								●							

SOURCE Reprinted with the permission of Exxon Corporation, "Telura® Industrial Process Oils," Exxon Corporation 1984.

6. **Special Grade** Telura 797 is a heavy, black residual oil used in newsprint ink formulations.

Oil Contamination

One of the main causes of wear and breakdown of machinery is contamination of lubricating or hydraulic oils by grit, water, and other unwanted materials. Regular inspection of industrial oils can save a great deal of breakdown time. This can be done by means of a fluid analysis kit (Figure 13). Examples of SAE oil contamination classification are also provided in the kit. Oil can also be contaminated with hazardous chemicals of which polychlorinated biphenyls (PCBs) are the most common, especially around electrical apparatus such as older transformers. PCBs are toxic and are thought to be a cause of cancer. Field tests for PCB spills or contaminated oil can easily be made with portable testers like the one shown in Figure 14.

Water

Aside from our municipal water systems, industrial water supply is of major importance. Water that has been used for industrial purposes is often contaminated with heavy metals and other materials that must be removed before this waste water is returned to a river or other outflow. Various kinds of tests are made

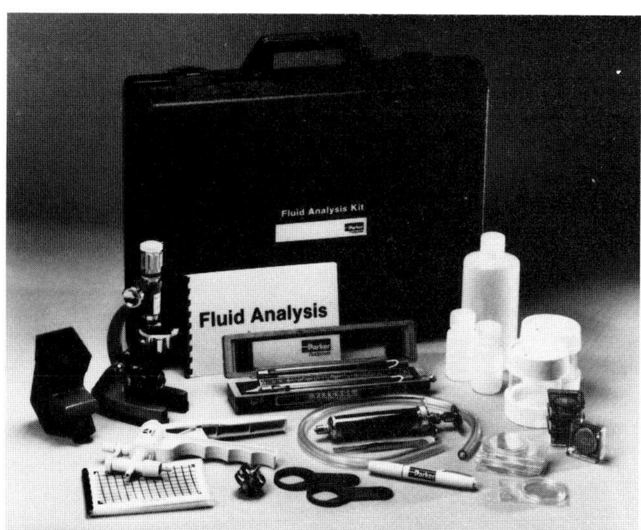

Figure 13
Fluid analysis kit. Oil samples in field, plant, and laboratory hydraulic systems can be easily checked for contaminants with this portable kit (Courtesy of Parker-Hannifin Corporation).

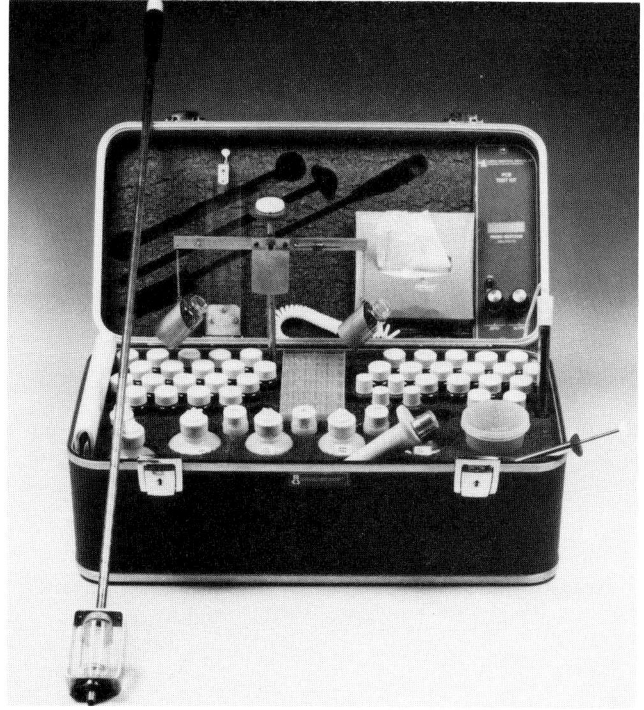

Figure 14
PCB field test kit for screening oil and soil. The soil collector is shown removed from the case and assembled (Courtesy of McGraw-Edison Company, Power Systems Division, Pittsburgh, PA).

to assure the purity of waste water; among them is an automatic metals monitor (Figure 15) that determines the level of cadmium, copper, lead, and zinc in waste water.

Figure 15
Water analyzer for testing water for heavy metals (Alewijnse Electrical Engineering, Environment Systems Technology Division, Nijmagen, Holland).

Heavy metals may pose a serious threat to public health if they occur in sources of potable water such as rivers, lakes, and groundwater. The metals monitor, a completely automatic electrochemical installation, can detect and analyze all types of water without any human supervision.

Besides potable water, the instrument can monitor sea water, rain water, and industrial waste water. It can also be used to detect corrosion in power stations and water works, to find leaks in pipes, and to analyze the effluents of the metal finishing, plating, and chemical industries. Besides cadmium, copper, lead, and zinc, other metals can be detected by the monitor if the computer program and treatment of the sample are modified.

All of the control, monitoring, and calculating functions are performed by a microprocessor. The data are recorded four times an hour on a printer and an LED display. In addition, daily averages of the metal concentrations are printed once a day. If it is connected to a telemetric recording circuit, the metals monitor will analyze concentrations ranging from .0005 to 1 mg/l.

MATERIALS APPLICATIONS

Material selection for a particular manufactured product is an involved and sometimes complex process requiring a broad understanding of the nature and behavior of materials. Many factors other than material strength and durability need to be considered. In view of domestic and foreign competition, the cost of materials is of prime consideration, in addition to the manufacturing cost. Product warranties and customer service requirements are to be considered as well as product liability and patent infringement. Many new materials for products are chosen because of the necessity for weight reduction, especially for energy savings in the field of transportation. However there is an inherent risk in making a sudden change in material for a product without extensive evaluation of its long-term performance. Many hasty substitutions of materials have resulted in product liability cases. Some of the factors involved in making a material choice are tensile strength, compressive strength, torsional strength, yield strength, fatigue strength, hardness, impact resistance, wear resistance, ductility, machinability, formability, hardenability, weldability, corrosion resistance, availability, and cost. Of course, there are many other factors that relate to the specific selection, such as the prevailing conditions—temperature, atmosphere, and vibration—to which the products will be subjected. Reference handbooks on materials are helpful when making selections. Some of these are the *Metals Handbook* and the *Source Book on Materials Selection,* both published by the American Society of Metals, and the *Cast Metals Handbook,* published by the American Foundryman's Association.

Figure 16
Pivot irrigator covers 160 acres in a circular pattern (Pierce Corporation, Eugene, OR).

Figure 17
After just a few years in the field, the section of galvanized pipe (*a*) shows the roughness and discoloration typical of a "worn-out" galvanized coating. The pipe with the electrostatically applied epoxy coating (*b*) still looks like new and is smooth to the touch even after several seasons of use (Pierce Corporation, Eugene, OR).

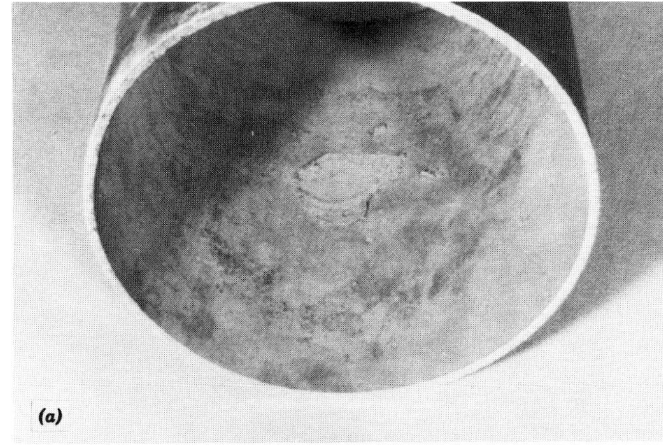

(*a*)

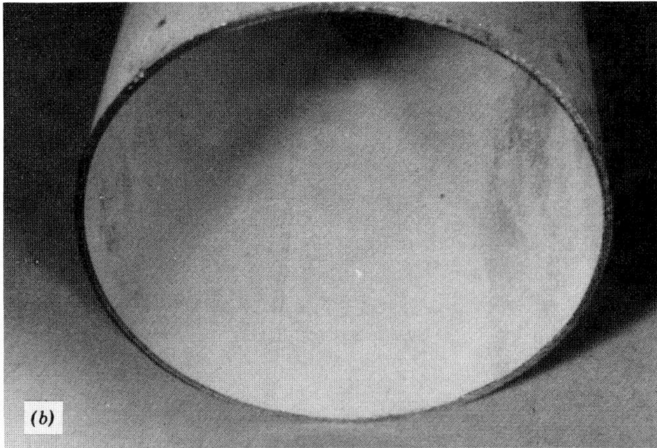

(*b*)

Example: Galvanized versus Plastic-Coated Pipe

A West Coast manufacturer of irrigation equipment, OEM (Original Equipment Manufacturer), had been using galvanized steel pipe to make pivot irrigators. These devices slowly move a pipe around a center tower in a circular pattern spraying water that often contains liquid fertilizer (Figure 16). Because the fertilizer is acidic, uncoated steel pipe would quickly corrode. Since zinc is a sacrificial metal, it soon begins to roughen inside causing a turbulence that requires more energy to move the water through the pipe (Figures 17a and 17b). A higher head (pressure) is required to move the water through the galvanized sprinkler pipe than through a plastic pipe, for example, in order to maintain sufficient pressure at the outer sprinkler heads, requiring more electrical energy.

A search was made by the company engineers for a substitute pipe material to solve this energy loss problem. Although plastic had less friction than the other alternatives, it could not substitute for two reasons. It would tend to deteriorate from ultraviolet rays in sunlight and it has insufficient tensile strength to support the weight of water in a suspended pipe. Aluminum, besides being expensive, would not substantially improve the energy loss problem. The Hazen–Williams formula of the coefficient of friction, in pipe, in which higher numbers indicate lower friction, shows black steel pipe with a number of 100. Galvanized pipe and aluminum are both 120. Since weight was not a factor and higher strength was required, steel was still the preferred pipe material. The solution was to coat steel pipe with a suitable plastic material. Steel pipe was cleaned by shot-blasting and electrostatically sprayed inside and outside with an epoxy powder that was then heat dried. This epoxy coating is inert to the low acid in the fertilizer. It is very smooth and its Hazen–Williams number is 150, the same as that of plastic pipe. This investigation led to an amazing revelation. A substantial cost savings in electric power for the farmer's water pumps could be realized by reducing the friction in the pivot irrigation pipe. With the cost of electricity in the Midwest being 3 to 6¢/kwh, a ten-machine operator (each machine covers 160 acres) over a ten-year period would realize a savings of $30,000 to $50,000 or more. Thus, the switch to epoxy-coated steel pipe gives the machine purchaser, the farmer, a competitive edge; it also will undoubtedly increase sales for that OEM company's product, the pivot irrigator.

REVIEW QUESTIONS

1. Which is the most common classification system used in the manufacturing industry for machine and alloy steels, the SAE–AISI or the ASTM specifications?
2. What kind of steels are designated by the ASTM specifications?
3. Name the quenching medium used for W1 and O1 tool steels.
4. Which steel shafting is more expensive, cold rolled (CR) or ground and polished (G and P)?
5. Which steel bar having the same alloy content is more expensive, hot rolled (HR) or cold rolled (CR)?
6. State the greatest advantage of HSLA steels over plain carbon steels for use in modern automobiles.
7. One type of stainless steel that is often used for cutlery and surgical equipment can be hardened by heating and quenching. What number series is it classified under?
8. As gray cast iron solidifies, it is unable to hold more than 2 percent carbon in solution, yet it may contain 4 percent carbon. What happens to the remaining 2 percent?
9. Would you make a part from gray cast iron if it were to be subjected to high impact loads? Why?
10. What kind of cast iron might be acceptable for the part described in Question 9, particularly if the impact loading were moderate?
11. Why is a thin sheet of pure aluminum sometimes sandwiched (clad) on tough alloy aluminum?
12. In what two ways can aluminum be hardened?
13. What is the most widely used type of copper for electrical conductors?
14. Of all the means of identifying the alloy and carbon content of metals, which is the most accurate although the most expensive method?
15. If only low and high plain carbon steels are present, what simple method could be employed to distinguish between them?
16. A manufacturer of a consumer product wishes to make a material change from metal to plastic for the purpose of weight reduction and appearance. What should be done before the new product goes on the market?

CASE PROBLEMS

Case 1
Separating Aluminum and Magnesium Bars

A manufacturing plant regularly received shipments of aluminum bars of various sizes on which machining and welding operations were performed. On one shipment the supplier had discovered the bars had been sent out unmarked and several magnesium bars had accidentally been mixed into the lot. There was no time to send for and receive a new shipment before running out of material. The two metals had to be separated, not only because magnesium was an unwanted material, but because it is a pyrophoric metal, tending to catch fire and burn fiercely when finely divided, as in machining chips, and from welding, when proper precautions are not taken. Therefore, it constituted an extreme hazard in this case where they were not normally using magnesium. The metal bars all looked identical and they varied so much in size that weight differences could not be readily used to identify them by unskilled help. A quick low-cost method was needed to separate these look-alike metals, one that an unskilled person could use. How would you solve this problem, considering what you have learned in this chapter?

Case 2
Cost of Steel Shipment

You receive a load of steel bars on a truck. The shipping order does not include the cost, which is on the invoice that will be mailed later. You need to know the approximate cost immediately in order to make a cost estimate for parts to be made from the steel bars. The load is made up of 20-foot-long steel bars: 10 flat bars (rectangular), $\frac{3}{4} \times 2\frac{1}{2}$ in., and 15 round bars $1\frac{1}{4}$ in. diameter. Metal shapes are sold by weight and different metals have different weights per unit volume. Steel weighs 487 lb/ft^3. The price of the steel in this shipment is .27¢/lb for the flat bars and .31¢/lb for the round bars. Weight in cubic inches of a steel bar is found by multiplying the cross-sectional area by the length in inches, multiplied by the weight of 1 in.^{3}. There are 1728 in.^{3} in 1 ft^3. What is the total cost of the shipment?

SURVEY OF MANUFACTURING PROCESSES

PART II

Steel plate is rolled through the finishing unit of the 160-in. sheared plate mill at Bethlehem Steel Corporation's Burns Harbor, Indiana, plant. The four-high finishing mill can produce plates from $\frac{3}{16}$ to 15 in. in thickness, up to 150 in. in width, and up to 120 ft in length (Bethlehem Steel Corporation).

Materials such as metals, plastics, and ceramics are made into useful articles and consumer products by many and varied means. Metals are cast into molds in a number of different ways to form intricate small shapes or massive machine parts. Metals are also rolled, pressed, and hammered into dies or forced through dies by extrusion to make special shapes. For example, iron and steel are heated to a high temperature so they can be easily formed by forging (hammering and squeezing). Although forging was once only a hot-metal operation, today cold upset forging is practiced even on steel. At intermediate (warm) temperatures, a metallurgically superior product can be produced for some purposes. For example, in warm forming, tough materials like SAE 52100 steel are routinely manufactured into high quality parts with the lower production costs of cold forming.

A large segment of manufacturing is devoted to the processing of sheet metal in cold working operations such as blanking, stamping, and forming. Home appliances, automobiles, and many other products depend on sheet metal processing.

Machinery of all kinds that requires precision parts depends on the machining and machine tool industry which could be considered the foundation of modern manufacturing. The tool and die industry, though little known and understood, is the backbone of modern industry.

Virtually every manufacturing process is in some way dependent upon tool and die shops. Stamping, pressing, injection molding, and die casting are just a few of the processes for which tool and die services are absolutely necessary.

Manufacturing of plastic and composite materials is steadily growing and is replacing many products formerly made of metal, leather, and wood. However, many of these manufacturing industries are interdependent. For example, plastic injection molds require special tool and die machining processes; the manufacture of the molding machine is a machine tool process.

CHAPTER 7 PROCESSING OF METALS: CASTING

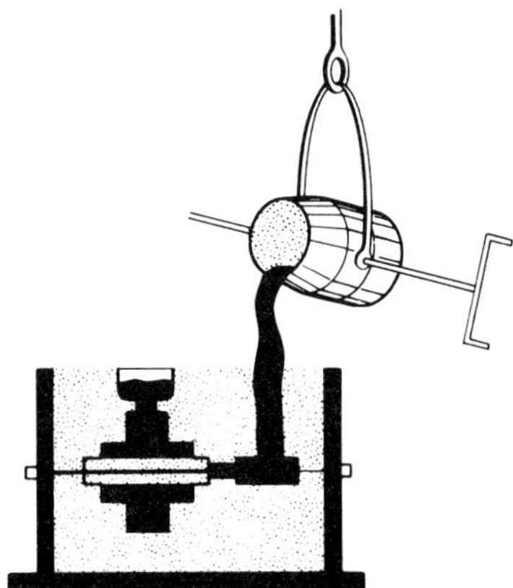

The process of casting metals is done by pouring or injecting molten metals under pressure into a mold or cavity having the desired shape. When the metal has solidified, the mold is stripped off, leaving the metallic shape. Virtually any shape can be produced by this method, often with such precision that subsequent machining is not required. However, a machining operation is usually required to finish castings for use in machinery.

Casting is one of the oldest methods of manufacturing metals. Prehistoric humans made tools by pouring molten metal into open molds made of baked clay or stone. Cast objects over 4000 years old have been found dating from ancient Assyrian, Egyptian, and Chinese cultures.

When designing a metal part to be manufactured, an engineer must choose a method of production. The part can usually be made by one or more processes including machining from solid metal, welding fabrication, powder metallurgy, pressing and cold forming, hot forging, or casting. The chief advantage of casting over these other processes is that smooth, flowing designs are possible, either for practical or decorative purposes. Also, the metal can be placed only where it is required. Therefore, the economy of using less metal for a part (especially when a very expensive metal is used), the eye-pleasing appearance (such as in machinery housings), and the possibility of producing intricate shapes are factors that should be considered when designing for manufacture. Sand casting is not a rapid method of production, but die casting is a relatively rapid process. Other casting processes lend themselves to production, but none should be considered to have high production rates as compared to punch press work or powder metallurgy.

Casting processes involve a large segment of the metals industry. These range from the tiniest precision parts to huge castings for machinery sections weighing many tons (Figure 1). Some metals that are too hard to machine after casting, such as those used for aircraft turbine impeller blades, can be cast to a precision shape not requiring any subsequent machining. Other soft metals, such as die cast and aluminum, are used to form articles such as carburetors for automobiles. It would take many hours of machining to make a complicated carburetor part from solid metal in a machine shop, but it takes only seconds in a die casting machine.

Several requirements are needed for the production of good castings, whatever the method used.

1. A method of melting the metal to the correct temperature.
2. A mold cavity of the desired shape, with sufficient strength to contain the metal without distorting or having too much restraint on the molten metal as it solidifies, which can cause cracking. Also, the mold must be designed to avoid internal porosity and cracking.
3. Molds must be arranged so that when molten metal is introduced into the mold, air and gases can escape so the casting will be free from defects.
4. Any mold material in internal cavities must have a provision for its removal. Finishing operations are often required to remove any excess material from the casting.

OBJECTIVES

This chapter should enable you to:

Figure 1
Centrifugal casting, 131,000 lb, 5 ft. diameter $35\frac{1}{2}$ ft. long, to be used for a suction roll shell in a paper machine. Here it is being machined to size (Copyright 1976, 1978, 1984, Sandusky Foundry & Machine Co., Sandusky, Ohio).

1. Show how sand is prepared and used to make molds into which molten metal is cast.
2. Explain the use of patterns for sand casting.
3. Describe the principles involved in the processes of shell molding and investment casting.
4. Describe and evaluate the processes of centrifugal casting, permanent molding, and die casting.
5. State the difference between several types of melting furnaces.
6. Show how good casting design can prevent many casting problems.

SAND CASTING

In sand casting, a specially prepared sand is used for the mold materials. The sand is **rammed** or packed around a **pattern** that has the shape of the desired casting. A series of steps in producing castings in a small foundry can be seen in Figures 2 through 7. Sand has the advantage of being highly refractory (can resist high temperatures without melting), so metals like cast iron and steel can be cast by this method. Sand casting accounts for the greatest tonnage of all types of castings produced.

Casting Sands

Sand is easily molded and capable of holding accurate detail and it can be reused. Two basic types of sand molds are **green sand molding** and **dry sand molding**. *Green* refers to moisture content. The sand should contain enough moisture to cause it to adhere together in a lump when squeezed in the hand. Moisture content is usually determined by one of several types of testing devices. Green sand is used for ramming around patterns to form molds for general casting purposes. Dry sand molding is often used to make cores that form holes and hollow places in green sand molds. A resin is usually put into the sand to harden it. Molding sand is a natural silica sand (SiO_2). Not all natural sand deposits are ideal for foundry work; however, less ideal sands may be used with a binder of volcanic ash or a synthetic binder. When very large castings are poured, there are tremendous pressures on the mold; sometimes a small amount of portland cement is used as a binder. Sand for molds must be carefully conditioned with additives and by proper mixing.

Preparing the Sand

Molding sand must have several characteristics:

1. **Cohesiveness** (bond) The ability to be packed (rammed) in a mold and retain its shape.
2. **Refractoriness** The ability to withstand high temperatures.
3. **Permeability** Porosity that allows gases to escape through it.
4. **Collapsibility** The ability to allow freedom for the solidifying, shrinking metal to move without cracking and to allow the cast part to be easily broken out of the mold.

Figure 2
Casting sand is being pushed around on a jolt-squeeze machine (Eugene Aluminum & Brass Foundry).

Figure 3
Drag (bottom half) of sand mold after it has been formed around the match-plate pattern (Eugene Aluminum & Brass Foundry).

Figure 4
Core being placed in mold (Eugene Aluminum & Brass Foundry).

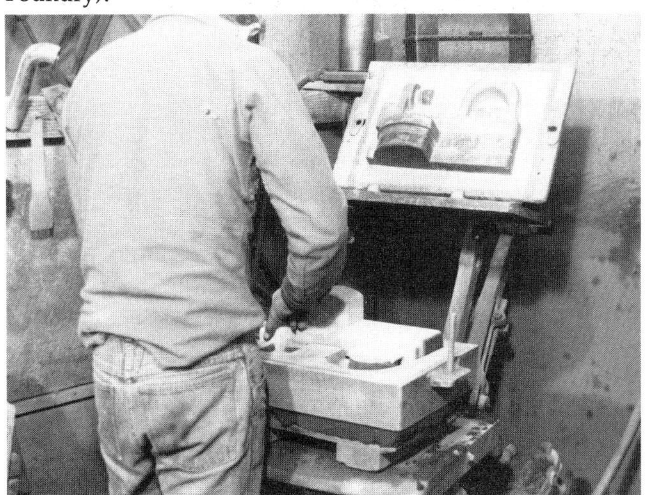

Figure 5
Cope (top half) being placed on drag flask (Eugene Aluminum & Brass Foundry).

Figure 6
Molten aluminum being poured into the mold (Eugene Aluminum & Brass Foundry).

Figure 7
Solidified casting broken out of molds (Eugene Aluminum & Brass Foundry).

To achieve good results, the sand grains or particles must be of the right size, clay must be added in the right proportion to give the required cohesive quality to the sand, and moisture content must be correct. The mold must be of a certain hardness, strength, and permeability if consistently good castings are to be obtained.

Some foundries add about 4 percent sodium silicate to the sand which remains soft until it is exposed to carbon dioxide (CO_2) gas, after which the sodium silicate hardens the mold in a few seconds. The gas is applied to the mold by forcing it through a small-diameter tube inserted in various places in the sand.

The two halves of a sand mold are prevented from sticking together by dusting them with a parting sand or a commercially prepared powder. The inside of a sand mold is often sprayed with foundry facing, a graphite solution, to prevent the sand from melting and sticking to the casting. It also provides a smooth surface on the mold face.

Molten metal is poured into the mold, where it solidifies. The mold is broken to remove the casting which is placed in a **shake-out**, a vibrating machine that removes the sand sticking to the casting and the cores. The sand from the broken-up mold and from the shake-out is conveyed to another machine that reconditions it and prepares it for reuse by crushing, screening, magnetically separating, and impinging of sand grains against a wear plate at high velocity. The fines (dust) are vacuumed away and the clean sand is transported to hoppers to be used again.

The sand mixture is placed in a muller (Figure 8) to thoroughly mix the ingredients with water to obtain the proper consistency. The muller consists of a large tub in which an arm swings around forcing rollers over the sand. Scrapers on the arm remove sand from the sides of the tub and force it in front of the rollers. The sand is again ready for molding.

Patterns

The first requirement in making a sand casting is to design and make a pattern (Figure 9). Patterns are usually made of wood when only a few castings are needed. For larger quantities, patterns are made of metal: aluminum, magnesium, or brass. Hard, tough plastics are also sometimes used for patterns.

Shrink Rate When metals solidify from the molten state, they shrink a given amount per foot. This is called the shrink rate. Each metal has a different shrink rate

Figure 8
Three types of mulling machines used for the conditioning of molding sand. (*a*) Continuous muller. (*b*) Conventional batch muller. (*Metals Handbook*, Vol. 5, 8th ed., American Society for Metals, 1970, p. 163. With permission.)

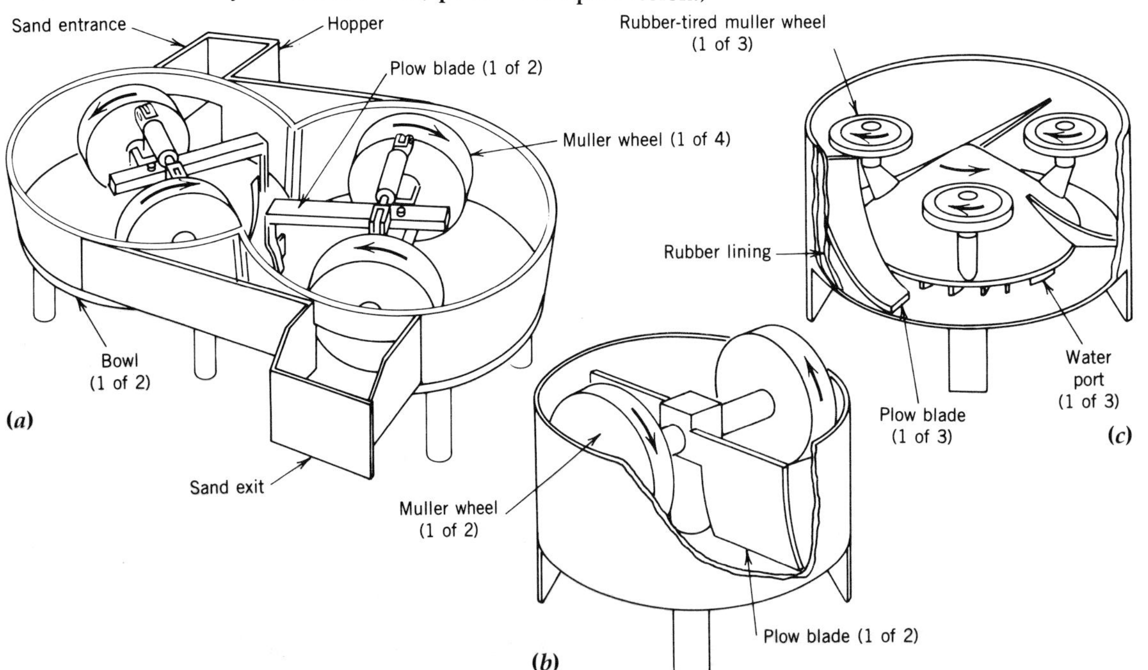

Figure 9
Metal match-plate patterns (Eugene Aluminum & Brass Foundry).

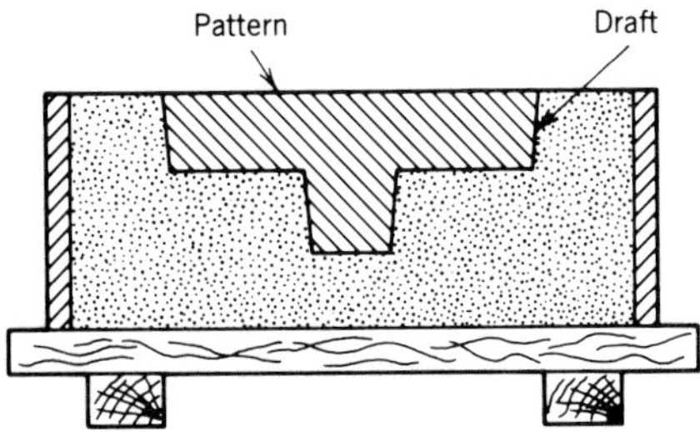

Figure 10
All sand patterns must have draft so they can be removed from the mold.

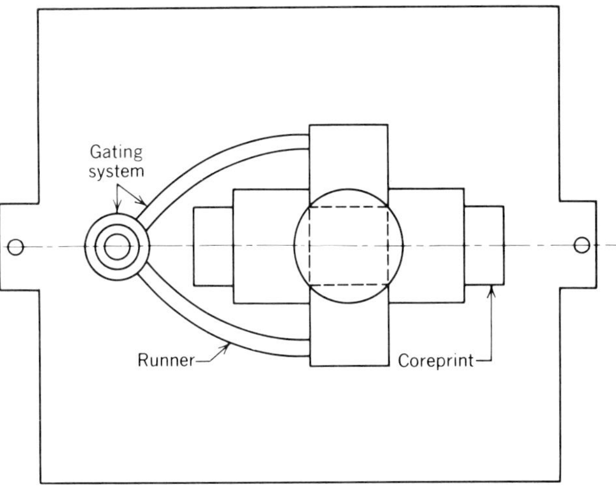

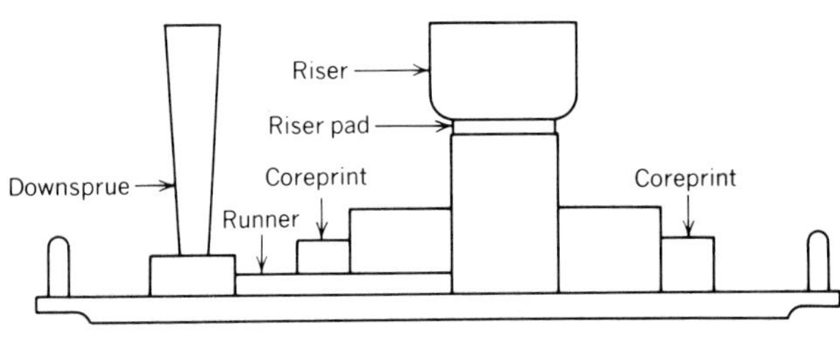

Figure 11
Top view and side view of cope match plate pattern. The down sprue and riser patterns are removed from one side and the cope pattern is removed from the other side (White, Neely, Kibbe, Meyer, *Machine Tools and Machining Practices,* Vol. I, © 1977 John Wiley & Sons, Inc.).

so the patternmaker uses a shrink rule when making measurements on the pattern. Shrink rates in in./ft for some metals are:

Cast iron	$\frac{1}{8}$
Steel	$\frac{1}{4}$
Brass	$\frac{3}{16}$
Aluminum	$\frac{5}{32}$
Magnesium	$\frac{5}{32}$

Draft Patterns must have draft or taper to permit removal from the mold (Figure 10). Draft is usually about $\frac{1}{8}$ to $\frac{1}{4}$ in./ft. Without draft on the pattern, pieces of the sand mold would be broken away as the pattern was pulled out. Previously made castings can sometimes be used as a pattern if additional shrinkage is not a factor. If there are areas on the casting that will have machined surfaces, the patternmaker will leave additional metal for removal. How much extra thickness is necessary depends on the roughness or accuracy of the finished casting. Sufficient material must be left for subsequent machining in order to cut under sand inclusions on the surface of the casting.

Green Sand Pattern

Although many kinds of patterns are used, such as split patterns, single piece, and loose piece types, patterns for green sand molding are usually made in two halves and are called **match plate patterns** or **cope** and **drag** patterns. *Cope* refers to the top half (Figure 11) and *drag* refers to the bottom half (Figure 12). These

Figure 12
Drag match-plate pattern (White, Neely, Kibbe, Meyer, *Machine Tools and Machining Practices,* Vol. I, © 1977 John Wiley & Sons, Inc.).

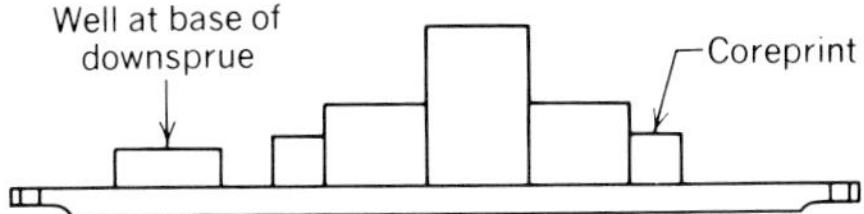

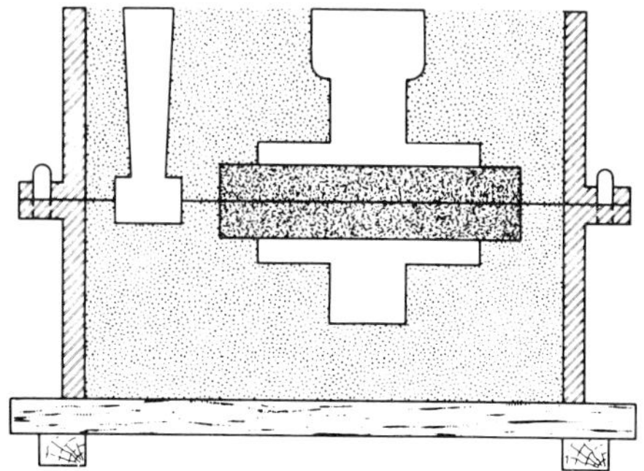

Figure 13
Sectional view of cope and drag with core in place on the mold board. At this point it is ready to make the casting (White, Neely, Kibbe, Meyer, *Machine Tools and Machining Practices,* Vol. I, © 1977 John Wiley & Sons, Inc.).

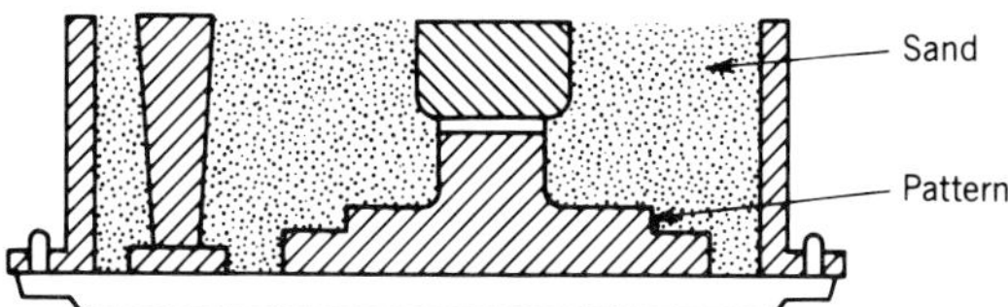

Figure 14
Sectional view of cope pattern in flask with the sand rammed in place and struck off.

match plate patterns are placed in a **flask** that is also made in two halves (Figure 13). The cope half of the pattern is placed on the mold board with the cope half of the flask and the sand is rammed into place and struck off (leveled off even with the top of the flask with a straight edge, Figure 14). The pattern also has an extension called a **coreprint** if a core is used. This space in the sand mold provides a support for the ends of the sand core. One or more holes are provided for pouring the metal. This hole, called a **sprue**, is connected to the mold cavity by the gate. When the molten metal is poured into a downsprue, it fills the mold cavity and other holes in the top of the mold, called **risers**. The risers are used as reservoirs to feed liquid metal to the mold cavity as it shrinks while solidifying. These vents also provide for escaping gases and allow impurities to float to the riser and out of the casting. The half-pattern is made in segments that can be separated, such as the riser and downsprue which are removed from the pouring side of the cope half of the flask, and the pattern is removed from the **parting line** side after turning it over. The drag half of the flask and pattern is also rammed and the pattern is removed in the same manner (Figure 15). Cores are installed, the two halves are placed together, and the mold is ready for placing the core and preparing for casting (Figure 16).

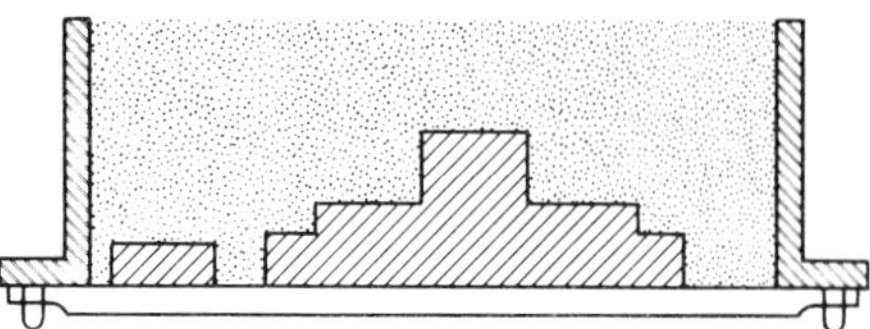

Figure 15
Sectional view of drag pattern in a flask with the sand rammed in place and struck off (White, Neely, Kibbe, Meyer, *Machine Tools and Machining Practices* Vol. I, © 1977 John Wiley & Sons, Inc.).

Figure 16
Completed casting (White, Neely, Kibbe, Meyer, *Machine Tools and Machining Practices,* Vol. I, © 1977 John Wiley & Sons, Inc.).

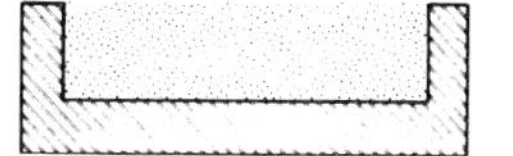
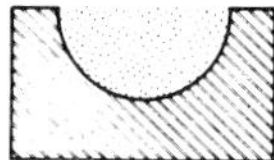

Figure 17
Sectional view of core box with the core (White, Neely, Kibbe, Meyer, *Machine Tools and Machining Practices,* Vol. I., © 1977 John Wiley & Sons, Inc.).

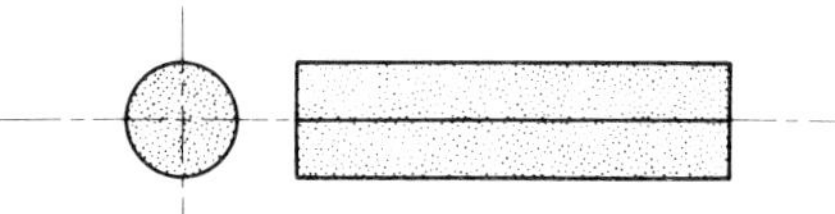

Figure 18
Two halves of core fastened together after being removed from core box and baked in oven (White, Neely, Kibbe, Meyer, *Machine Tools and Machining Practices,* Vol. I, © 1977 John Wiley & Sons, Inc.).

Cores

Dry sand molding is used to make cores. Cores are made by ramming sand into a simple core box or by forcing sand into the mold using a core blowing machine (Figure 16). These green cores are sometimes baked in an oven. The organic binder of linseed oil, cereal, or resins causes the core to harden, yet allows for some permeability. Most cores are now made with **no bake sands** in which a resin and a catalyst are com-

Figure 19
Large sand mold being gravity cast with molten cast iron, © 1983 Pottstown Machine Company, Pottstown, PA).

bined to achieve the desired strength and porosity with the baking process. Some cores are made by the sodium silicate and carbon dioxide process mentioned earlier. When it can be done conveniently, the core can be part of the green sand mold itself (Figures 17 and 18). The core is placed in the sand mold at the parting line in the coreprint. The mold is now ready for pouring the molten metal (Figure 19). Some cores are now made hollow using a "hot box," thus using less sand, and they are lighter to handle and more precise.

The Advantages and Disadvantages of Sand Casting

The greatest advantage of sand casting is that almost any metal can be used and there is almost no limit on size, shape, or weight of the part. Sand casting has the most direct route from pattern to casting. Tooling costs are low and the gravity casting process is economical. Among the limitations involved in sand casting is the need for machining to finish the castings, especially large ones having rough surfaces. Long, thin projections are not practical to make using sand castings and certain alloys will almost always produce defects.

Sand castings are extensively used for machine tool housings, bases, slideways, and other parts, and they are used also for the automotive industry. Castings cannot have porosity or be too hard or too soft for these applications, and they must be consistent from casting to casting. Statistical process controls (SPC), quality controls for castings, are rapidly becoming a survival issue for foundries that provide these items since scrap losses must be kept to an absolute minimum—there is strong foreign competition in these industries.

The Evaporative Casting Process

In this process the pattern, sprue, and riser are made of foamed polystyrene that remains in the sand mold during the pouring of the metal. When the metal is poured, the heat vaporizes the polystyrene pattern almost instantaneously, leaving the mold shape intact as it is being filled with metal (Figure 20). This process

Figure 20
Evaporative casting process. This photo depicts the way the intense heat of molten aluminum vaporizes a foam pattern, leaving behind a high-quality casting (Photo courtesy of Ford Motor Company).

Figure 21
Polystyrene (styrofoam™) pattern for a bronze casting. The pattern is made in sections (Eugene Aluminum & Brass Foundry).

Figure 22
Bronze casting called *Coast Spirit*, permanently located in front of the library at the University of Victoria, British Columbia (Eugene Aluminum & Brass Foundry).

can be used for castings of any size and shape since the patterns do not need to be removed prior to casting; also, for the same reason, no draft is required on the pattern (Figures 21 and 22). But patterns made by cutting and shaping the polystyrene are economical for only one or two castings. So only if the patterns themselves can be mass produced in a mold can this method be used for production runs. The economy of this method lies in the inexpensiveness of the pattern. Ordinary patterns are usually quite costly and would usually be prohibitive for only one part. The sand is prepared in the same way as for other green sand molding and is packed around the polystyrene pattern by ramming or blowing. Care must be taken not to shift this lightweight pattern while ramming the sand around it.

STEEL CASTING

Although steel is somewhat more difficult to cast than gray cast iron, it is, to some extent, replacing cast iron for parts requiring higher strength. These range in size from small parts weighing a few ounces to huge hydraulic press castings weighing many tons. Steel is more difficult to cast because it has a higher melting point of 2500 to 2900° F (1371 to 1482° C) than cast iron (2100° F or 1149° C) and has less fluidity, that is, ability to flow into small crevices. Most steel castings are of a medium carbon steel with manganese and some other elements added to ensure soundness. The mechanical properties of steel castings approach those of wrought steels. Toughness, ductility, and tensile strength are much greater than that of cast iron. However, the cost of steel castings is greater than that of cast iron and where these mechanical properties are not important, cast iron is a better choice. Also, cast iron is better for machine frames and beds because it tends to maintain its shape better than steel does. That is, it has less tendency to "creep" or warp because of internal stresses gradually being relieved. Because of the graphite flakes, gray cast iron provides a better bearing surface than steel when machine parts slide together and it also provides better vibration damping.

Since steel has a high melting point, it must be cast in a refractory material such as sand. Therefore, most steel is sand cast. Also, the shell molding process and investment-shell process are both used for precision steel casting since the mold is made of a refractory substance.

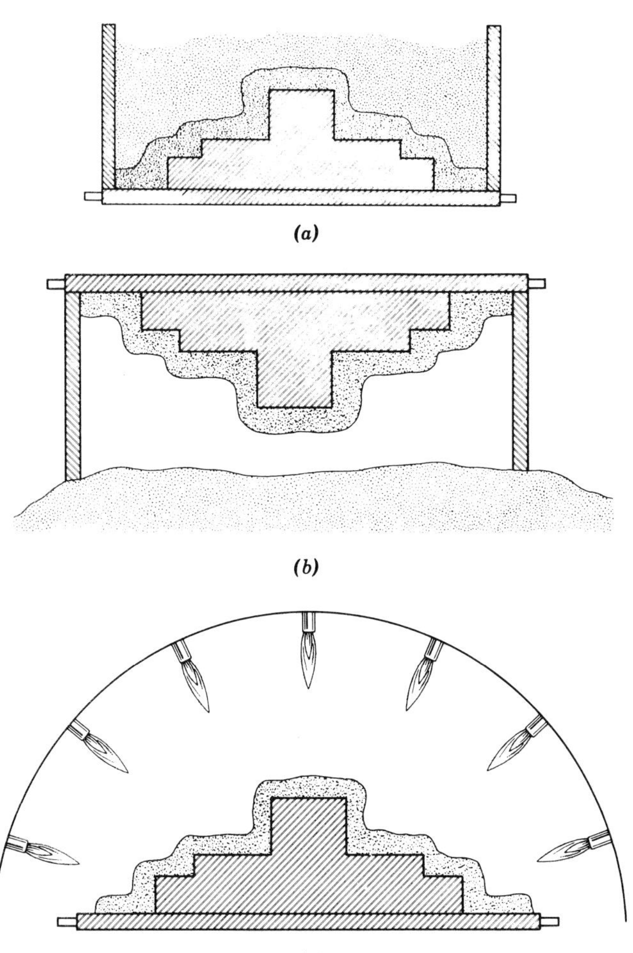

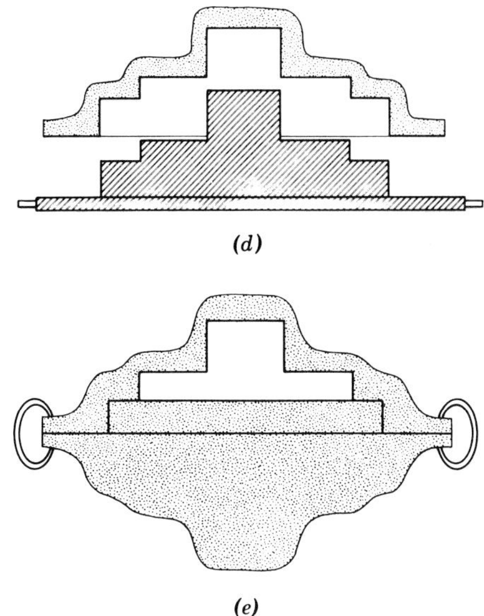

Figure 23
Shell molds. (*a*) Heated metal pattern in dump box with sand. (*b*) Dump box is turned, leaving some sand adhering to the pattern. (*c*) Pattern is placed in an oven for 1 or 2 min to harden it. (*d*) Half-shell mold is stripped from pattern. (*e*) Mold halves are clamped together with core. The top half-shell is shown as a sectional view (White, Neely, Kibbe, Meyer, *Machine Tools and Machining Practices*, Vol. I, © 1977 John Wiley & Sons, Inc.).

SHELL MOLDS

The shell molding process is a type of sand casting that provides a finer detail and smoother finish because the sand is finer and it is combined with plastic resin to make a smooth mold surface. The process requires less labor than sand casting and it can be automated for mass production. This method requires the use of metal patterns that are heated to temperatures up to 450° F (232° C), coated with a silicon release agent, and put into a dump box. A phenolic resin mixed with the sand is poured over the heated pattern. Some of the sand adheres to the pattern and solidifies, forming a shell around it. The thickness of the shell is determined by the length of time the pattern is in contact with the sand, and ranges from $\frac{1}{8}$ to $\frac{1}{4}$ in. The dump box is inverted to remove the loose sand (Figure 23). The pattern and adhering sand are placed in an oven and heated to a temperature of 600° F (316° C) for one or two minutes. The half-shell is then removed from the pattern and the two shells are glued or clamped together to form a complete mold. This thin mold is strong enough for casting small or thin parts, but heavier castings require the use of a backup material such as shot or sand in a pouring jacket.

The cost of the metal patterns is relatively high and the gate and runners must be included on the pattern. The saving in machining time and labor costs plus the high productivity make shell molding an economical process for quantity production. Many shapes can be made by this process, but since there is a parting line and the half-shell must be removed from the pattern, many complex shapes cannot be made by this process and the size of the part is limited. Shell mold casting has good repeatability and most ferrous and nonferrous metals can be cast by this method.

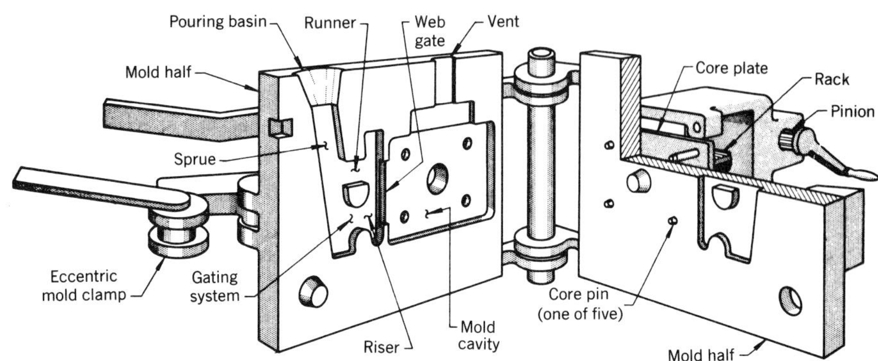

Figure 24
Book-type manually operates mold-casting machine, used principally with molds having shallow cavities (*Metals Handbook*, Vol. 5, 8th ed., American Society for Metals, 1970, p 266. With permission).

PERMANENT MOLDS

The greatest disadvantage of sand casting is that a new mold must be made for each casting, and some dimensional inaccuracies gave rise to the development of a more permanent and accurate kind of mold. Despite the name, permanent molds can be reused at most for only a few thousand castings, after which they lose their true shape and must be scrapped. Most permanent molds are made of gray cast iron or steel. Graphite molds are often used for casting higher temperature metals. The molds are made by machining processes and are hand finished or polished. A refractory wash is applied to the mold prior to casting. When cores are needed, they are made as in sand casting and are not reused. The mold halves are hinged so they can be opened and closed quickly and accurately (Figures 24 and 25).

In permanent mold casting, the metal is poured from a ladle, using gravity to fill the mold. Cast iron and nonferrous metals are cast in this manner. The molds are heated at the beginning of a run and the temperature is maintained by the casting rate. This is necessary to avoid chilling the metal too quickly, which can produce laps or cold-shuts in the casting. These are cracks where two adjacent portions of metal do not solidify together. Usually only simple shapes are cast by this process because of the rigidity of the mold and shrinking of the metal when it solidifies. Since these molds are not permeable, they must be vented at the joint where the two mold halves come together. In this, permanent molds are similar to those used in die casting in which the molten metal is forced into the mold with high pressures and vents are provided at the parting line to remove gases. Some permanent molds are also pressure poured but they use lower pressures than die casting does.

Slush casting is also used with permanent molds to make a shell of metal in the mold. The molten metal is poured into the mold and allowed to solidify to a certain wall thickness against the mold, and the remainder is dumped out, thus producing a shell. Toys, lamp bases, and ornamental objects are made by this process.

Good surface finish and dimensional accuracy plus

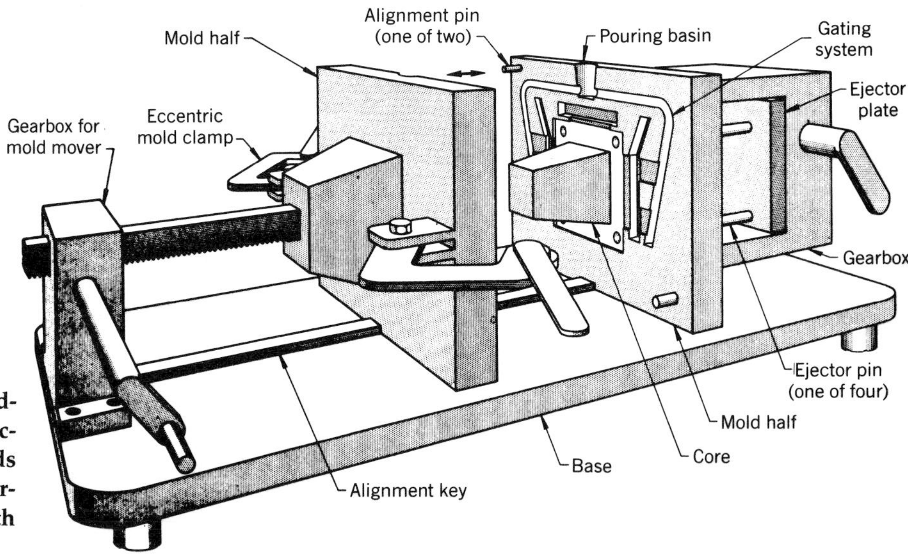

Figure 25
Manually operated permanent mold-casting machine with straight-line retraction, required for deep-cavity molds (*Metals Handbook*, Vol. 5, 8th ed., American Society for Metals, 1970, p.266. With permission).

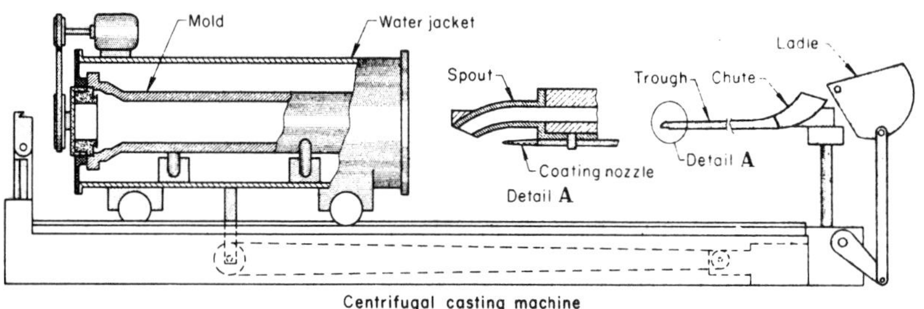

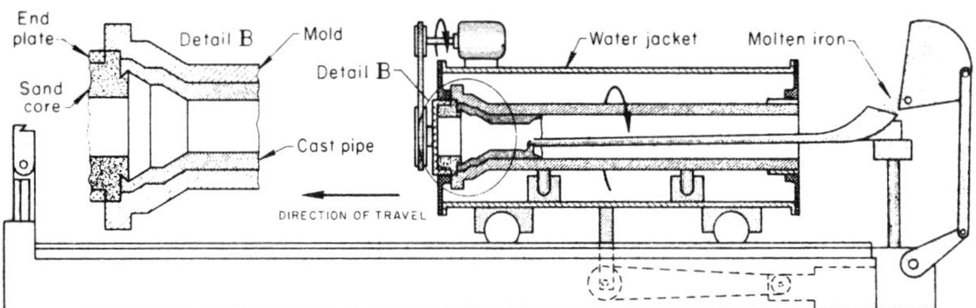

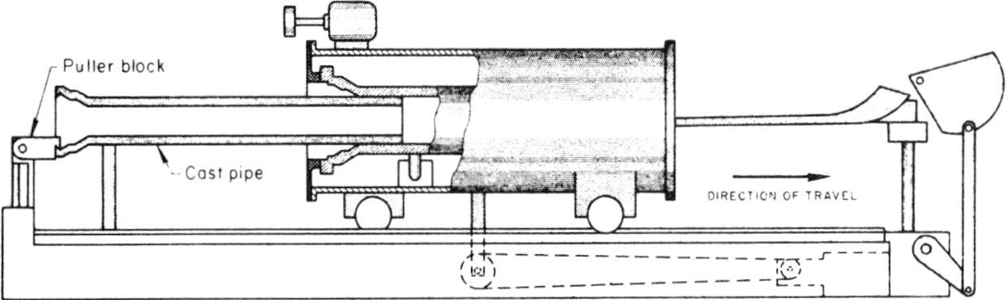

Figure 26
Machine for centrifugal casting of iron pipe in a rotating water-jacketed mold (*Metals Handbook*, Vol. 5, 8th ed., American Society for Metals, 1970, p. 267. With permission).

Figure 27
Large centrifugally cast tube (Copyright 1976, 1978, 1984, Sandusky Foundry & Machine Co., Sandusky, Ohio).

low porosity are obtained with permanent molds. Repeated use of molds and a rapid production rate with low scrap loss make this casting process ideal for moderate production runs of a few thousand pieces. Some disadvantages are high mold cost, limited intricacy of casting shape, and that the method is unsuitable for high temperature melting metals such as steel.

CENTRIFUGAL CASTING

Centrifugal casting is a process in which molten metal is poured into a rapidly revolving mold. The liquid metal is forced to conform to the shape of the mold by centrifugal forces many times the force of gravity. In this process the mold rotates about either a horizontal or a vertical axis. No core is needed to make an inner surface since this process naturally produces a hollow shape such as pipe (Figures 26 and 27). The thickness of the mold controls the cooling rate and therefore the grain structure of the cast part. Also, if a hard outer wear surface is needed with an inner machinable soft metal, two dissimilar metals can be used; the outer one to harden when solidified and the inner one to remain relatively soft. For example, a thin wall of stainless steel can be poured first and then less expensive low carbon steel can be poured, giving the pipe a corrosion re-

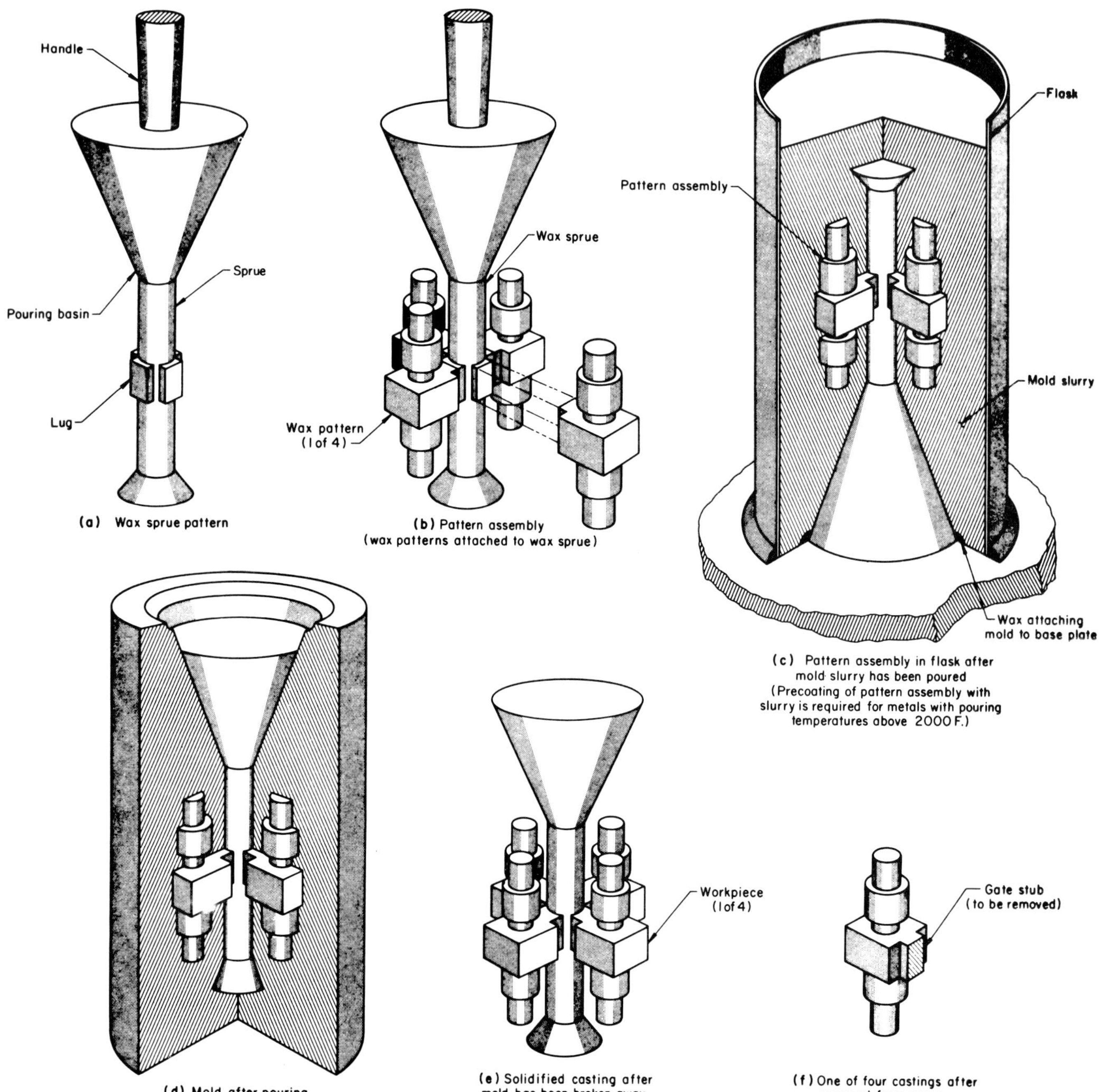

Figure 28

Steps in the production of a casting by the solid investment molding process, using a wax pattern. (*a*) Wax sprue pattern. (*b*) Pattern assembly (wax patterns attached to wax sprue). (*c*) Pattern assembly in flask after mold slurry has been poured. (Precoating of pattern assembly with slurry is required for metals with pouring temperatures above 2000° F.) (*d*) Mold after pouring. (*e*) Solidified casting after mold has been broken away. (*f*) One of four castings after removal from sprue. (*Metals Handbook,* Vol. 5, 8th ed., American Society for Metals., 1970, p. 239. With permission).

sistant jacket at low cost. Lighter elements, impurities, and slag collect on the inner wall of a centrifugal casting and can be removed later by machining.

Vertical or semicentrifugal casting is the rotation of a mold about its vertical axis. The metal is poured into a central reservoir that allows the liquid metal to flow outward into the spinning mold by centrifugal force. This process provides a denser structure in the metal than ordinary gravity casting does. It is ideal for steel wheels, brake drums, and cylinder barrels. A similar process, called centrifuging, rotates a group of molds or an entire mold away from the center of rotation. A pouring basin allows the metal to flow outward into the molds. This method produces superior quality castings.

INVESTMENT CASTING

One of the oldest methods of casting metals is the lost wax process. The pattern is made of wax and is used only once, but the wax is not really lost since it is melted and poured out of the mold and reused for another pattern. This method is used for dentistry, for making jewelry, and for one-of-a-kind art forms. When it is used for industrial purposes, significant production rates are possible.

Plaster of paris is generally used for the mold material in the lost wax process (Figure 28). The mold must be broken and destroyed to remove the cast part. Plaster of paris molds can be used only for the lower temperature metals such as aluminum, zinc, tin, and some bronzes.

When the pattern has sufficient draft and is shaped so half-shells can be removed, ordinary wood or metal patterns may be used in plaster mold casting. The mold halves are removed and baked to remove moisture. Obviously the cost of such molds is very low. Smooth surface finishes and dimensional accuracy are obtained by this method. But mold making time is somewhat long and the size of objects made by this method cannot be too large, so it is limited to being used for prototypes and for low-volume production with nonferrous metals.

Investment Shell Process

Metals having higher pouring temperatures, such as cast iron and steel, are precision cast by a similar process known as precision investment casting or the **investment shell** process. Since this is a production casting process, a method of mass producing the wax patterns is needed. This is done by injecting the molten wax into a split mold in the same way plastic objects are rapidly made in injection molding machines (Figure 29). These wax patterns are assembled in groups on a sprue in clusters resembling a Christmas tree (Figure 30). The wax assembly is dipped into an investment or slurry of refractory material. This is done a number of times to develop a shell with a thickness of up to $\frac{1}{4}$ in. The shell is then heated in an oven to melt out the wax which is later reused. The mold is then heated in a

Figure 29
Automatic wax pattern molder. Large quantities of duplicate wax patterns can be produced on this machine (Courtesy of (Mueller-Phipps International, Inc., Poughkeepsie, NY).

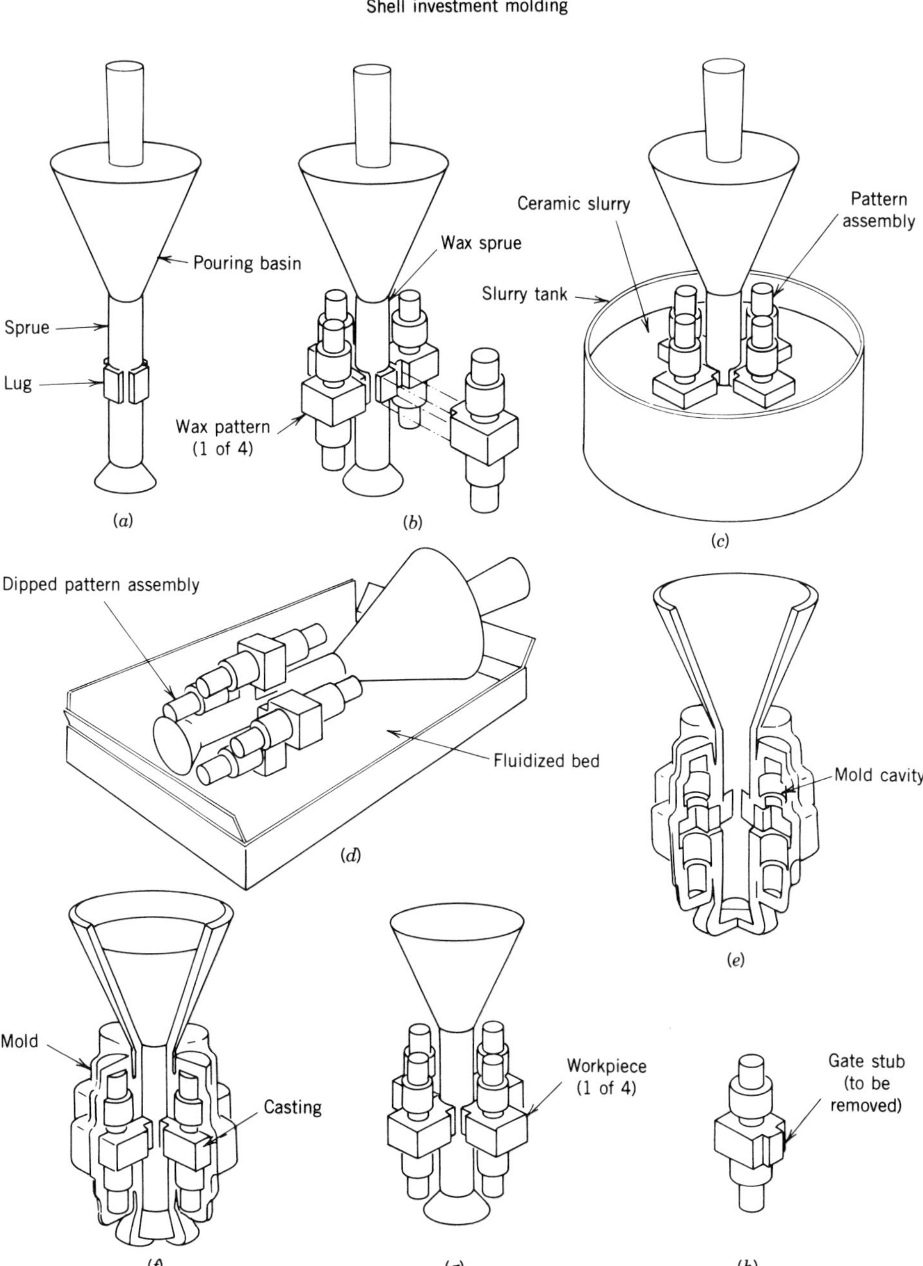

Figure 30
Steps in the production of a casting by the investment shell molding process, using a wax pattern. (*a*) Wax sprue pattern. (*b*) Pattern assembly (wax patterns attached to wax sprue). (*c*) Pattern assembly dipped in ceramic slurry. (*d*) Pattern assembly stuccoed in fluidized bed. Dipping (*c*) and stuccoing (*d*) are repeated until required wall thickness of mold is produced. (*e*) Completed mold after wax pattern has been melted out (mold shown in pouring position). (*f*) Mold after pouring. (*g*) Solidified casting after mold has been broken away. (*h*) One of four castings after removal from sprue. (*Metals Handbook*, Vol. 5, 8th ed., American Society for Metals, 1970, p. 239. With permission.)

furnace to 1600° F (871° C) and then placed in a flask where a heavy mold material is poured around it for additional support. The molten metal can be put into the mold by gravity, pressure, or vacuum.

This method can be automated to some extent and is widely used in the automotive and aerospace industries. Some of the advantages of investment shell casting are:

1. Unusual and nonsymmetrical shapes that would not allow withdrawal of an ordinary pattern are easily produced.
2. Smooth surfaces and high accuracy can be obtained. Machining time can be reduced or eliminated.
3. Nonmachinable alloys can be cast with high precision.

Most parts made by this process are limited to a small size and there are other limitations such as in the use of holes and cavities. The process is more involved than some other casting processes, such as permanent molding and die casting, and is therefore more expensive per part.

The Shaw Process

In this process, a rubbery jelling agent and a slurry of refractory aggregate is poured over the pattern. This rubbery mold hardens sufficiently to be stripped off the pattern and will return to the exact shape of the pattern. The mold is ignited to burn off the volatile elements, and it is then placed in a furnace and brought to a high temperature. The mold is then ready for pouring. The greatest advantage of this mold over plaster molds is its permeability and good collapsibility, which allow for production of more delicate and intricate shapes with fine detail and higher quality castings.

DIE CASTING

Die casting is similar to permanent molding in that a metal mold made in two halves is used. The difference is that the metal is not gravity poured into the mold (die), but instead the metal is injected under high pressures from 1000 to 100,000 PSI. This requires massive machines which are generally operated hydraulically to exert the hundreds of tons of force necessary to hold the two halves of the die together when the molten metal is being injected (Figure 31).

The dies are usually made of alloy or tool steel and are quite expensive to make. Some have one or two identical mold cavities for larger parts and others may have several cavities. Some dies are more complicated and have sections that move in several directions. Grooves or overflows around the cavity on the parting face provide for gases to escape. However, overflows of excess metal must be trimmed off after the casting is removed from the mold by a secondary operation. This is done with trimming dies that also remove the sprues and runners. The mold must also have provisions for water cooling so a constant operating temperature can be maintained. Knock-out pins provide for ejection of the part when the die is opened (Figures 32 to 34). When cores are used, they are made of metal and are usually drawn out before the die is opened. Cores are retracted, either in a straight line or in a circular motion.

It can be easily seen that these complex dies are quite expensive. Their costs can range from $5000 to more

Figure 31
Die casting machine (TVT Die Casting & Mfg. Inc.).

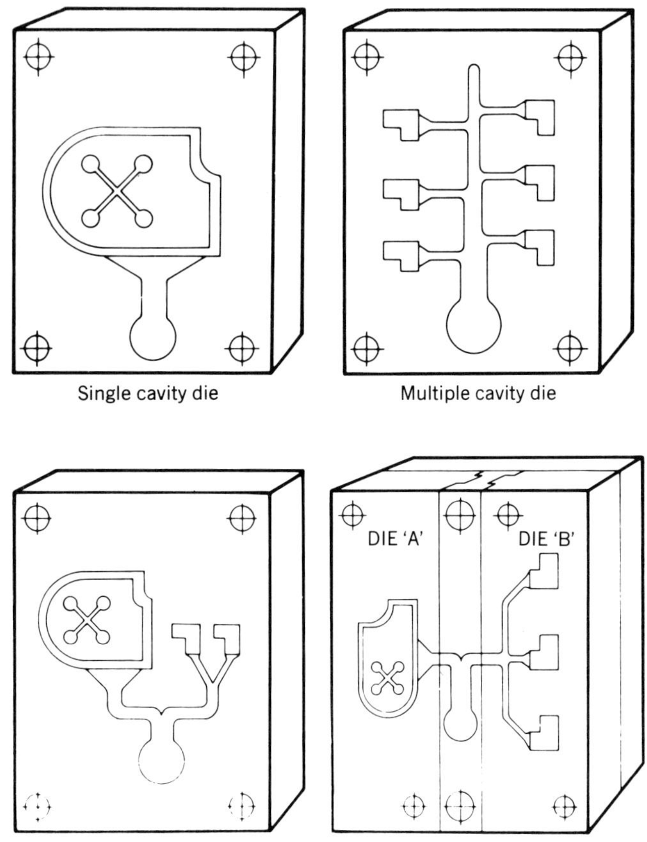

Figure 32
Die casting molds showing four different types of die cavities (Courtesy of American Die Casting Institute).

Figure 33
Part of a die showing mold cavity at center, raw material on the left, and product at the right (TVT Die Casting & Mfg., Inc.).

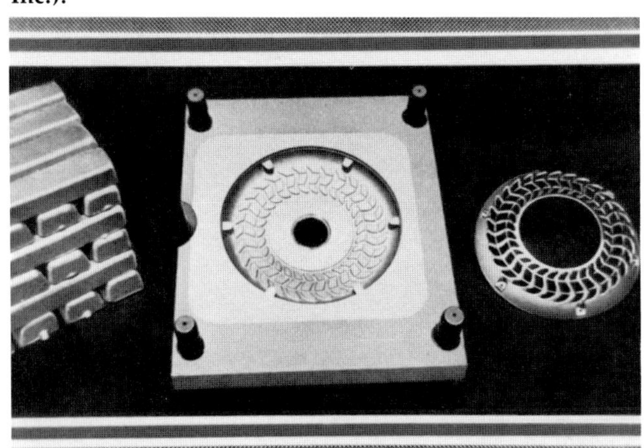

Figure 34
Some of the great variety of parts made by die casting (TVT Die Casting & Mfg., Inc.).

than $10,000, depending on complexity and size. Obviously, this process is not suitable for small quantity production or for making very large parts. The large investment for dies and machines will pay off only when very large quantities (20,000 to millions of parts per year) are required.

Die casting is a means of producing castings of lower temperature alloys at a relatively high rate. These castings are usually thin walled, smooth, and highly accurate. The process is highly adaptable to the manufacture of small parts, such as automobile door handles, carburetors, wiper motors, kitchen appliance parts, and thousands of small items we use every day.

The quality of die castings is high because of a rapid cooling rate that produces fine grains in the metal. The surfaces tend to be harder than the interior as a result of the chilling actions of the metal die. Porosity is sometimes a problem as a result of entrapped air, but with proper venting this can be overcome.

The most common die cast alloy is zinc based. Others are brass and bronze, aluminum, and magnesium. Zinc alloys have the lowest melting point, about 700° F (371° C), and so have a less destructive effect on dies. Aluminum and magnesium alloys melt at about 1100° F (593° C) and brass and bronze melt at about 1700° F (927° C). Therefore the dies using these alloys have a shorter life because of thermal shock which causes crazing (microcracking) on the die surfaces.

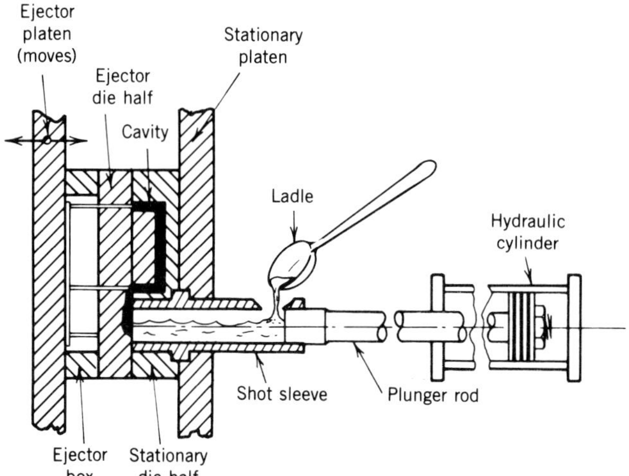

Figure 35
Cold-chamber machine. Diagram illustrates die, cold chamber, and horizontal ram or plunger in charging position. (Courtesy of American Die Casting Institute).

Die Casting Machines

The two basic systems and machines used for die casting are the **cold-chamber** machines and the **hot-chamber** machines. Cold-chamber machines are more commonly used and utilize the principle shown in Figure 35. Cold-chamber machines have a separate melting and holding furnace. For each casting cycle, the molten metal is fed into the cold chamber and then is forced into the die by the plunger (Figure 36). These machines are often used for casting the higher temperature metals because the chamber remains relatively cool and there is less tendency for spalling or cracking than in the hot-chamber machine. Although the cycle time is somewhat slower than that of the hot-chamber machine, the productivity is still high, as much as 100 cycles per minute.

A typical hot-chamber machine with a submerged gooseneck injector and plunger is illustrated in Figure 37. This type has an oil-fired or gas-fired furnace with a cast iron pot for melting and holding the metal. The plunger and cylinder are submerged in the molten metal that is forced through the gooseneck and nozzle into the die at each cycle. This type of die casting machine

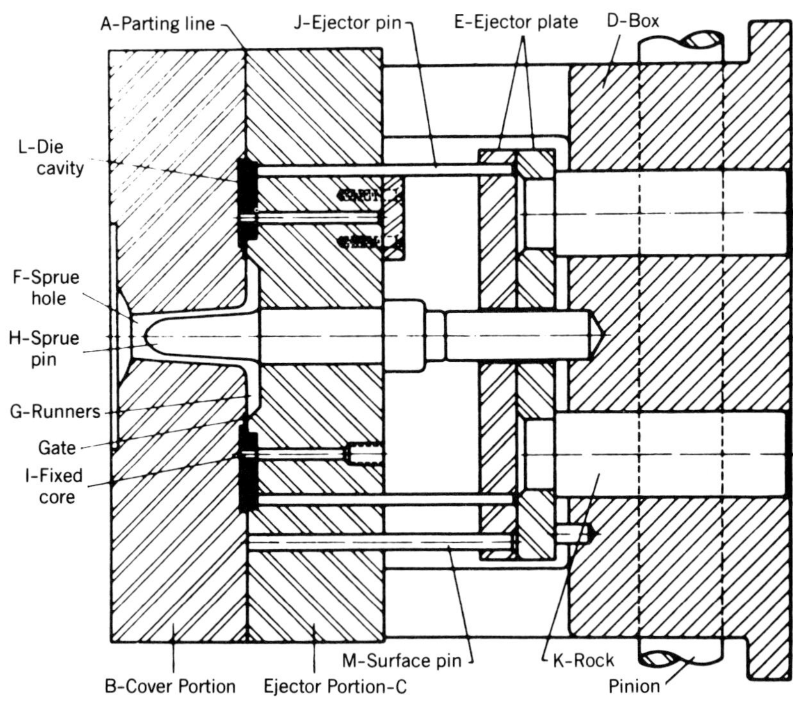

Figure 36
Typical die (Courtesy of American Die Casting Institute).

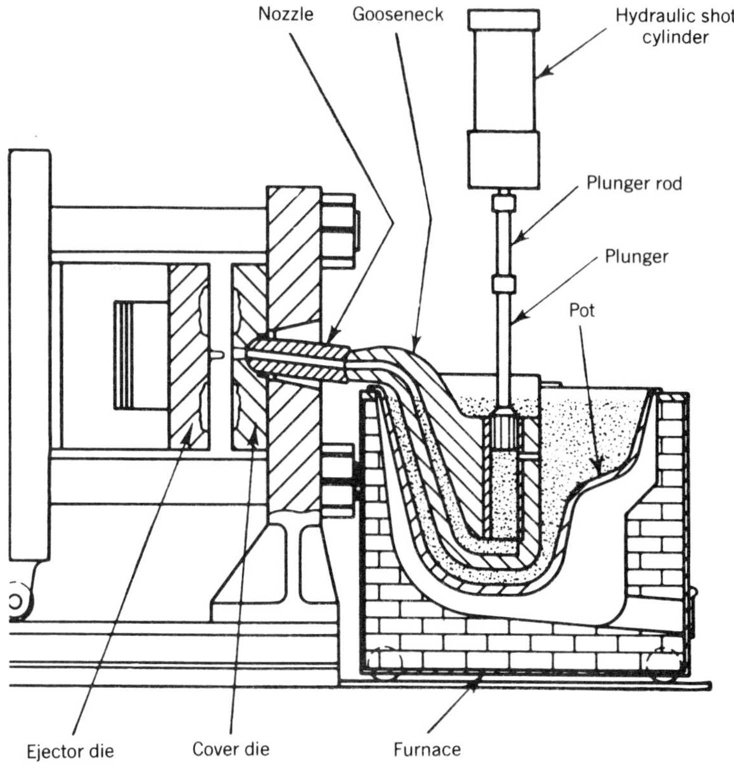

Figure 37
Hot-chamber machine. Diagram illustrates plunger that is submerged in molten metal, gooseneck, nozzle, and dies. Modern machines are hydraulically operated and are equipped with automatic cycling controls and safety devices (Courtesy of American Die Casting Institute).

can only be used for the low melting point zinc-based and tin-based alloys. This machine is much faster in operation than the cold-chamber machine.

Several problems and difficulties have plagued the die casting process, especially when attempts were made to speed up or fully automate die casting systems. The major problems were **flash** (Figure 38), sticking, and soldering (molten metal adhering to the die). Many kinds of automatic unloading and loading devices were developed and die sprayers were used to control these problems with mixed results. Other problems such as surface defects, dimensional inaccuracies, and internal porosity were addressed. Increasing the injection plunger velocity and reducing the total filling time was found to reduce these problems.

Figure 38
Die cast parts (clutch arm) showing flash on the left and flash removed on the right (Consolidated Metco, Inc.).

FURNACES AND HOT METAL HANDLING

The **cupola furnace** (Figure 39) has been of primary importance in the melting of cast irons for foundry work. Coke (produced from coal) is used for producing heat in the cupola furnace. The cupola furnace is a circular steel shell lined with a refractory material such as firebrick. It is equipped with a blower, air duct, and wind box with tuyeres for admitting air into the cupola. A sand bottom keeps the molten metal from burning through and is sloped so the iron will run out when the tap hole is opened. The process is begun by starting a wood fire on the sand bottom and placing coke on top. When the coke bed is thoroughly ignited and is about two to four feet thick, iron scraps or pigs are added and alternated with coke plus small amounts of limestone for a flux. The blower is turned on and the iron begins to melt at the top of the coke bed. The

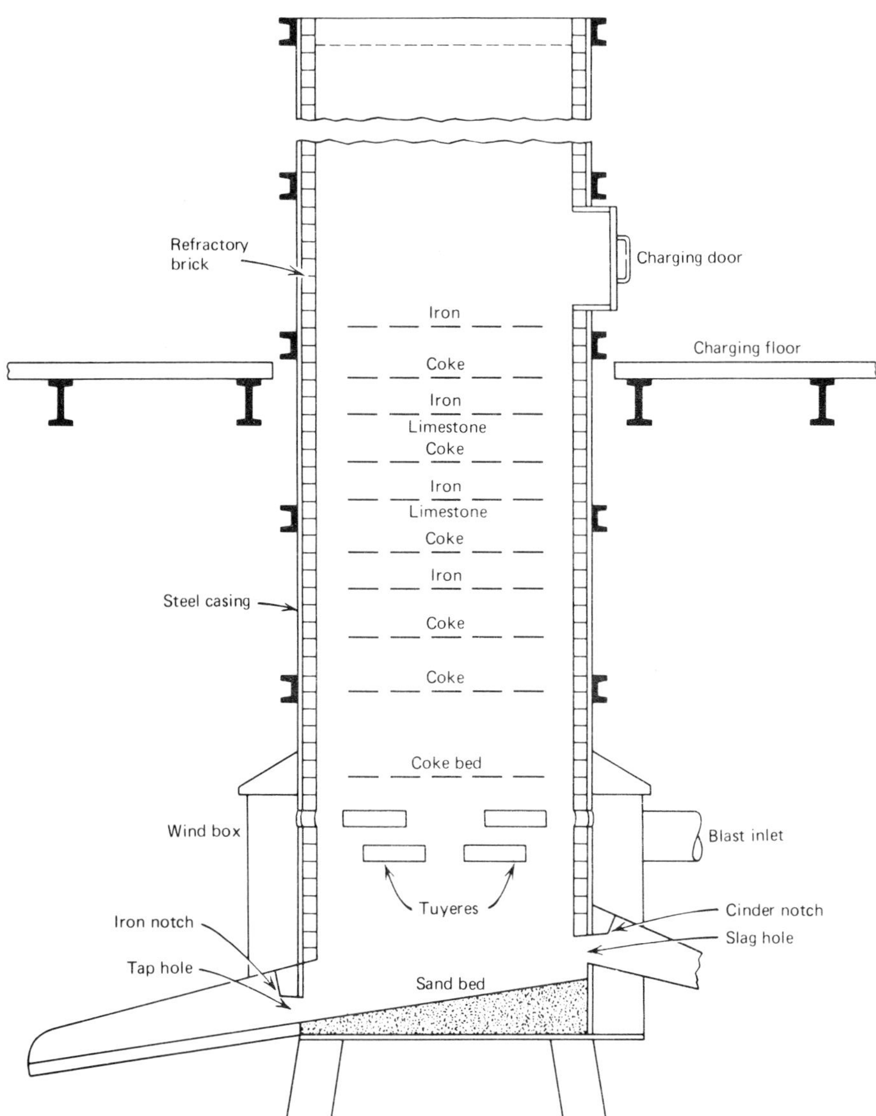

Figure 39
The cupola furnace (Neely/*Metallurgy* 2 Ed., © 1984 John Wiley & Sons, Inc.).

molten iron is drawn off into a ladle after which suitable metallurgical tests are made. Metallurgical control of the cast iron is made at the furnace before the ladle is moved to the casting floor. These tests reveal the amount of graphitization and other conditions. Certain alloy-rich materials or silicon to control graphitization are added in the ladle. Also, if nodular iron is desired, elements such as magnesium or cerium are added to the ladle to cause the graphite to form nodules (spheres) instead of flakes. Nodular iron is much tougher and stronger than gray cast iron which contains graphite flakes. When the tests are completed and the additions are made in the ladle, the molten metal is ready to be poured into the prepared molds.

When fossil fuels such as coke are burned, considerable smoke and fumes are released into the atmosphere. In areas where air pollution is a critical problem, these furnaces are fitted with dust collectors and other air pollution equipment; where cheap electricity is available, the cupola furnaces are replaced with electric furnaces. Electric furnaces for melting cast iron can be either direct arc, electric induction, or electric resistance types (Figures 40 to 42). Sometimes reverberatory or hearth furnaces are used for melting cast iron for foundry work, but they are less economical in that they have no regenerative system; that is, they have no air blast preheaters to utilize the hot air from the furnace, unlike most modern cupola furnaces.

Figure 40
Molten metal being poured from an induction furnace into a ladle. The furnace is tilted to make the pour (Inductotherm Corp.).

Figure 42
Front view of 250KW, 1KHz phasor and left swing furnace (Courtesy of Ajax Magnethermic Corp., Warren, Ohio).

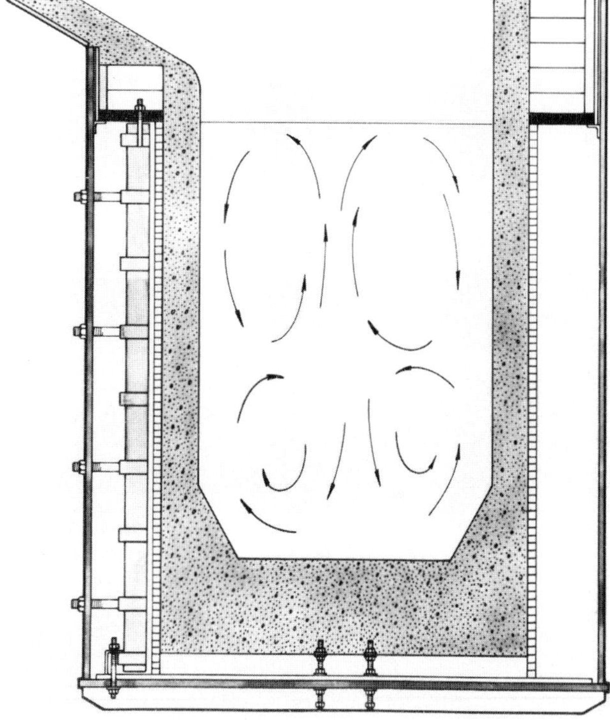

Figure 41
Cross section of a 60-cycle coreless furnace with a rammed lining (Courtesy of Ajax Magnethermic Corp., Warren, Ohio).

Pouring Practice

Ladles are usually constructed of steel and lined with firebrick or other refractory material (Figure 43). The interior of the ladle is heated or kept hot while in use so as not to cool the molten metal. In small foundries, a hand-held, shank-type ladle is often used to pour small quantities (Figure 44). This type requires two persons. The teapot ladle can contain more metal and is supported on an overhead monorail or crane. The handwheel is turned to tilt the ladle and make the pour. This and the bottom pour types keep the slag and oxidized metal from going into the mold. In some operations the molds are placed on a pouring floor, whereas other operations are automatic or semiautomatic, the molds being carried along a conveyor to the ladles where a measured amount of molten metal is poured into the mold.

Casting Clean-up

After the solidified castings are removed from the mold, they are cleaned in a shake-out and small parts are often put in a tumbler. The risers and gates must be knocked off (often this can be done on gray cast iron because of its brittleness) or cut off by means of sawing or use of an abrasive cut-off wheel.

Figure 43
Tea pot ladle.

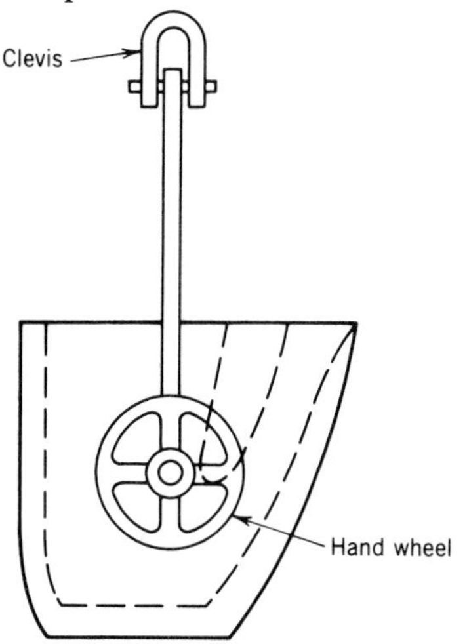

Figure 44
Shank-type ladle.

CASTING DESIGN AND PROBLEMS

Since there are many factors that influence the quality and performance of a casting, careful attention must be given to several design requirements if the best results are to be obtained. The more complex castings require more considerations than the simple ones do. The drafter or engineer should understand the problems involved or confer with the foundry for suggested alterations. Sufficient allowance of metal should always be provided when machining operations are to be performed. Size of the casting, roughness, and thickness of surface hardness are involved in this decision. The location of the parting plane is also very important since the pattern must be extracted at this plane without disrupting the sand mold.

Other factors to be considered are:

1. Weight of the casting and mold strength.
2. Effective gating and sufficient riser.
3. Number of cores and their placement.
4. Required dimensional accuracy.
5. Radii, thickness of sections, and amount of shrinkage.

Cracking can occur at the sharp corners and where thick sections join with thin sections (Figure 45). Proper radii can reduce this problem. Also, the cooling rates are greater for the thin section than for the thick one; this also causes cracking at the juncture. Cooling rates may be increased in a thick section by using "chills," metal sections embedded in the mold to absorb heat (Figure 46). The shape of casting should be designed to have as uniform thickness as possible, but in many products this is not possible. Parts that have ribs or spokes are subject to this kind of cracking, but often the ribs can be staggered (Figure 47). When parts intersect, such as in a 90 degree turn, interior corners (fillets) should be provided, but if they are too generous, hot spots may develop causing shrinkage areas (Figure 48). This can cause porosity or shrinkage voids. Holes provided by cores can be used to eliminate these weak points in a casting. If risers are inadequate to supply molten metal when the casting is shrinking during solidification, shrinkage cavities can result (Figure 49).

Figure 45
Casting problem: thin to thick section. The spokes, being smaller, cool more rapidly than the thick rim, causing them to break.

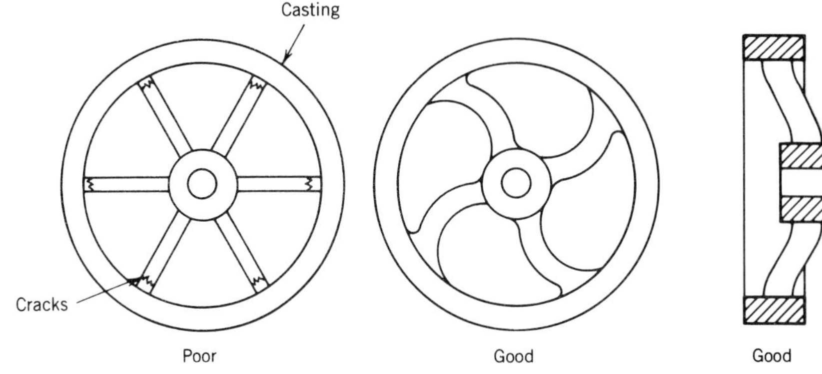

Figure 47
Staggered ribs.

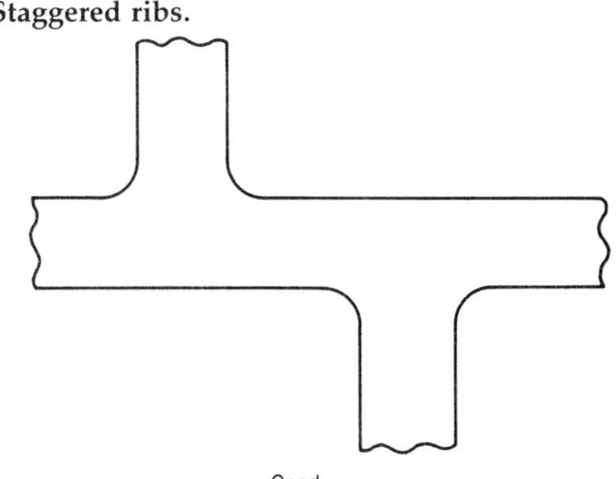

Figure 46
Chill in a mold around the rim of a wheel. This equalizes the cooling rate between the thick rim and the thin spoke. Odd numbers of spokes are often used to reduce cracking.

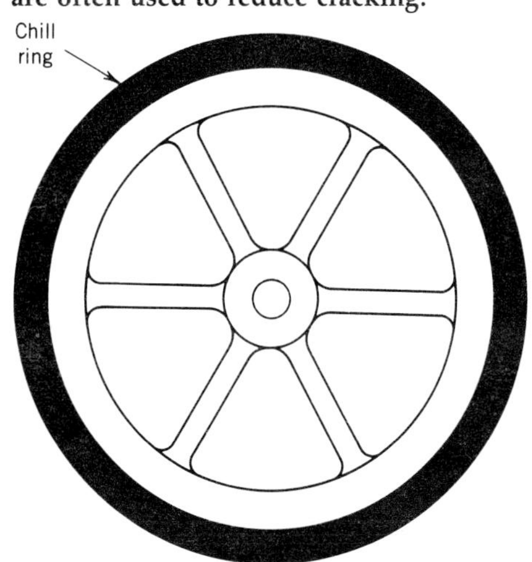

Figure 48
Shrinkage areas, hot spots cause voids.

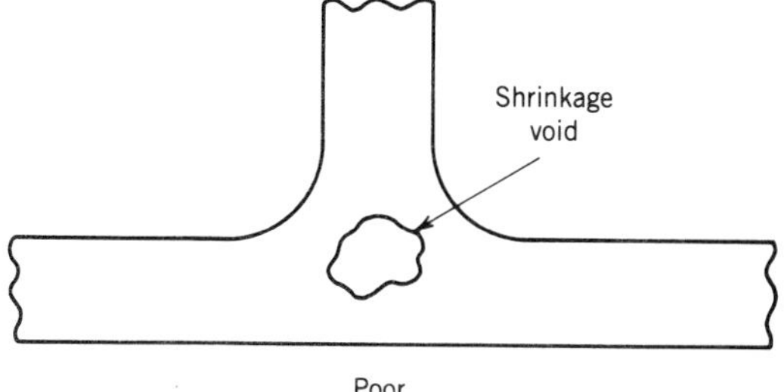

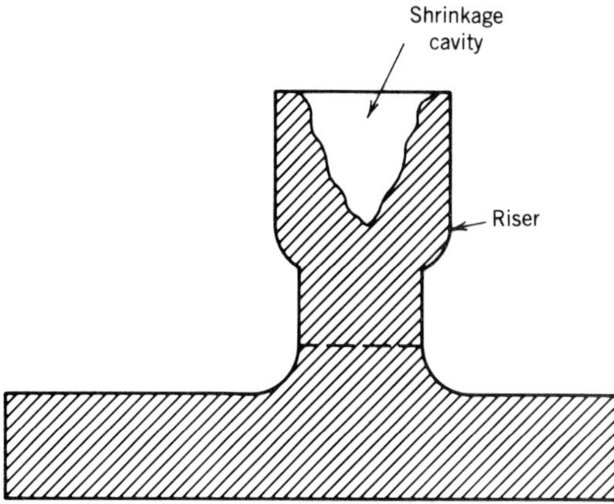

Figure 49
Shrinkage cavities.

REVIEW QUESTIONS

1. When precision surfaces are required on a casting, which one of the several types of castings is most likely to need a machining operation subsequent to casting? Which two types are least likely to need subsequent machining?
2. What are patterns for sand casting made of? What material is used to make patterns for shell molds? Which kind of pattern volatilizes when the metal is poured?
3. What characteristic of sand casting patterns is most necessary for removal of patterns?
4. What is shrink rate and what has this to do with patternmaking?
5. What are cores and of what materials are they made?
6. If you had to choose between a steel casting and a gray cast iron casting of a crane hook, which would you choose? Why?
7. A gear housing for a tractor is to be sand cast. Which metal would you choose for this casting, gray cast iron or steel? Why?
8. You are required to manufacture 1000 simple, small parts that are to be cast in aluminum. Would using a permanent mold be less expensive per part than preparing 1000 sand molds? There is no time limit. Explain your reason.
9. What is slush casting and what is it used for?
10. Name a relatively rapid method of casting pipe and tubing.
11. In which type of casting is the pattern removed by heating and melting it? List three advantages in using this casting process.
12. Name the most rapid method of producing small cast parts. Is this system practical to use for only 20 parts? For 100? Explain the reasons for your answer.
13. Which fuel does the cupola furnace use? What metal is usually melted in it? Name one other type of furnace used for the same purpose.
14. How does the molten metal get from the furnace to the mold?
15. What happens to the sand castings after they have solidified in the mold and cooled sufficiently to handle?
16. A cast iron casting has failed in use by breaking at a T joint where one leg meets the other at the joint. On inspection, a large shrinkage cavity is found in the joint which obviously caused the failure. How could this have been prevented?
17. In a sand casting, the top edge near the riser is full of impurities and porosity; also, a large shrinkage cavity is present. What could be the cause and how can it be corrected?
18. What is the major difference between a hot-chamber and a cold-chamber die casting machine? Which one is used mostly for low-temperature-melting alloys?

CASE PROBLEMS

Case 1
Hydraulic Cylinder Head Problem

A design for a metal cylinder calls for a metal cylinder head (Figure 50) that must withstand considerable forces. Zinc-based alloys, aluminum, and magnesium cannot be used because they are not sufficiently strong for the size limitations of the design. Cast iron, bronze, or steel may be used since they all have sufficiently high tensile strengths. These metals vary somewhat in cost and in ease of casting. The least expensive material and process to be used to manufacture these cylinder heads of about 100 parts per 8 hours, for a total run of 1000 heads, must be selected. Only 20 days are given to produce them. The options are:

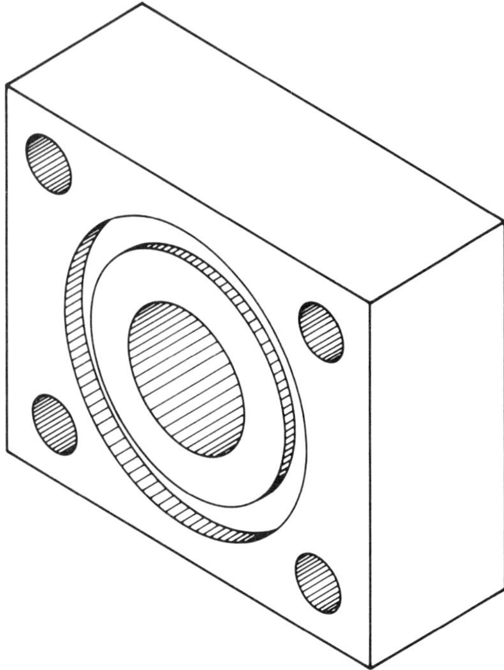

Figure 50
Metal cylinder head.

1. Machine all holes, cavities, and surfaces from solid bars of steel. These come in 20-ft-long bars that are sawn off in suitable lengths for machining.
2. Make sand castings of steel and finish by machining.
3. Make sand castings of cast iron and finish some surfaces by machining.
4. Cast the parts in cast iron by investment casting.
5. Use permanent molds and cast them in bronze.
6. Use the process of die casting with bronze and eliminate machining altogether.

Which of these six methods would produce the least costly parts and get the job done on time?

Case 2
Manufacturing a Small Bronze Gear

A small bronze gear must be made at a high production rate. Machining is ruled out as being too slow, and although extrusion (forcing metal through a die) of these gears could be a competitive option, it is also ruled out because no extrusion machine is available to the manufacturer. The order requires 10,000 gears to be delivered within 30 days. It will take about 18 to 20 days to make either sand casting patterns, permanent molds, or die casting dies. This leaves 10 to 12 days for production.

This means that about 1000 parts per day must be manufactured, allowing two days for packaging and shipping. The manufacturer will receive payment of $150,000 for the complete order. Which casting method should be used?

CHAPTER 8 PROCESSING OF METALS: HOT WORKING

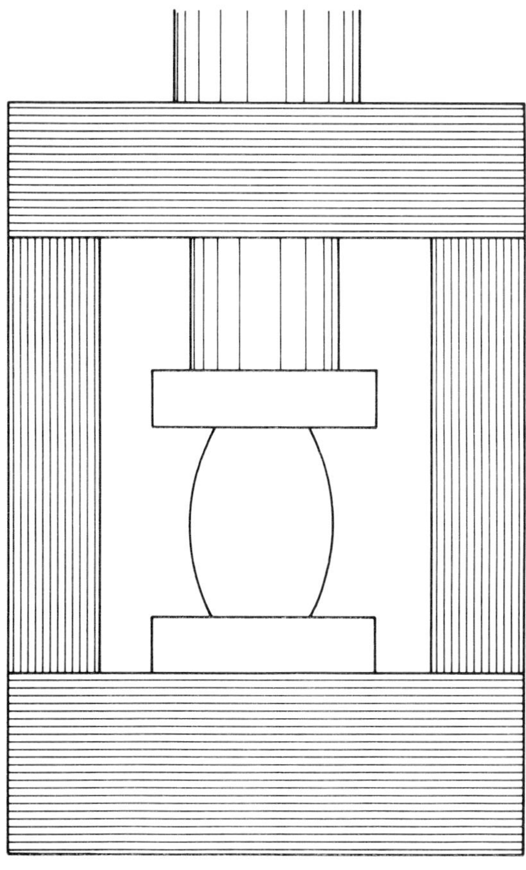

In earlier times, wrought iron and steel were forged by shaping a piece of red hot metal on an anvil by pounding on it with a hammer. The hammering process not only produced a desired shape, it also improved the grain structure of the metal and therefore increased its strength and durability. When the blacksmith needed to make a round bar from a square one, a half-round die block, called a swage, was inserted in a square hole in the anvil and an upper half-round die was held in the hand by a handle. The square bar was heated red hot in a forge and placed in the lower die while the upper one was hammered down on the bar. This process was repeated all along the bar, thus producing a length of round stock. This is roughly the process used for hot forging various shapes today, using upper and lower dies and powerful presses. Of course, steel bars of various useful shapes and dimensions are usually rolled in a steel mill. However, larger sections and other shapes are often forged. Modern forging processes account for a large part of basic metalworking and often a forged or rolled shape is subjected to a secondary process of metalworking such as stamping or machining.

OBJECTIVES

This chapter should enable you to:

1. Explain how molten steel is formed into industrial shapes.
2. Show the significance of recrystallization and grain structure of hot rolling and forging.
3. Describe several methods of hot forming metals and explain their advantages.
4. Describe two methods of hot forming pipe and tubing.

HOT ROLLING

In Chapter 2 you learned how raw materials, ores,

Figure 1
Powerful tongs lift an ingot from the soaking pit where it was thoroughly heated to the rolling temperature (American Iron & Steel Institute).

Figure 2
(*a*) Blooms, (*b*) slabs, and (*c*) billets are formed from ingots.

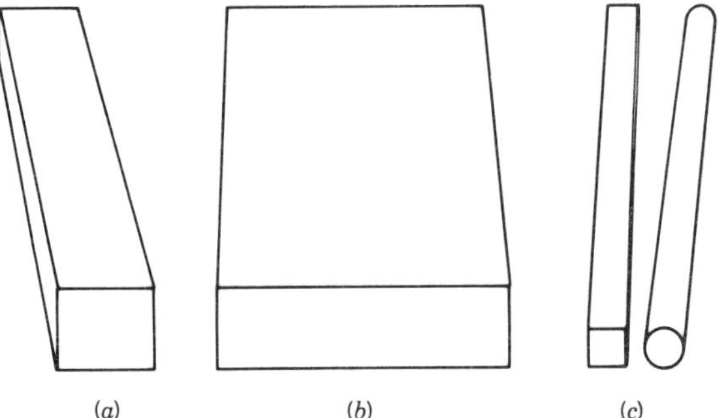

coke, and limestone are used to extract pig iron and how steel is produced in furnaces. This molten steel must be formed into useful shapes in the steel mill; this process begins in two ways. In the conventional process by which much steel is still processed, the molten metal is teemed or poured into cast iron ingot molds. These molds are then moved to the stripper where they are removed from the now solidified ingot. Before the ingot can be rolled, it must be reheated uniformly to about 2200° F (1204° C) throughout. This is done in a soaking pit (Figure 1). The soaking period requires from four to eight hours.

Figure 3
Reversing two-high roughing mill (American Iron & Steel Institute).

Figure 4
Operator at console of reversing two-high roughing mill (Courtesy of Bethlehem Steel Corporation).

Figure 5
Diagram of sequences in rolling hot ingot into a slab on a reversing two-high rolling mill (American Iron & Steel Institute).

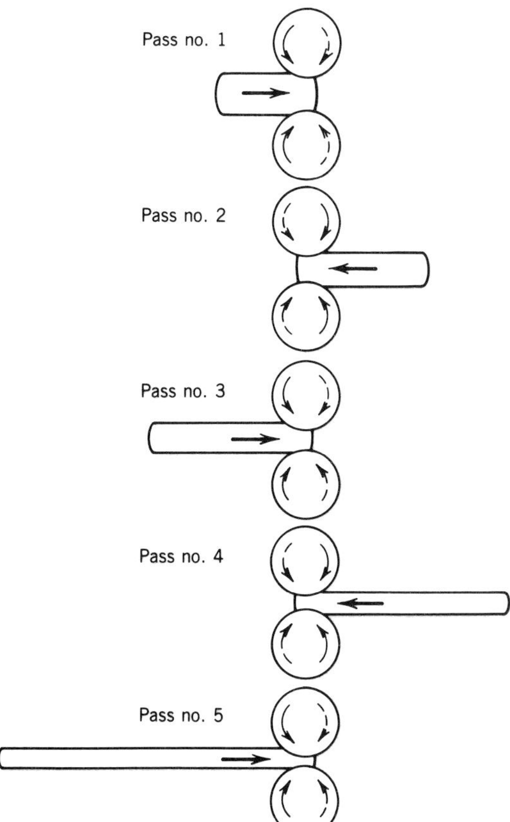

Figure 6
Structural steel and rails are rolled from blooms. Standard shapes are produced on mills equipped with grooved rolls. Wide flange sections are rolled on mills that have ungrooved horizontal and vertical rolls (American Iron & Steel Institute).

Rolling Mills

The heated ingot is sufficiently plastic to be taken to the rolling mills to be shaped into workable forms of blooms, slabs, and billets (Figure 2). This is done on roughing mills. Blooms and slabs are most frequently rolled on a reversing two-high rolling mill (Figures 3 to 5).

Blooms are forms of semifinished steels that are either square or rectangular in cross section. They are further processed in structural rolling mills (Figure 6) into railroad rails or shapes such as I beams, angles, wide flange beams, zees, tees, and H-piles. They are made of plain low carbon steel which can be drilled, heated, or welded without such complications as work hardening, embrittlement, or cracking. Blooms are sometimes further reduced in size to billets on billet mills (Figure 7) before being processed into finished products such as bar and rod stock. Bar and rod (from which wire is made) can be made of either low carbon steel or higher carbon and alloy steel.

Figure 7
Steel billet emerges from 35-in. blooming mill. The billet has been rolled from an ingot. The product of this mill will eventually be made into bars. Dials on the two-stand mill indicate size of billet being rolled. Alloy and tool steel ingots are processed in this particular mill (Courtesy of Bethlehem Steel Corporation).

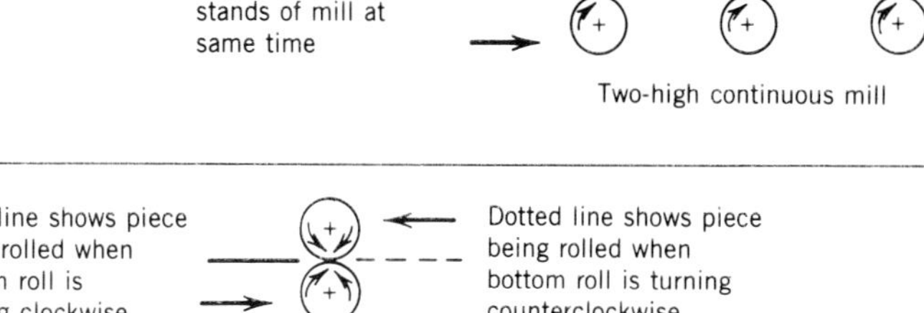

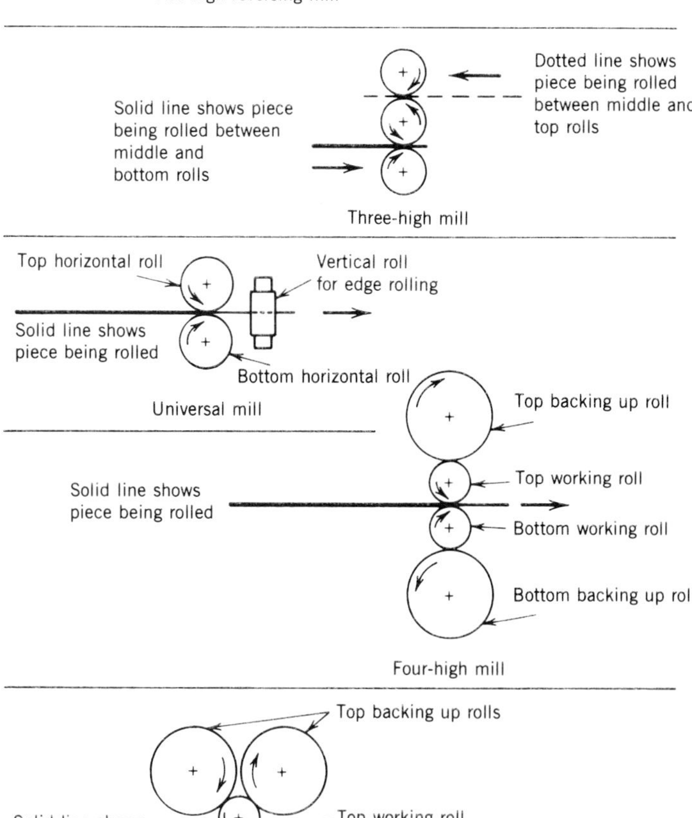

Figure 8
Diagrams demonstrating various arrangements of rolls and rolling procedures used in steel mills. Operating under extreme heat and pressure, rolls must frequently be changed, and a supply of new or refinished mill rolls is always maintained from the mill's own machine shop (American Iron & Steel Institute).

Figure 9
Hot saw cuts finished rolled shapes to customer length after delivery from the 48-in. finishing mill (Courtesy of Bethlehem Steel Corporation).

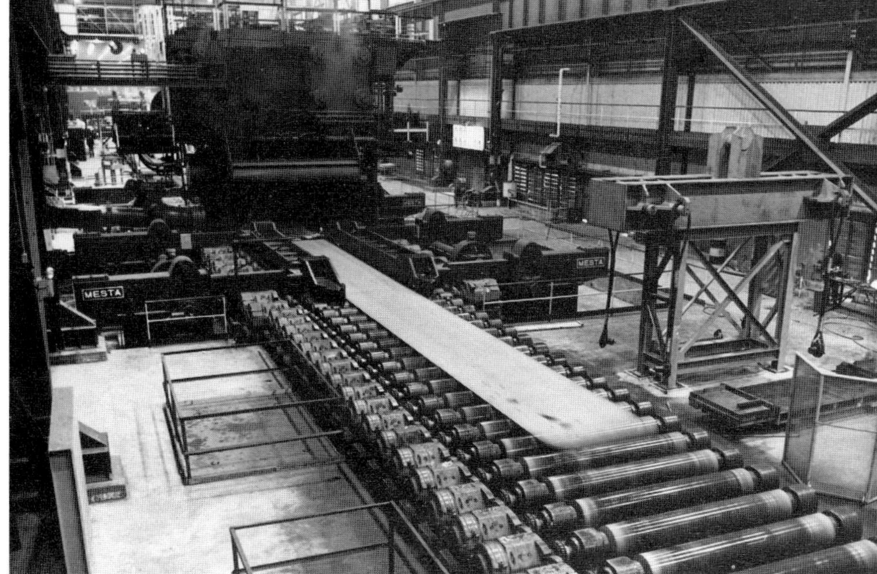

Figure 10
Plate enters the four-high finishing stand in the 110-in. sheared plate mill. The plate is rolled to desired length and thickness by being passed back and forth through the finishing stand rolls. The mill has AGC (automatic gage control) and is completely computerized (Courtesy of Bethlehem Steel Corporation).

As the ingot is carried through the massive rolls, the hot plastic steel is squeezed to a smaller thickness. The rolls are brought closer together on each pass, causing the ingot to get longer and thinner as it passes between the rolls (Figure 8). Rolled shapes become quite long and must be cut to standard customer-length (Figure 9). Finishing mills for plates, strip, and sheets are often continuous, having a number of mill stands, each adding to the final shape. Sometimes this is done on 3- or 4-high reversing mills, passing the hot steel forward between the lower rolls and then shifting the slab to the upper rolls to return back through the same mill stand (Figure 10).

All hot rolled steel forms a black or gray mill scale on the surface that is caused by oxidation of the red hot steel in the presence of oxygen in the air. The oxide scale helps protect the steel from corrosion after it has cooled.

Figure 11
Diagram of continuous casting (American Iron & Steel Institute).

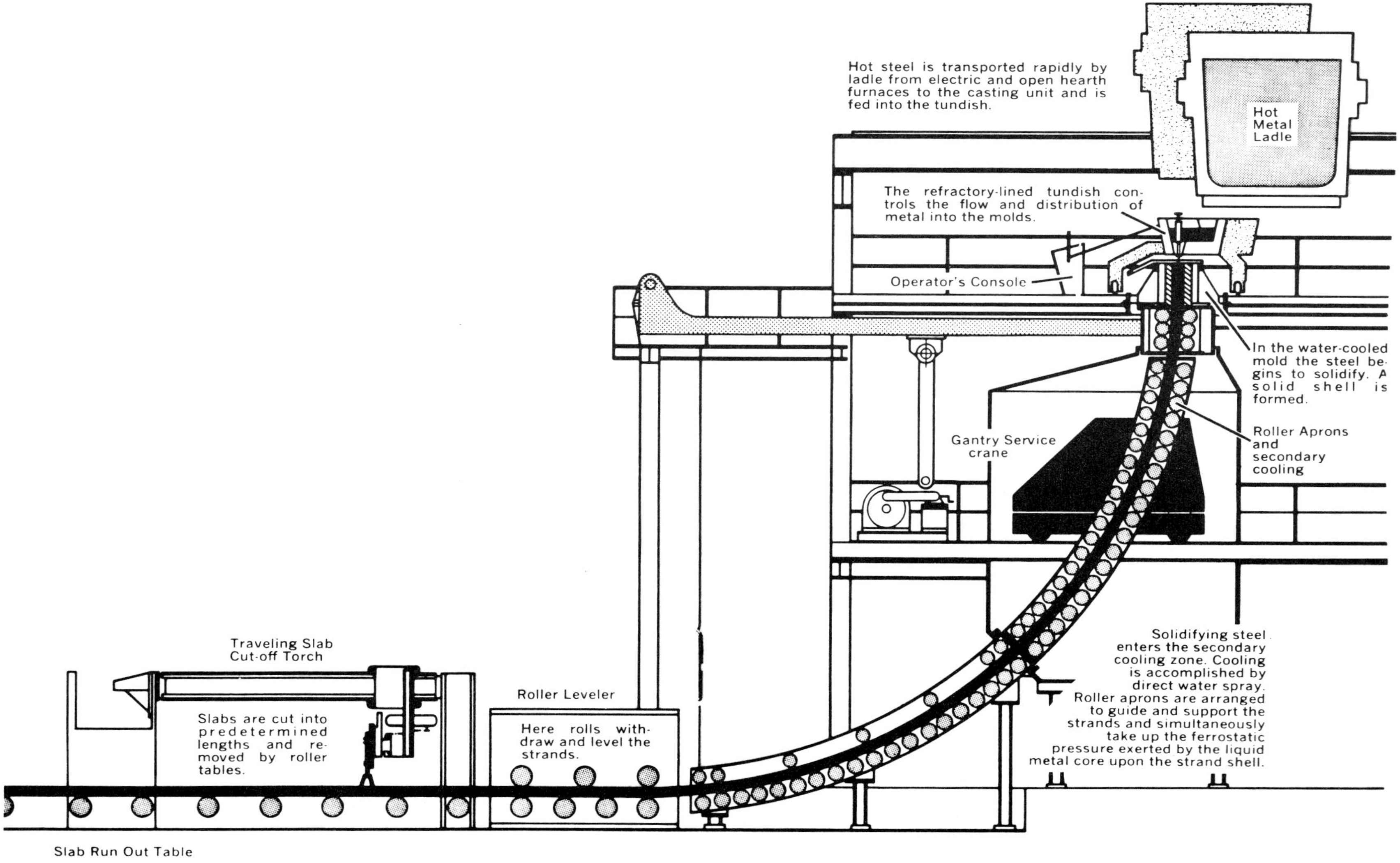

Figure 12
Molten steel is poured from a ladle into the tundish which regulates the flow of liquid steel into two water-cooled copper molds for the slabs to be continuously cast (Courtesy of Bethlehem Steel Corporation).

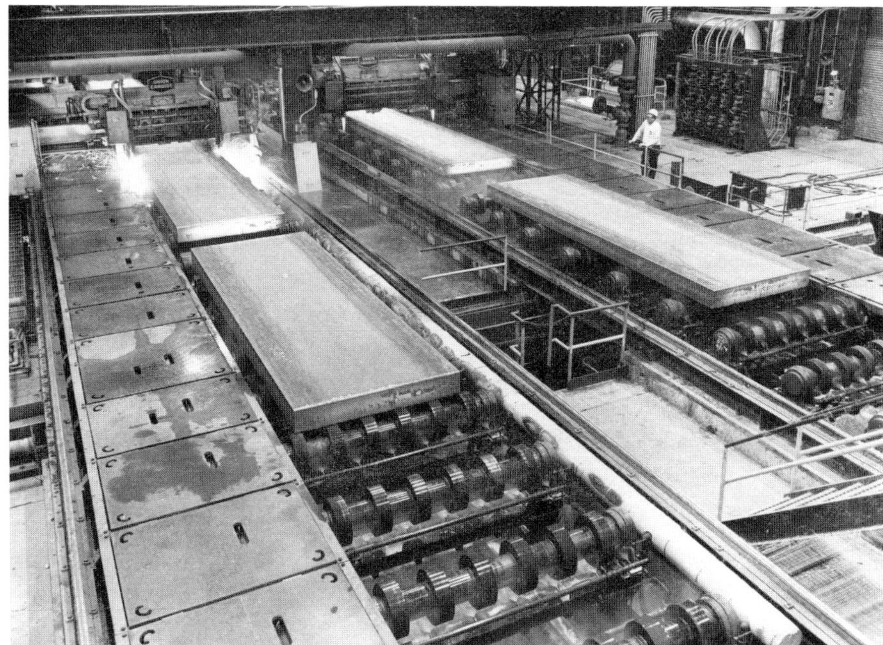

Figure 13
Perfectly formed slabs of steel are cut to required lengths by an automatic flame torch after passing through the continuous slab caster (Courtesy of Bethlehem Steel Corporation.

STRAND CASTING

A second procedure of steel processing that bypasses ingot teeming, stripping, soaking, and rolling in roughing mills is called strand casting (Figure 11). It is also called continuous casting, because the molten metal is continuously supplied to a tundish or reservoir (Figure 12) where it is fed through a water-cooled copper mold from which it emerges as a continuous ribbon of steel in the desired cross section as slabs or billets. Sprays of water under high pressure cool and harden the metal still further. The ribbon of metal is cut into suitable lengths by a traveling torch as the steel moves along on rolls toward the mill stands (Figure 13). In some strand casters, the descending column of metal is cut to the desired length while it is still in the vertical position, then the cut length falls forward onto the rolls and is carried away. Of course the metal is at a rolling or forging heat, just under the temperature of solidification. Both ingot and strand casting procedures are primarily designed for high-tonnage production. The continuous strand casting unit at Bethlehem's Burns Harbor, Indiana plant can convert 300 tons of molten steel to solid slabs in about 45 minutes, whereas a minimum of 12 hours is needed for conventional processing. Smaller batches are made in foundries in which cast steel, for example, can be further processed by forging.

There are some similarities in the two methods of casting metals for hot rolling. The continuous-cast ribbon of steel, like the ingot, cools from the outside toward the center, forming large, columnar, dendritic crystals (Figure 14). Impurities and gas pockets tend to migrate toward the center and to the still liquid area. These undesirable grains and gases can be removed or altered by the process of hot working—rolling, drawing, extruding, or forging. Also, the trapped impurities tend to elongate and flow along the steel as stringers or fibers in the metal in the hot working process. These fibers are usually beneficial, giving the steel more resistance to transverse (crosswise) cracking. Whether the slabs or billets are made by casting ingots or by the continuous-strand casting process, the subsequent rolling processes in mill stands remain the same.

Figure 14
Large dendritic crystals form in solidifying steel which must be reformed into smaller uniform grains by the rolling process.

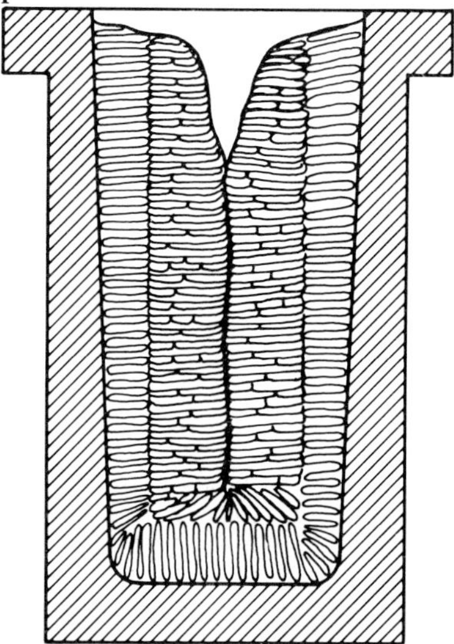

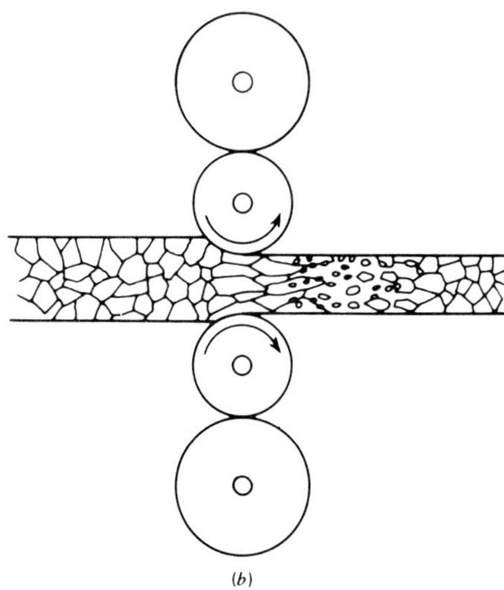

Figure 15
Recrystallization in hot rolling (Neely/*Metallurgy* 2 Ed., © 1984 John Wiley & Sons, Inc.).

RECRYSTALLIZATION

Ingot steel, with its typically coarse columnar grain structure, is quite unsatisfactory for any use where strength is required. A part made directly from ingot steel can shatter when a working force and shock loads are applied. The columnar grains in a cast ingot must be **recrystallized** to give the steel the needed strength. This is done by hot working, such as forging or rolling processes (Figure 15).

Factors that influence the grain size obtained from hot deformation are:

1. Initial grain size.
2. Amount of deformation.

Figure 16
This illustrates the fibrous quality of rolled steel, called anisotropy. Since the fibers are in the direction of rolling, the metal is stronger along one axis than in the other (Neely/*Metallurgy* 2 Ed., © 1984 John Wiley & Sons, Inc.).

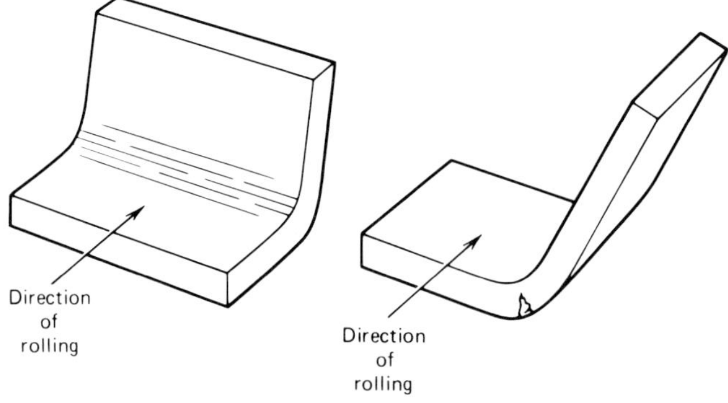

3. Finishing temperature.
4. Rate of cooling.

Because the metal is at a high temperature, the reformed crystals begin to grow again into large grains, but they are not as irregular and large as before. Now as the hot working progresses as the metal is cooling, each deformation results in smaller, more uniform grains that are flattened to some extent, giving the metal a condition called anisotropy or "grain"; that is, the metal is more ductile and can be deformed to a greater extent in one axis than the other without cracking or splitting (Figure 16).

This characteristic of anisotropy of rolled or hot- (and cold-) worked metals is very important in welding and cold-forming processes. For example, the metal can be bent to a smaller radius in the direction of rolling than it can at 90 degrees from it. Although steel ingots are not an acceptable form for making steel products until they are hot worked, steel castings contain alloying elements and are subjected to heat treatments that give them a good metallurgical structure so they do not require subsequent hot work to strengthen them.

FORGING PROCESSES

In ancient times, copper and other nonferrous metals were sufficiently soft and ductile so they could be shaped while cold by hammering. Iron and steel were not sufficiently ductile while cold to be hammered into a shape unless the piece was very small or thin. Heating metals to forging temperatures greatly increases their plasticity and workability. Therefore, hot forging was and still is a useful method of forming steel articles, both large and small.

Steel bars and sheets were repeatedly heated and hammered to produce armor, swords, and agricultural implements. It was in 1783 that Henry Cort of England built a rolling mill that had grooved rolls; this rolling mill made shaped bars of uniform cross section possible. As early as the thirteenth century, the tilt hammer came into use in which water power was used to raise or tilt a lever arm to which a weight was attached. When it fell, a heavy blow was delivered to the hot iron. In the mid-seventeenth century at the Saugus Iron Works in Massachusetts, a 500 pound weight on the end of a wooden beam was used as a forging hammer in which a cogged wheel driven by a waterwheel periodically raised and dropped the weight onto an anvil. Thus it was possible to make large forgings such as heavy ship's anchors (Figure 17).

Forging Machines

Today large forgings are produced in massive hydraulic presses (Figure 18). Smaller articles are forged by a variety of methods and machines. **Open-die drop forging** (smith forging), **impression-die drop forging, press forging, swaging,** and **upset forging** are the most commonly used types in manufacturing. When choosing a particular forging operation, there are several considerations to be made. Force and energy required, size of the forging, repeatability, speed and workability, and metallurgical structure of the part are major factors. However, the amount of time hot metal is in contact with the die is important to die life and is also a consideration when choosing a forging machine for a particular forging operation.

Figure 17
Woodcut showing a large wrought iron ship's anchor being forged in eighteenth century France. The arm and the shaft of the anchor are being forge welded together in this view (Dover Publications).

Figure 18
A large forging to be used in a large aircraft. Subsequent machining operations will be performed on it (Lockheed Corporation).

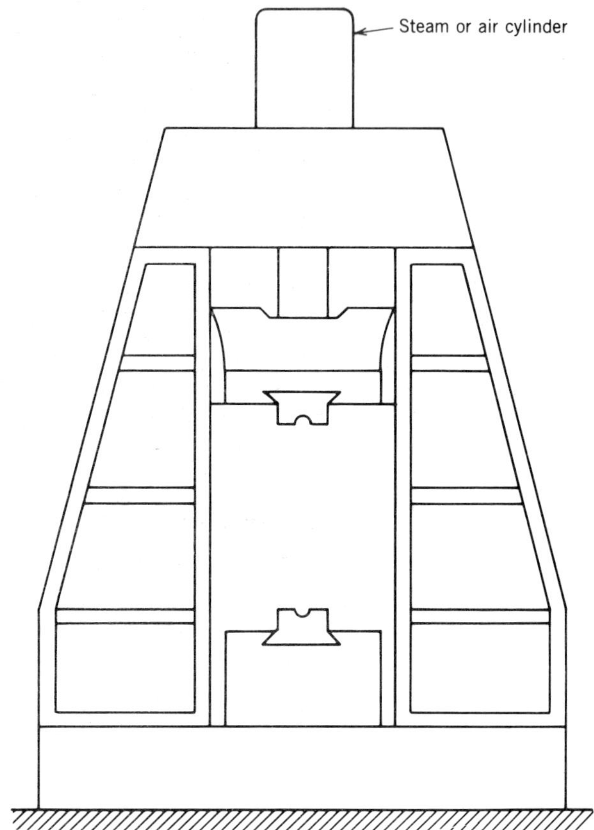

Figure 19
Drop forging hammer. Steam or air pressure is used to raise the hammer and force it down on the heated metal.

The Drop Hammer The drop hammer (Figure 19) is a development of the old tilt hammer. Drop hammers today are usually steam or air assisted to increase the force of the hammer blow and to raise the weight after each blow. Large articles such as machinery shafts that are to be finished by machining are usually open-die forged, using repeated blows while turning and moving the hot metal with tongs or a manipulator. This type of forging is essentially the same principle as that used in ancient times; the metal is squeezed or hammered to increase its length and decrease its cross section, or it is upset, that is, pressed on the end to enlarge it, increasing its diameter and shortening the length. There is a size limit for drop forgings, but, as mentioned earlier, very large forgings can be formed in hydraulic presses by the open-die method. In open-die forging, drop hammer operators develop a skill in giving the hammer just the right amount of force. When an upper and lower die are used, it is called impression-die forging. This method increases production rates and ensures repeatability. Often two or more progressive dies are used, each contributing to the final shape (Figure 20). Drop hammers require repetitive blows to bring the workpiece to size, whereas mechanical and hydraulic presses can often make a forging in one or two strokes. However, both types of forging machines are adaptable to automated forging processes.

Mechanical Presses These machines make use of a heavy flywheel that stores energy for the forging stroke. Mechanical presses (Figures 21 and 22) of this type have

Figure 20
Progressive dies used with a mechanical press. Here a round bar at forging heat will be moved into the die by means of mechanical fingers. At each stroke kick-outs move the partially formed forging to the next die. The bar is converted to two connecting rod forgings and the last die operation is that of removing the flash (Illustration courtesy of National Machinery Co.).

Figure 21
Automatic mechanical presses are high production forging machines (Illustration courtesy of National Machinery Co. Maxipres® is a registered trademark of National Machinery Co.).

Figure 22
Mechanical press with a cutaway view showing the driving mechanism. Maximum die used in a press depends on shut height, which is the vertical opening with the ram down and adjustment up (Photo courtesy of the Minster Machine Company).

Figure 23
Many years ago parts were made one at a time by an operator, often under dangerous conditions, unlike automatic press systems today (Graebener Press Systems, Inc., Providence, RI).

Figure 24
High energy forging press (Photo courtesy of Siempelkamp Corporation).

an eccentric shaft or crank that moves the ram down and back to the starting position when a clutch is actuated. The greatest pressure is exerted at the bottom of the stroke. One problem is the possibility of jamming the ram near bottom dead center, but most mechanical presses have a stall release that frees the stuck ram and upper die when this happens. These machines are much quieter in operation, are more accurately controlled, and can exert far more force and energy in each stroke than drop forging machines. For this reason, the mechanical press is often chosen when a part must be made to closer tolerances. During the past 25 years, there has been a gradual trend toward mechanical press forging. Fewer blows are required per forging and less operator skill is needed with these machines. Mechanical presses squeeze rather than impact the hot metal with a shorter forging-to-die contact time, thus extending die life.

A modern version of an old principle (Figure 23) is a high-energy type of mechanical forging press (Figure 24) that utilizes a vertical screw instead of a crank or eccentric shaft to transfer the flywheel energy to

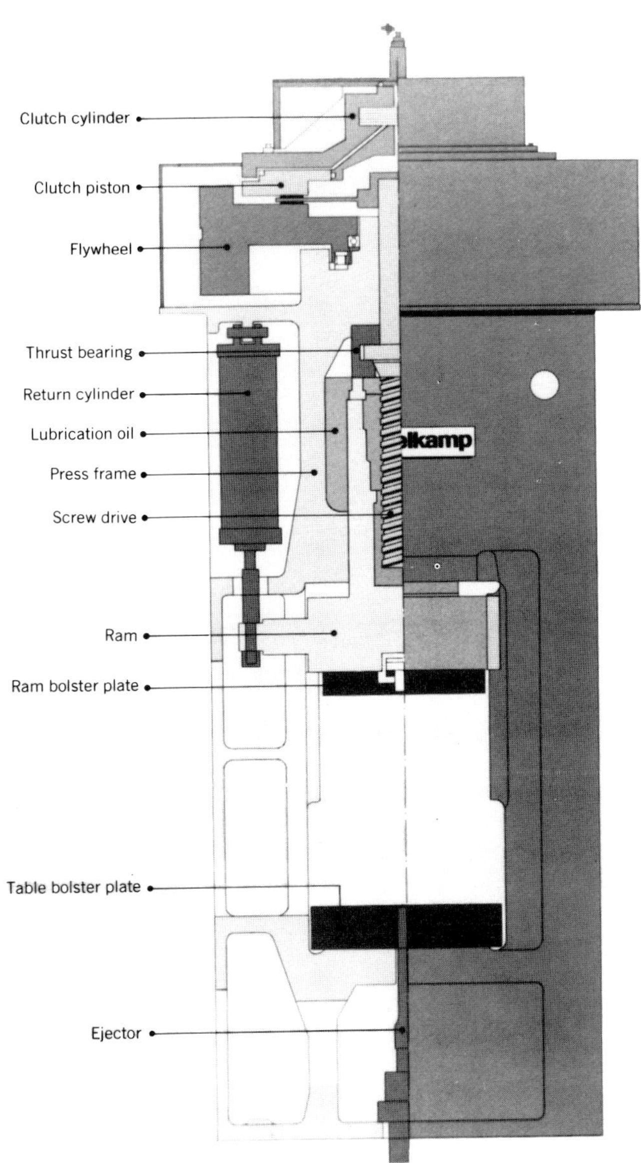

Figure 25
Drawing of the parts of a high energy screw press (Photo courtesy of Siempelkamp Corporation).

Figure 26
Preform in one-blow forging on a high energy press. (*a*) Preforging or descaling. (*b*) One-blow finish forging (Photo courtesy of Siempelkamp Corporation).

the ram. This arrangement provides uniform force throughout the stroke, unlike crank presses in which the major tonnage is confined to the end of the stroke (Figure 25). Like the drop hammer, the screw press does not have a bottom dead center. A continuously rotating flywheel imparts motion to a screw when a clutch is engaged. It appears that the screw press combines two of the best features of the drop hammer and the crank press, high impact and a squeezing action. Even with hard blow forging, the press frame and mechanical components are not subject to high stresses and the die is in contact with the hot metal a very short time. Often only one or two blows are needed to complete a single forging (Figure 26). High energy forging presses provide high flexibility since they can be used for either open or closed die forging operations.

Hydraulic Presses Hydraulic presses (Figure 27) are the slowest types, but their advantage is that they can exert very high tonnages for large forgings. Instead of mechanical linkages and cranks, the press ram is powered with one or more large hydraulic cylinders. Oil is pumped at high pressure into the cylinders and the ram moves down at a constant rate. Smaller cylinders

Figure 27
Large open-die hydraulic forging press with manipulator positioning the forging ("Forging Presses," Schirmer-Plate-Siempelkamp, Hydraulische Pressen GmbH, Krefeld, West Germany).

Figure 28
This huge forging press is used to dish heavy plate for pressure tank heads ("Pressing for All Applications in Metalforming," Schirmer-Plate-Siempelkamp, Hydraulische Pressen GmbH, Krefeld, West Germany).

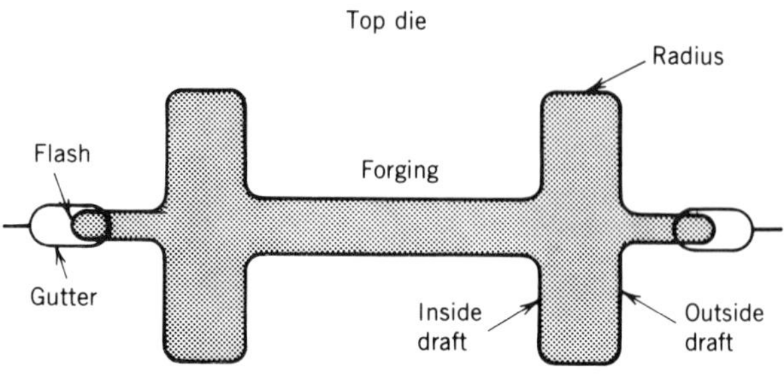

Figure 29
Die showing parting line and gutters for flash overflow.

Figure 30
Typical automatic progressive die set for use with a 2000-ton press. Guide pins at the rear align the upper half of the die. Parts at the front show the progressions in making the part; the last one to the right shows the sheared-off flash (Illustration courtesy of National Machinery Co.).

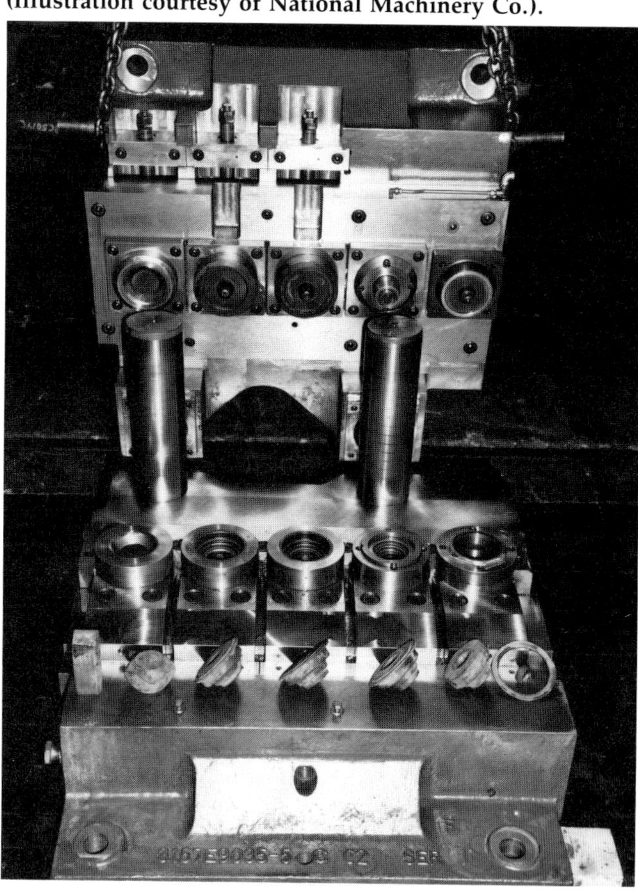

return the ram to its upper position. Large hydraulic presses can exert thousands of tons of pressure, making them ideal for some of the largest forgings (Figure 28). Virtually every metal can be hot forged within its forging temperature range.

Forging Dies

The design of forging dies is similar to that used for die casting. There is a parting line where the two halves of the die meet and gutters into which the overflow or flash can move. There must be adequate draft so the part can be removed from the upper and lower die halves without sticking. Generous corner fillets and corner radii are necessary to allow for a smooth flow of material (Figure 29). A forging die must be able to withstand extremely high temperatures while the red-

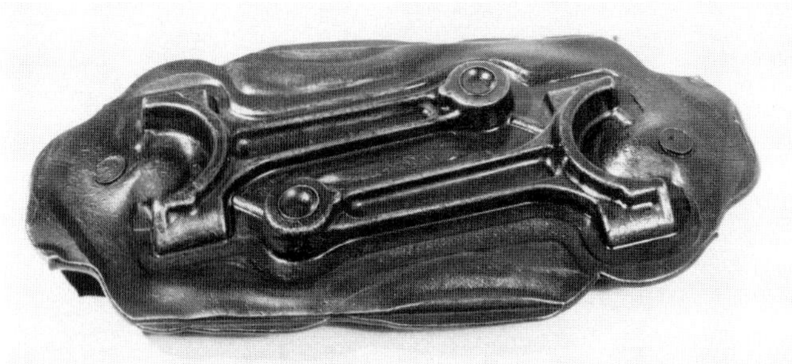

Figure 31
Forged connecting rods showing flash which must be removed, either in the automatic die set or as a secondary operation in a separate die (Illustration courtesy of (National Machinery Co.).

Figure 33
Hot crankshaft forging in a die after the last blow. It will now be removed from the die and the flash will be sheared off (Illustration courtesy of National Machinery Co.).

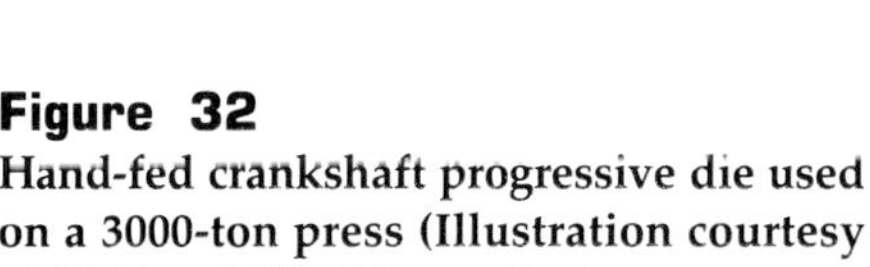

Figure 32
Hand-fed crankshaft progressive die used on a 3000-ton press (Illustration courtesy of National Machinery Co.).

Figure 34
Stages in the formation of a crankshaft by hot forging from bottom to top (Illustration courtesy of National Machinery Co.).

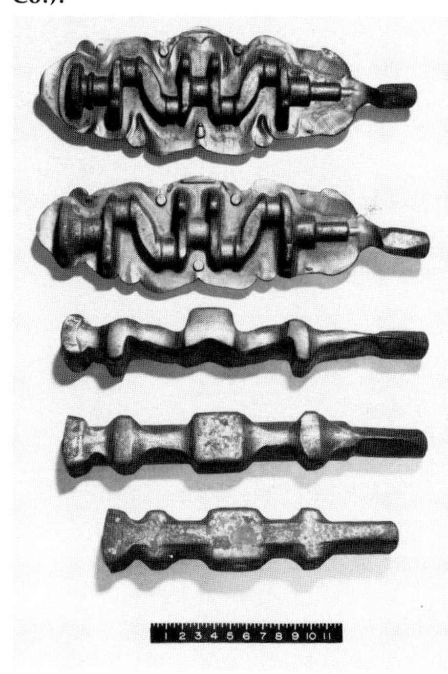

hot metal is in contact with it. Special hot-work tool steels are used for this purpose to prevent spalling and cracking of the dies. Normally, only simple designs can be made in forging dies (Figure 30). Forgings all have a "flash" of excess metal (Figure 31) which must be removed in a secondary operation, usually in a small press that shears it off. Sometimes a series of dies is necessary to preform the bar so it will conform to the shape of the next set of dies until the final shape is made (Figures 32 to 34). Computer aided design (CAD) is quite useful in designing these multiple dies. From the starting shape of the bar stock, each step altering its shape prior to the finished shape can be displayed on the screen. The determination can be made as to the amount of deformation and its required force for each step so it will be within the capacity of the equip-

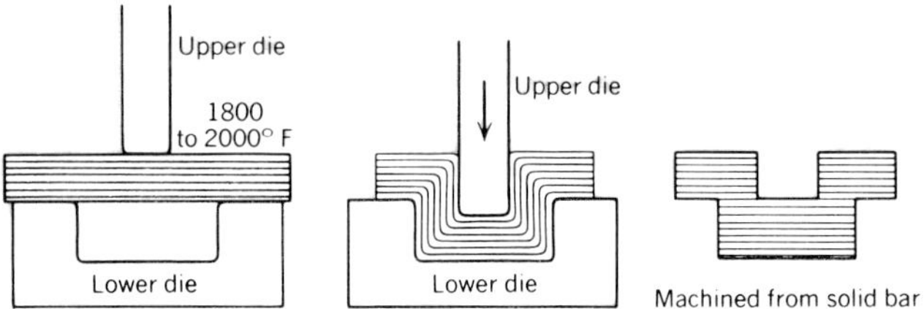

Figure 35
Grain flow in a solid bar as it is being forged, compared to a machined solid bar (White, Neely, Kibbe, Meyer, *Machine Tools and Machining Practices* Vol. I, © 1977 John Wiley & Sons, Inc.).

ment with the least strain on the dies. Cast preforms are often used in automated forging operations. Powder metal preforms are also used.

Forging gives metal good strength and resistance to failure by metal fatigue and shock loads. This is why most good tools are forged and critical parts such as automobile wheel spindles and axles are forgings. Forged parts are often machined to provide bearing surfaces, splines, keyseats, and so forth, but the essential strength of the forging remains. However, when bar stock is used to make a part by machining it from the solid material, the grain flow (anisotropy) is not always in the most advantageous direction and does not flow around shoulders and corners the way forgings do. Therefore, machined parts from stock rolled bars are not as strong or resistant to failure as forgings are (Figure 35).

Hot Upset Forging The term *hot upset forging* refers to high production forging of small shapes. Its prime application is for forgings needed in large quantities. High production hot forging as an automated process has been applied to making nut blanks and other similar products (Figure 36) for years. Nut blanks are also cold formed in similar operations as is discussed in Chapter 9.

Hot forming machines are massive and complex automatic feeding, shearing, forging, and punching systems capable of very high output, 150 parts per minute (Figure 37). The process of nut making takes four steps on a three-die hot former (Figure 38). The heated metal stock is fed into a stop where it is sheared off (Figure 39). The first die upsets the blank; the second die forms the blank to a hexagonal shape and partly forms the hole; and the third die punches out the hole and the blank nut is ejected. One of the advantages of hot forming is that a large percentage of the hot blank can be displaced to make various shapes. Bolt heads, flanged axles, and similar shapes are made in a process called **heading** which is similar to the nut making process. Heading can be done by either a hot forging or a cold upset process. The trend seems to be toward the cold upset forging process and away from hot work upsetting. However, shapes such as bearing caps and front wheel hubs are hot formed, whereas more symmetrical shapes such as wrist pins and gear blanks are cold formed. Heading and the cold upset process are fully covered in Chapter 9, "Processing of Metals: Cold Working."

Figure 36
These are typical shapes made in hot formers—short, round, symmetrical, and with a hole (Illustration courtesy of National Machinery Co.).

Figure 37
A hot former is an automatic feeding, shearing, forging, and punching system capable of very high output. This machine makes 150 parts per minute (Illustration courtesy of National Machinery Co.).

Figure 38
The tooling arrangement for a three-die hot former progressing from right to left shows the steps in nut making—upset, form, and punch out the hole (Illustration courtesy of National Machinery Co.).

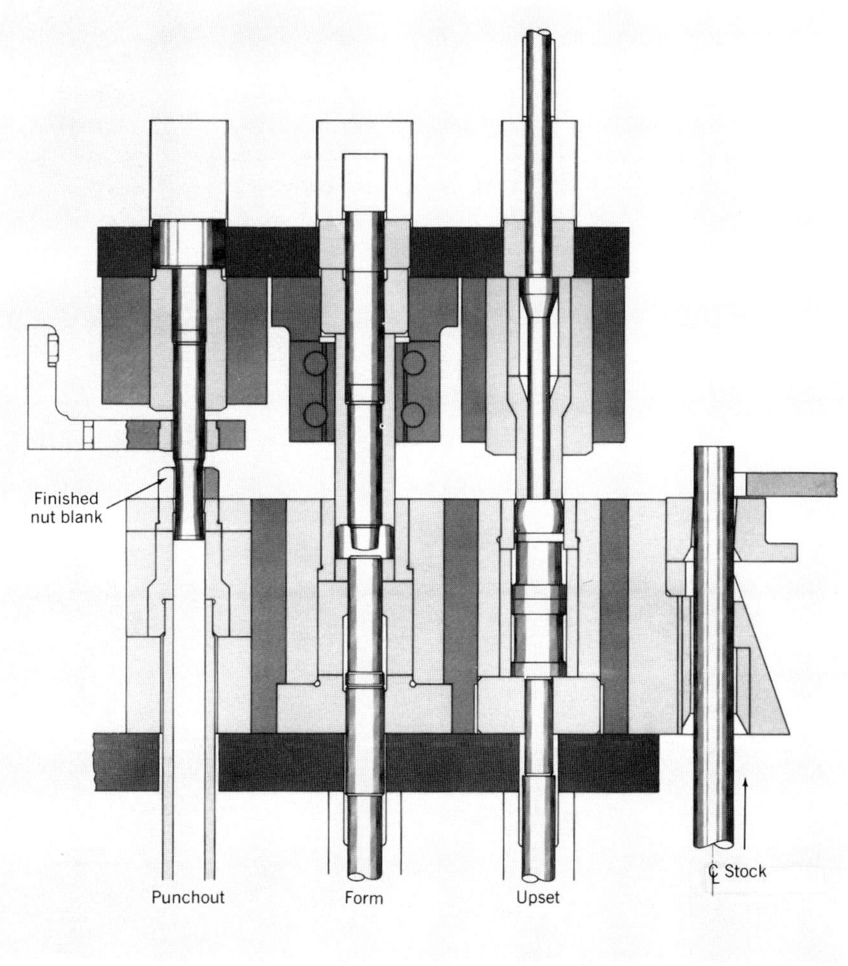

Figure 39
Hot shearing. A computer measures bar end movement into the hot former, then signals feed rolls and cutoff transfer to cycle at preset times (Illustration courtesy of National Machinery Co.).

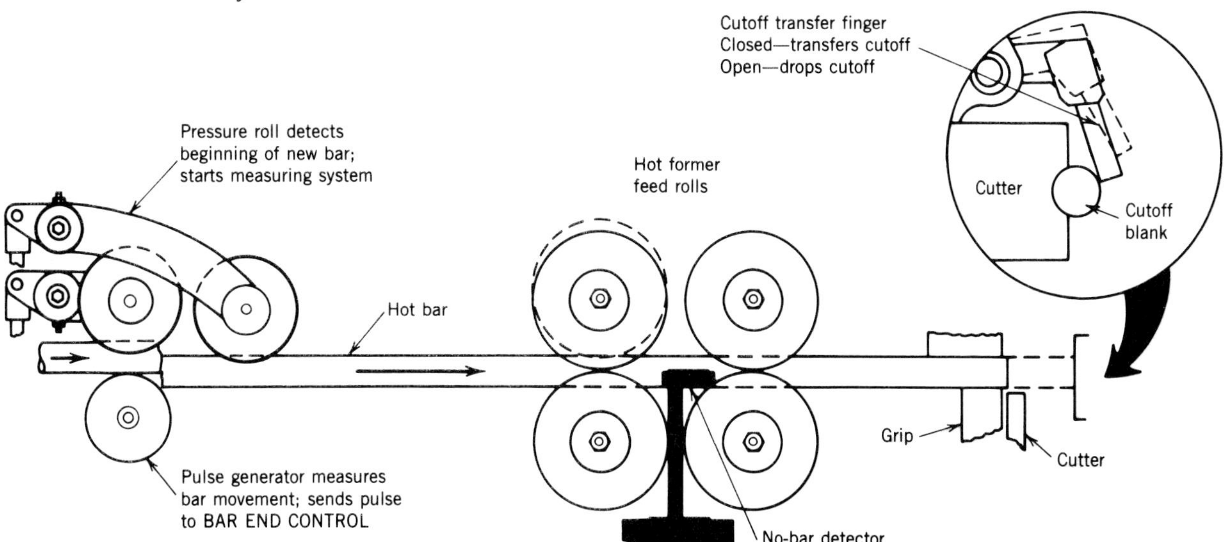

Figure 40
Press system for the coining of bicycle bearing races. Parts are oriented and conveyed up the incline on the left, then fed down the tube escapement device that admits them individually into the open die. The blanks continue through a five-station progressive die with the operation steps shown in Figure 33 (Graebener Press Systems, Inc., Providence, RI).

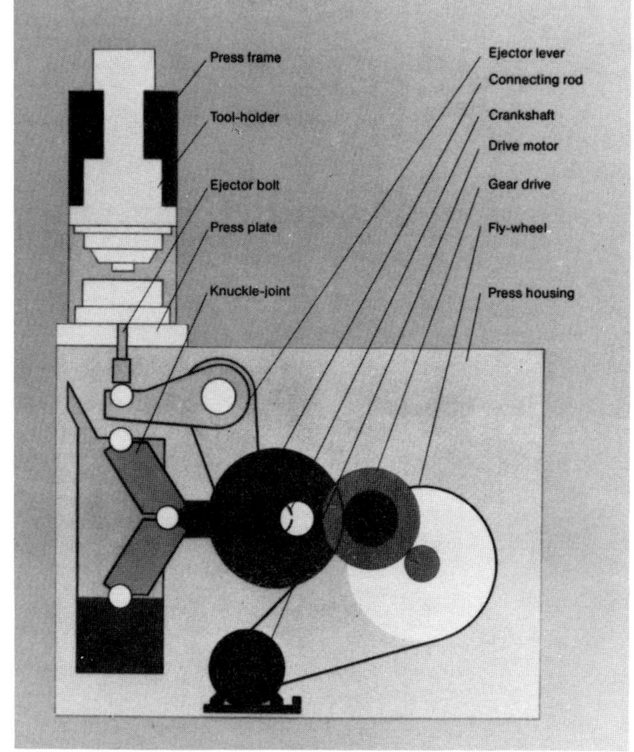

Figure 41
Principle of the Graebener knuckle-joint press system (Graebener Press Systems, Inc., Providence, RI).

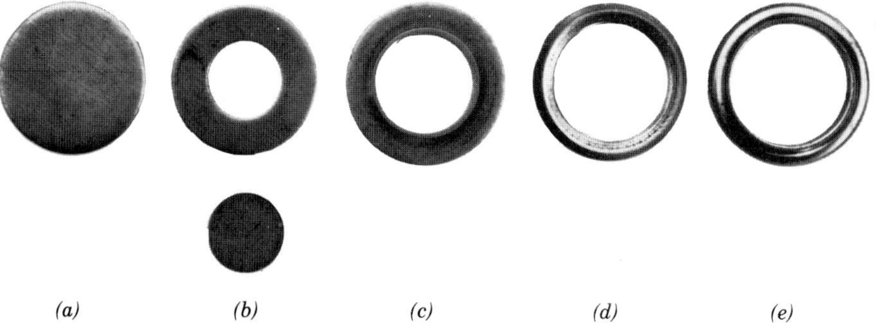

Figure 42
Sequence in producing bicycle bearing races: (*a*) Material. (*b*) Blanking. (*c*) Upsetting. (*d*) Forming. (*e*) Finishing. (Graebener Press Systems, Inc., Providence, RI).

Another type of very high production forming system that is in sharp contrast to the older methods (as shown in Figure 23) is the knuckle-joint press principle (Figures 40 and 41). These machines use progressive dies in an advanced automated system. Small parts such as socket wrenches and bicycle bearing races (Figure 42) are produced with these presses. Hot, cold, or warm forming can be done on knuckle-joint presses, depending on the requirements of the design and the toughness of the metal. Open-end and box-end wrenches are perfect examples of parts that require a forging process. For example, an end wrench made from steel strip by blanking would be a very poor product, tending to bend or break in use. Forgings are much stronger. Combination box-end and open-end wrenches are produced on two presses that are linked by a conveying belt (Figure 43). Together they produce about 1800 pieces per hour. Since the material is a chrome–vanadium alloy, it can only be hot forged. Combination wrenches are produced in a sequence of operations (Figure 44). Wrenches are carried out of the machine after the final finishing operation (Figure 45). The wrenches collect in containers (Figure 46) for transfer to other operations.

Figure 43
Interlinked press line for the production of $\frac{7}{8}$-in. combination wrenches: (left) Line 1 - GK 360 and (right) Line 2 - GK 800 (Graebener Press Systems, Inc., Providence, RI).

Figure 44
Sequence of operation for $\frac{7}{8}$-in. combination wrenches: (*a*) On a GK 800: (1) Setting. (2) Straightening of head. (3) Forming. (4) Trimming. (*b*) On a GK 360. (1) Piercing. (2) Flat coining. (3) Edge coining. (Graebener Press Systems, Inc., Providence, RI).

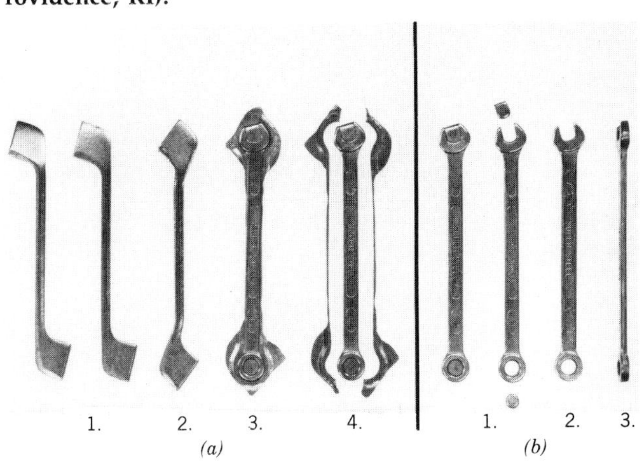

Figure 45
Wrenches emerging from automatic machine (Graebener Press Systems, Inc., Providence, RI).

Figure 46
The finished product—$\frac{7}{8}$-in. combination wrenches (Graebener Press Systems, Inc., Providence, RI).

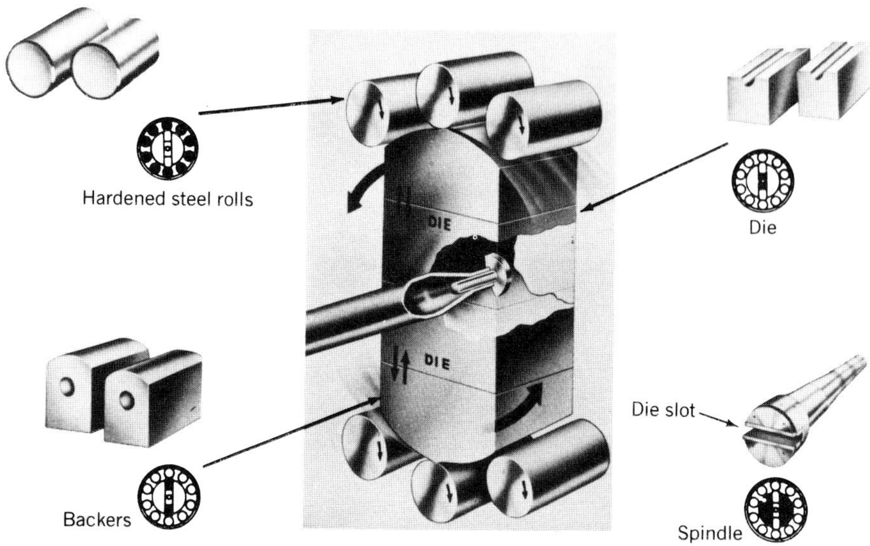

Figure 47
Tubing being necked down by the swaging process (Courtesy: Aluminum Company of America).

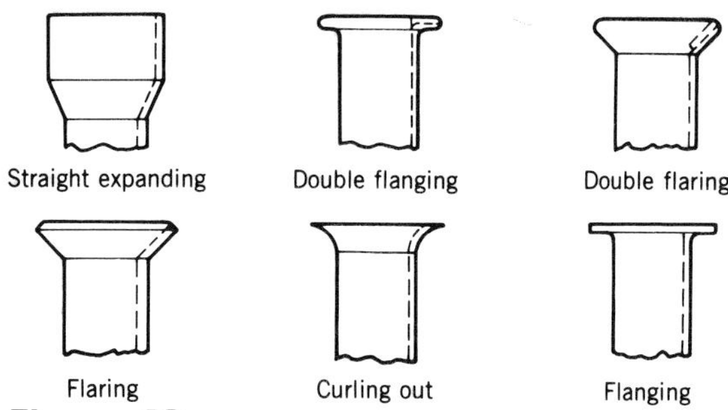

Figure 48
Types of expanded tube ends (Aluminum Company of America).

SWAGING

Swaging, like nut making and bolt heading, can be either a hot- or a cold-forming operation. Considerably more force and more massive machines are needed for cold forming than for hot forming. Therefore, larger, heavier parts are most often hot formed. Swaging involves forcing, hammering, or rolling a rod or tube into a smaller or necked-down shape (Figure 47).

Swaging as a cold-working operation is often performed in a rotary swaging machine. Typical swaging types are shown in Figures 48 to 50. The two-die swaging machine delivers blows at a rate of 1800 to 4000 per minute by action of the backers alternately striking the work rolls and then flying outward because of centrifugal force. Thus the metal is formed around a die to produce the required shape. Some machines are hand fed and others operate automatically.

Figure 49
Types of reduced tube ends (Courtesy: Aluminum Company of America).

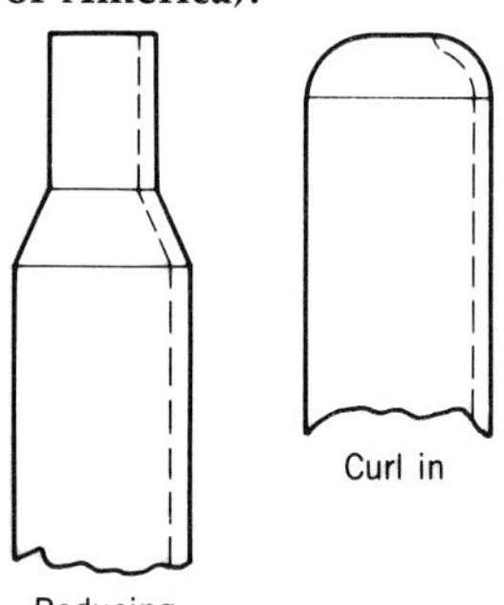

Figure 50
Types of beaded tubes (Courtesy: Aluminum Company of America).

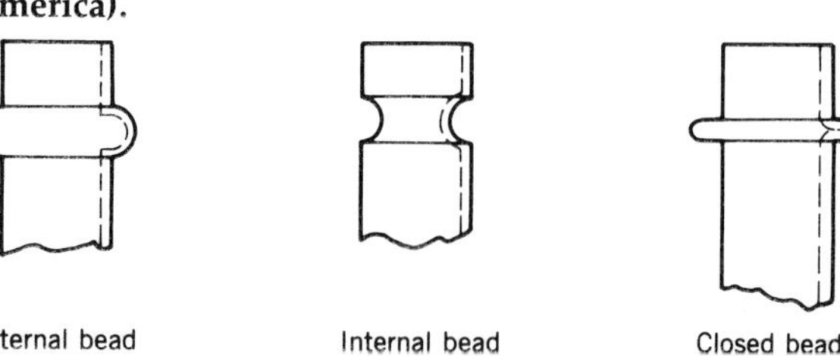

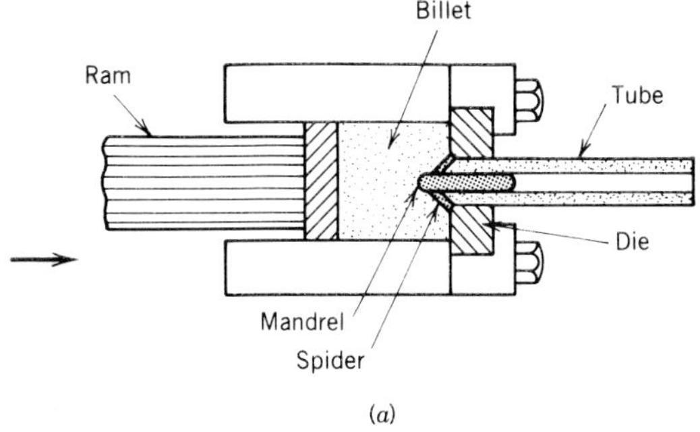

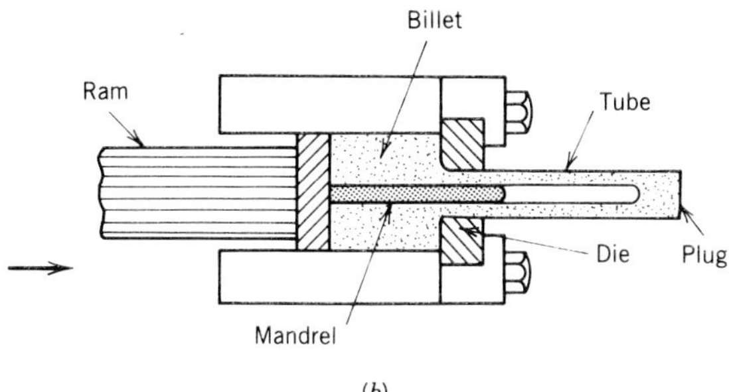

Figure 51
Forward extrusion. (*a*) Use of the spider mandrel to produce hollow extruded shapes. (*b*) Mandrel fastened on the ram producing a hollow shape.

HOT EXTRUSION

Like nut forming and heading operations, extrusion can be done with hot, warm, or cold metals. Extrusion is a process of forcing metal through a die, not unlike squeezing toothpaste out of a tube. Nonferrous metals are usually extruded cold or warm, whereas ferrous metals usually need to be at a forging heat to make them sufficiently plastic to be extruded. Square and round tubular products, structural shapes, and round, square, or hexagonal solid shapes are some examples of extruded pieces. There is almost no limit to the kinds of intricate cross-sectional shapes that can be made by this process. Extrusion produces surfaces that are clean and smooth with accurate shapes that can be held within close tolerances. Tubular products made by direct extrusion require a spider mandrel (Figure 51) that allows the metal to flow around it and around the spider which holds the mandrel in place. The great advantage of the extrusion process is that it allows such intricate shapes to be made. For example, various size bronze gear shapes can be extruded in lengths of 20 to 40 ft, and later cut off in short lengths to make inexpensive gears. However, in cases where a product can be rolled instead of extruded, the choice should be a rolled product if large amounts are needed because rolling is a less expensive method for shaping large quantities of metal. For short runs, extrusions may be more economical since tooling costs are much lower.

HOT DRAWING

Drawing is a process of forming sheet, plate, or solid metal into a hollow, cup-shaped part. Cold drawing uses relatively thin metal and changes the thickness of the metal only slightly, whereas, on the other hand, hot drawing deforms the metal to a very great extent. Hot drawing is used to make thick-walled parts of simple shapes that are usually cylindrical (Figures 52 to 54). Heavy duty hydraulic cylinders, artillery shells, and oxygen tanks are made by the hot drawing process.

HOT SPINNING

Although most metal spinning operations are carried out on cold metal, very large, tough metals are spun while hot. Domed heads for pressure vessels are often made by hot spinning. This process consists of shaping flat or preformed metal disks over a rotating form. Pressure applied by the spinning tool causes the hot metal to flow over the form. Since simple tooling is needed for spinning, it is a less expensive method of forming circular and cylindrical shapes than the use of drawing or stamping dies.

Figure 52
Hot drawing press, called a push bench, for the production of thick-walled hollow bodies and pressure vessels ("Presses for All Applications of Metalforming," Schirmer-Plate-Siempelkamp, Hydraulische Pressen GmbH, Krefeld, West Germany).

Figure 53
After the mandrel has forced the hot metal through the draw ring, a stripping device removes the forging as the mandrel is drawn back ("Presses for All Applications of Metalforming," Schirmer-Plate-Siempelkamp, Hydraulische Pressen GmbH, Krefeld, West Germany).

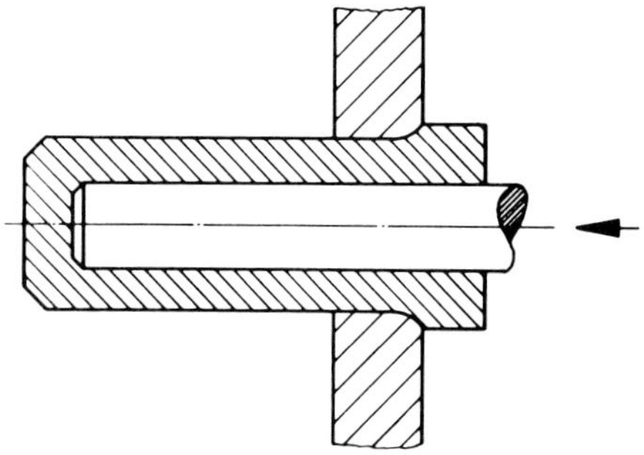

Figure 54
Diagram showing the principle of producing thick-walled steel bodies ("Presses for All Applications of Metalforming," Schirmer-Plate-Siempelkamp, Hydraulische Pressen GmbH, Krefeld, West Germany).

Figure 55
Piercing of solid billet to make seamless tubing (White, Neely, Kibbe, Meyer, *Machine Tools and Machining Practices*, Vol. I, © 1977 John Wiley & Sons, Inc.).

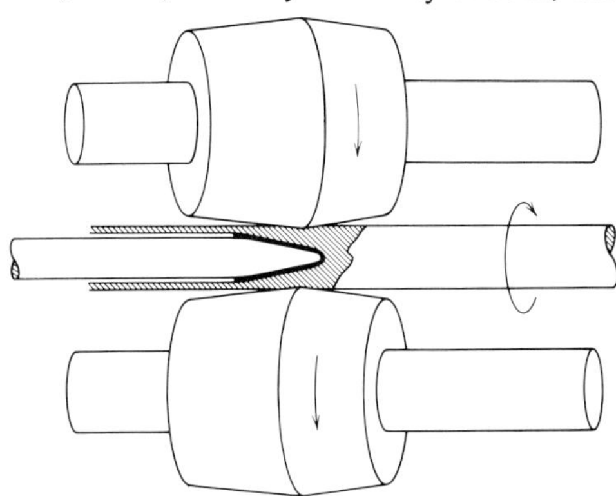

SEAMLESS TUBING

Seamless tube is hot formed by the process of roll piercing (Figure 55). A heated, cylindrical billet of steel is passed through rotating conical rolls whose axes are in different planes. This action causes a "pipe" or opening in the center of the billet that flows over a conical mandrel that shapes and sizes the thick-walled tube. A following operation in a rolling mill with grooved rolls and a mandrel further elongates the tube and reduces its wall thickness. Seamless tube is often cold drawn as a finishing operation. See Chapter 9 for cold drawing.

PIPE WELDING (ROLL FORMING)

Hot formed pipe is made of **skelp**, a low-carbon semifinished steel strip. Essentially, the hot strip is drawn through a series of rolls that gradually curl the flat strip into a cylinder (Figure 56). As the pipe is formed, the edges are forced together, and at a white heat they make a weld bond. This is called butt welding. Since the skelp is unwound from a continuous coil, the pipe must be cut to lengths. This is done with a flying saw, one that follows the moving pipe as it cuts it to length. Small diameter pipe, up to 3 in., is butt-welded. For larger diameters it is lap welded to give it greater strength. Hot-formed pressure welded pipe is much less expensive than seamless tubing and it is widely used for underground water systems, household plumbing, and for many industrial uses. However, because it is pressure welded (butt welded), it is not acceptable for high pressure applications such as in hydraulic machinery where the seamless pipe is used.

Figure 56
Small-diameter pipe formed by continuous butt welding. Skelp is formed and welded into pipe as it passes through sets of rolls. A flying saw cuts the pipe into lengths (Courtesy of Bethlehem Steel Corporation).

REVIEW QUESTIONS

1. By what means are white hot ingots formed into slabs, blooms, and billets?
2. What modern method of steel manufacture bypasses the ingot stage and produces slabs or blooms directly?
3. Large dendritic crystals that weaken the metal form in cast metals such as ingots and strand casting. How are these removed from the metal?
4. What is anisotropy in metals?
5. Why are very large forgings heated although smaller forming operations are sometimes carried out cold? Explain.
6. What kind of forging machine would you choose to use for a very large open-die operation? Which would you choose for making a run of 10,000 high-precision parts?
7. Hot metal tends to stick to forging dies. What design characteristic helps to prevent this from happening?
8. What is "flash" and how is it removed?
9. If you were required to design a machine part, such as an automobile axle, that will be subjected to high shock loads, would you choose a forging, a casting, or would you machine it from solid bar stock? Why?
10. How are bolt and nut blanks formed?
11. What kinds of shapes can be made by hot extrusion?
12. A tank end having a hemispherical shape must be made of $\frac{3}{4}$ in. thick flat steel. Name two methods by which this operation can be accomplished.
13. Briefly explain how seamless tubing is made and how common steel water pipe is made.

CASE PROBLEMS

Case 1
Failure of a Tractor Spindle

A front wheel axle spindle on a tractor became damaged from a bearing failure. A replacement part would take six weeks to get if it were ordered from the factory and the farmer needed the tractor for his spring planting. A new spindle was made at a local machine shop from solid steel of the same alloy as the original. After two weeks of use, it broke in half and the wheel fell off. Considering what you have learned in this chapter, what do you think caused the failure of the machined spindle?

Case 2
Hot Pressing

Open die forging on flat anvils usually requires a pressure of 3 to 5 tons per square inch on the tool faces for mild steel at a forging temperature. Harder alloy steels may require two or three times that amount, and up to 15 tons per square inch are needed for the very hardest steels. Pressures may need to be doubled if swages or dies are used.

A 3-in. square billet, 10 ft long, of hot rolled mild steel is to be open die forged into a lifting arm for a hoist. Upper and lower swaging dies will be used to form half the length of the bar into an elliptical shape from the square. Each of the lower and upper swages has a surface area of 20 in.^{2}. A 300 ton hydraulic forging press is available. Will it be sufficient for the job? Approximately how much pressure will this forging operation require?

CHAPTER 9 PROCESSING OF METALS: COLD WORKING

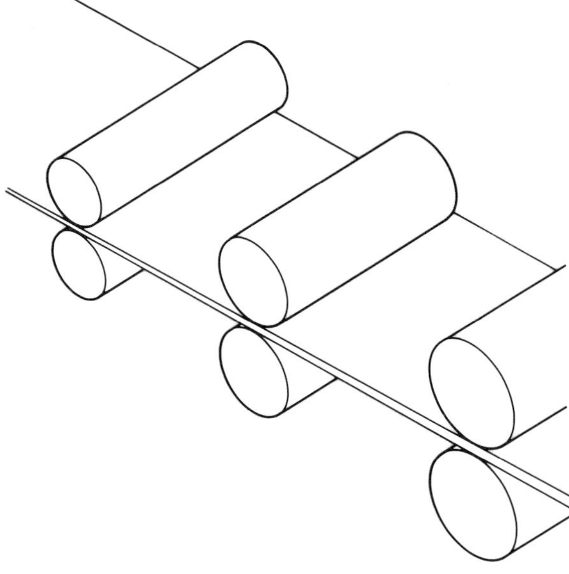

In ancient times, cold working was accomplished by hammering on soft metals such as gold, silver, and copper for jewelry and other ornamentation. Iron wire was made in the Middle Ages by pulling iron rods through progressively smaller holes in a steel plate. Later, the improved properties of cold-drawn products were discovered and a new industry was born. After being hot formed in a steel mill, steel shapes such as bars, sheets, and tubes could be descaled and further shaped by the process of cold forming. Today, cold-formed products range from very fine hypodermic needles to huge pipeline tubes and from tiny hair-size filaments to propeller shafts for ships. Almost any conceivable shape can now be made by one or more processes of cold forming. Greater strength-to-weight ratios, better finishes, and closer dimensional tolerances are some of the advantages a cold-worked metal has over a hot-worked metal.

OBJECTIVES

This chapter should enable you to:

1. Explain the effects of cold working on metals.
2. Describe how hot-rolled steel is prepared and cold finished in steel mills.
3. List a number of cold-forming operations and explain their principles, advantages, and uses.
4. Account for the difference between cold-rolled and machined parts and select one process over another for a particular product.
5. Make a choice among cold-forming operations for a particular product.

FACTORS IN COLD WORKING

One of the most valuable characteristics of metals is known as **plasticity**, the ability of metals to be deformed permanently in any direction without cracking or splitting. Higher temperatures as in hot working tend to make metals more plastic, but recrystallization at high temperatures causes them to retain their soft condition with its lower strength. Higher strengths and hardnesses are obtained by deforming the metal at normal temperatures. Permanent plastic deformation can occur only at stresses above the elastic limit or yield point (Figure 1). As the cold work continues, more force is required and the hardness increases until a certain stress is reached; at this point the metal ruptures or splits apart. This is called the **breaking point** and it is often slightly lower than the **ultimate strength**. When metals are cold worked to a certain point, the next operation requires forces greater than those previously applied to deform the metal further. Each operation brings the particular metal closer to its ultimate strength

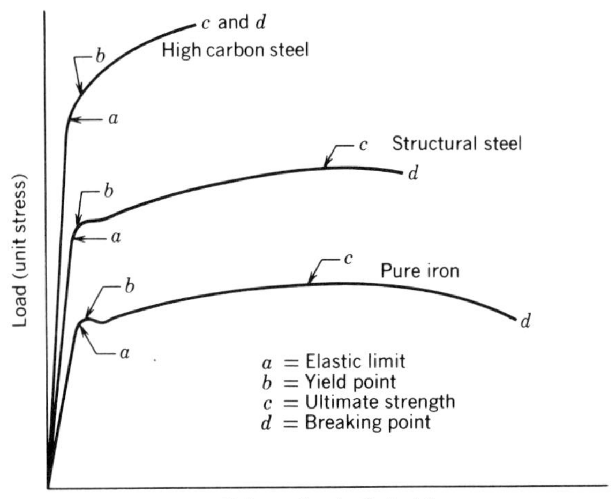

Figure 1
Above the yield point more stress is required as the cold work continues. The *d* on the diagram indicates where failure occurs. (White, Neely, Kibbe, Meyer, *Machine Tools and Machining Practices* Vol. II, © 1977 John Wiley & Sons, Inc.).

and point of rupture. The object in cold working metals, therefore, is to stop well short of failure.

The degree of deformation (amount of cold working) determines the level of toughness, strength, hardness, and remaining ductility. Different specifications for a manufactured product can thus be obtained. In the case of sheet steel, for example, it is possible to produce from one-quarter-hard steel to full-hard steel. Quarter-hard steel can be bent back 180 degrees without breaking, whereas half-hard steel can be bent only 90 degrees and full-hard steel can be bent only 45 degrees on a radius of about the material thickness. If more deformation is needed, then a **process anneal** is used to restore plasticity which at the same time reduces the strength of the metal and lowers the force or stress required to deform it further. Process anneal on the cold-worked steel is often carried out in a closed container of inert gas to avoid scaling problems. When this is done, it is called **bright anneal**. See Chapter 4, "Heat Treatments of Metals," for a discussion of annealing processes. Some of the advantages of cold working as compared to hot working are:

1. Closer dimensional tolerances.
2. Better surface finish.
3. Better machinability.
4. Superior mechanical properties.
5. Enhanced directional properties.

Some of the disadvantages of cold working are:

1. The metal is less ductile.
2. Annealing is sometimes required.
3. It may contain residual stresses, causing warping when machined.
4. More massive and powerful machinery is required.
5. Subsequent heating or welding will change the cold-work properties and weaken the metal.

Elastic Recovery

When a metal is placed under stress within its elastic range, it will return to its former shape when the load is removed. If the metal is made to deform or take on a permanent set by loading it beyond the elastic limit into its plastic range, it will be permanently deformed, but will bounce back to some extent because of its elastic properties. This characteristic of metals, called **springback** (elastic recovery) (Figure 2), is a design consideration for forming and bending dies and fixtures. Parts are thus overbent several degrees, ironed, or bottom beaded to counteract springback.

Ductility and Malleability

Ductility is the property of a metal to deform permanently or to exhibit plasticity without rupture while under tension. Any metal that can be drawn into a wire is ductile. The ability of a metal to deform permanently when loaded in compression is called **malleability**. Metals that can be rolled into sheets or upset cold forged are malleable. Most ductile metals are malleable, but some very malleable metals such as lead are not very ductile and cannot be easily drawn into wire. Some ductile metals are steel, aluminum, gold, silver, and

Figure 2
Cold working operation showing springback. When sheet metal is formed as in (*a*) and is removed from the die as in (*b*), it will partially revert back to its original shape.

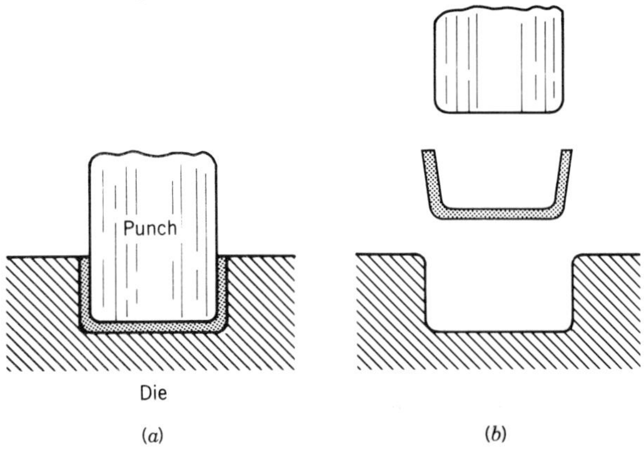

Punch
Die
(*a*)
(*b*)

Figure 3
Straightening rolls. The rolls alternately bend the metal slightly beyond its elastic limit in two axes so that when it emerges it is straight.

nickel. A few nonferrous alloys such as brass and Monel are also quite ductile, but most alloys of steel are less ductile than plain carbon steel. Some metals and alloys, such as stainless steel, high manganese steel, titanium, zirconium, and Inconel, tend to "work harden," that is, to quickly increase hardness as cold working progresses. However, most metals tend to work harden to some extent. Austenitic stainless steels will tend to remain ductile until a very high hardness is achieved. This is an engineering property that is valuable when higher hardness as well as toughness is a requirement for a cold-formed product. Annealing will restore a softer condition as well as restoring the ductility of these work hardening metals. Work hardening is often a troublesome difficulty in machining operations.

Preparing Hot-Rolled Steel for Cold Rolling

Hot-rolled steel bars and plates must be sufficiently oversize because the cold-finishing process reduces them in cross section. These are also sometimes straightened in a set of straightening rolls (Figure 3). The surface must be clean and free from scale. The hot-rolled bars or sheets are placed in a hydrochloric or sulfuric acid dip which removes the scale; this is called **pickling.** The acid is washed off and the steel is dipped in lime water to remove any acid residue. Sometimes the hot rolled steel requires full annealing to make it as soft and ductile as possible prior to the cold-working operation.

COLD ROLLING IN THE STEEL MILL

The grain or crystal structure of the metal is permanently altered, flattened, and lengthened by the process of cold rolling (Figure 4). After a process anneal, the recrystallized metal can be further cold worked since the soft condition of the grains allows for more deformation to take place.

The coils of pickled hot-rolled steel pass through a series of rolling mills at a high rate of speed, 1 to 2 miles/min is not uncommon. The strip begins with a thickness that is a little less than $\frac{1}{8}$ in. and a length of about $\frac{3}{4}$ mile. Two minutes later it is squeezed to the thickness of two playing cards and it is more than 2 miles long (Figure 5). After annealing, the strip passes through a temper mill. There it is given the desired thickness, flatness, and surface quality, and the desired temper or hardness. Sometimes the full coil is shipped to customers, other times the coiled strip is cut into lengths for shipping flat.

Figure 4
Cold rolling flattens and lengthens the grains in the direction of rolling (Neely/*Metallurgy* 2 Ed., © 1984 John Wiley & Sons, Inc.).

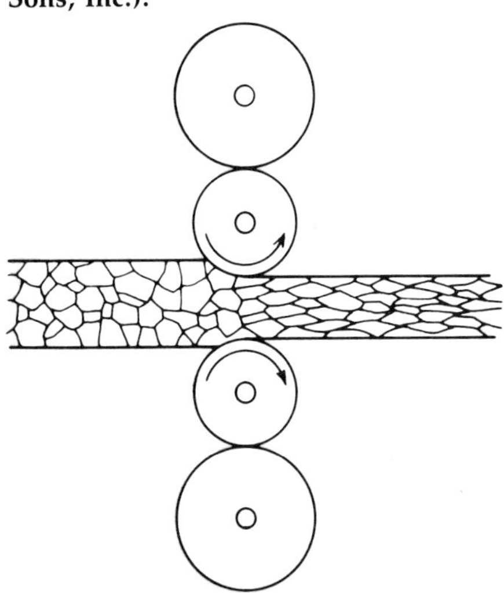

Figure 5
Cold strip mill showing the operator at his console in the foreground and the rolling mill stands in the background (American Iron & Steel Institute).

Figure 6
Sendizmer mill for rolling steel strip to close tolerances. Only the two small central rolls touch the metal; the outer cluster of rolls is for backup and bearings. These mills are used to make thin strip and foil (American Iron & Steel Institute).

Thin-gage, high-precision steel sheets were made possible by the **Sendizmer mill** (Figure 6). It was first developed to produce thin sheets for light weight radar equipment used in aircraft. Today these thin sheets are needed for space programs. This mill also produces steel foil of .003 in. thickness or less for packaging and other uses. The work rolls on these mills are quite small in diameter, from 1 to $2\frac{1}{2}$ in., and they are sometimes several feet in length. These are supported by a cluster of back-up rolls. High surface quality of steel strip can be obtained by planishing and polishing with abrasives. Stainless steel sheet and strip are sometimes given a mirror finish before being shipped to the customer.

Metals other than steel, such as aluminum, copper, and titanium, are also rolled into bars, plates, and sheet stock, using similar methods. Both hot and cold rolling processes are used for most nonferrous metals.

Surface Coatings on Sheet Steel

Since iron and steel have a tendency to rust in the presence of oxygen and moisture, many protective coatings for them have been developed. One of the first of these was the coating of iron with tin by hammering it onto steel sheets. This method can be traced back to the thirteenth century. Tin does not corrode in the presence of moisture and will readily bond to clean iron by hot dipping or by plating by the electrolytic method—the way most tin plate is produced today (Figure 7). The tin plating on steel may be less than

Figure 7
Steel coated with tin by the electrolytic process leaves the plating line (Courtesy of Bethlehem Steel Corporation).

Figure 8
Minimized spangle steel coils await shipment. These coils are produced on the plant's modernized galvanized line (Courtesy of Bethlehem Steel Corporation).

Figure 9
Blanking die set in a press showing a punch making round blanks (Courtesy: Aluminum Company of America).

.001 in. thick. Since available supplies of tin are dwindling and the demand for it is so great worldwide, steel companies began to search for substitutes for tin-coated steel. Chromium plate is sometimes used as a substitute and other coatings are now being used for some purposes. Galvanized sheet steel is used for corrugated roofing material, culvert pipe, and many small articles such as steel pails (Figure 8). Manufactured steel parts for boats, street lighting fixtures, and other exposed metal objects are often galvanized. Although zinc can be deposited by the electrolytic process, by far the most prevalent plating method is the hot-dip process. In the sheet steel mill this process is continuous and often includes a final coating of paint which is dried in bake ovens as the steel moves through them. These galvanized and painted sheets are capable of being formed or bent without destroying their protective coatings. Many other coatings are also put on steel to protect it from deterioration or to enhance its appearance. These will be covered in Chapter 16, "Corrosion and Protection for Materials."

BLANKING AND PRESSING

One of the most versatile forms of metal working is that of converting flat sheet and strip metals into useful articles. Sheet metal can be pierced, punched, or blanked to produce flat shapes. It can be drawn or cupped to form such items as pots and pans. It can be rolled into cones or cylinders or pressed into automobile body shapes; plus it can be shaped by many other forming operations.

Although the cutting of a blank is actually a shearing operation, it is included here because it is the first step in forming a product from sheet stock. **Blanking** is the operation of cutting a flat shape out of a strip of sheet metal in which the hole material is saved for further operations. If the hole material is scrap, it is called **punching**. By this process any shape hole can be made. **Piercing** is the operation of cutting (usually small) round holes in sheet metal; if the holes are small and close together, it is known as **perforating**. Blanking with a conventional punch and die (Figure 9) is usually employed where large quantities are required. It is usually used only for thin sheet, but thicker material can be punched or blanked, depending on such factors as press size, blank size, and material thickness (Figure 10). Some blanks may be more economically cut by the circular shears (Figure 11) or nibbling machines (Figure 12). Where square or rectangular shapes are needed, blanks may be cut with sheet metal shears (Figure 13). Blanks may be sawed in a vertical band saw, several pieces stacked together (Figure 14). Hand-held power shears can also be used to cut blanks (Figure 15). It is obvious that using any of these methods to make blanks is much slower than using a die set and they are only useful when a few parts are to be produced.

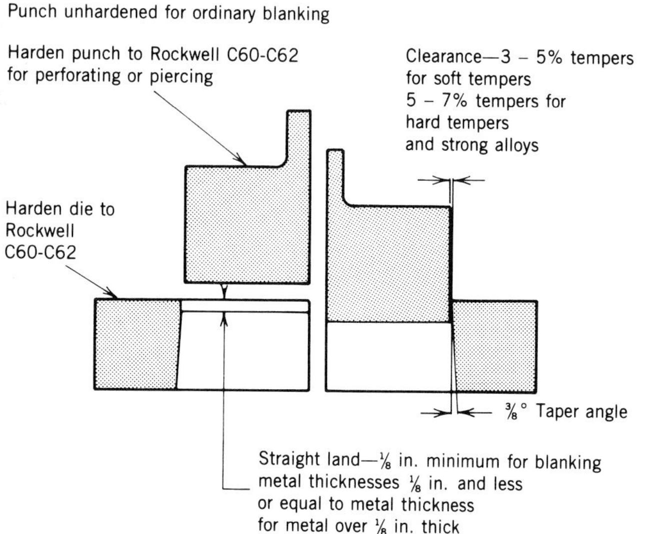

Figure 10
Hardness and clearance of blanking tools may vary with the type of sheet metal used (Courtesy: Aluminum Company of America).

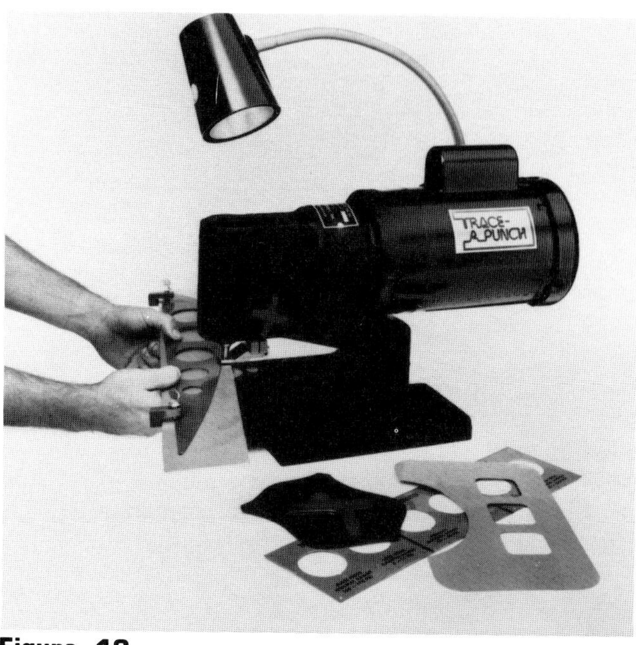

Figure 12
Duplicating nibbler. This machine cuts sheet metal by means of a small oscillating punch. Here it is being moved along a template to duplicate that shape (Heck Industries, Inc.).

Figure 11
Circular shears cutting a large disk (Niagara Machine & Tool Works).

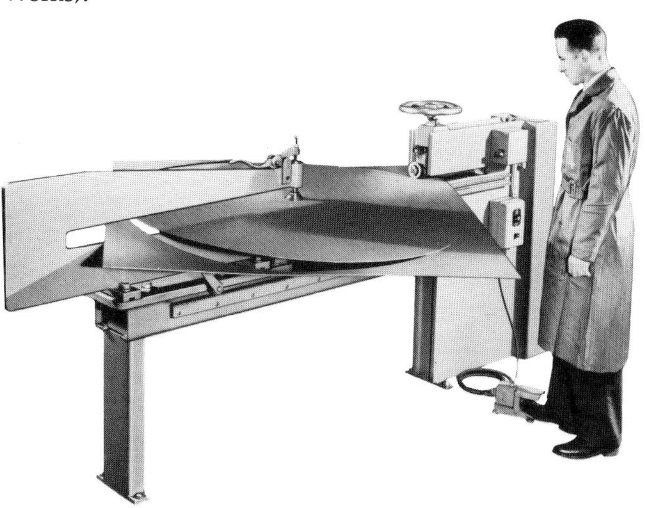

Figure 13
Guillotine shear for shearing sheet and plate.

The basic components of piercing and blanking die sets, as shown in Figure 9, are a **punch**, a **die**, and a **stripper plate**. The punch holder (top plate) of the die is attached to the press ram which moves the punch in and out of the die. The die holder of the die set (bottom plate) is attached to the bolster plate of the press. Guide posts in the die set on which bushings slide maintain precise alignment of the cutting members of the die. The stripper plate removes the material strip from around blanking and piercing punches. The material strip is advanced after each punching stroke by a feeding mechanism.

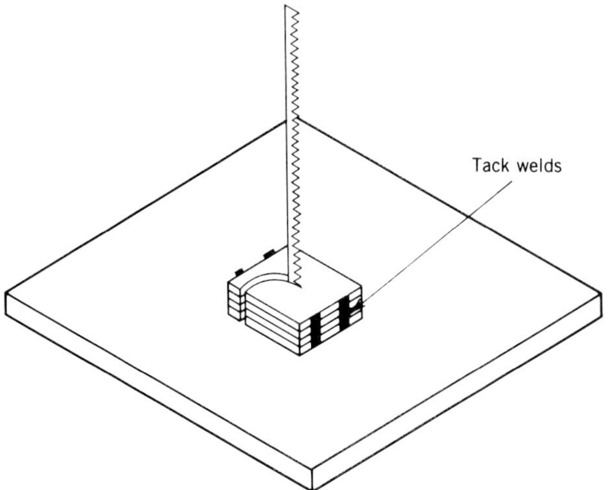

Figure 14
Band sawing a stack of blanks.

Figure 15
A portable hand-held power shears for cutting sheet metal blanks (Robert Bosch Power Tool Corp.).

An alternate method of blanking is the use of steel rule dies (Figure 16). This method was first used for cutting softer materials such as cardboard, cloth, and plywood, but the process is now used for metal. A steel band is fastened on edge in a groove in the upper die plate. Neoprene rubber pads take the place of the usual stripper plate. Steel rule dies have an expected life of about 100,000 parts. Conventional blanking dies may produce three times that many parts before re-sharpening is required. An open lower die can be used or a solid maple or urethane slab can be used to cut against, lessening the cost.

When mechanical presses (Figure 17) are fitted with

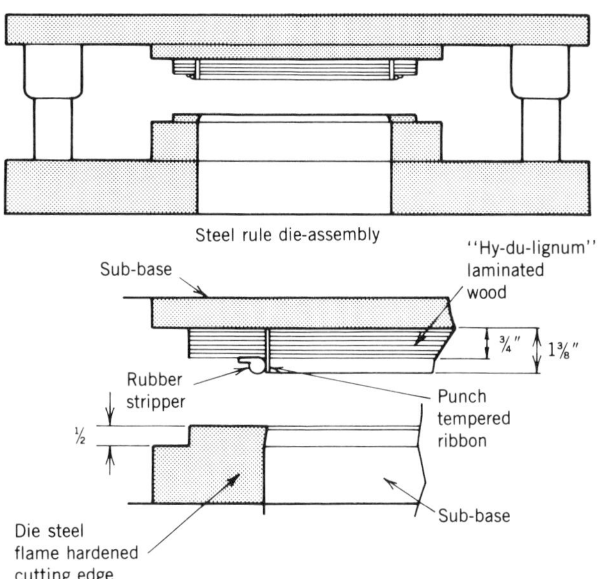

Figure 16
Steel-rule die blanking technique (Courtesy: Aluminium Company of America)

Figure 17
Mechanical presses such as this one are extensively used for punching, blanking, and forming of sheet metal strip (Photo courtesy of The Minster Machine Company).

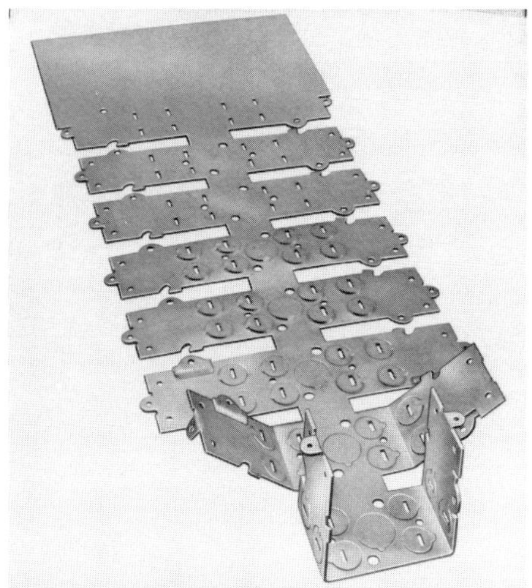

Figure 18
Electrical outlet boxes are punched and formed with a progressive die (Photo courtesy of The Minster Machine Company).

progressive punching and forming dies, a continuous manufacturing of products is possible. Parts for electrical outlet boxes are formed from steel strip (Figure 18).

DRAWING, FORMING, AND EXTRUDING METAL

It is easy to imagine molten metal flowing by gravity into molds and steel softened by heating to a white heat being hammered or forced into shape. Also we are familiar with the hardness and toughness of cold metals. It is more difficult to understand how cold metal can be made to flow when sufficient pressure is applied. Virtually every operation performed on hot metal can also be done on cold metal, but the limit to the size of these parts is in the massiveness and power of the machinery. Since the cost of heating metal is eliminated in cold forming, it is a less expensive method, especially for the smaller piece parts, and there is no thermal damage to dies.

Combination hot, cold, and/or warm forming has the advantage of reducing equipment costs where several operations are involved and often reduces the need for secondary machining and operating personnel. Larger parts can be hot formed and later finish formed by cold forming methods. There is a new trend toward warm forming in which metals that are difficult to cold work, such as titanium, stainless steel, and some alloy steels, can be formed without the great pressures and stresses of cold forming. For example, bearing races are now being warm formed of tough alloy steel that will subsequently be hardened and ground. Steels that contain more than .40 percent carbon are difficult to cold form but can easily be formed while warm, that is, below the transformation temperature but at or above the recrystallization temperature, about 900° F (482° C) for steel. Warm forming can substantially increase production rates for some materials.

Coining and embossing are stamping operations that form the surface of metal. Impressions of letters, figures, and patterns are formed by pressing them onto the metal. Usually knuckle-joint presses are used for coining and embossing because these machines are of rigid construction, have rapid short strokes, and exert intense pressure during the last part of the stroke as a result of the knuckle-joint action. See Chapter 8 for a description of knuckle-joint presses.

Drawing Plate, Sheet, and Foil

This kind of drawing process is one in which a flat piece of metal is formed into a hollow shape by apply-

Figure 19
Forces applied to a blank during cold drawing operation (Courtesy: Aluminium Company of America).

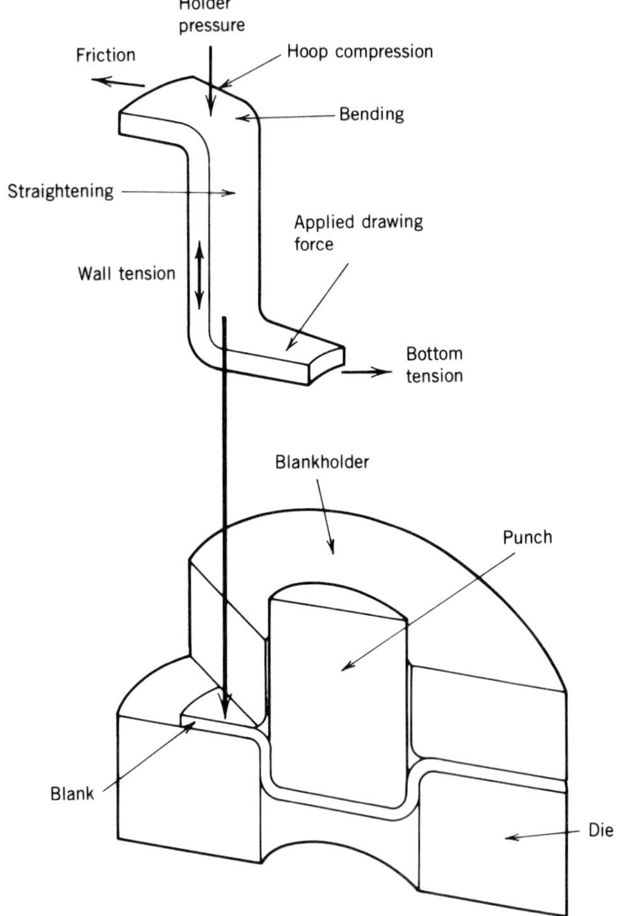

Figure 20
Blanked and drawn parts showing progression of drawing operation (Courtesy: Aluminum Company of America).

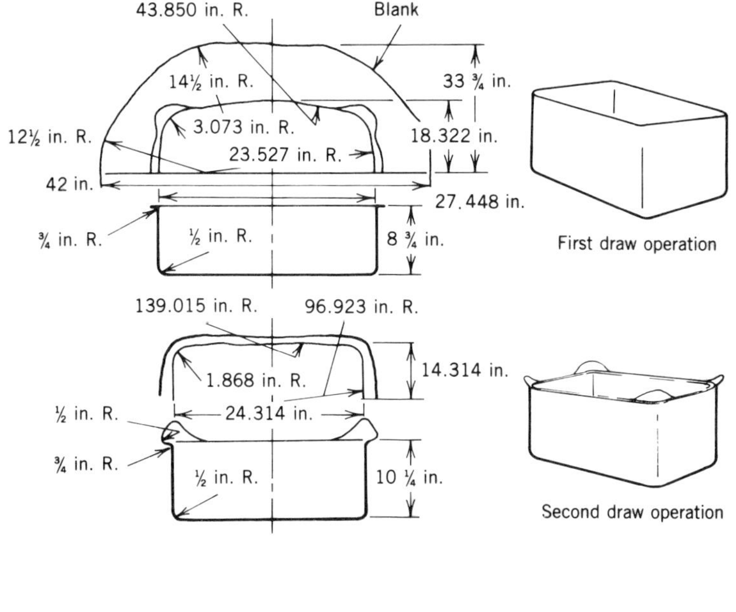

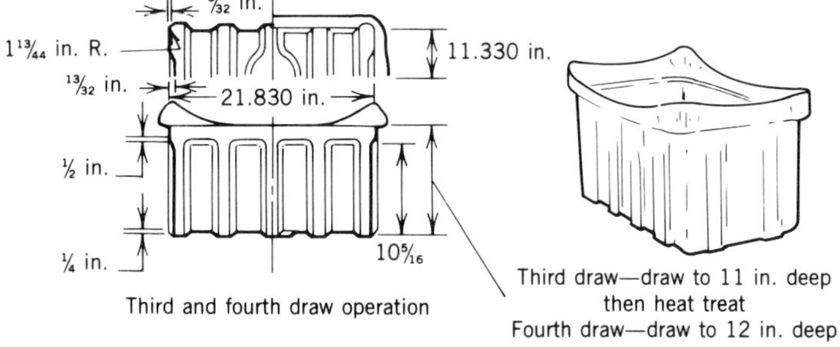

Figure 21
Sequence of draw reductions for 6061 aluminum alloy rectangular case (Courtesy: Aluminum Company of America).

ing force with a punch to the center portion of the metal. The punch draws the metal into a die cavity (Figure 19). In this operation, the metal is stretched on the sidewalls and made to assume an exact thickness. Sheet metal drawing is usually performed in a vertical press.

Blanking and piercing are sometimes combined in one operation for the purpose of continuous manufacture of small parts such as ash trays and bottle caps. This is usually done with progressive dies (Figure 20). Deep drawing is often done with several presses, each having a different die that contributes to the final shape (Figure 21). When the wall thickness must be further reduced and the parts lengthened, as in the manufac-

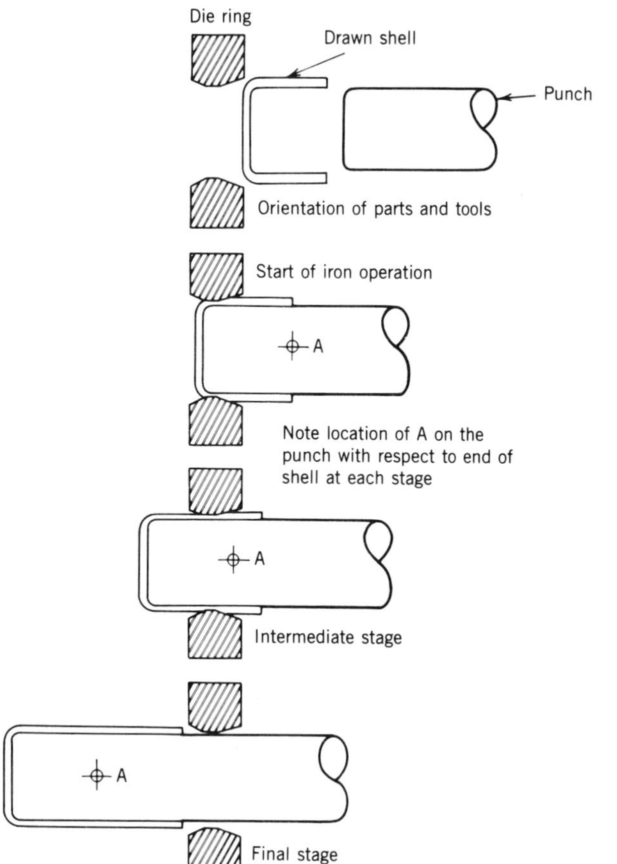

Figure 22
Ironing a drawn shell (Courtesy: Aluminum Company of America).

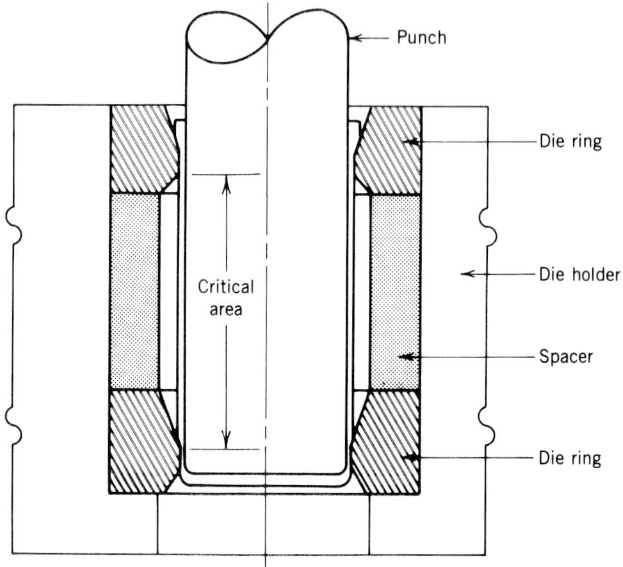

Figure 23
Multiple reduction with stacked ironing dies (Courtesy: Aluminum Company of America).

ture of cartridge cases, an ironing process is used (Figures 22 and 23). Sometimes an anneal is required between pressings.

Bar, Tube, and Wire Drawing

Hot-rolled bars, after pickling and cleaning to remove the scale, are reduced in cross section by pulling them through a die that is slightly smaller than they are. This operation is performed on a draw bench (Figure 24). The drawing process hardens the metal and gives it a smooth finish suitable for machinery shafting.

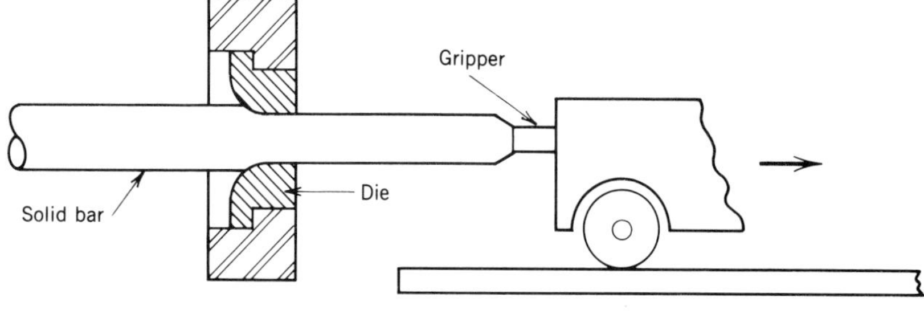

Figure 24
Draw bench for solid-bar stock.

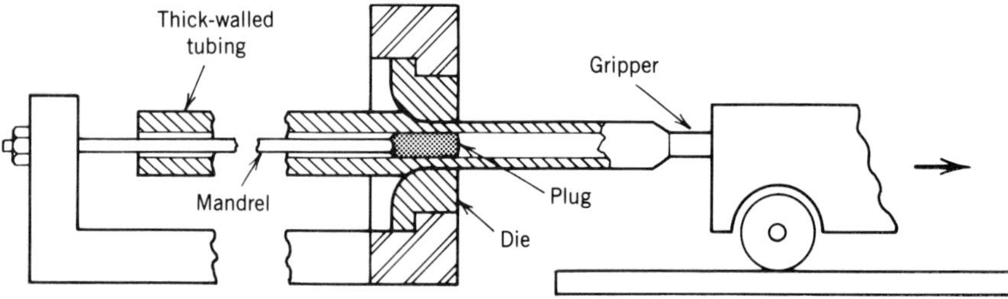

Figure 25
Draw bench for seamless tubing.

Figure 26
Battery of modern wire drawing machines. Rod enters at the left, is reduced in size as it passes through successive dies, and is coiled at the right (Courtesy of Bethlehem Steel Corporation).

Seamless tubing is also cold drawn on a draw bench; the only difference in the operation is that there is a mandrel inside the tube to thin the walls and provide an internal finish (Figure 25). Because of this finish operation and the higher cost of pierced seamless tubing, cold-drawn seamless tubing is considerably more expensive than butt welded black pipe. Seamless steel tubing, after a honing operation, is widely used for hydraulic cylinder manufacture and smaller diameters are used for high pressure pipes that carry oil or other fluids.

Figure 27
Enlarged cross section of wire drawing through a die (White, Neely, Kibbe, Meyer, *Machine Tools and Machining Practices,* Vol. I, © 1977 John Wiley & Sons, Inc.).

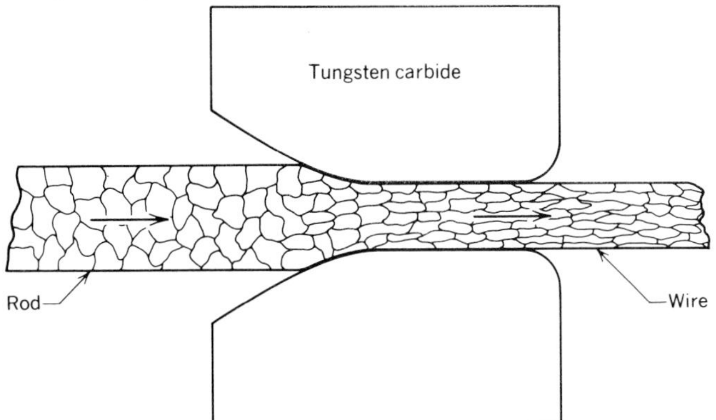

Wire drawing is basically the same process as bar drawing except that it involves much smaller diameters of metal and it is a continuous process (Figures 26 and 27). The wire is drawn through a series of dies, each one slightly smaller in diameter than the previous one. The wire is pulled by a rotating capstan or drum that is located between each set of dies. Coolant, acting as a lubricant and to cool the die, is flooded over the dies since this is a cold-forming operation and unwanted heat is developed in the operation. The finished wire is wound on a reel.

Cold Forming

Also called cold heading or cold forging, this is a metal upsetting process carried out on machines designed for rapid production of small parts from wire stock (Figure 28). Upsetting machines are frequently rated by maximum diameter cutoff capacity. For example, a $\frac{5}{16}$ in. cold header can shear alloy steel up to $\frac{5}{16}$ in. in diameter. Generally, only small parts such as screw and bolt blanks, rivets, and ball bearing blanks are produced on cold upset machines, but some machines are capable of upsetting much larger parts (Figure 29). The largest single use of upsetting machines is in forming heads on fasteners such as rivets and screws. Threads are subsequently rolled on the blanks giving them a much greater strength and fatigue resistance than cut screws (Figure 30).

In the past, most threaded fasteners were made on automatic screw machines that turned the part to size from bar stock and then threaded it with a die, followed by a cutoff operation. Using automatic screw machines is necessary when high precision and complex shapes are involved, but they are much slower than cold-head-

Figure 28
Wire upsetting machine (Illustration courtesy of National Machinery Co.).

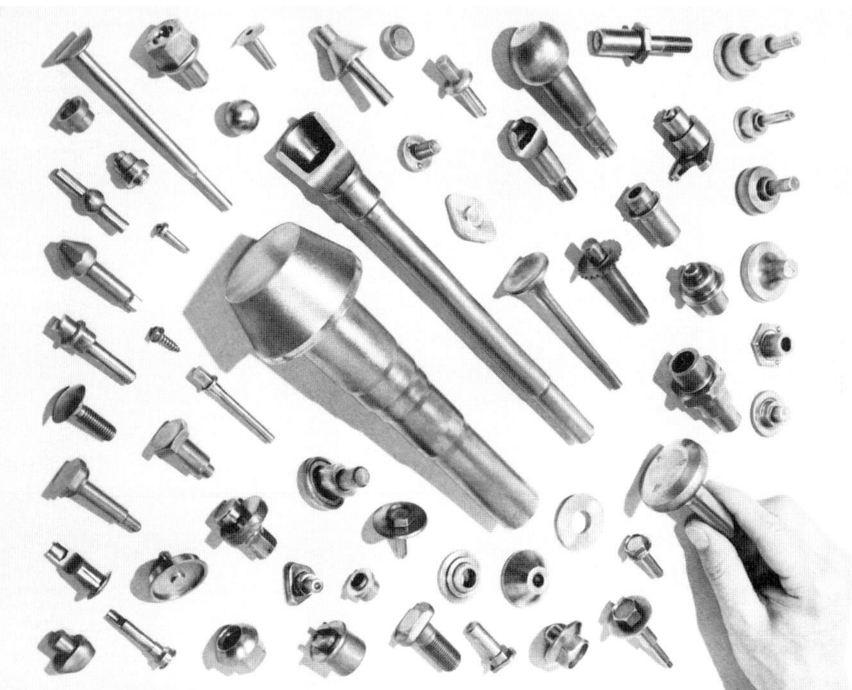

Figure 29
An array of cold upset parts (Illustration courtesy of National Machinery Co.).

Figure 30
Grain structure of a cold formed bolt blank (Illustration courtesy of National Machinery Co.).

ing machines; a small part made in 10 sec is acceptable on an automatic machine. However, production rates on upsetting machines can be as high as 36,000/hr for small unpierced rivets, and No. 8 size screw blanks can be made at 27,000/hr. Bolts $\frac{3}{8}$ in. in diameter can be headed, pointed, and threaded at a rate of 15,000 per hour. Another advantage of forming parts is that nothing is wasted; almost no scrap is cut away as in machining. There are also fewer rejects with the upsetting process.

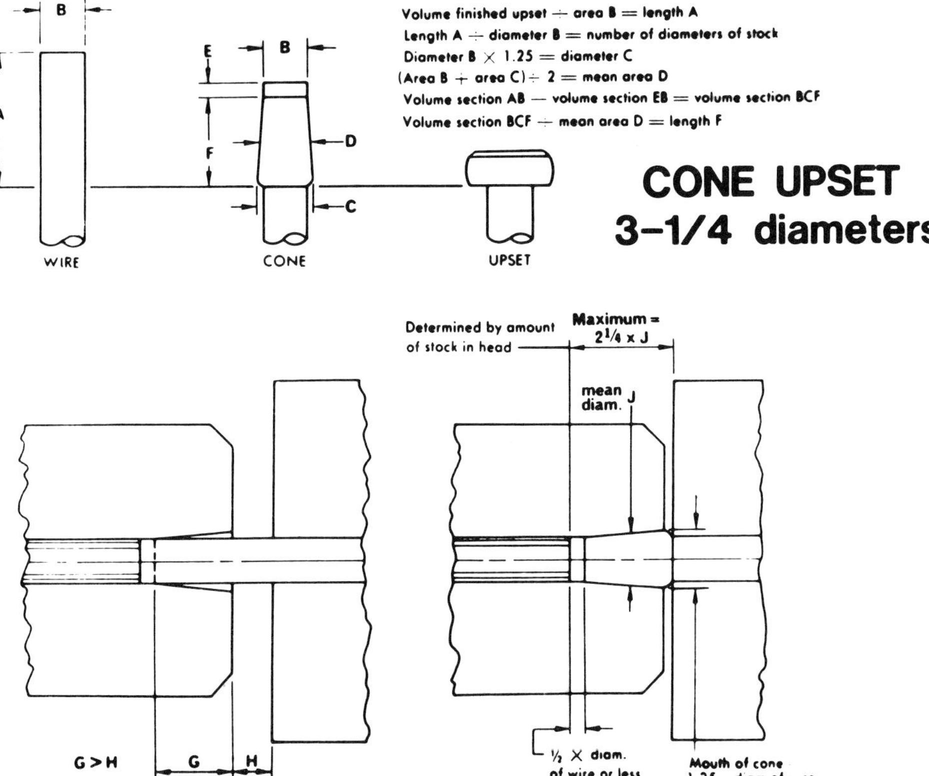

Figure 31
Cone upsetting. Sequence of operations in upsetting a bolt head (Illustration courtesy of National Machinery Co.).

Figure 32
Modes of upsetting (Illustration courtesy of National Machinery Co.).

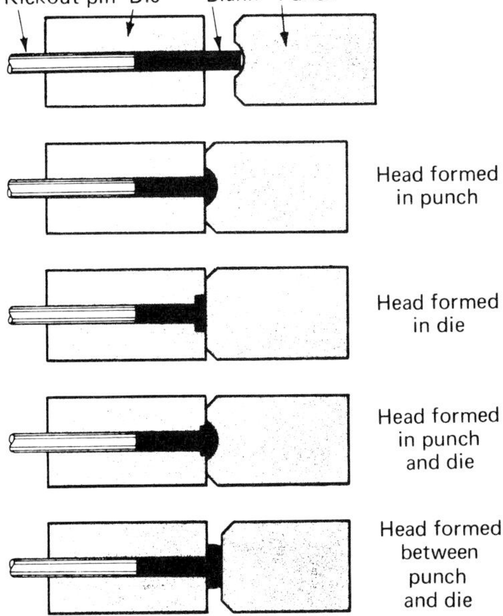

The process and sequence of operation for a cold heading operation is shown in Figure 31. The wire or bar is first formed into a cone before making the head because the unsupported length tends to buckle if the heading operation is done first. Several modes of upsetting are shown in Figure 32; some parts are upset in the punch, some in the die, some in both punch and die, and others between the punch and die. The five basic operations performed in cold-forming machines are shown in Figure 33. Combinations of these give cold forming great versatility. An example of this is the spark plug shell (Figure 34). Figure 35 shows the progression of upsets and extrusions used to make the spark plug center post. This sequence of operations and the tooling involved is shown in Figure 36.

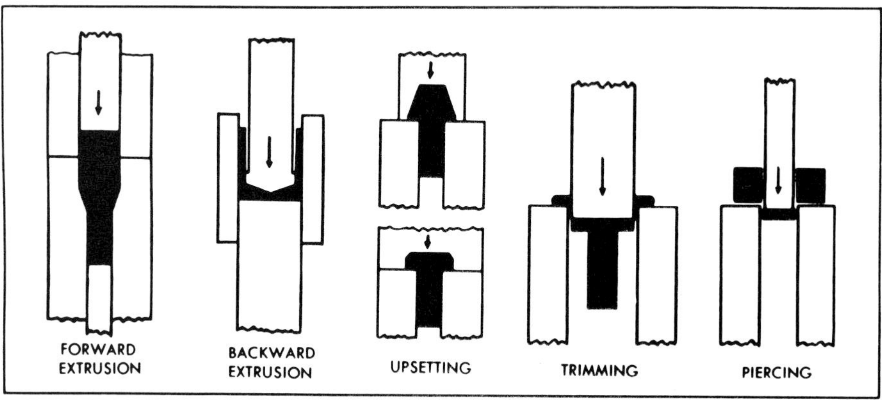

Figure 33
Five basic operations performed in cold-forming machines. Combinations and variations of these operations give the cold former a wide range of application (Illustration courtesy of National Machinery Co.).

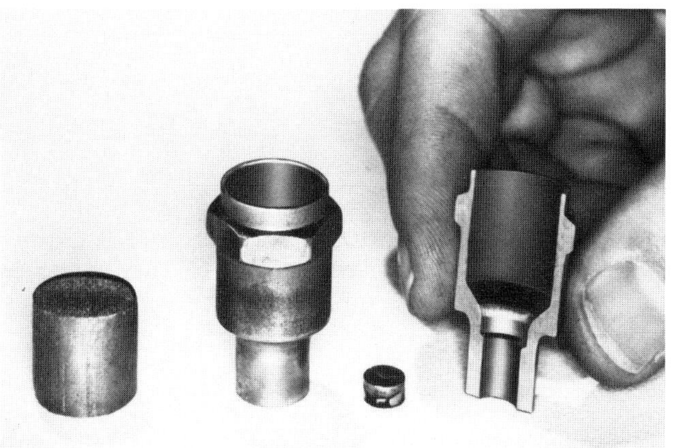

Figure 34
From blank to finished spark plug shell in a six-forming die machine (Illustration courtesy of National Machinery Co.).

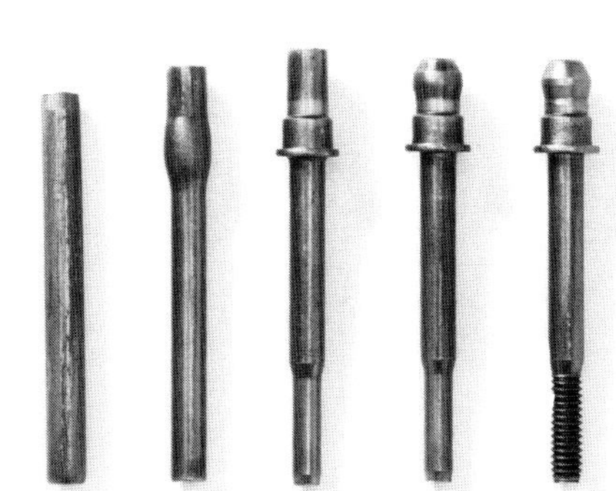

Figure 35
Progression of upsets and extrusions used to make spark plug center post (Illustration courtesy of National Machinery Co.).

Figure 36
(*a*) Air-loaded pin punches the spark plug center post blank into die. (*b*) Die forces inserts into closed positions. (*c*) Continued advance of pin upsets metal into cavity formed by inserts. (*d*) As punch case withdraws, inserts open to allow them to clear largest upset diameter (Illustration courtesy of National Machinery Co.).

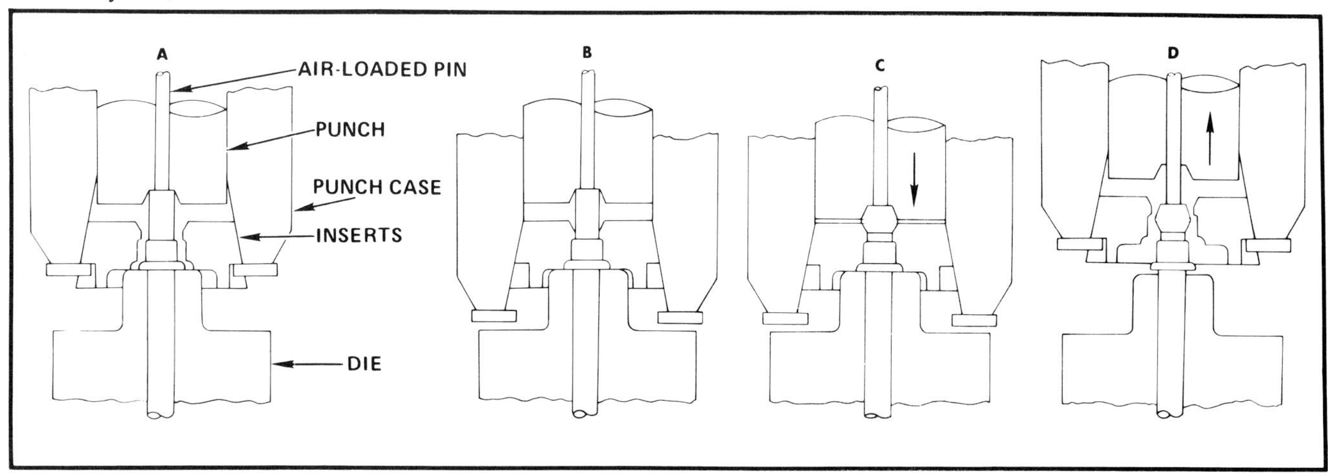

Figure 37
Four-die nut former (Illustration courtesy of National Machinery Co.).

Figure 38
Four-die nut forming. The top row of operation shows forming done by the punches; the bottom row shows forming done in the dies. The first blow upsets the blank and starts the hex formation. The second blow displaces metal away from the center of the blank toward the corners of the hex. The third blow finishes both faces and marks the hole for punch out. The final blow punches the slug, forming the hole. This method retains controlled metal flow for high-quality nuts (Illustration courtesy of National Machinery Co.).

Nut blanks are made in four- or six-die nut formers (Figure 37). The operation is similar to upsetting and to the nut blank hot formers discussed in Chapter 8. As in other cold upset operations, the advantage over hot forming is the higher production rate and lower cost. A sequence of four-die nut forming is shown in Figure 38.

Cold Forming Threads, Worms, and Gears

Although thread and gear forming is not a new idea, it is well suited to be a part of a new trend toward the continuous processing of metals. Ideally, the process would be from the raw ore to continuous casting to rolling mills to upset and roll forming to the finished product. In such a system, waste would be reduced and transferring and shipping of materials eliminated.

External thread forming by the process of cold rolling screw threads is a follow-up operation after cold heading screw blanks. However, threads may also be cut as a machining operation on cold formed blanks. The decision whether to machine threads or roll them can be based on several factors.

1. **Plasticity** The amount of deformation required is directly related to the plasticity of the metal. Some grades of aluminum, gray cast iron, and die cast metals cannot be rolled because of their insufficient elongation and reduction of area.
2. **Work hardening** Most stainless steels and high manganese steels rapidly work harden when cold worked. These metals do not readily lend themselves to cold rolling.
3. **Workpiece design** Some parts are not adaptable for rolling; those with threads that are too close to a shoulder (Figure 39), for example. Some tapered threads and threads on inaccurate blanks should be cut instead of rolled.

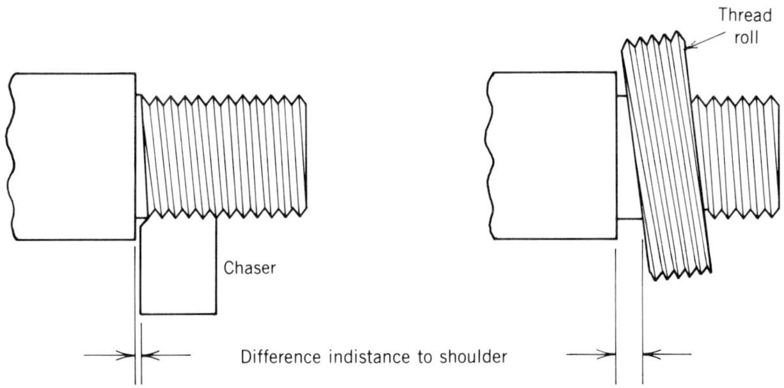

Figure 39
If threads must be made close to a shoulder, cutting is better than rolling.

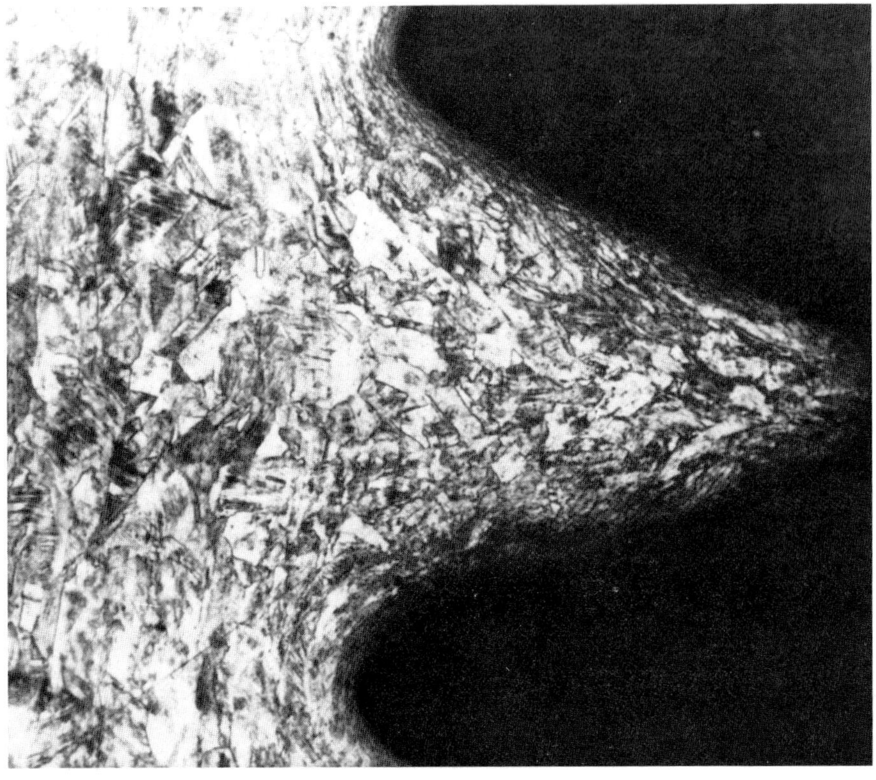

Figure 40
Longitudinal section of a type 304 stainless steel fastener that was cold worked by being thread rolled in the annealed condition. Cold worked austenite (and probably some martensite) is in the area immediately below the surface of the rolled thread (50×) (*Metals Handbook,* Vol. 7, 8th ed., American Society for Metals, 1972, p 134. With permission).

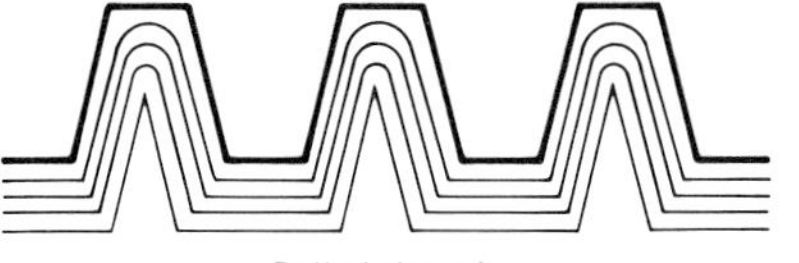

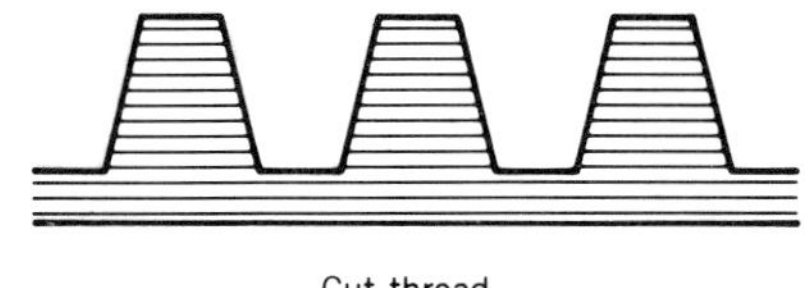

Figure 41
Rolled threads have a harder surface than cut threads and are generally stronger.

Figure 42
Production tapping of nuts is usually done with bent shank taps to make it a continuous process.

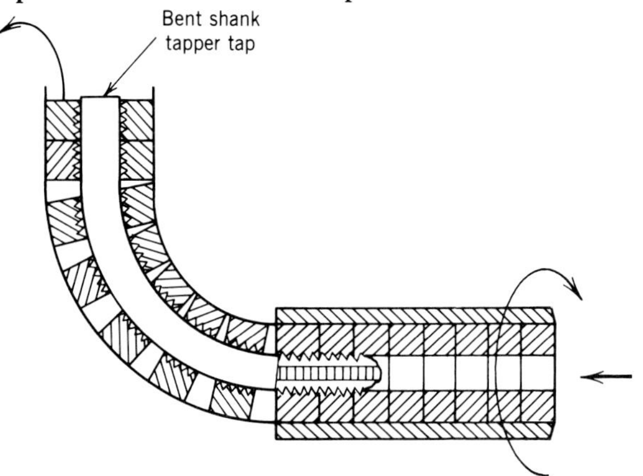

Figure 43
Thread forming taps (Courtesy of DoAll Company).

Style B, Bottoming Series

Style P, Plug Series

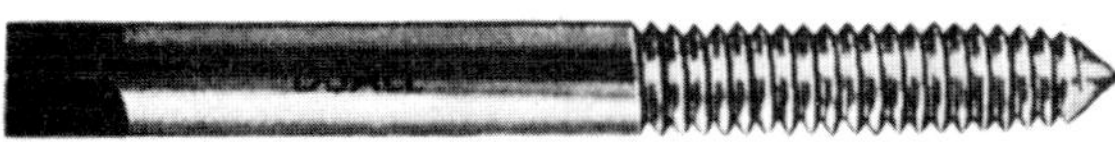

4. **Finish** Rolling is far superior to cutting if the finish is a consideration.
5. **Surface hardness** The rolling process produces a greater surface hardness (Figure 40) that wears well and helps to resist metal fatigue at thread roots.
6. **Higher speeds** Generally, thread rolling is a somewhat faster operation than cutting threads with dies.
7. **Tensile strength** Rolled threads are stronger than cut threads because of grain flow (Figure 41).

Thread rolling machines have two or three hardened, threaded rolls between which threads are formed on blanks. Several methods are used for forming threads, splines, worms, and gears: infeed forming, throughfeed forming, the reciprocal method, the infeed–throughfeed–out method, and the automatic continuous method.

Internal threads are often cut using taps, but they may also be formed using thread forming taps (Figures 42 and 43). As in thread rolling, metal is displaced to form the threads, work hardening and smoothing it. Both cutting and forming methods for internal threads are very rapid processes and the choice of method can be based on the same factors as those of external roll forming of threads.

Gearing

A **worm** is a spiral or helical groove, similar to a screw thread that engages with a worm wheel or worm gear to impart motion. Worms have traditionally been machined on lathes or milling machines since they are highly precision parts. However, it is possible to achieve the required precision by the cold-rolling process with the added advantage of the part receiving a very smooth finish and work hardened surface. This reduces friction and provides a longer service life. In this process, a blank is placed in a rolling machine and moved between rolls in the infeed process where hundreds of tons of pressure is applied as the rolls and blank turn and form the shape of the worm.

Helical and spur gears are formed exactly as worms are formed. Smaller gears are cold formed, but larger gears with coarser teeth are hot formed. For high precision gears, a follow-up operation is necessary. Even machine-cut gears require a finishing operation of gear shaving or grinding. This will be further discussed in Chapter 12.

Cold Extrusion

This is a process of producing long lengths of various shapes such as rod, tubing, and structural forms. Usually only soft metals with low yield strengths are extruded cold and tough metals such as steel are extruded hot.

Methods of Extrusion There are basically three methods to form extrusions: direct, indirect, and impact. In the process of extrusion, metal is forced through a die that has the desired shape at extremely high pressures, causing the metal to deform plastically and to flow through the die taking its shape. Soft metals such as lead can be extruded with low pressures of about 20 tons/in.^{2}, but higher pressures are required for aluminum, brass, and other nonferrous metals.

Direct extrusion (Figure 44) is more commonly used for both hot and cold extrusions. Direct or forward extrusion is widely used to produce special shapes that are not available from rolling mills. However, the cost of rolled products is considerably less than that of extruded forms and the surface finish is better. Since the cost of extrusion dies is less than the cost of rolls, extrusions are more practical for short runs.

Indirect extrusion (Figure 45), also called backward extrusion, has the advantage of creating less friction, but the restricted length of extrusion and extra equipment limits the usefulness of this method.

Figure 44
Direct or forward extrusion. The flow lines show how the metal is pushed out of the die by the use of great forces. Softer metals are usually extruded cold, whereas harder metals are brought to a forging heat before being extruded (Neely/ *Metallurgy* 2 Ed., © 1984 John Wiley & Sons, Inc.).

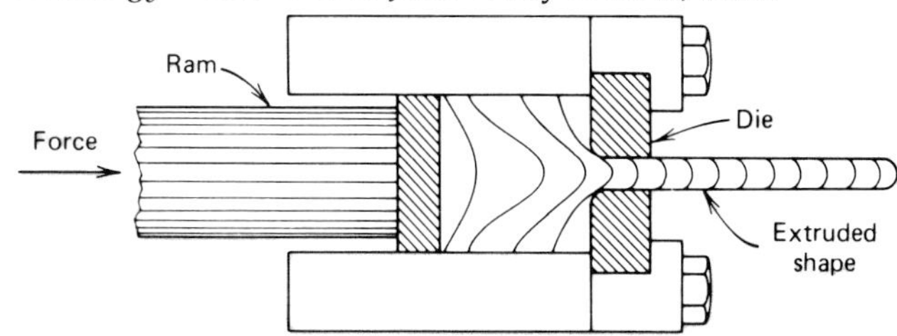

Figure 45
Indirect or backward extrusion.

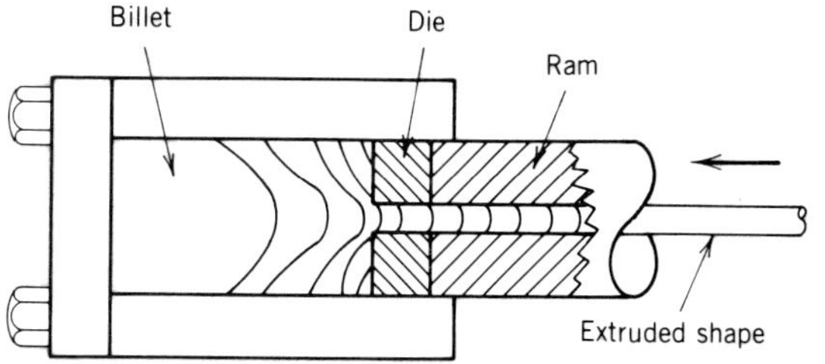

Figure 46
The method of forming thin-walled tubes by impact extrusion. A flat blank is placed in the die (*a*) and the punch is brought down rapidly with a single blow (b). The material in the blank then "squirts" upward around the punch. When the punch is withdrawn, a stipper plate removes the tube (Neely/*Metallurgy* 2 Ed., © 1984 John Wiley & Sons, Inc.).

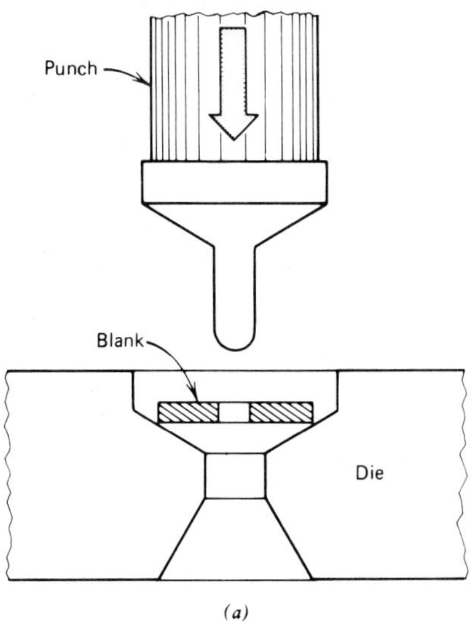

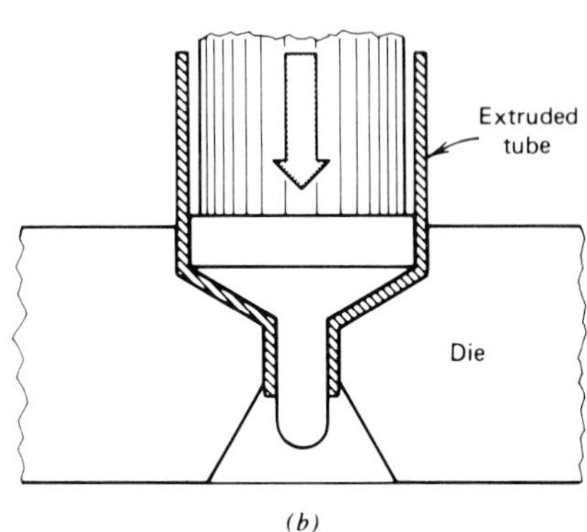

Figure 47
Simple bend in one axis.

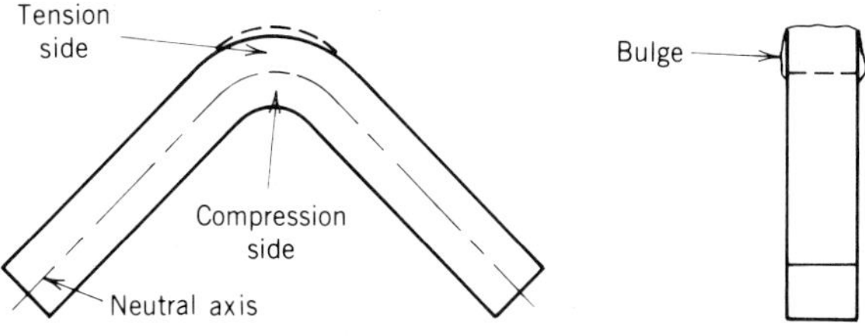

Figure 48
Bar folder. These devices are used for bending light sheet metal.

Impact extrusion (Figure 46) is used to produce products such as collapsible tubes (toothpaste) and soft drink cans from soft metals, usually aluminum. In the process of impact extrusion, a thick slug of metal is placed in the cavity of a die. The slug is struck by a punch having the size and shape of the inside of the part. The impact of the punch is sufficient to cause the metal in the blank to flow very quickly backward through the space between the punch and die cavity. The part adheres to the punch and is removed by a stationary stripper as the punch is withdrawn. Impact extrusion is a very rapid production method of forming cold metal.

BENDING, STRAIGHTENING, AND ROLL FORMING

Bending refers to a simple bend in one axis. When two or more bends are made simultaneously, as with a die, the process is usually called forming or drawing. Bending metals causes a plastic deformation about a linear axis, making little or no change in the surface area of the bend. The neutral axis of the bend (Figure 47) is not located at the center, equidistant between the outer and inner surfaces of the bend, because the yield strength

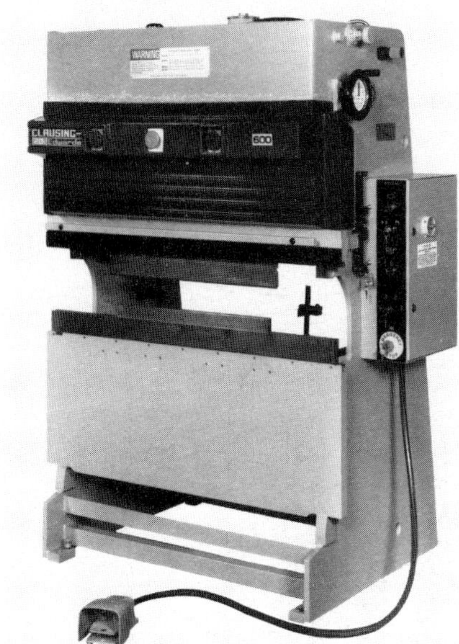

Figure 49
Vertical press brake (Clausing Machine Tools).

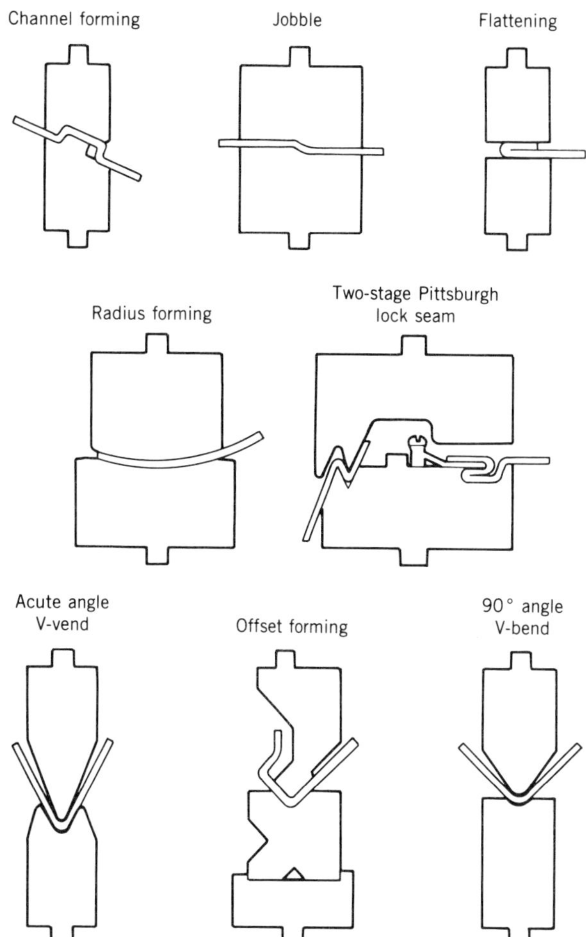

Figure 50
Typical press brake tools (Courtesy: Aluminum Company of America).

of metals in compression is somewhat higher than in tension and this causes the inner compression sides of the bend to bulge, whereas the outer tension side is thinned and reduced in width. Also, as in other cold-forming operations, there is a tendency of the metal to unbend; this varies considerably in metals. For example, lead has almost no springback, soft aluminum has very little, but 301 stainless steel has considerable springback.

Machines that make relatively sharp bends or small radii in sheet metal are called bar folders (Figure 48). These tools are used to bend sheet metal for architectural shapes, furnace and air conditioning pipes, and sheet metal products. When bends in heavier sheet metal are required or more complex bends are needed, a press brake is used (Figures 49 to 54). These machines are designed to make short strokes to a predetermined depth, and they are powered either with hydraulic cylinders or in much the same way as a crank punch press.

Roll bending (Figures 55 to 57) is used to form curved shapes, cylinders, or rings. The metal used can range from thin sheet metal to massive bar or plate stock. Plates, bars, and structural shapes are formed by this method.

Figure 51
Press-brake forming of bead in two operations (Courtesy: Aluminum Company of America).

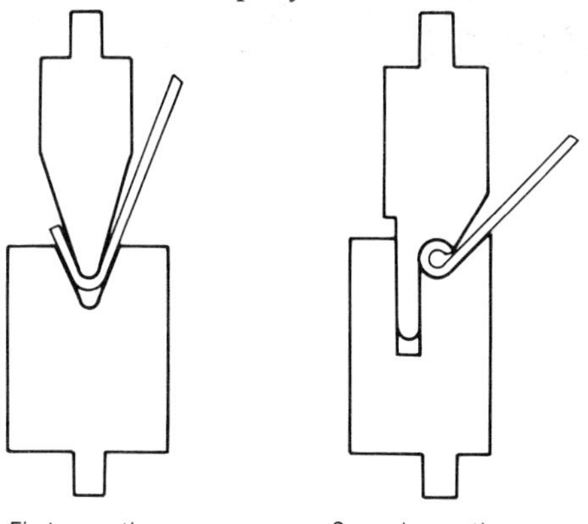

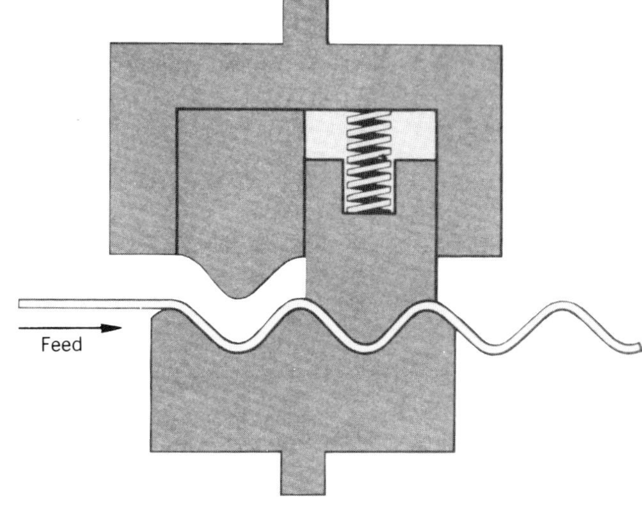

Figure 52
Press-brake forming of a corrugated sheet (Courtesy: Aluminum Company of America).

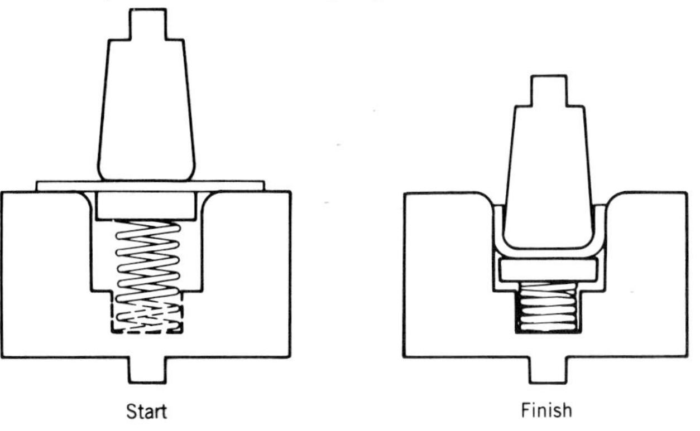

Figure 53
Brake tool with spring-loaded pressure pad in bottom die (Courtesy: Aluminum Company of America).

Figure 54
Brake tools used with urethane die (Courtesy: Aluminum Company of America).

Figure 55
Roll bending is used to form cylindrical parts (Courtesy: Aluminum Company of America).

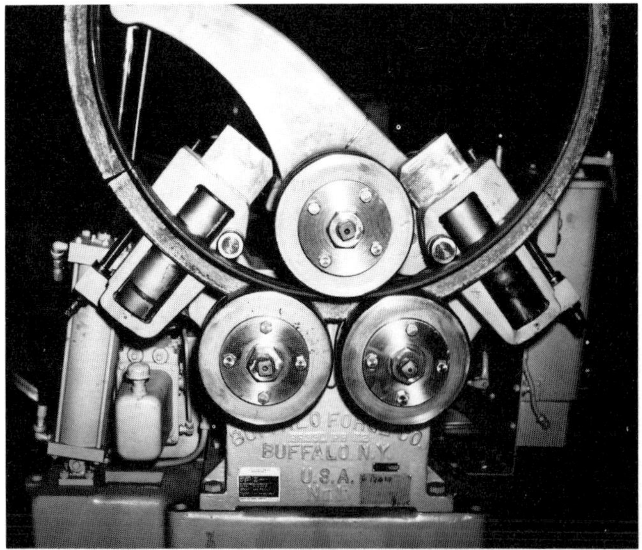

Figure 56
Vertical roll-bending machine. Heavy bars may be rolled into rings with this machine (Buffalo Forge Company).

Figure 57
The pinch-type roll bender is used to form shapes, such as channels, into rings (Buffalo Forge Company).

Figure 59
Typical roll-formed shapes (Courtesy: Aluminum Company of America).

Figure 58
Roll-forming machine (Courtesy: Aluminum Company of America).

Cold-roll forming (Figure 58) consists of bending a flat strip into a complex shape by a series of rolls, each contributing to the final shape. Channels, moldings, building trim, and rain gutters are formed by this method (Figure 59).

Straightening or **flattening** is a process designed to remove unwanted bends from metal. Sheet or bar stock can be straightened by being passed through a series of rolls, as shown in Figure 3. The metal is bent back and forth to a point slightly beyond its elastic limit, thus removing previous permanent deformations. Heavy sections are often straightened in a press by applying pressure on the bend while the part is supported at two points.

Stretch forming light-gage sheet metal can be done by stretching it over a contoured form or die block. This is done on a machine that has gripping jaws that are mechanically or hydraulically powered (Figure 60).

Stretch-wrap forming (Figure 61) is similar to stretch forming. The blank is first stretched beyond the yield point and then wrapped around a form block. Stretch-draw forming (Figure 62) is a shallow stamping process in which bottom and top dies are brought together to form a stretched sheet. The principal advantages of

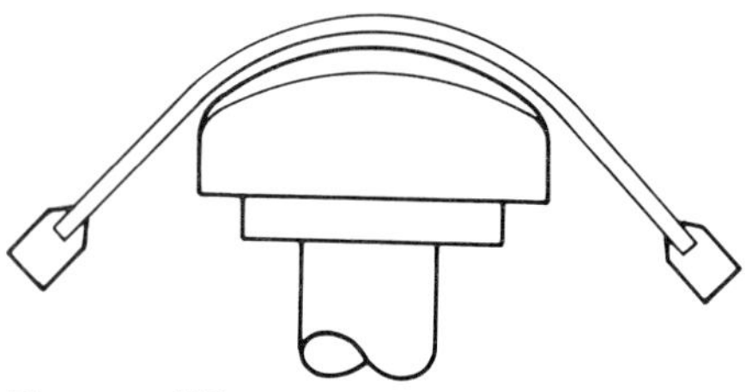

Figure 60
Stretch forming (Courtesy: Aluminum Company of America).

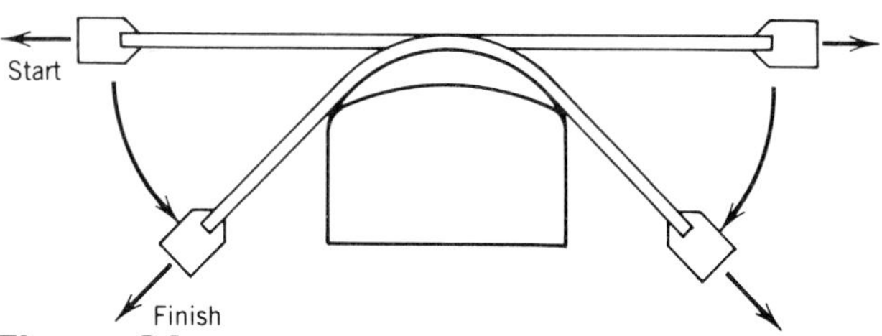

Figure 61
Stretch-wrap forming (Courtesy: Aluminum Company of America).

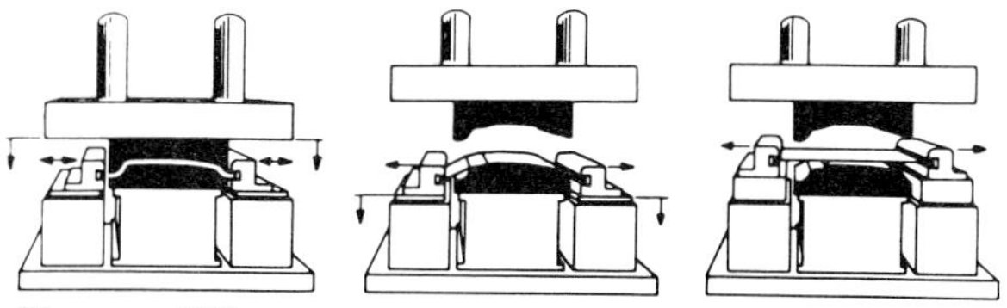

Figure 62
Stretch-draw forming (Courtesy: Aluminum Company of America).

stretch forming are that it greatly reduces springback, virtually eliminates wrinkles, and reduces tooling cost.

High energy rate forming processes (HERF) involve a high rate of workpiece deformation. The most important of these techniques are chemical (explosive), electrohydraulic, electromagnetic, and high energy mechanical forming.

Explosive forming (Figure 63) is used for operations such as shaping metal parts, cladding, joining, and forming powdered metal parts.

Electrohydraulic metalworking (Figure 64) is similar to explosive forming. Instead of a chemical explosion to produce shock waves, an electric arc along a bridgewire immersed in a liquid provides the high energy needed to form the metal.

With explosive forming there is virtually no limit to the size or thickness of plate that can be formed. The process is relatively slow but the original investment is also low. Springback is reduced. Large dome shapes and curved gore segments are produced by this means.

Magnetic pulse compression and expansion (Figure 65) is used for metals that are good electrical conductors. A magnetic field acts like a compressed gas to deform the metal. It is used to blank, bulge, compress, perforate, dimple, swage, flange, and assemble parts.

METAL SPINNING AND FLOW FORMING

Some metals such as titanium alloys do not draw well in press forming. However, they can be formed into cylindrical, tapered, or curved shapes by metal spinning (Figure 66). Metal spinning is a process in which

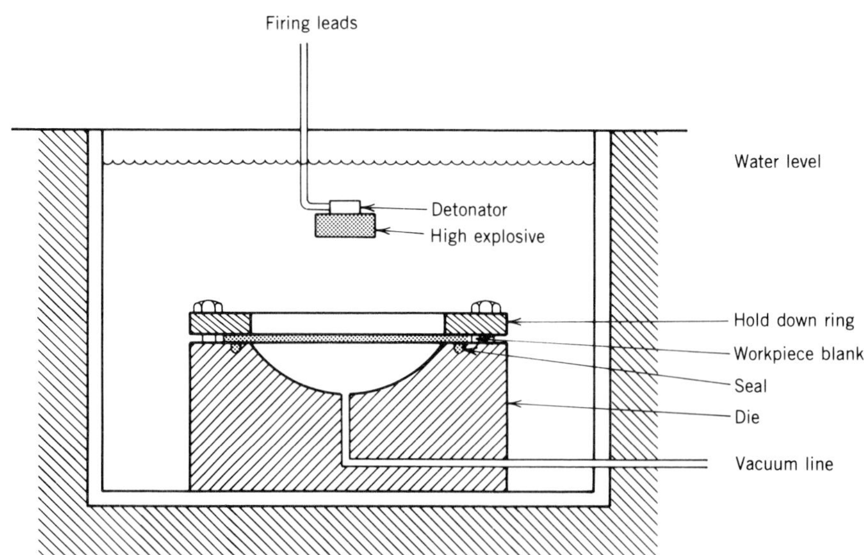

Figure 63
Explosive sheet and plate forming.

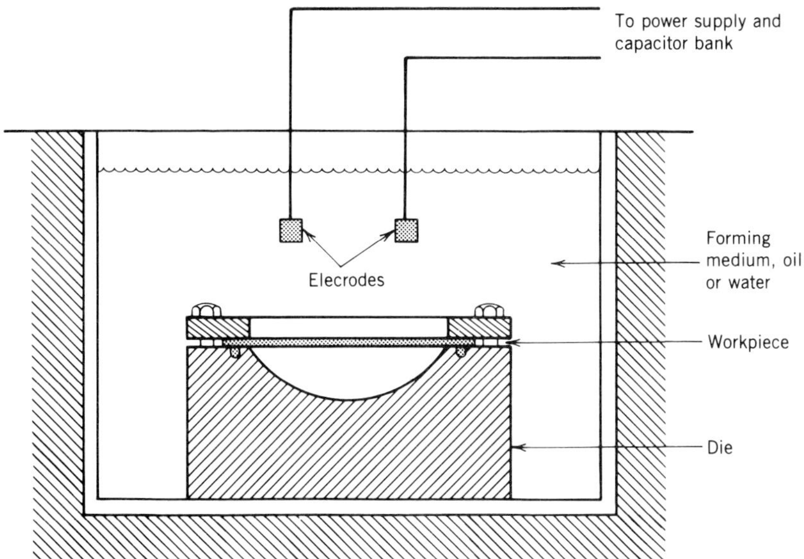

Figure 64
Components in electrohydraulic metal-working.

a disk of metal is rotated and forced against a form, called a chuck (Figure 67), with no intended reduction of wall thickness. In contrast, in flow forming (Figure 68) the blank is intentionally reduced in thickness while being lengthened. Objects such as seamless tubes and oxygen tanks and many industrial products and consumer items can be cold formed with this method. Shear spinning (Figure 69) is similar to flow forming in that the cross section of the blank is reduced. In shear spinning the blank is usually a disk instead of a tube.

Shearing Operations

Shearing is a sheet metal cutting operation that makes no chips. It consists of two straight- or square-edged hardened blades, the bottom one stationary and the top one fastened to a ram. Other forms of metal working that also shear metal are blanking, trimming, piercing, notching, and shaving. These shearing operations differ from simple shearing in that the cutting edges are curved, whereas shear blades are always straight.

Slitting is a shearing process in which rolls of sheet metal are cut into narrower widths. This is done by means of rotary shears in which the larger diameter slitting blades on one cylindrical roll fit into grooves on the mating roll.

Figure 65
Typical applications of magnetic pulse compression, expansion, and flat forming coils (Courtesy: Aluminum Company of America).

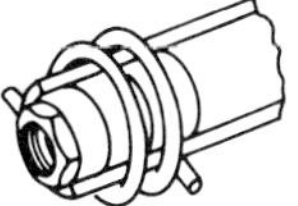

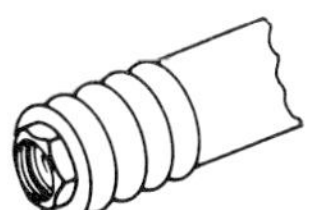

Figure 66
Products formed by metal spinning (Courtesy: Aluminum Company of America).

Figure 67
Diagram of metal spinning. Flat circular blanks are often formed into hollow shapes such as photographic reflectors. In a spinning lathe, a tool is forced against a rotating disk, gradually forcing the metal over the chuck to conform to its shape. Chucks and follow blocks are usually made of wood for this kind of metal spinning (Neely/*Metallurgy* 2 Ed., © 1984 John Wiley & Sons, Inc.).

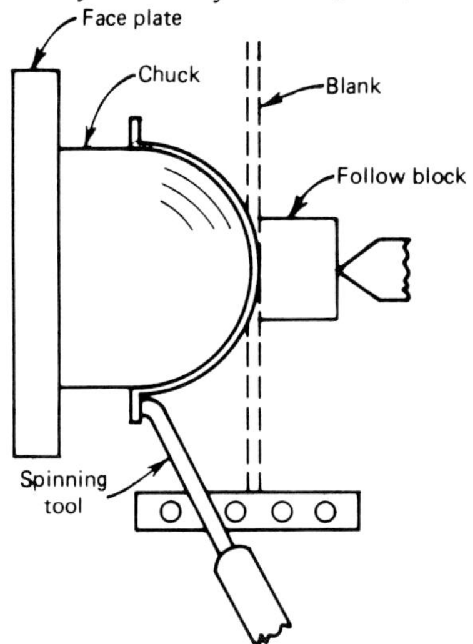

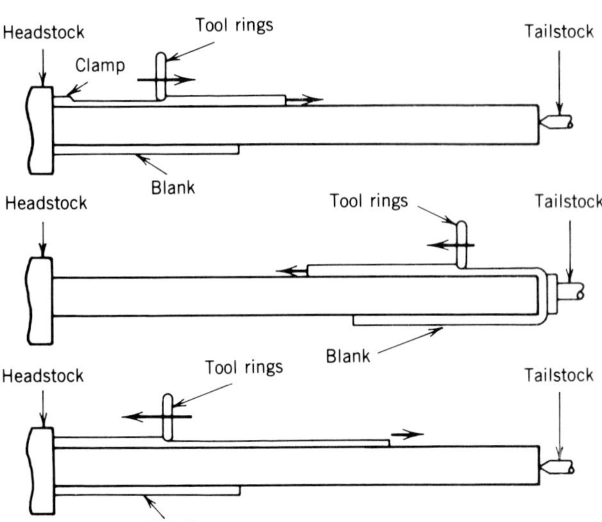

Figure 68
Flow forming diagram—spinning tubular parts (Courtesy: Aluminum Company of America).

Figure 69
The principle of the shear spinning process (Courtesy: Aluminum Company of America).

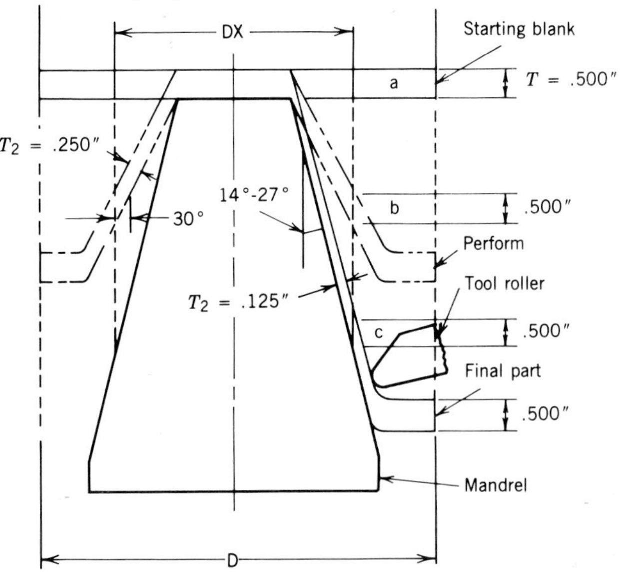

Figure 70
In the electric-resistance welding process, skelp coming through the forming rolls is welded automatically (Courtesy of Bethlehem Steel Corporation).

Figure 71
Pipe forming press-crimping tool ("Presses for All Applications of Metalforming" Schirmer-Plate-Siempelkamp, Hydraulische Pressen GmbH, Krefeld, West Germany).

FORMING PIPE AND TUBING

Pipe and tubing are formed with rolls in a manner similar to that of cold-roll forming. Smaller sizes of square, rectangular, and round tubing are rolled to shape and the edges are electric-resistance welded together (Figure 70). This process makes use of the natural resistance of the metal when an electric current is applied close to the edges to be joined to generate sufficient heat for welding. Larger-diameter pipes, such as those used for gas and oil pipelines, are made in special hydraulic presses (Figures 71 to 73). A final O-

Figure 72
Second stage in large pipe forming—U-canning transport ("Presses for All Applications of Metalforming," Schirmer-Plate-Siempelkamp, Hydraulische Pressen GmbH, Krefeld, West Germany).

Figure 73
Steps in formation of large steel pipe ("Presses for All Applications of Metalforming," Schirmer-Plate-Siempelkamp, Hydraulische Pressen GmbH, Krefeld, West Germany).

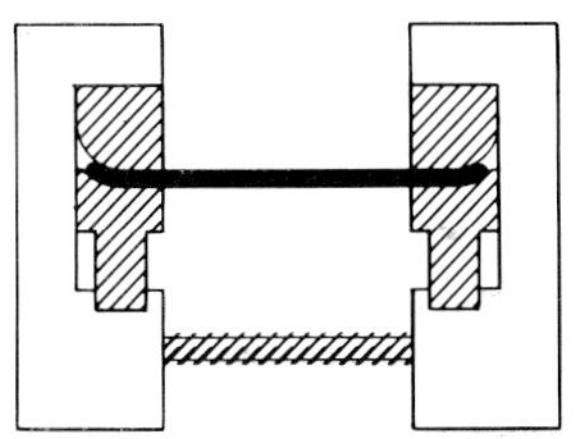

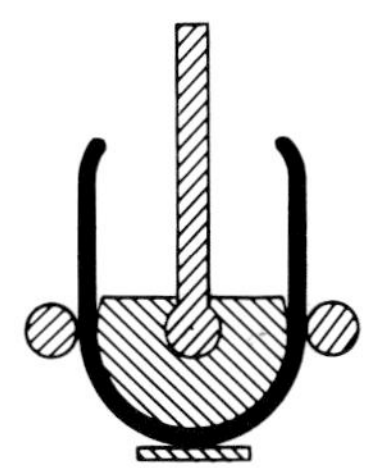

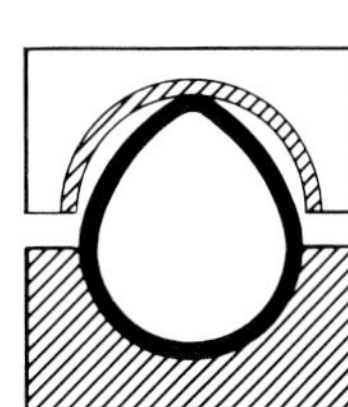

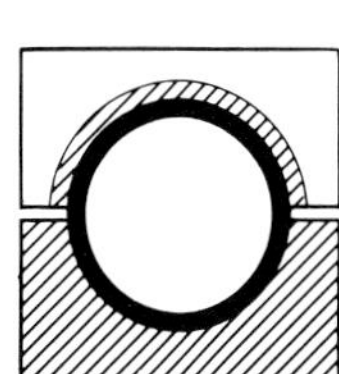

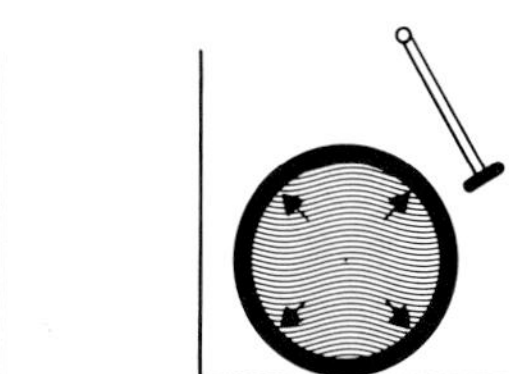

Figure 74
O-forming press gives the pipe its final shape ("Presses for All Applications of Metalforming," Schirmer-Plate-Siempelkamp, Hydraulische Pressen GmbH, Krefeld, West Germany).

Figure 75
The final step in making a large diameter steel pipe is welding the curved plates together by the submerged arc process (Courtesy of Bethlehem Steel Corporation).

Figure 76
Hydrostatic pipe tester ("Presses for All Applications of Metalforming," Schirmer-Plate-Siempelkamp, Hydraulische Pressen GmbH, Krefeld, West Germany).

ing press makes the cylindrical shape (Figure 74). The edges are joined by arc welding processes (Figure 75) and the pipe is tested hydrostatically by capping the ends and pumping in water under pressure (Figure 76). For some uses the pipe is galvanized or covered with a corrosion resistant coating.

RUBBER PAD PRESSES

Certain products that do not require deep draws can be produced on rubber pad presses (Figure 77). Forming is done with an elastic pad that presses sheet metal against forms made of metal or of dense laminated wood (Figure 78). The elastic pad is usually made of polyurethane or similar material. The forming principle is shown in Figure 79. The major advantages of this method of sheet metal forming are that manufacturing costs for either short or long runs are low, dies are inexpensive, and large objects can be made, up to 5 by 10 ft plan area. This process is particularly useful for aircraft component manufacturers. Materials such as titanium and stainless steel can also be formed at slightly elevated temperatures by this method.

Figure 77
Rubber pad press showing forming tools on the press table ("Rubber Pad Presses," Schirmer-Plate-Siempelkamp, Hydraulische Pressen GmbH, Krefeld, West Germany).

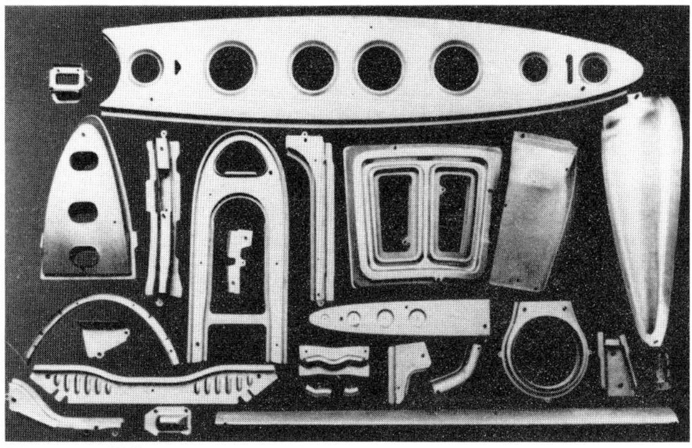

Figure 78
A large number of different components can be made simultaneously during one press cycle with rubber pad presses ("Rubber Pad Presses," Schirmer-Plate-Siempelkamp, Hydraulische Pressen GmbH, Krefeld, West Germany).

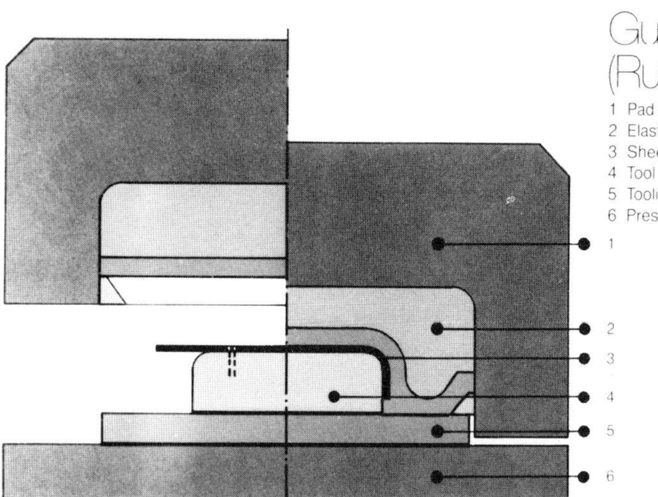

Figure 79
Principle of rubber pad forming process ("Rubber Pad Presses," Schirmer-Plate-Siempelkamp, Hydraulische Pressen GmbH, Krefeld, West Germany).

REVIEW QUESTIONS

1. What is the property of metals that makes possible the cold deformation of metals?
2. When considerable deformation is required in a metal, what can be done to avoid brittleness or rupture?
3. Are closer dimensional tolerances and better finish advantages of hot forming, or of cold forming?
4. In which type, cold forming or hot forming, is more massive machinery required to form a given size part?
5. What is springback?
6. What must be done to hot-rolled steel in the mill before it can be cold finished?
7. Cold-rolled steel strip is the material used for many products such as auto bodies and kitchen appliances. How is it made?
8. Aluminum foil is used every day in our homes. How is it made?

9. What is the difference between blanking and punching?
10. How can a sheet metal blank be formed into a cup shape?
11. How is the process of making wire similar to that of cold drawing bars?
12. The cold-forming upsetting machines can make parts from wire stock at a high rate of production, as much as 36,000 parts per hour. What kind of parts are usually made on these machines?
13. Larger, more complex products than those of Question 12 are made from bar stock on cold-forming machines that do combined operations in a sequence of operations. Name two products made on these machines.
14. How are threads, worms, and gears cold formed?
15. Threads cannot be rolled on gray cast iron, die cast metals, and 301 stainless steel. Why is this so?
16. In what ways are rolled threads superior to cut threads?
17. Would extrusion or cold rolling be more practical for producing various odd-shaped cross sections from soft metals such as aluminum in short runs? Why?
18. When press brakes or bar folders are used, what is the process called: forming, shaping, bending, or drawing?
19. What is the difference between metal spinning and flow forming?
20. Large steel pipes are electric arc welded after being formed. How are small round, square, or rectangular tubes joined at the seams after being formed?
21. If thin aluminum panels, 4 by 8 ft, needed a shallow, ribbed form to give them rigidity and 300 were required, which would be more economical to use, standard drawing dies, or a rubber pad press?

CASE PROBLEMS

Case 1
Cold Working versus Hot Forming

Automotive wrist pins were once made only on automatic screw machines, an operation that turned 40 percent of the starting stock into scrap. Wrist pins are usually less than 1 in. in diameter, about 3 or 4 in. long, and hollow. These have been formed of steel for the past 20 years. They must have a good finish and accurate dimensions. Which method, hot forming or cold forming, do you think the manufacturers chose for making this part?

Case 2
Stainless Steel Nuts

A manufacturer of fasteners uses a four-die nut former to produce blanks. They received an order for 80,000 stainless steel Type 305 hex nuts. A trial run was made with the stainless steel bar stock and the piercer punch began to break down after only a few hundred nuts were produced. What change in the operation can be made so the manufacture of these nuts can be made profitable with no breakdowns?

Case 3 (Part 1)
Punching Pressure

The rule for finding the approximate pressure for punching circular holes in sheet steel is:

$$P = D \times t \times 80$$

where

D = diameter of punched hole
t = thickness of sheet steel
P = pressure (force in tons)

You are given a die set having four punches: two are $1\frac{1}{4}$-in. diameter and two are $\frac{3}{4}$-in. diameter. You are asked to determine the press tonnage required to punch holes in $\frac{1}{2}$ in. sheet steel. What size press will be required for this operation?

Case 3 (Part 2)
Punching Pressure

When a punched part is not circular, the same formula as in Part 1 may be used but instead of the diameter of the hole, one-third the length of the perimeter or outline of the hole is used. You are asked to determine the pressure required to punch a rectangular hole in $\frac{1}{4}$-in. steel plate, of which the length is $3\frac{1}{2}$ in. and the width is $2\frac{1}{2}$ in. What size press will be needed?

Case 4
Force Required to Shear Plate

Squaring shears have an upper and lower blade or knife (Figure 80). The lower one is in a level position and is

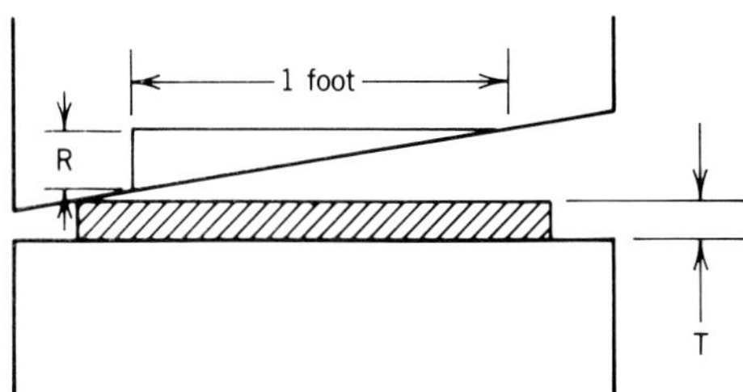

Figure 80

stationary, whereas the upper one moves up and down and has rake to allow for a gradual cutting action. Most plate shears have $\frac{1}{2}$ in./ft rake and scrap shears have up to 2 in./ft rake because distortion of the cut piece is not important and larger rakes require less force. One formula to determine the force required to shear mild steel plate is:

$$\text{maximum } KP = \frac{174{,}000 \times T^{1.6}}{R^{0.79}}$$

where

KP = knife pressure, lb
R = rake, in./ft
T = plate thickness, in.

Using this formula, calculate the pressure required to shear $1\frac{1}{4}$ in. steel plate with $\frac{1}{2}$ in./ft rake, and with 2 in./ft rake. Also determine KP for $\frac{1}{2}$ in. plate with $\frac{1}{2}$ in./ft rake.

CHAPTER 10 POWDER METALLURGY

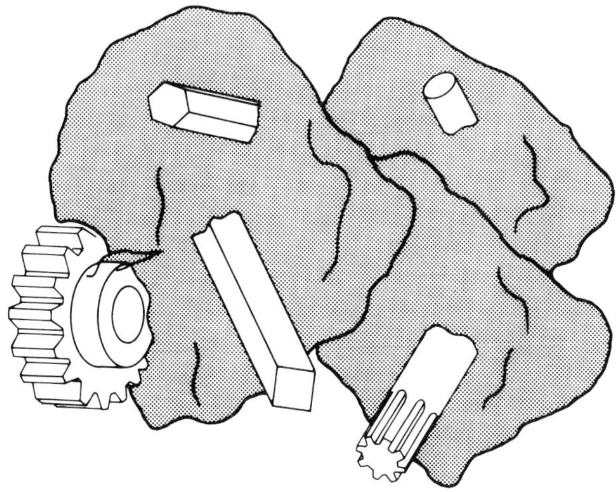

Powder metallurgy, commonly known as P/M, is essentially the compression of finely divided metal powder into a briquette of the desired shape which is then heated but not melted to form a metallurgical bond between the particles. It is one of the four major methods of shaping metals. The other three are machining, hot or cold plastic deformation, and casting.

Oil impregnated P/M bearings have been in use in automobiles since the late 1920s, but acceptance of this method has grown substantially in recent years. Technological advances in P/M are growing so rapidly that some consider it to be a new process of forming metals despite its having been in use for several decades. Even sheet stock is being produced by P/M and products that are difficult if not impossible to produce by other means are being manufactured with P/M at high production rates.

Over one-half the P/M products manufactured are used in transportation (automobiles and trucks). Parts used in farm and garden equipment and household appliances are also made by P/M. Many new applications will no doubt be found in the future for this unique method of forming metals.

OBJECTIVES

This chapter should enable you to:

1. Understand and explain the processes involved in simple die compaction.
2. Explain the methods by which metal powders are produced.
3. Describe the metallurgical principles involved in bonding of powders in the sintering process.
4. State the advantages of several advanced powder metal processes.
5. Give reasons for deciding to use P/M processes for manufacturing a part.

HOW P/M PARTS ARE MADE

The basic traditional process of making P/M parts consists of three steps: **compacting** or molding, **sintering**, and **secondary operations** (Figure 1). In the molding sequence, loose powder (or a blend of powders) is placed in a die and compacted between punches at room temperature. The ejected part, called a green compact or briquette, is now a solid shape, but it can easily be broken or chipped by mishandling (Figure 2). The briquette is then heated in an appropriate atmosphere to a temperature high enough to cause the powder particles to bond together by a solid state diffusion Figure 3) and to homogenize any alloy constituents in the powder. Melting does not normally occur. The P/M part is now ready for use unless other finishing operations are needed.

Secondary operations may include sizing (cold coining, for example), machining or grinding, heat treating, tumble finishing, plating, or impregnating with oil, plastics, or liquids. Secondary operations increase the cost of the finished part; therefore, a goal of the designers in P/M is to limit the use of secondary operations and, if possible, complete the article in the first two steps of compacting and sintering. However, the sintering process tends to deform and shrink the shaped briquette slightly. Such parts as precision gears should have a finishing operation in which they are forced through a sizing die.

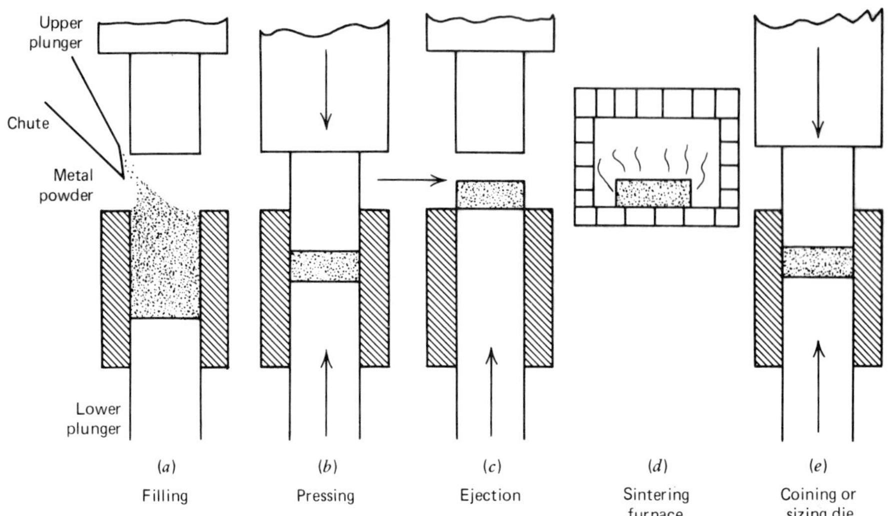

Figure 1
Sequence of operations for powder metallurgy. The depth of die cavity and the length of plunger stroke is determined according to the density required. (*a*) A measured amount of metal powder is placed in the die cavity. (*b*) Pressure is applied. (*c*) The briquette is ejected from the die cavity. (*d*) The parts are sintered at a specified temperature for a given length of time. The parts are now ready for use. (*e*) If more precision is needed, they can be sized in a coining die (Neely/*Metallurgy* 2 Ed., © 1984 John Wiley & Sons, Inc.).

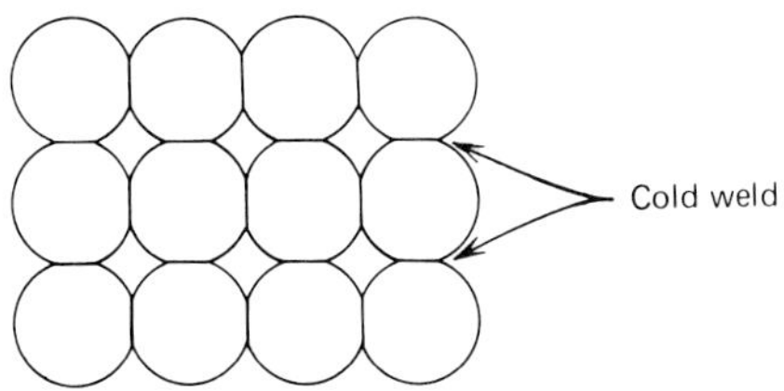

Figure 2
The compressed particles in the briquette are cold welded together at this stage with very weak bonding (Neely/*Metallurgy* 2 Ed., © 1984 John Wiley & Sons, Inc.).

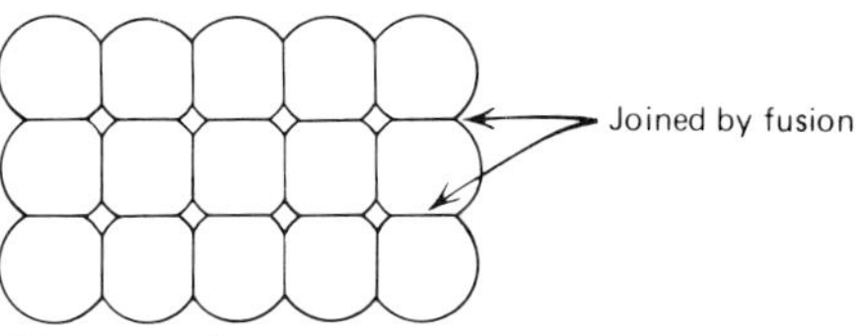

Figure 3
After sintering, the particles in the briquette become fused together (Neely/*Metallurgy* 2 Ed., © 1984 John Wiley & Sons, Inc.).

METAL POWDERS

Some of the powders from which P/M products are made are iron, alloy steel, stainless steel, copper, bronze, tin, lead, nickel, cobalt, tungsten carbide, titanium carbide, and tantalum carbide. The three most important methods of producing metal powders are **reduction of oxides**, **electrolysis**, and **atomization**. By far the largest quantities of powder produced are of iron. The raw material may be either a high grade iron ore or mill scale. The ore is mixed with coke and limestone and heated at 1950° F (1066° C) in ceramic containers for several days. The resulting sponge iron is crushed, impurities are separated out by magnetic separation, and then it is further reduced with hydrogen in a continuous belt furnace. Iron powder from electrolysis is characterized by its high purity and high compressibility. Iron is deposited as a brittle sheet on the cathode of an electrolytic cell in an electrolyte of iron sulfate or iron chloride. The sheet is stripped off the cathode, washed, dried, and ground into powder of the desired fineness.

The advantage of the atomization process is that it uses scrap steel to make iron powder. The steel is melted and carburized in a cupola and the molten metal is sprayed or atomized in a nozzle with compressed air. This produces hollow spherical particles containing carbon with a coating of iron oxide. This oxide and carbon containing material is then reduced and annealed in a continuous furnace.

In all of these processes, the powders must be ground further, usually in a ball mill (Figures 4 and 5) to the desired fineness. Metal powders are screened and larger particles are returned for further crushing or grinding. The powders are classified according to particle size and shape in addition to other considerations such as carbon content, presence of impurities, density, and metallurgical condition of the grains. Particle sizes range from about .002 in. to less than .0001 in. in diameter. Test sieves are used to determine particle size; the smallest is the 325-mesh sieve which has an opening of 44 microns. This method of testing has been standardized throughout the industry. Various powders are often blended by tumbling or mixing. A lubricant such as graphite is added to keep the punch and die from seizing or scoring.

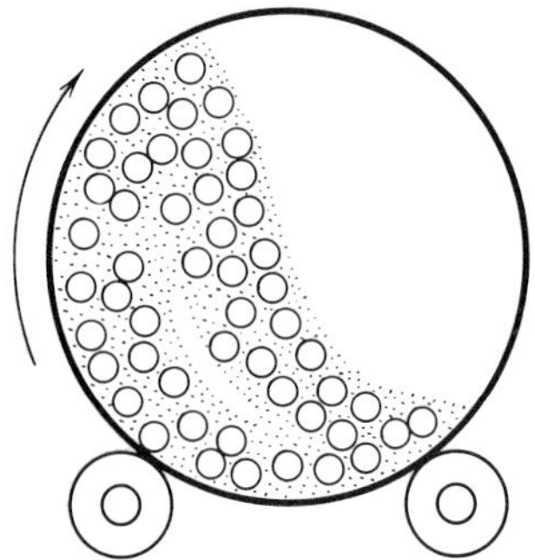

Figure 4
Action of the ball mill is shown as a continuous grinding as the drum rotates (Neely/*Metallurgy* 2 Ed., © 1984 John Wiley & Sons, Inc.).

Figure 5
Blending metal powders (Pennsylvania Pressed Metals Inc.).

Figure 6
Custom-blended powder is precision metered into a die cavity and compacted under high pressure (Pennsylvania Pressed Metals Inc.).

Figure 7
Green compacts of gears are produced and stacked ready to be taken to sintering furnaces (Pennsylvania Pressed Metals Inc.).

COMPACTION OF POWDERS

Compacting or pressing gives powder products their shape. In the basic process of compacting, the powder is pressed from the bottom upward and from the top downward, in two dimensions, not from the sides. Unlike liquids, which flow in all directions under pressure, powders tend to flow mainly in the direction of the applied pressure (Figures 6 and 7).

Mechanical presses (Figure 8) are favored when the applied load is not too high. Eccentric (crank) presses rarely exceed 30 tons (60 tons would be considered quite large). Toggle-type presses may reach a 500-ton pressure. The advantage of mechanical presses is that they have a rapid rate of production, as much as 50,000 parts per hour. Hydraulic presses (Figure 9) can exert very high forces, as much as 5,000 tons or more. They

Figure 8
Automatic presses have controlled molding speeds up to 50,000 parts per hour (Pennsylvania Pressed Metals Inc.).

Figure 9
Large hydraulic press used for briquetting metal powders or sponge for experimental purposes (Bureau of Mines).

have the advantage of a long stroke and an easily adjustable stroke length, but their operating speeds are much slower than those of mechanical presses. Simple die compaction of powders with these various presses has such advantages as speed, simplicity, economy, and reproducibility. It produces a strong, dimensionally accurate, and relatively inexpensive product. However, it does have limitations. Length-to-diameter ratios must be relatively small; long parts will have uneven densities, being denser nearest the punches, and so they would have uncertain properties and should not be made by die compaction. Grooves or undercuts or parts with thin sections cannot be made by simple die compaction. Not every part is a good candidate for powder metallurgy. However, some of these limitations are overcome by alternative forming techniques such as split die techniques to provide undercuts, isostatic pressing, and densification methods.

Advanced Processes

Metal powder forming can be separated into two divisions, cold compaction and hot compaction. Die compaction can be done either hot or cold. Engineering properties such as tensile and compressive strength depend to a great extent on the **density** of the pressing. **Hot pressing**, in which the powder is pressed in the die at a high temperature, produces a density approaching that of rolled metal.

Since dies can only compact powder in one axis, some shapes such as hollow hemispheres, long parts, and internal threads cannot be made by this process. But there is a method in which pressure can be applied from all directions: isostatic pressing. In cold isostatic pressing (CIP), the powder is loaded into molds made of rubber or other elastomeric material and subjected to high pressures at room temperature (Figure 10). Pressure is transmitted to the flexible container by water or oil. The parts are removed and sintered, followed by secondary operations if needed. With hot isostatic pressing (HIP), an inert gas such as argon or helium is used in a pressure chamber to provide the squeeze. This gas is reclaimed between each batch of pressings. Hot isostatic pressing provides more density and achieves a finer microstructure than the cold process. Powders are often preformed to an oversize shape prior to placing them in the isostatic chamber. Heat is applied to the preform by induction for a short time while the gas pressure compacts the preform. Temperatures may be as high as 1600 to 2000° F (871 to 1093° C) with pressures of 15,000 PSI or higher. Isostatic pressing is useful only for certain special applications. CIP is a comparatively slow process, 120 parts per hour would be considered a very high rate. The HIP is an even slower process. However, its advantages are low cost

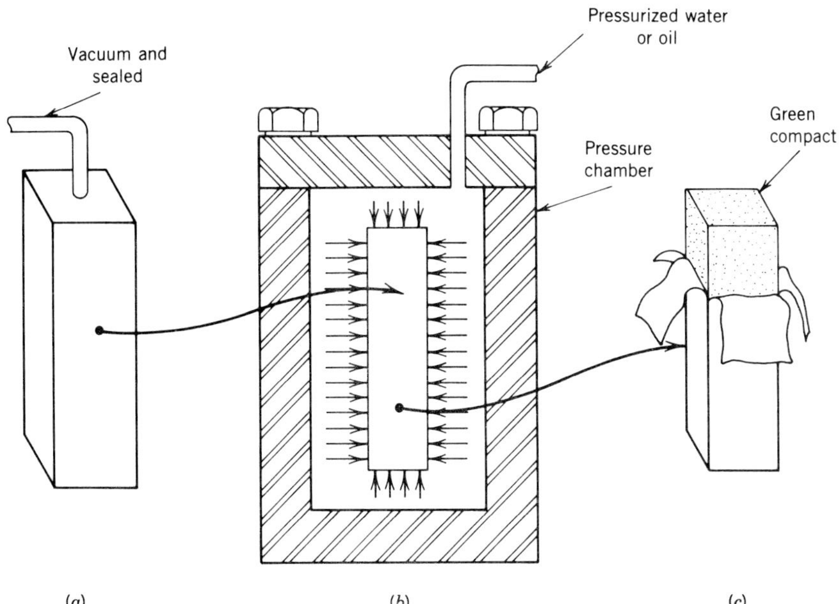

Figure 10
Isostatic pressing. (*a*) Prepared powder is placed inside a flexible container or mold. A vacuum is drawn in the mold and it is then sealed. (*b*) The powder and mold are then placed in a pressure chamber into which water or oil is pumped under pressures of 15,000 PSI or more. (*c*) The green compact is removed from the pressure chamber and the container is stripped off.

tooling versus expensive dies. Parts made by either CIP or HIP are not limited by the shape constraints of rigid tooling.

Powder Forging

Fully dense P/M parts equalling or surpassing those of wrought products in mechanical properties such as tensile and compressive strength are being produced in commercial quantities by powder forging (P/F). The green compact or preform is made in a conventional press and it is sintered, followed by a restrike (forge) that brings it to the final density. Mechanical properties may sometimes exceed those of wrought metals because a more uniform composition is achieved in P/M processes. Fatigue and impact strengths are particularly high in powder forgings. P/F bearing races have been shown to outlast wrought steel races by a factor of 5 to 1.

Metal Powder Injection Molding

A new P/M technology that borrows a plastic injection molding (I/M) process shows great promise for production of precision small parts. In fact, some variations of this new process can use plastic molding machinery. In order to inject powders into molds, the particle size must be much finer than that used for conventional P/M processes. This "dust" is combined with a thermoplastic binder. The molding step is performed at injecting pressures of about 900 PSI and about 325° F (163° C). The result is a green compact that is sintered in the conventional fashion after being debinderized in an oven at about 400° F (204° C). This removes the plastic binder as an evaporative process. Thin walls, high densities, unsymmetrical shapes, and accurate dimensions are possible with this method. Of course, injection molding is not a high-production method compared to simple die pressing.

Metal Powder-to-Strip Technology

The direct rolling of metal strip from a powder slurry has been under development for many years and is now beginning to enter commercial production. Since conventional production of strip requires large amounts of energy and large investments in rolling mills, there is a need for a low-cost method of producing short runs of metal strip. The powder-to-strip process produces thin strip directly without hot or cold rolling. In this process, an appropriate powder mix is blended with water and a cellulose binder to form a fine slurry. The slurry is deposited on a moving band as a continuous film (Figure 11). After drying, the moving strip is compacted between rolls and then sintered, first to remove the binder and then to bind the particles. It is rolled a second time and resintered to remove porosity. As in all of these advanced P/M processes, metals or alloys not possible to form in any other way can be produced with powdered metals. Bimetal alloys can be produced in strip and high strength titanium strip can now be produced for the aircraft industry.

Powder Extrusion

Metal powders can be hot extruded with or without presintering. Metal powders are placed inside a copper can which is evacuated and sealed. The assembly is

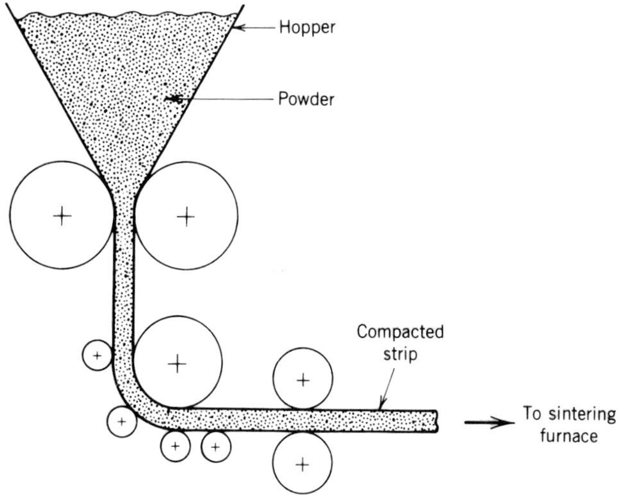

Figure 11
Powder rolling can produce this compacted strip of difficult-to-work, refractory, or reactive metals.

heated and extruded as a single unit. Metal billets and tubing from powder are made by this process.

SINTERING

In **solid phase sintering**, which is most common, two or more powders (constituents) having different melting points are combined. The green compact part must be heated from 60 to 80 percent of the melting point of the lowest melting constituent. This usually requires from one-half to two hours in a sintering furnace to produce metallurgical bonds. **Liquid phase sintering** is carried out above the melting point of one of the constituents or above the melting point of an alloy formed during sintering. The major constituent powder never melts.

The important changes that take place during the process of solid phase sintering are:

1. **Diffusion** This takes place on or near the surface of the particles as the temperature rises. For example, any carbon present in the voids (spaces between the particles) will diffuse (penetrate) into the metal particles.
2. **Densification** Particle contact points increase considerably. Voids decrease in size, therefore lowering porosity. As a result, there is an overall decrease in the size of the part during the sintering process. A green compact, therefore, must be made larger to allow for this shrinkage.
3. **Recrystallization and Grain Growth** Sintering is usually carried out well above the recrystallization temperature of metals and grain growth can occur between particles in a contact area. Methods of inhibiting excessive grain growth are often used since large grains tend to weaken metals.

When one of the blended metal powders has a melting point below the sintering temperature, a liquid phase of that metal fills the voids between the particles that do not melt. Density and strength are increased and porosity is reduced by this method.

In liquid phase sintering, **infiltration** is a process in which the pores or voids of a sintered or unsintered compact are filled with a metal or alloy of a lower melting point. For example, an iron or steel compact is heated in contact with a molten copper alloy at a temperature lower than the melting point of the steel compact. The molten copper is drawn into the pores of the compact and ideally fills the entire void area. This process increases densities and tensile strengths considerably.

High density and low porosity are not always desireable. Porous filters or prelubricated bearings are produced by loose sintering or by combining the powder with a combustible or volatile substance, such as sawdust, that is later removed by sintering after the green compact is made. Very large parts are sometimes formed with very low pressures or none at all, and then loose sintered (called pressureless sintering) followed by a cold forging operation.

Sintering furnaces on production lines are usually of the continuous type (Figure 12). Furnace atmospheres are usually of a hydrocarbon gas. However, with certain metals or alloys other gases may be used. Some manufacturers use a nitrogen gas atmosphere for both ferrous and nonferrous metals.

Figure 12
Molded briquettes are sintered in controlled atmosphere furnaces to bond metal powder particles at temperatures up to 2200° F (1204° C) (Pennsylvania Pressed Metals Inc.).

Figure 13
Precision powder parts such as gears must be finished by resizing to increase their accuracy and density (Graebener Press Systems, Inc., Providence, RI).

SECONDARY OPERATIONS

For many products the slight dimensional changes that take place during sintering are acceptable. Where close dimensional tolerances must be maintained, such as in self-lubricating bearings, the product must be resized after sintering (Figure 13). Sizing is similar to coining; the part is forced into a sizing die of the correct shape by a sizing punch.

Small electric motors and many other mechanisms have sintered, porous, prelubricated bearings that last the lifetime of the motor. In some operations in the food and textile industries, extra oil from periodic lubrication could contaminate a product. These porous bearings are ideal for these uses. Self-lubricating bearings are able to absorb as much as a third of their weight in oil, which is stored in all of the pores of the metal. A thin film of oil covers the shaft and, when the temperature rises from rotation of the shaft, more lubricant is supplied. When the shaft stops and the temperature drops, the oil is reabsorbed by capillary action, thus none is lost by dripping or leakage.

Articles can be finished by plating, but because of the porous quality of sintered products, precautions should be made to avoid trapping the plating electrolyte in the pores. This could cause corrosion and ultimate failure of the part. Sintered parts that are infiltrated with lower melting point alloys are less likely to have this problem. These articles can be painted or coated using the same processes used for rolled or cast metals.

Machining and precision grinding are common secondary operations. Drilling and threading holes and reaming operations are often necessary for plain die compacted products because they cannot be done in the die. Cylindrical or flat surfaces are often ground because many P/M parts are relatively hard.

Depending on the content, P/M parts can be heat treated by the same methods as metals formed by other processes. Steel parts can be hardened by quenching and then tempered. Annealing is sometimes necessary before machining can be performed.

P/M PRODUCTS AND THEIR USES

A wide array of small parts made by the P/M process can be seen in Figure 14. Manufacturers often choose P/M over such methods as forging, cold forming, or machining to produce these products because of the following characteristics of P/M processed parts.

1. Superior engineered microstructures and properties with precise control.
2. Consistent properties and qualities.
3. Controlled porosity for filters and self-lubrication.
4. Very low scrap loss.
5. Wide variety of shape designs.
6. Unlimited choice of alloys and composites.
7. Low-cost, high-volume production.
8. Good surface finishes.
9. Close dimensional tolerances.
10. Little or no machining required.

A single part can be made having different properties in different areas of the part. A part can be made in one piece by the P/M process whereas several pieces would be needed with other methods. Internal and

Figure 14
These are some of the thousands of intricate shapes and designs that are produced by the powder metallurgy process (Pennsylvania Pressed Metals Inc.).

external gear teeth that end at a shoulder can easily be formed by P/M methods.

Surprisingly, P/M steels are almost as strong (in terms of tensile strength) as machining steels, and powder forging processes bring the tensile strength up further, to the same level or above. The versatility of the P/M process allows some parts to be designed lighter, and objects of very hard metals, such as tungsten carbide cutting tools for machine tools, to be made. Friction materials in the form of bimetal powder materials that are bonded to a steel base and aluminum-based anti-friction materials containing graphite, iron, and copper in small percentages are made by the P/M process. Copper–nickel powders are often formed as a layer on steel strip and then sintered. The sintered strip is impregnated with babbitt metal and formed into insert bearings for automobile and aircraft engines.

Some disadvantages of P/M are found in the conventional cold die compacting and sintering process. Since these products are somewhat porous and present a larger internal surface to any corrosive atmosphere, they have a somewhat lower corrosion resistance than solid metals. These particular products tend to have slightly poorer plastic properties (ductility and impact strength) than conventionally produced metals.

FACTORS FOR DESIGN OF P/M PRODUCTS

When parts are made by the cold, plain die compaction process, several elements of design should be observed. Thin sections and feather edges should be avoided (Figure 15). Generous fillet radii should always be provided in a die (Figure 16), and internal holes should have rounded corners (Figure 17). External corners should be chamfered (Figure 18), and narrow deep slots should be avoided (Figure 19). Splines or keyseats should have rounded roots.

Using secondary machining processes is common practice for making precision P/M parts. Holes, tapers and drafts, countersinks, threads, knurls, and undercuts usually must be machined after the parts are sintered. However, holes in the direction of pressing are readily produced in P/M parts. Round holes are easiest to produce but shaped holes, keys, splines, hexagonals, squares, and any blind holes are easily made.

Holes that are not in the direction of pressing generally have to be machined later. Some redesign often can eliminate secondary machining (Figures 20*a* and 20*b*). The purpose of the undercut must be kept in mind. In this case, it is simply a relief for a sleeve whose internal edge should not contact any radius on the flange. A preferred arrangement that will still do the job and be even stronger is shown in Figure 20*b*. Vertical (straight) knurls and splines can be made in the

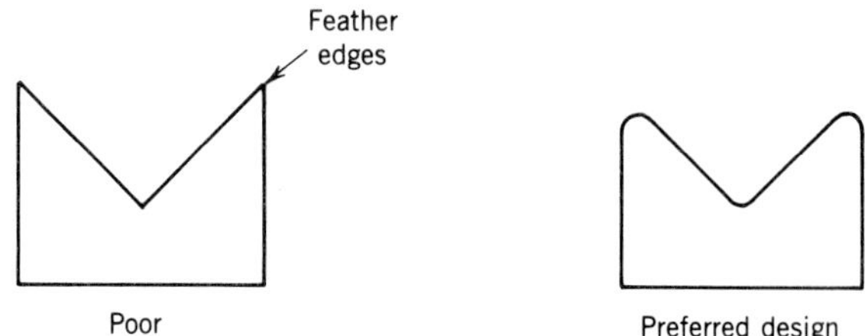

Figure 15
Metal powders do not fill dies with sharp feather edges well. A preferred design is one with rounded corners.

Figure 16
Sharp corners on shoulders should be avoided. A fillet radius should be used.

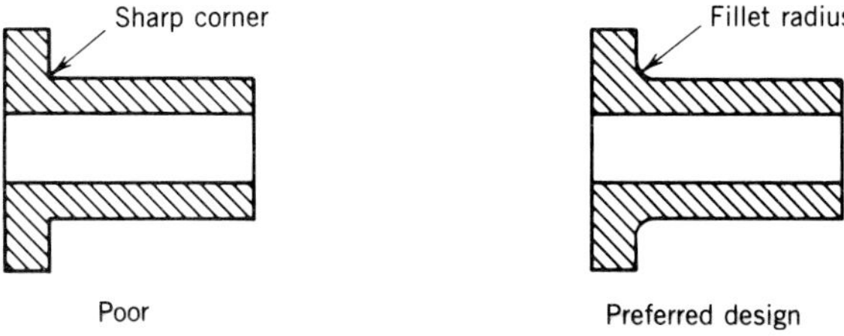

Figure 17
Avoid sharp corners. Internal shapes should have rounded corners.

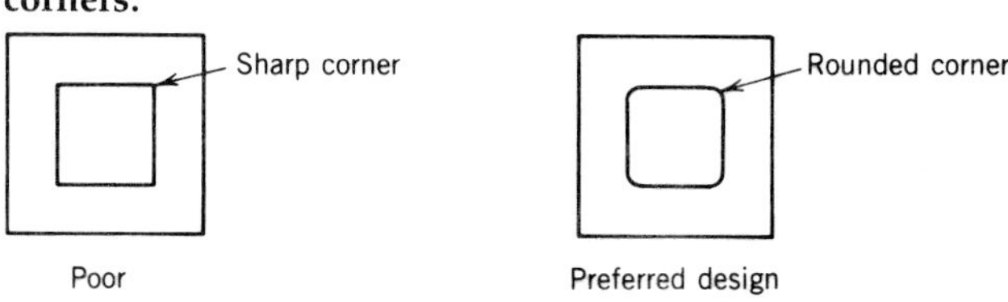

Figure 18
Both of these chamfers are used, but the 30-degree angle is the better form.

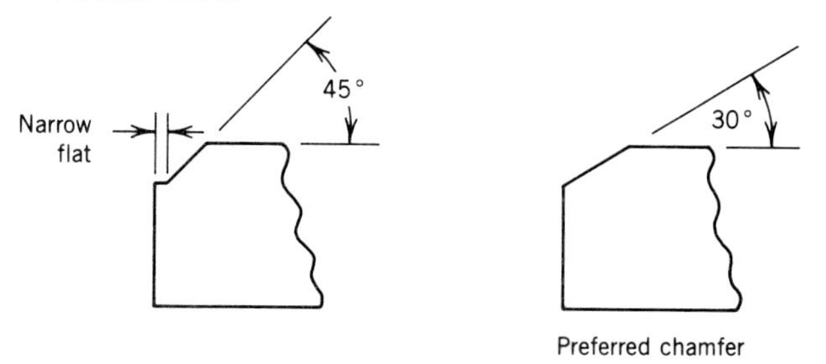

Figure 19
Deep slots are not easily made by the P/M process. Shallow grooves with rounded bottoms are better.

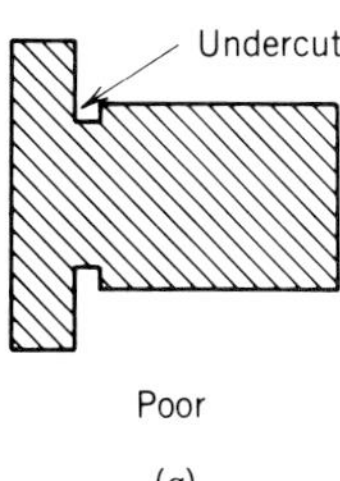

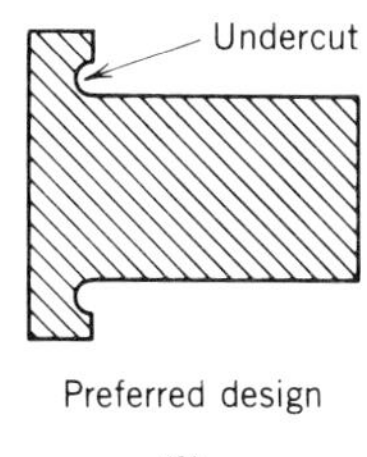

Figure 20
Undercuts as in (*a*) are difficult to make, requiring special expensive dies. The design for an undercut at (*b*) serves the same purpose and is easily done with simple pressing dies.

direction of pressing but diamond or angled knurls cannot be pressed because they would interfere with part ejection.

P/M is the ideal way to make cams and gears. Combinations of stepped or clusters of gears or gears and cams together are easily produced by die compaction. Gears with blind corners, both internal and external, are very difficult to make by other processes, such as machining or forming, but are easily produced by P/M processes. Helical gears with angles up to 45 degrees can theoretically be made, but gears with angles up to 30 degrees are manufactured in quantity. Hubs for gears and sprockets should be located as far as possible from the root diameters of the gear; that is, the hub should be as small as possible while retaining sufficient strength.

Probably the greatest advantage in adopting the P/M process over other methods of manufacture is that it allows for redesign that makes use of the great versatility of the P/M process. Many small mechanisms such as those found in pneumatic drills, electronic printers and sequencers, door locks, firearms, and sewing machines have a number of small parts that can often be combined into one piece by redesigning for powder metallurgy. Of course a great savings in production can be made and a simplification of the mechanism can be made as well. P/M is not suitable for every metal product, but its use should always be investigated when making a decision for designing a part to be manufactured.

REVIEW QUESTIONS

1. In simple die compacting, what are the three steps in producing a precision finished part?
2. Name the three most important methods of producing metal powders.
3. A heat-resistant alloy part that has a melting point higher than that of steel is to be used to make 20,000 small symmetrically shaped parts. Both P/M and die casting processes have high production rates and making the dies would cost about the same for each. Which method would you choose? Why?
4. Why would a bushing or sleeve of 1 in. diameter and 4 in. in length have uncertain mechanical properties and probably fail in use if it were produced by simple die compaction?
5. Name two methods of increasing the density of P/M products.
6. Simple die compaction presses metal powders in only one axis, making undercuts, threads, and cross holes impossible. By what two methods of compaction can these operations be done?
7. Name three processes of metalworking other than die compacting of powders that have been adopted for P/M.
8. In liquid-phase sintering, does all of the metal powder reach the melting point, at least momentarily? Explain.
9. How does infiltration increase density?
10. What method is used to make metal filtration devices and prelubricated porous bearings?
11. Is it possible to harden and temper a P/M part to make a wear resistant mechanical part?
12. A manufacturer must tool up to produce 20,000 metal handwheels with a knurl on the outer edge of the wheel. The parts will be produced with P/M simple die compaction. The type of knurl is not specified. What kind of knurl, diamond or straight, should be used? Why?

CASE PROBLEM

Case 1
Improving a Good Product

A drive hub and flange has been successfully produced by the P/M process from iron powder. However, under some shock load conditions, the hub failed. It was found to have considerable porosity that contributed to its somewhat low impact strength. How can this product be made more dense with increased impact strength without changing the process from plain die compaction and without changing present die size?

CHAPTER 11 PRINCIPLES OF MACHINING PROCESSES

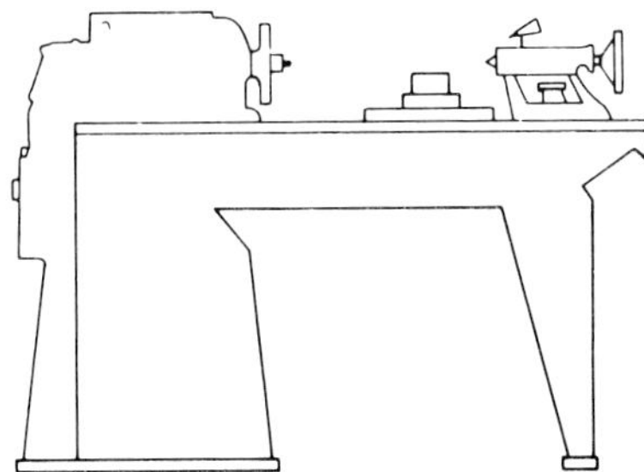

Machining is essentially the process of removing unwanted metal from bar stock, forgings, or castings to produce a desired shape and dimension. It is one of the four major processes for forming metal. The other three are deformation of hot or cold metal, casting of molten metals, and powder metallurgy. Machining is done by shaving away the metal in small pieces, called chips, using very hard cutting tools and powerful, rigid machine tools. The tool material can be held stationary and moved across a rotating workpiece as on a lathe, or a rigidly held workpiece moves into a rotating or moving cutting tool as on a milling machine. There are several variations of these principles. Cutting metal with saws and metal removal by grinding are also considered to be machining operations.

Historically, machining processes probably began with Besson in 1569 when he invented the screw cutting lathe and later when a practical version was built by Henry Maudslay in 1800. Horizontal milling machines first appeared in 1820 when Eli Whitney used them for the manufacture of firearms, and the vertical milling machine first appeared in 1860. Of course these primitive machine tools continued to be improved and subsequently developed into the manufacturing marvels we have today, many of which are programmed by a computer to operate through a complex series of machining steps without the constant attention of a machinist. Today there are many varieties of machine tools; some specialized machines are designed primarily for manufacturing at high production rates. The more versatile machines are designed to make one-of-a-kind prototype parts for research and development (R and D), tool and diemaking, or to repair existing machinery. These versatile machines are usually found in tool rooms or local machine shops and are operated by general machinists or tool and die makers. Machine operators usually work only with production machinery.

If one word were used to describe machining as a form of metalworking, it would be **precision**. No other form of metalworking can approach the accuracy obtainable by the various processes of machining with the possible exception of some powder metallurgy processes. When turning metals with diamond tools, it is possible to achieve accuracies within ±.0001 in. This kind of precision is also possible with grinding processes. Large heavy sections such as those used for dies are made to highly precise dimensions, a virtual impossibility for other kinds of metalworking. Also, only machine tools can make other machine tools.

These, then, are the major reasons for choosing a machining operation to produce a particular product, even though the machining process may, in some cases, be more costly. For example, gears for an alarm clock may be entirely acceptable if stamped out of sheet metal or produced by plastic injection molding, which would be far less expensive than machining them; but gears for an automobile transmission must be hard, strong, and precisely made so they will not be noisy or fail under heavy shock loads. These transmission gears must be made by machining processes and subsequently heat treated and precision ground.

OBJECTIVES

This chapter should enable you to:

1. Describe some principles and standards of measurement, and explain the uses of most measuring instruments used in machining.
2. Understand the principles of metal removal in machining and explain the characteristics and uses of various tool materials.
3. Show why cutting fluids are used and explain their special characteristics.

MEASURING SYSTEMS

Since machining operations require varying degrees of precision and different accessibilities (for bores, offsets, slots, and outside diameters), a number of different measuring instruments are needed by the machinist and quality control inspector. These range from semiprecision tools, such as steel rules and spring calipers, to highly precise measuring instruments, such as height gages, gage blocks, and surface plates.

Standards of Measurement

The industrial world has, for the most part, adopted the metric system of measurement, known as the *Systéme International d'Unites* or the *International Metric System* (SI). This system is based on the meter (m), which is 39.37 in. long. In manufacturing, almost all metric linear dimensions are in millimeters (mm) even when the dimension is more than 1 m. The United States is committed to conversion to the metric system from the English system of feet and inches. However, this conversion is being only very gradually accepted in the United States and most manufacturing plants and machine shops still use the English system of measurement and probably will for some time to come. Because both measuring systems are in use in the United States, machinists and mechanical engineers often need to refer to conversion tables or use conversion factors such as the following:

1 in. = 25.4 mm or 2.54 cm
1 mm = .03937 in.

Measurement in machining processes is normally confined to size or dimension of length, that is, linear measure. Width, depth, and diameter of circles are other expressions of linear measure. Squareness and degree of angularity as expressed in degrees, minutes, and seconds of arc are other very important kinds of measure used in machining operations. In addition, terms such as *flatness, straightness, concentricity* (or eccentricity), and *surface finish* are elements of metrology found in manufacturing operations that require special measuring devices.

Measuring Tools

Tools and instruments used in machining operations for measurements may be separated into two major categories: **direct-reading instruments** having a graduated scale, and **indirect-reading instruments**, sometimes called transfer measuring tools. These do not have any graduated scale, but are instead used to adjust to and transfer the size of the dimension to a direct-reading instrument that measures it. Two other categories found in measuring instruments that pertain to the level of accuracy are precision and semiprecision measuring tools.

Semiprecision Measuring Tools Some measuring tools such as machinists' combination sets and steel rules could be considered to be semiprecision tools. The steel rule (Figure 1) can be read, with average eyesight, to an accuracy of $\frac{1}{64}$ in. and a metric rule of .5 mm. English rules are made in four fractional sets, $\frac{1}{8}$, $\frac{1}{16}$, $\frac{1}{32}$, and $\frac{1}{64}$; they are also made in $\frac{1}{100}$ in. and $\frac{1}{10}$ in. divisions on inch–decimal rules. Several combinations are usually found on each rule. Metric rules normally have 1 cm, 1 mm, and $\frac{1}{2}$ mm divisions. Machinist's rules are made in 6 in. or 150 mm lengths and longer. A 12-in. rule is generally used with a square head or combination set (Figure 2); however, other rule lengths are available. Squareness and angularity are measured with the square head and protractor on the steel rule of the combination set. The accuracy of the protractor is only to the nearest degree, not minutes or seconds, and the square may or may not be within a few minutes of accuracy.

Spring calipers (Figure 3) are sometimes used to transfer measurements from steel rules and are reliable for that use to about $\frac{1}{64}$ in. Thickness gages (Figure 4) consist of hardened steel leaves made to specific thick-

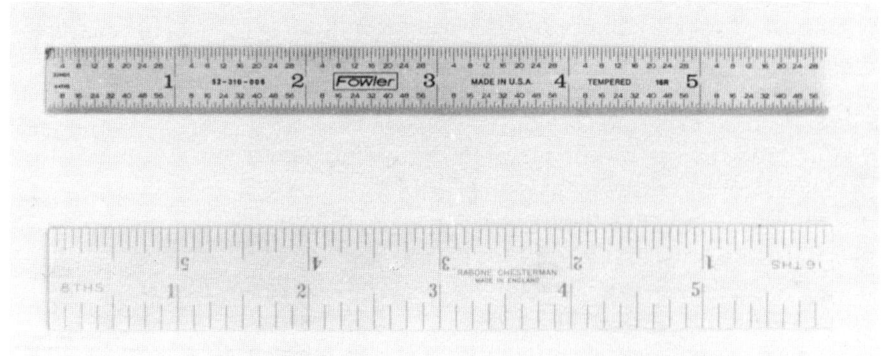

Figure 1
Steel rules.

Figure 2
Combination set. A square head, protractor, or centerhead may be mounted on a graduated blade for various uses (Compliments of Mitutoyo/MTI Corp.).

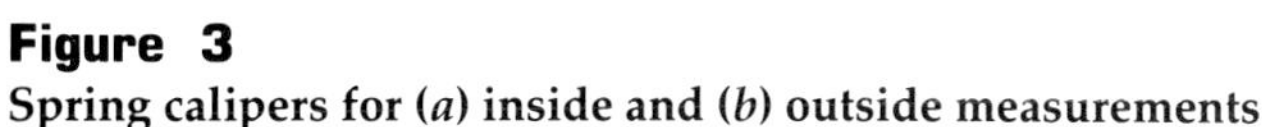

Figure 3
Spring calipers for (*a*) inside and (*b*) outside measurements.

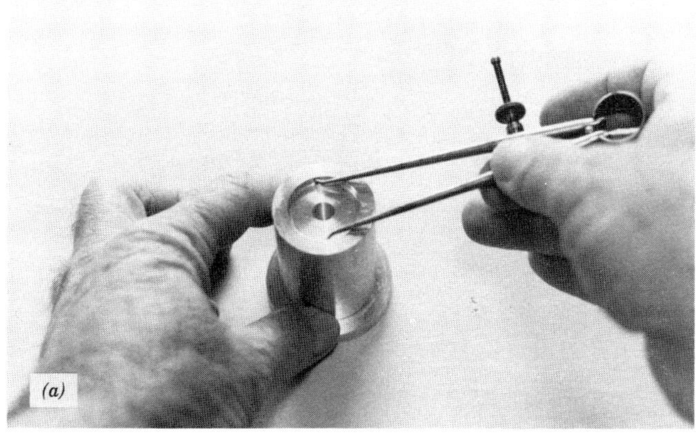

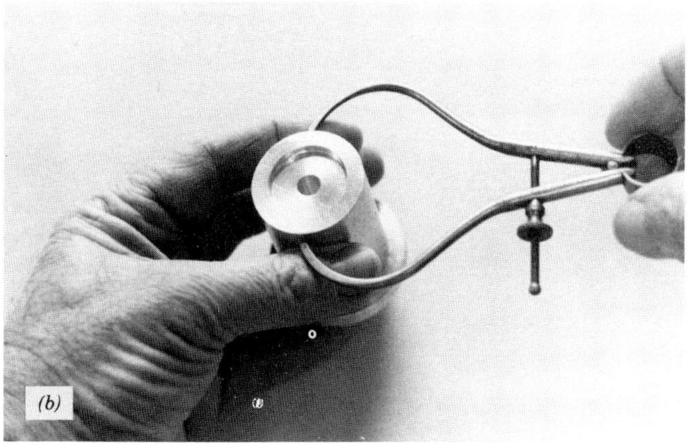

Figure 4
Thickness gages, also called feeler gages, have leaves of various thicknesses (Compliments of Mitutoyo/MTI Corp.).

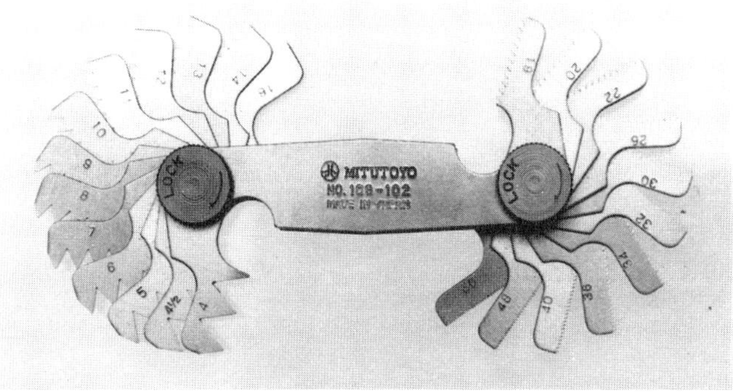

Figure 5
The screw-pitch gage is used to determine thread pitches (Compliments of Mitutoyo/MTI Corp.).

nesses which are marked on each leaf in either inch or metric measure. They are used mostly to determine the dimension of small spaces between machine members and are often used in conjunction with a straightedge to determine flatness on a metal part. Screw pitch gages (Figure 5) are used to determine thread pitches (distance between threads), and radius gages (Figure 6) are used to check the form of arcs of specific radii on machine parts. Rule depth gages (Figure 7) are used to determine the depths of grooves and offsets.

Measuring instruments are only as reliable as the user's skill. Accuracy can be affected by the following.

1. Line matching. This requires good eyesight.
2. Parallax in reading. That is, an angled view of matching lines or measured object which causes an erroneous reading.

3. Deformation of the workpiece or instrument. For example, forcing a micrometer when measuring a part instead of using the ratchet stop or friction thimble to get a true "feel" of measuring pressure.
4. Temperature effects. A workpiece or measuring tool can expand and contract with temperature changes, thus altering the reading.

Figure 6
Radius gages.

Figure 7
The rule-depth gage is a semiprecision measuring tool, used here to measure the depth of a counterbore.

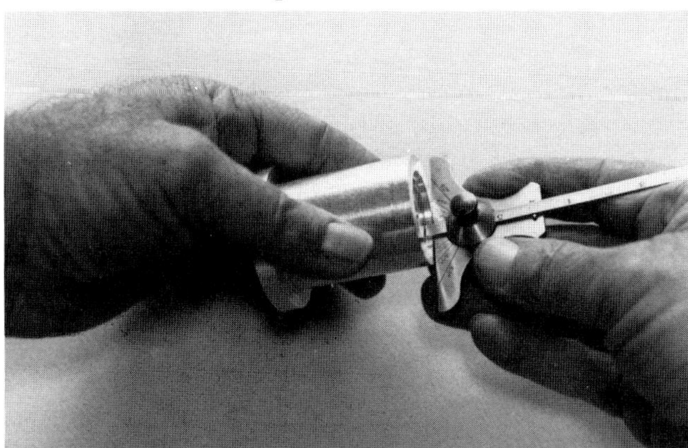

Figure 8
Small (0 to 1 in.) micrometer in use.

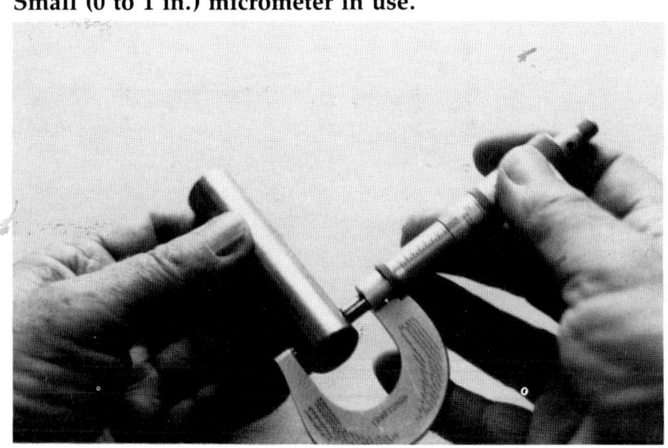

Figure 9
Large micrometer (The L. S. Starrett Company).

Figure 10
Combination inch/metric micrometer.

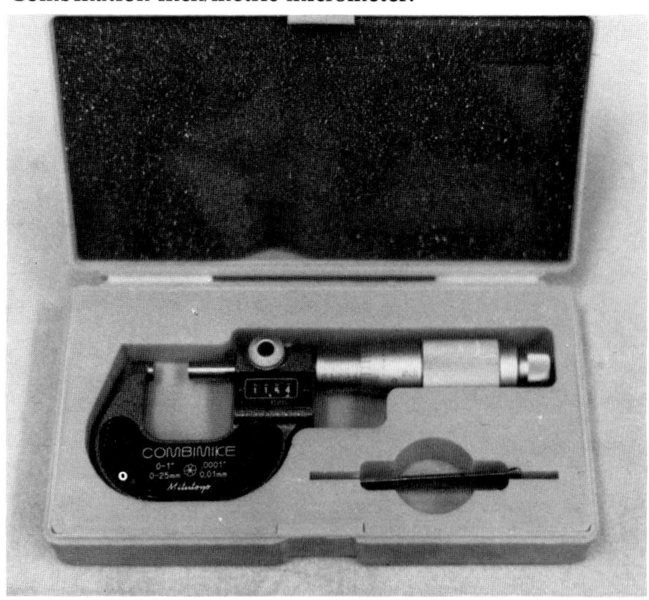

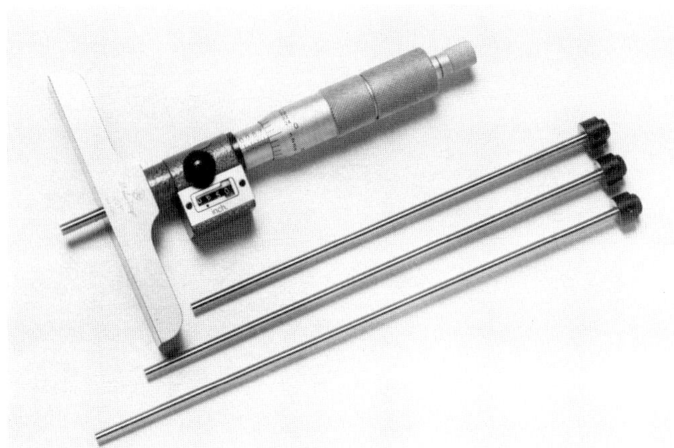

Figure 11
This depth micrometer reads in thousandths of an inch on the thimble and in a digital readout counter. It has interchangeable rods so the range can be extended (Compliments of Mitutoyo/MTI Corp.).

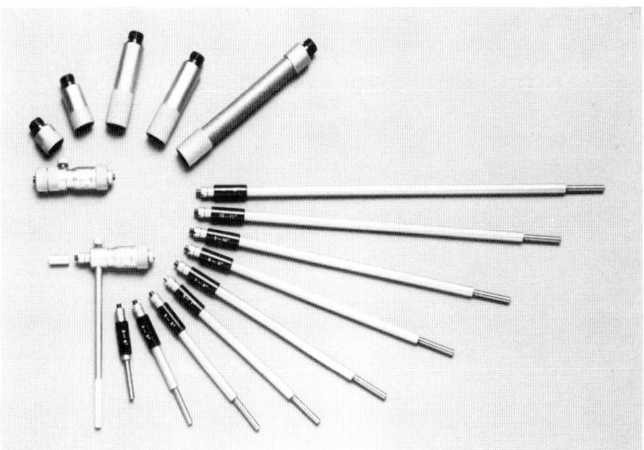

Figure 14
Inside micrometer set with interchangeable rods that give it a wide range of measurement (Compliments of Mitutoyo/MTI Corp.).

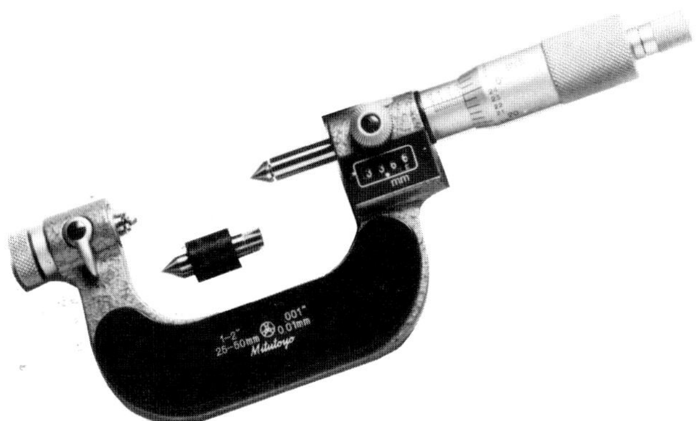

Figure 12
Screw thread micrometer which uses different interchangeable anvils (Compliments of Mitutoyo/MTI Corp.).

Figure 13
Measurement for thread pitch diameters is made on the flank of the threads.

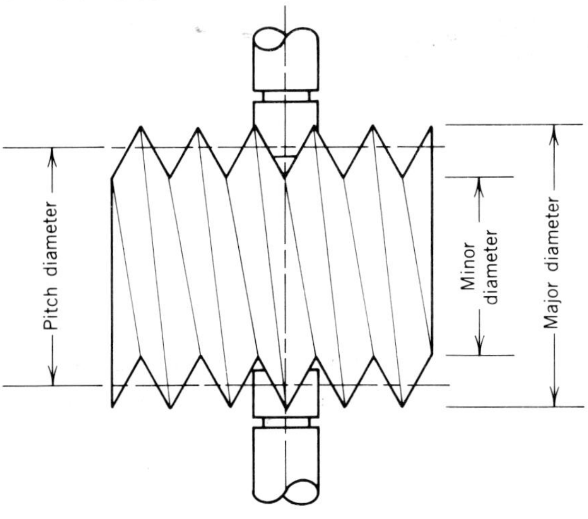

Precision Instruments These tools include micrometer calipers, vernier and dial calipers, depth and inside micrometers, screw thread micrometers, universal bevel protractors, small hole and telescope gages, snap gages, and adjustable parallels. Also used for highly precise measurement are dial indicators, precision height gages, dial comparators, optical comparators, sine bars, gage blocks, air gages, and surface texture analyzers.

Micrometer calipers (commonly termed mikes by machinists), perhaps the most widely used of all precsion measuring tools, are made in all sizes from 0 to 1 in. or 0 to 25 mm (Figure 8) to very large sizes that can measure several feet or more than 1 m (Figure 9). Because of the present slow transition from English to metric measure in the United States, some micrometers are made to read in both English and metric measure (Figure 10), often one reading is on the thimble graduations and the other is a digital reading. Some micrometers now have only a digital readout. Depth micrometers are used to make precision measurements of depth of slots and offsets on machine parts (Figure 11). Screw thread micrometers (Figure 12) have interchangeable anvils for different thread pitches and a 60 degree point on the spindle, making possible a precision measurement of 60 degree screw thread **pitch diameters** (Figure 13). The measurement is taken, not on the root or outside diameter of the thread, but on the flanks (sides). Inside micrometers (Figure 14) have the same barrel and graduated thimble as outside or caliper-type micrometers but are used to measure internal surfaces such as bores.

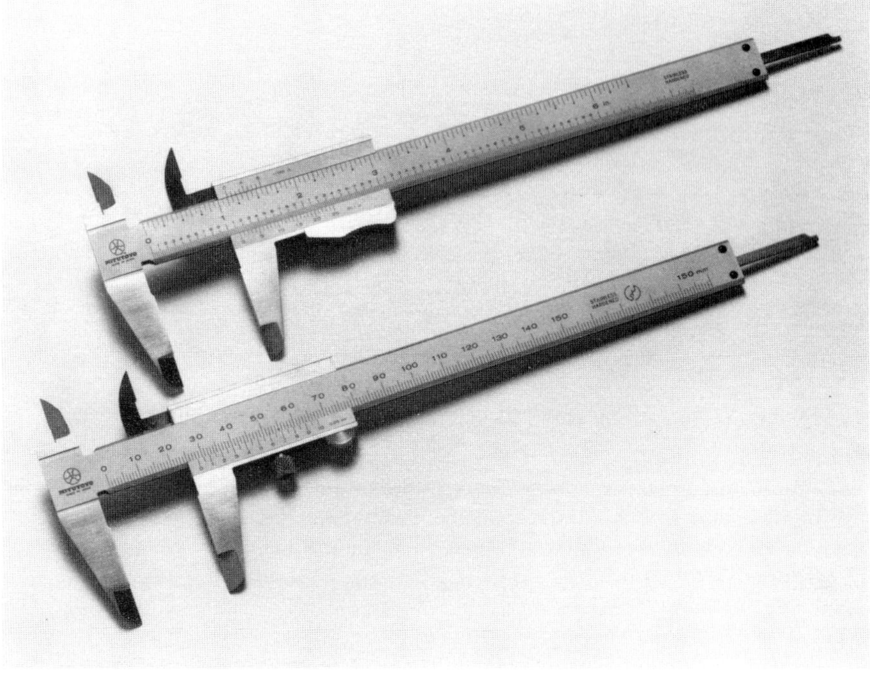

Figure 15
Vernier calipers, the upper one measures in inches and the lower one measures in millimeters (Compliments of Mitutoyo/MTI Corp.).

Figure 16
Vernier caliper measuring outside diameter. This vernier caliper measures both inches and metric.

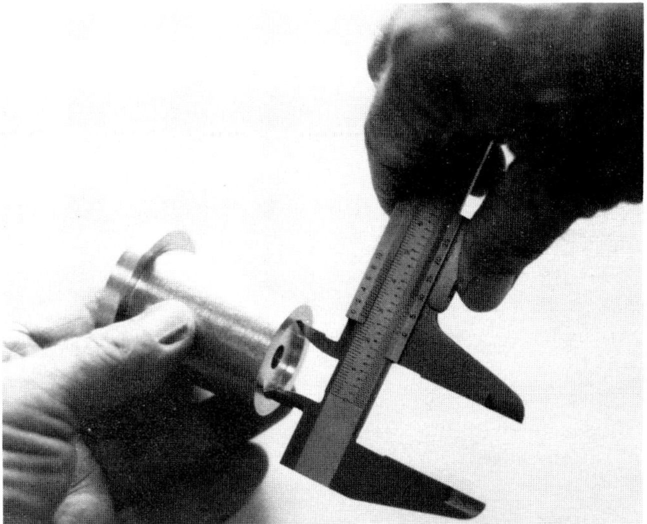

Figure 17
Vernier caliper measuring inside diameter.

Vernier calipers (Figures 15 to 18) have sliding jaws and are read by means of dividing .025 or .050 in. increments into thousandths of an inch, or .25 mm increments into two hundredths of a millimeter. This is done by means of a vernier plate that is inscribed with one less line than the number of divisions on the slide bar; that is, ten divisions may be obtained between two marks by using nine divisions on the vernier plate. A simpler method of reading a slide-type caliper is to use a dial-type or digital readout caliper (Figure 19). Vernier or dial slide calipers are capable of discriminating to .001 in. or .02 mm, and micrometer instruments can discriminate to .0001 in. or .01 mm. Most vernier and dial slide calipers have a provision to mea-

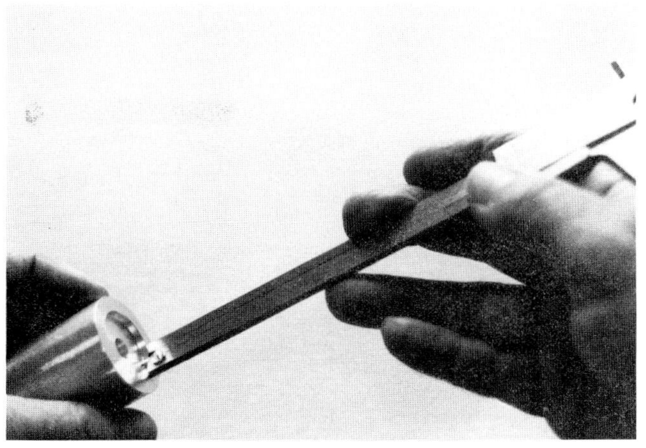

Figure 18
Vernier caliper measuring depth.

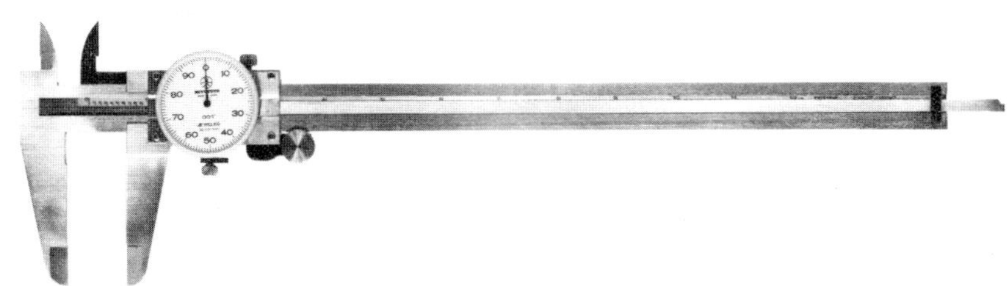

Figure 19
Dial caliper (Compliments of Mitutoyo/MTI Corp.).

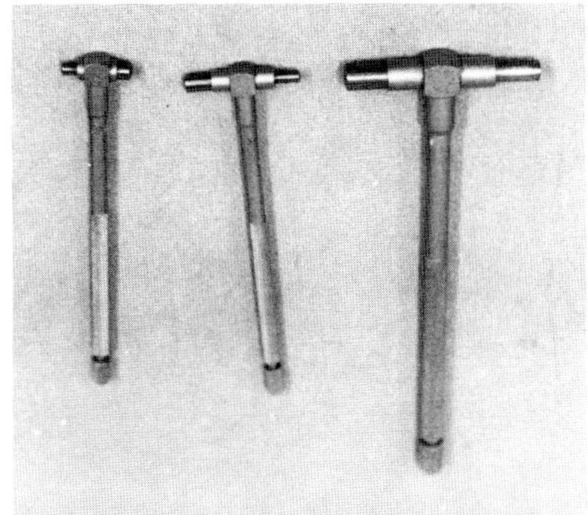

Figure 20
Telescoping gages.

sure depths of slots and offsets.

Transfer-type measuring tools do not make direct readings, they are adjusted to the internal dimension and then measured with other instruments such as micrometer calipers or vernier calipers. These include telescoping gages (Figure 20), small hole gages (Figure 21), and adjustable parallels (Figure 22). Dial indicators

Figure 21
Small-hole gage being measured with a micrometer.

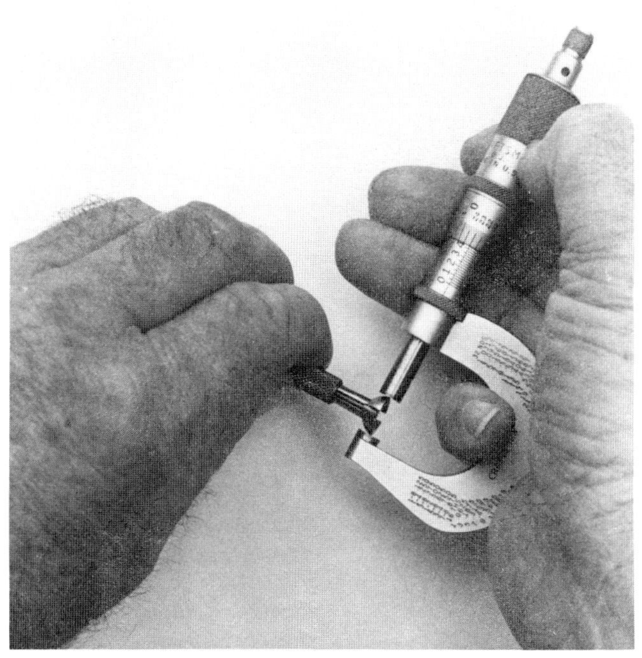

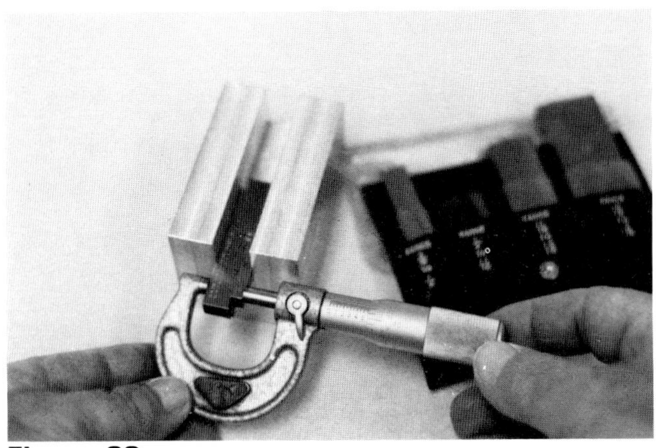

Figure 22
Adjustable parallel is measured with a micrometer.

Figure 23
Dial indicator (Compliments of Mitutoyo/MTI Corp.).

are precision instruments that detect minute movements or locations that can be read on a dial that can be calibrated in .001 or .0001 in. or in metric measure. Dial indicators have many uses, from measuring runout (eccentricity) in workpieces to transferring measurements on a surface plate (Figures 23 and 24). Dial indicators are often fastened to a surface gage or height gage and are used to transfer measurements from a

Figure 24
Dial indicators are often mounted on magnetic stands so that they can be fastened where necessary to machine parts for taking measurements (Compliments of Mitutoyo/MTI Corp.).

Figure 25
Test dial indicator (Compliments of Mitutoyo/MTI Corp.).

precision height gage (Figures 25 to 27). Measurements of this kind can have a consistent accuracy of .0001 in. or .01 mm. The universal bevel protractor is a precision tool for measuring angles (Figure 28). It is more accurate than the combination set protractor, which can measure only in degrees. The bevel protractor has a vernier plate which makes possible a reading within the range of 5 min of arc. Gage blocks (Figure 29) come in sets in several grades of accuracy in the range of a few millionths of an inch. Each block has a discrete dimension. The set has several increments of dimension, .100 in., .200 in., 1.0001 in., and so forth and adding or stacking blocks of different thicknesses makes possible a great variety of dimensions. The blocks are "wrung" together to remove the air to make a stack. They are used together with sine bars (Figure 30) to

Figure 26
Test dial indicators such as this one are often used for transferring precision measurements from workpiece to precision height gage or gage block stacks. To do this the height gage must be mounted on a surface gage or a standard height gage. These measurements usually take place on a granite surface plate (The L. S. Starrett Company).

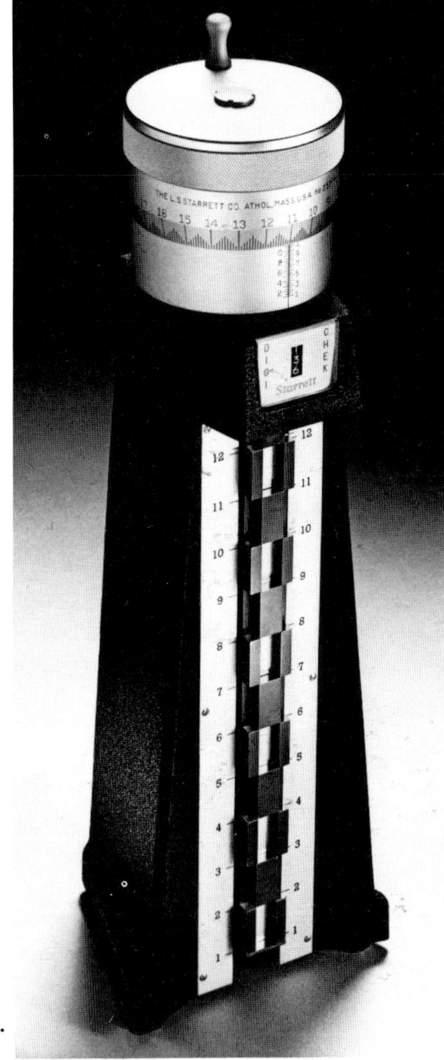

Figure 27
The precision height gage is capable of measuring to .0001 in. (The L. S. Starrett Company).

Figure 28
Universal bevel protractor (Compliments of Mitutoyo/MTI Corp.).

Figure 29
Gage block sets (Compliments of Mitutoyo/MTI Corp.).

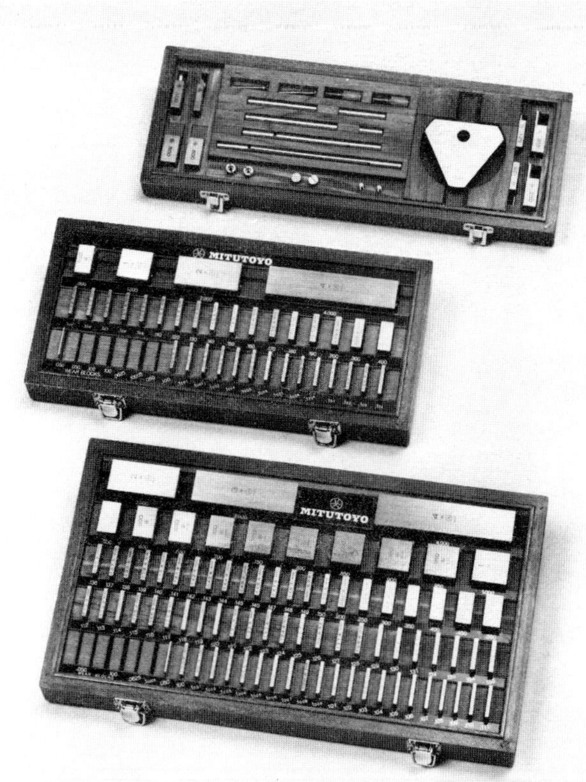

Figure 30
Precision sine bars (Compliments of Mitutoyo/MTI Corp.).

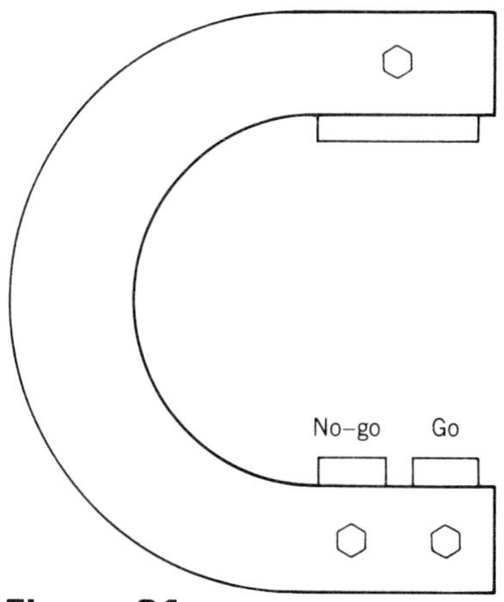

Figure 31
Snap gage. These tools are useful for making a quick check of dimensions.

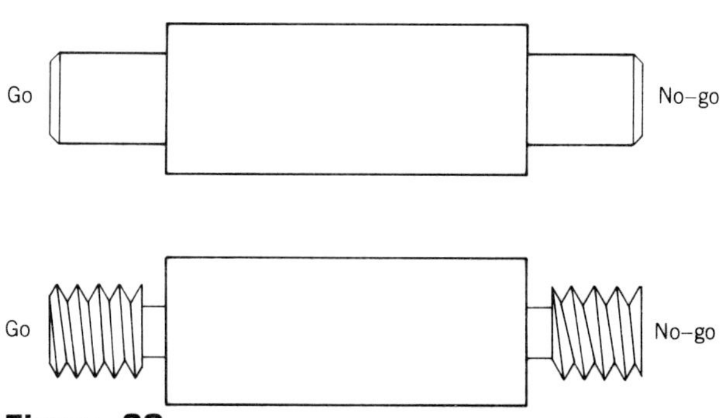

Figure 32
Plug and thread gages.

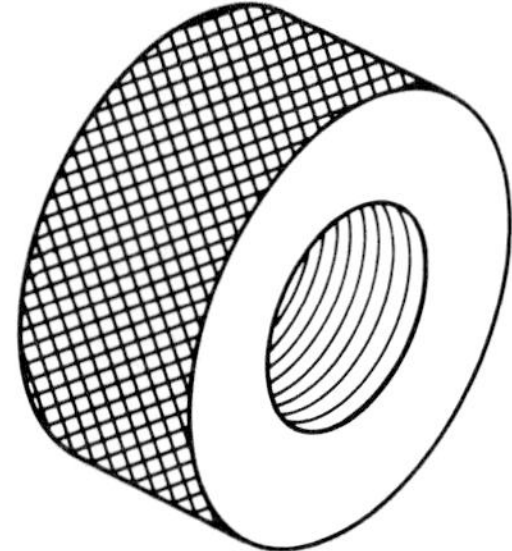

Figure 33
Ring gage for checking threads.

Figure 34
The air gage measures clearances between bores and nominal shaft sizes by means of air pressure. The setting ring is used for adjustment purposes (Federal Products Corporation).

Figure 35
Dial indicator comparator stand. Small parts can be quickly measured with this tool (Compliments of Mitutoyo/MTI Corp.).

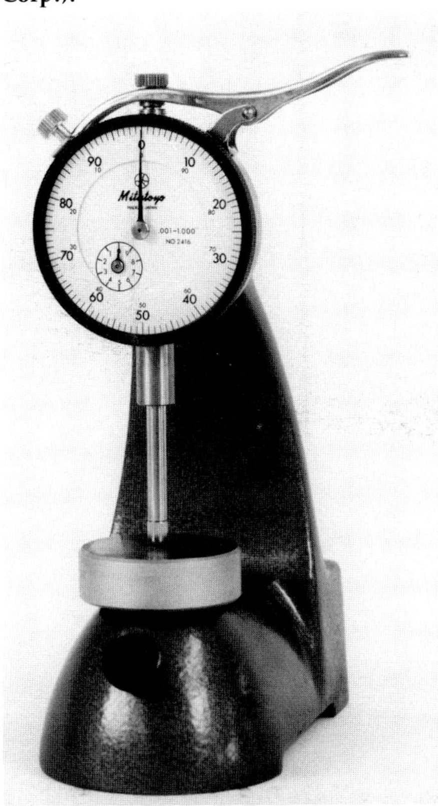

make very precise angles in the range of 1 to 5 seconds of arc. When gage blocks are used with a .0001 in. or .01 mm dial indicator on a surface plate to transfer measurement, the accuracy is somewhat greater than that of the dial indicator alone, possibly over .00001 in. or .0003 mm.

Snap gages (Figure 31) are used in manufacturing to determine quickly if the part is within tolerance. Go/no-go plug gages for bores and internal threads (Figure 32) are commonly used in manufacturing operations to check the accuracy of mass produced parts. Go/no-go ring gages are used for external dimensions and threads (Figure 33). Air gages (Figure 34) and dial indicator comparator stands (Figure 35) are used where many parts of the same size must be precisely measured.

Optical comparators (Figure 36) not only measure precise dimensions in areas not accessible to other in-

Figure 36
Optical comparator or profile projector can be used for precision measurement or comparing curves and radii (Compliments of Mitutoyo/MTI Corp.).

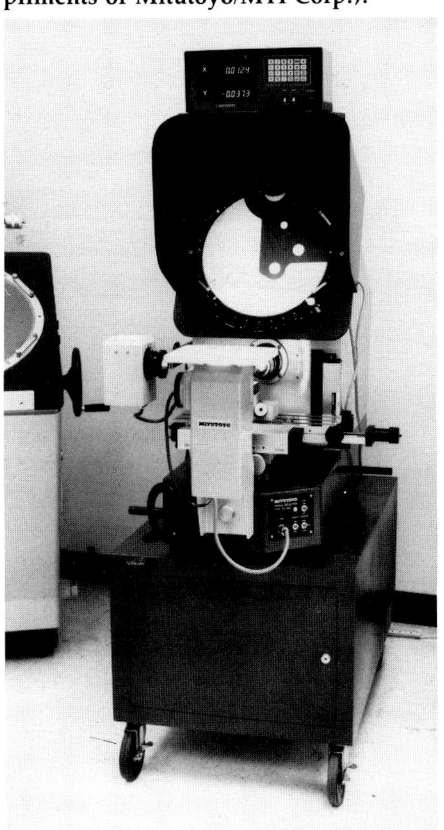

struments but allow profiles to be compared to a standard on a screen. Surface texture can be measured in several ways. Machining marks and grades of finish can be determined to some extent by using comparison standards of roughness, measured in microinches (one-millionth of an inch) and abbreviated as Mu in. (Figure 37). Tool finishes are measured in microinches, from a fairly smooth 30Mu to a coarse 250Mu. Ground or polished finishes are less than 30Mu, 4Mu being a mirrorlike polished surface. Special instruments are used for more accurate measurements of surface texture (Figure 38).

Figure 37
Roughness comparison standards.

Tolerances and Fits

If no method of quality control were used in modern manufacturing, mechanical parts would not fit, nor could replacement parts be obtained with any assurance that they would fit or have the same dimensions as the original. A system of tolerances and fits that all manufacturers use makes sure that a part made in California or Japan will fit its counterpart made in Ohio or Spain. This subject will be fully discussed in Chapter 20, "Quality Assurance and Control."

PRINCIPLES OF MACHINING AND METAL REMOVAL

Metal cutting could be analogous to whittling on wood with a sharp knife except that wood can split and metal normally does not. Basically, the mechanism of chip formation on metal with a metal cutting tool is one of **compression** and **plastic deformation** due to **shear strain.** This action results in the continuous shear failure of the work material just ahead of or at the point of the tool (Figures 39 and 40).

Three factors are involved in the process of chip formation.

1. The geometry and properties of the cutting tool.
2. The properties of the work material.
3. The interaction between the tool and workpiece.

Because of the high forces involved in metal cutting, considerable heat is generated at the point of the cutting tool. Most of the heat is carried off in the chips and the remainder goes into the tool and workpiece. Of course if coolants are used, much of the heat is removed by the coolant. The basic machining operations are shown in Table 1.

Figure 38
Surface texture analyzing instruments (Compliments of Mitutoyo/MTI Corp.).

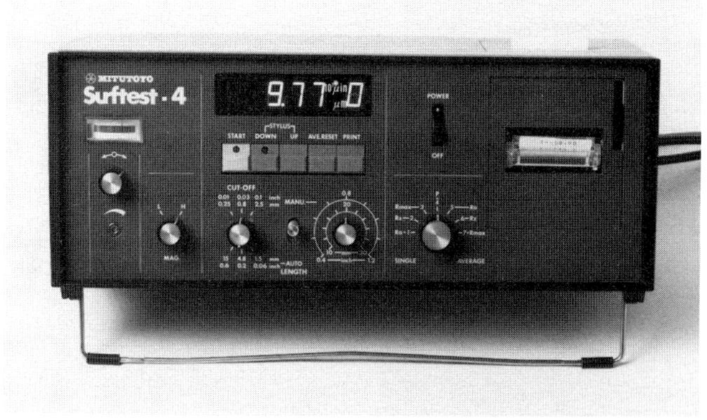

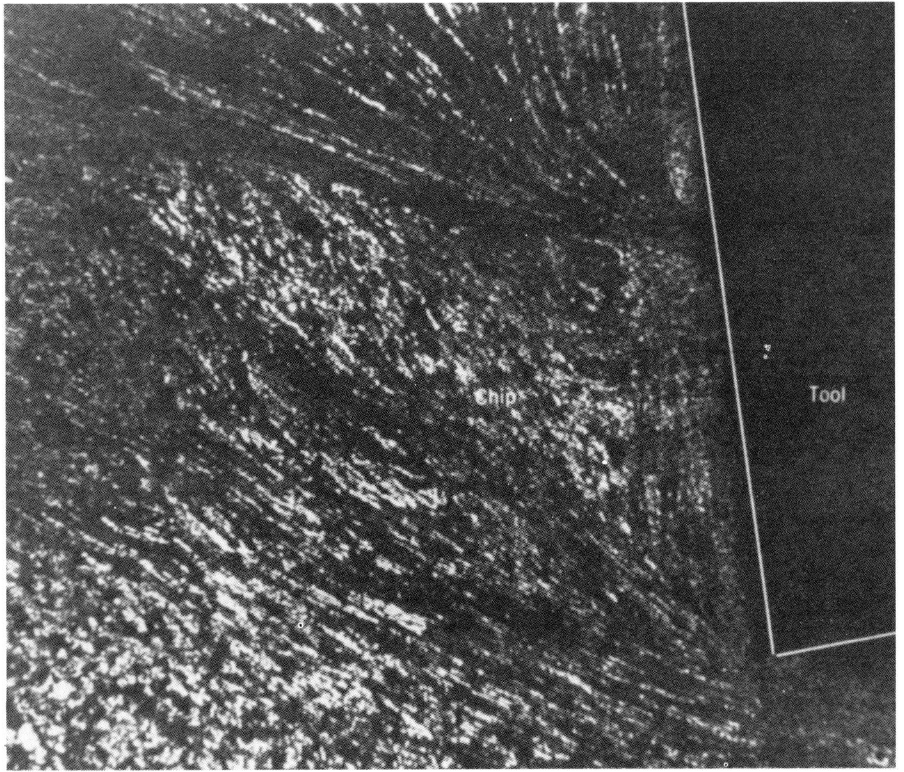

Figure 39
Micrograph of chip formation showing shear plane. Point of negative rake tool producing a discontinuous chip (100×) (Neely/*Metallurgy* 2 Ed., © 1984 John Wiley & Sons, Inc.).

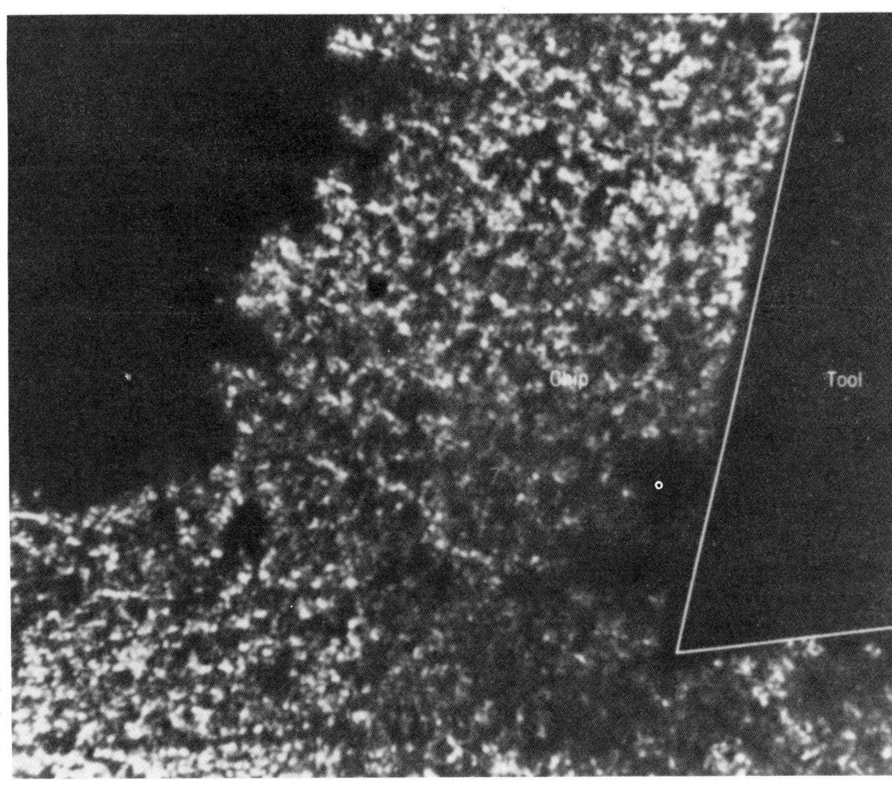

Figure 40
Positive rake tool showing less distortion ahead of the tool (100×) (Neely/*Metallurgy* 2 Ed., © 1984 John Wiley & Sons, Inc.).

Table 1
Machining Principles and Operations

Operation	Diagram	Characteristics	Type of Machines
Turning	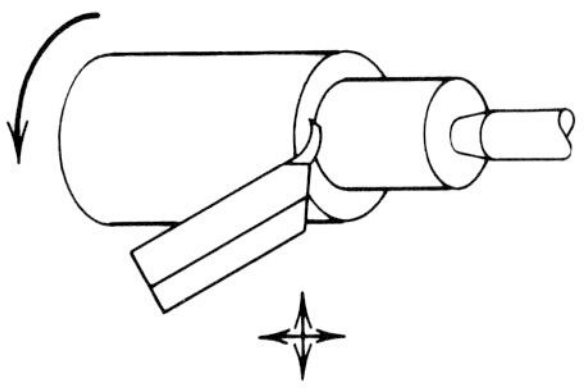	Work rotates, tool moves for feed	Lathe and vertical boring mill
Milling (horizontal)	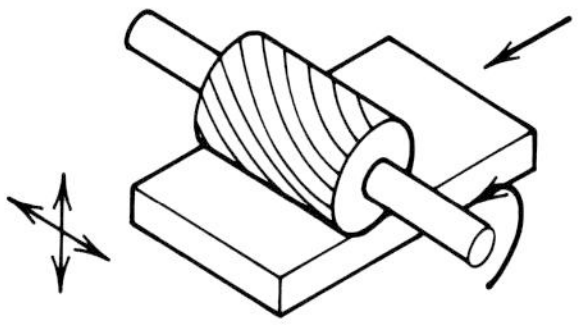	Cutter rotates and cuts on periphery. Work feeds into cutter and can be moved in three axes.	Horizontal milling machine
Face milling	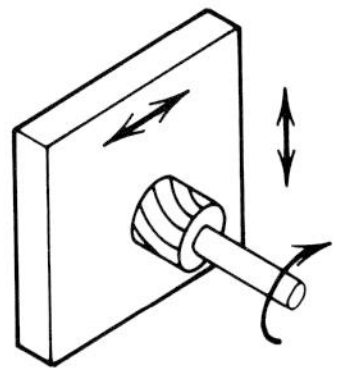	Cutter rotates to cut on its end and periphery of vertical workpiece.	Horizontal mill, profile mill, and machining center
Vertical (end) milling	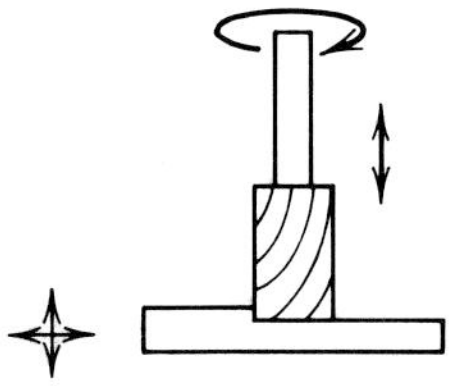	Cutter rotates to cut on its end and periphery, work moves on three axes for feed or position. Spindle also moves up or down.	Vertical milling machine, die sinker, machining center
Shaping	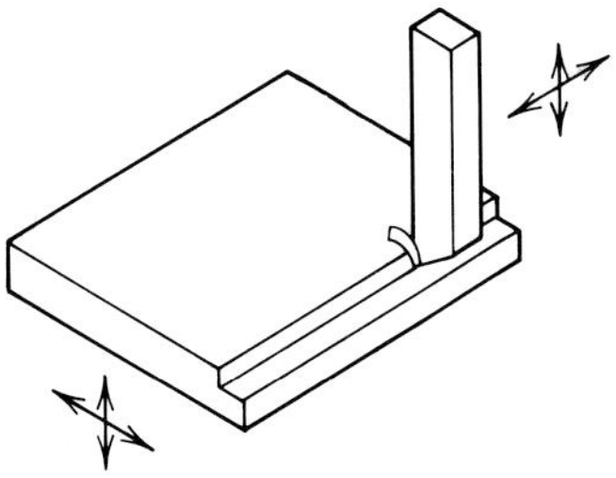	Work is held stationary and tool reciprocates. Work can move in two axes. Toolhead can be moved up or down.	Horizontal and vertical shapers
Planing	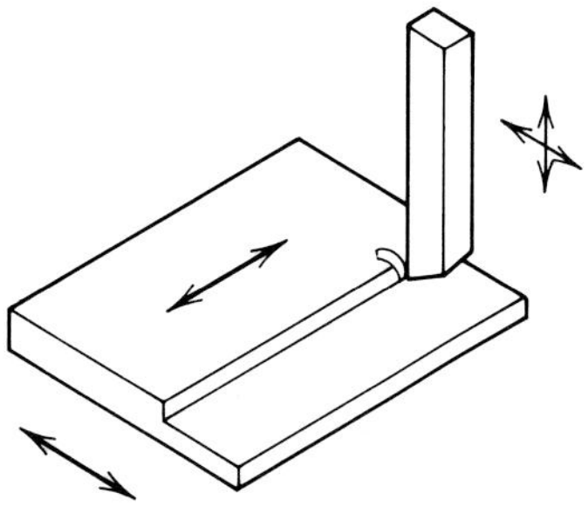	Work reciprocates while tool is stationary. Tool can be moved up, down, or crosswise. Worktable cannot be moved up or down.	Planer

Table 1
(Continued)

Operation	Diagram	Characteristics	Type of Machines
Horizontal sawing (cutoff)	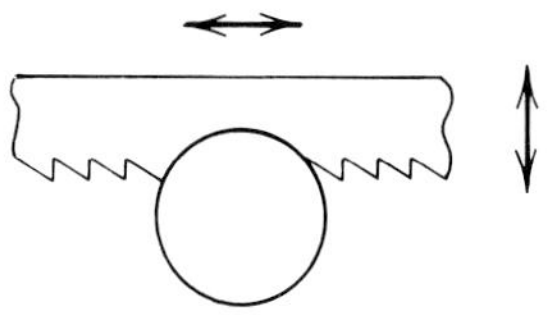	Work is held stationary while the saw cuts either in one direction as in bandsawing or it reciprocates while being fed downward into the work.	Horizontal bandsaw, reciprocating cutoff saw
Vertical bandsawing (contour sawing)	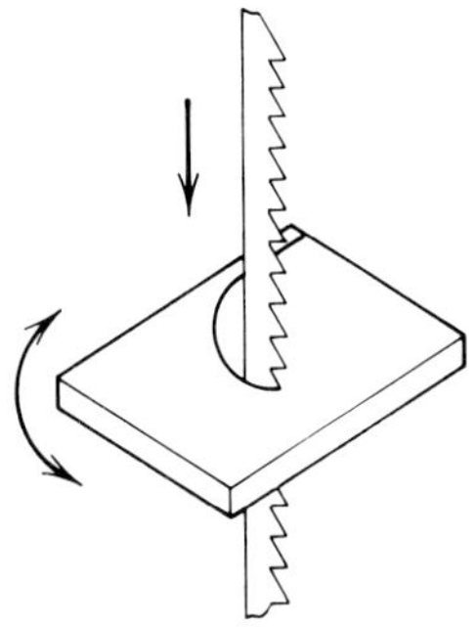	Endless band moves downward, cutting a kerf in the workpiece which can be fed into the saw on one plane at any direction.	Vertical bandsaw
Broaching	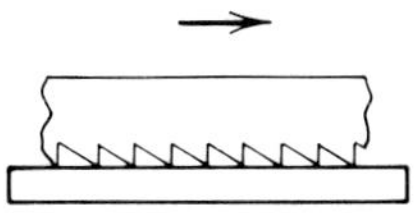	Workpiece is held stationary while a multitooth cutter is moved across the surface. Each tooth in the cutter cuts progressively deeper than the previous one.	Vertical broaching machine, horizontal broaching machine
Horizontal spindle surface grinding	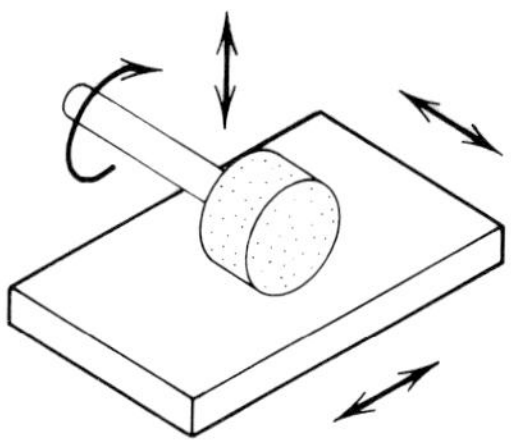	The rotating grinding wheel can be moved up or down to feed into the workpiece. The table, which is made to reciprocate, holds the work and can also be moved crosswise.	Surface grinders, specialized industrial grinding machines
Vertical spindle surface grinding	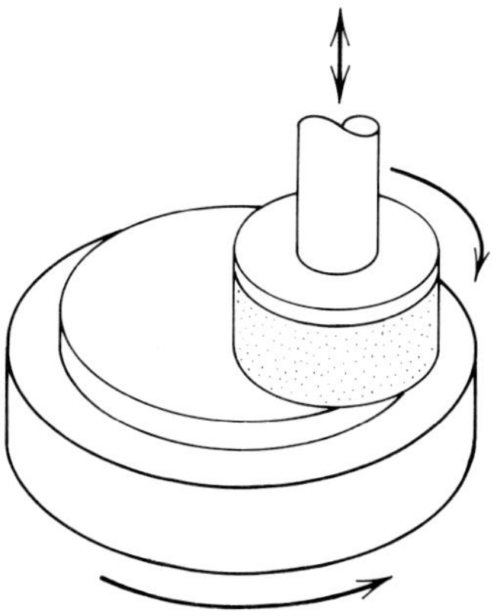	The rotating grinding wheel can be moved up or down to feed into the workpiece. The circular table rotates.	Blanchard-type surface grinders

Table 1
(Continued)

Operation	Diagram	Characteristics	Type of Machines
Cylindrical grinding	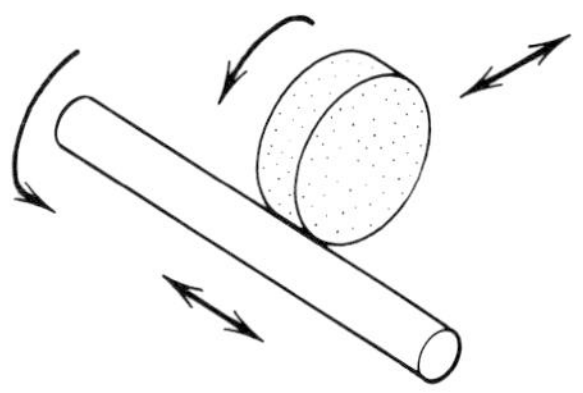	The rotating grinding wheel contacts a turning workpiece that can reciprocate from end to end. The wheelhead can be moved into the work or away from it.	Cylindrical grinders, specialized industrial grinding machines
Centerless grinding	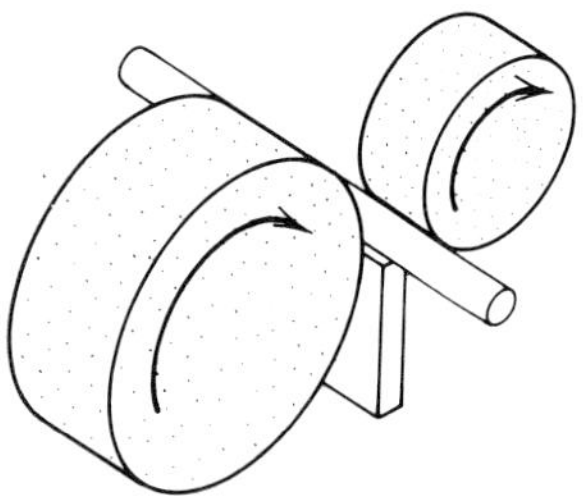	Work is supported by a workrest between a large grinding wheel and a smaller feed wheel.	Centerless grinder
Drilling and reaming	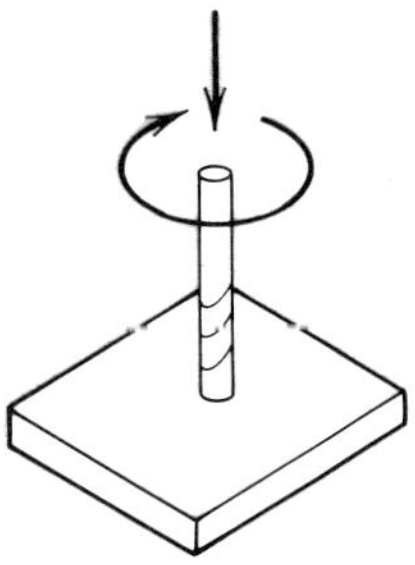	Drill or reamer rotates while work is stationary.	Drill presses, vertical milling machine
Drilling and reaming	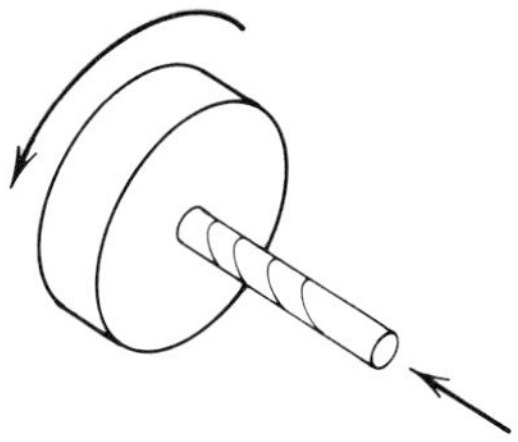	Work turns while drill or reamer is stationary.	Engine lathes, turret lathes, automatic screw machines
Boring	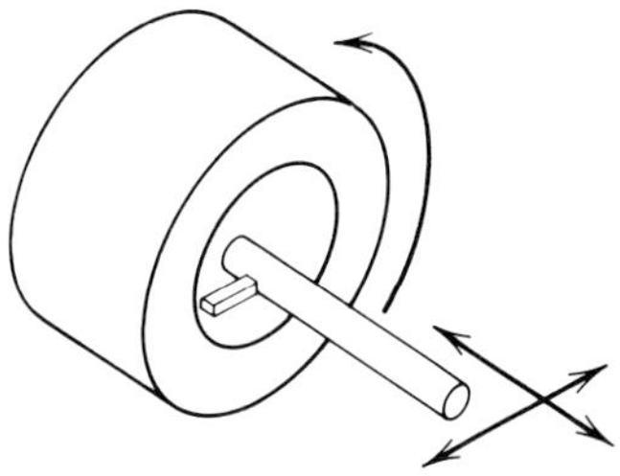	Work rotates, tool moves for feed on internal surfaces.	Engine lathes, horizontal and vertical turret lathes, and vertical boring mills. On some horizontal and vertical boring machines, the tool rotates and the work does not.

Tool Materials

Metals that are machined range in hardness from soft aluminum to heat-treated tool steels that are almost as hard as a knife blade. This shows the need for a great variety of cutting tool materials; some must be very hard and others must be very strong to withstand intermittent cutting action. Tool shapes also vary considerably for use with materials of different hardnesses and the shape of chipbreakers is vital to chip control.

Carbide Tools Most cutting tools used for machining operations involving chip removal from workpieces are made of tungsten carbide particles with a cobalt matrix. These powders are compressed into a briquette of the required tool shape and then sintered in a furnace, causing the insert to coalesce into a very hard, strong solid material. They are made in various grades ranging from tough and less hard for rough steel machining to very hard and somewhat brittle for cast iron and abrasive materials. Titanium and tantalum carbides are sometimes added to give them greater hardness or wear resistance. Carbide tools were made in the past by brazing a small sintered tungsten carbide insert onto a steel shank. That tool was sharpened by grinding it on a silicon carbide wheel. Virtually all carbide tools used today in manufacturing operations are throwaway inserts that have several indexable cutting edges (Figure 41). Various carbide insert holders are made, both for external and internal turning on lathes and other turning machines (Figures 42 and 43). Milling cutters with carbide inserts have long life and cutting edges can be replaced. Boring bars which are used for internal turning operations are also provided with carbide inserts. However, high speed steel tools are still used to some extent for small boring bars and insert boring tools.

Ceramic tools, often used for harder materials and where high finishes are required, are used at much higher speeds than carbide inserts but require very rigid setups and machines. They are normally used with no coolant. Diamond tools, like ceramic inserts, do not have high removal rates but can produce exceedingly high finishes and hold close tolerances. Ceramic coated tungsten carbide inserts that can take interrupted cuts without breakdown are now available. These are coated with titanium nitride, aluminum oxide, and/or titanium carbide.

High-Speed Tools Eighty to one hundred years ago hardened plain carbon tool steel was used for metal cutting tools on lathes, shapers, planers, and milling machines. Cutting speeds had to be very slow because carbon steel would permanently lose its hardness if it got too hot from machining. It is still used for some tools such as files, knives, and chisels since it can hold a very keen edge. A new type of tool steel developed around 1900 was called high-speed steel because it could

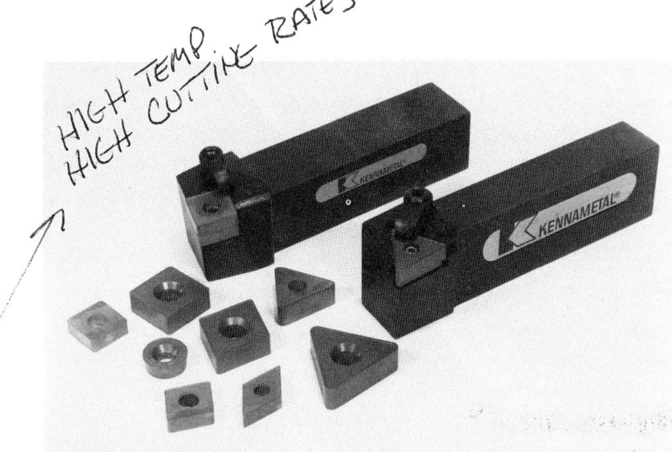

Figure 42
Carbide toolholders for turning machines (Photo courtesy of Kennametal Inc.).

Figure 41
Carbide inserts (Photo courtesy of Kennametal Inc.).

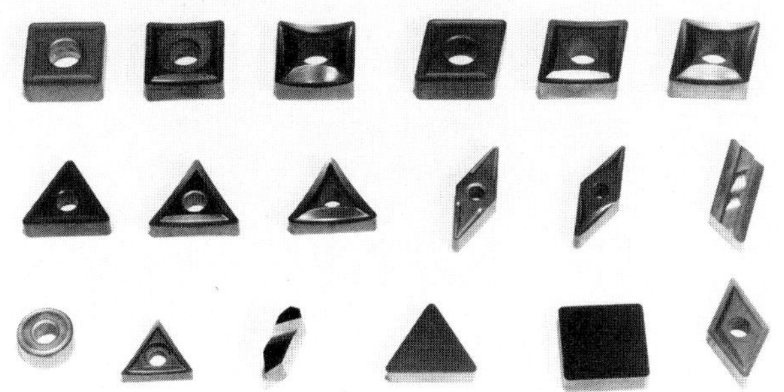

Figure 43
Boring bar holder with interchangeable heads for internal turning operations (Photo courtesy of Kennametal Inc.).

get hot, cool down, and still be as hard as before. High-speed steel drills, milling cutters, and lathe tools are still widely used. Some materials, such as some stainless steels, cut cleaner using high speed tools with less work hardening than they would with carbide cutters. Lathe tool blanks of high-speed steel are made in various square sizes in short lengths. These must be sharpened by hand on a bench grinder or precision sharpened on a tool and cutter grinder.

Tool Geometry Since high-speed tools have to be shaped by hand, tool geometry can be more easily explained using these tools rather than carbide tools as examples. The shape and angles of the cutting tool must be fairly accurate in order to have sufficient strength at the cutting edge, to ensure good chip formation and flow, and to provide sufficient relief (clearance) so the tool will cut into the work material and not rub. A commonly used industrial tool signature (see accompaning table below) with a drawing of the parts and angles of a high-speed tool is illustrated in Figure 44. The angles given are only examples and they could vary according to the application. This is an example of a right-hand cutting tool. There are many shapes of cutting tools, including left-hand, threading, and form tools, but all of them must have correct angles for relief and rake. Some tools, such as those used for cutting bronze, may have a zero or negative back rake to avoid gouging or digging into the work. Chip-breakers are sometimes needed on tools to control chip formation (Figures 45 and 46).

Figure 44
The parts and angles of a tool, and the tool signature. Angles given are only examples. (White, Neely, Kibbe, Meyer, *Machine Tools and Machining Practices*, Vol. I, © 1977 John Wiley & Sons, Inc.).

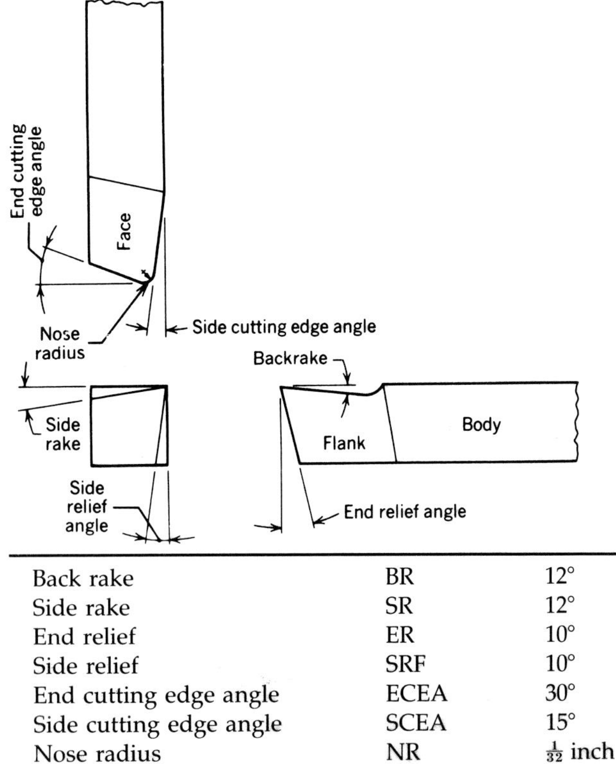

Back rake	BR	12°
Side rake	SR	12°
End relief	ER	10°
Side relief	SRF	10°
End cutting edge angle	ECEA	30°
Side cutting edge angle	SCEA	15°
Nose radius	NR	$\frac{1}{32}$ inch

Tool Signature

1. The tool shank is that part held by the toolholder.
2. Back rake is very important to smooth chip flow, which is needed to have a uniform chip and a good finish.
3. The side rake directs the chip flow away from the point of cut and it provides for a keen cutting edge.
4. The end relief angle prevents the front edge of the tool from rubbing on the work.
5. The side relief angle provides for cutting action by allowing the tool to feed into the work material.
6. The cutting edge angle (SCEA) may vary considerably. For roughing, it should be almost square out 5 degrees, but tools used for squaring shoulders or for other light machining could have angles from 15 to 32 degrees depending on the application. This angle may be provided by turning the toolholder or by grinding it on the tool bit or both. In finishing operations with a large nose radius + light cut, SCEA is not an important factor.
7. The side cutting edge angle, which is usually 10 to 20 degrees, directs the cutting forces back into a stronger section of the tool point. It helps to direct the chip flow away from the workpiece. It also affects the thickness of the cut.
8. The nose radius will vary according to the finish required. The smallest nose radius that will give its desired finish should be used.

Figure 45
Stringy chips.

Figure 46
Chips that are 9-shaped are best.

Chip Control Some manufacturing machines carry the chips away for disposal on a conveyor. Since stringy, wiry chips would entangle machine parts and eventually jam up the operation, chip control on automatic machinery is a necessity. This is done with chipbreakers (Figures 47 and 48) that break the chips by curling them against the workpiece so they break off (Figure 49). A C-shaped chip is often produced but a 9-shaped chip is the most desirable form for ease of handling and for safety's sake. Chips are very sharp, like knives or more often like saws. They should not be handled with bare hands since they can cause deep cuts or act as sharp slivers, especially if they are milling machine chips. Also chips, especially C-shaped chips, tend to fly from the work at high speed, sometimes as far as 20 to 30 feet away. They can easily penetrate the eye if safety glasses are not worn. Even when walking through a machine shop or manufacturing facility, safety

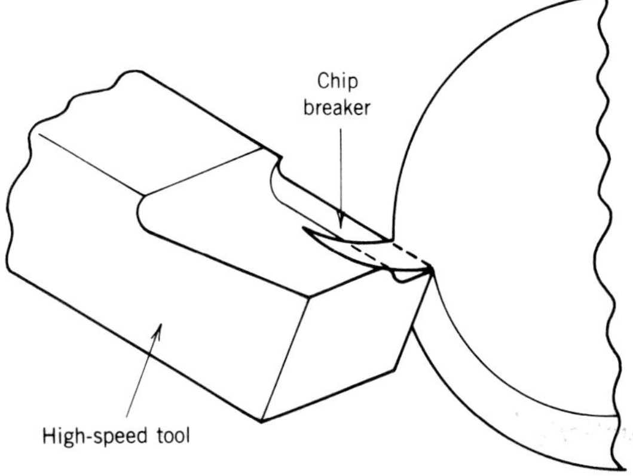

Figure 47
High-speed chip breakers.

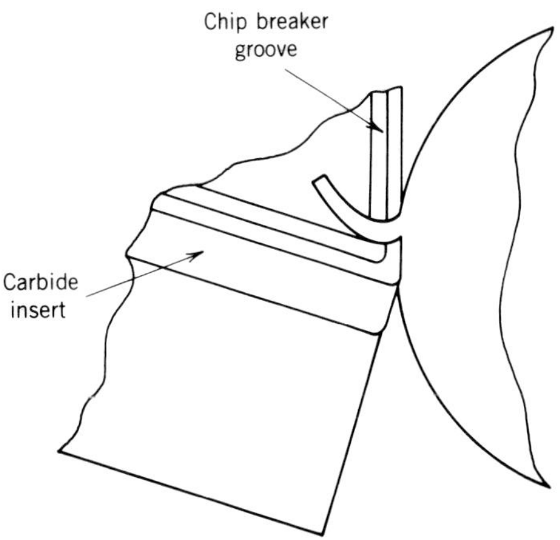

Figure 48
Carbide chip breakers.

Figure 49
How chips are broken. Chip flow with plain tool and with a chip breaker (White, Neely, Kibbbe, Meyer, *Machine Tools and Machining Practices*, Vol. I, © 1977 John Wiley & Sons, Inc.).

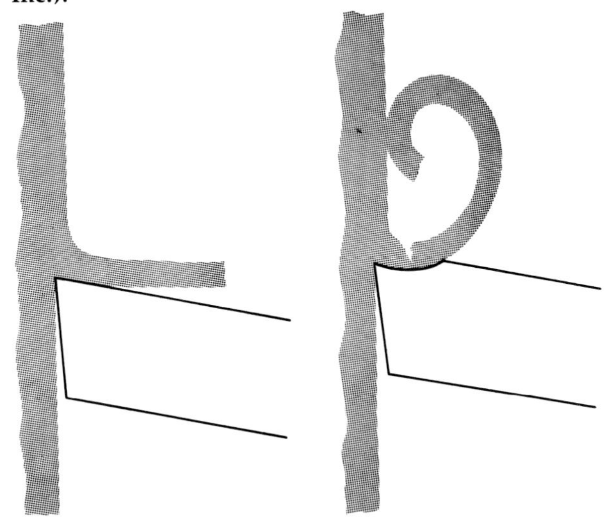

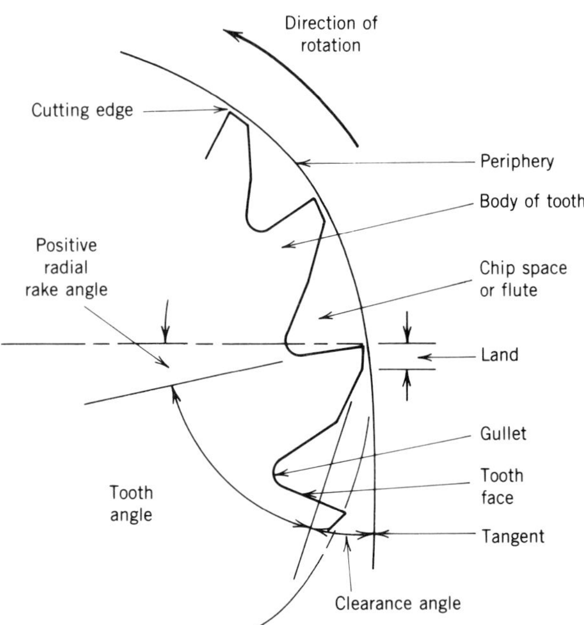

Figure 50
Milling cutter angles.

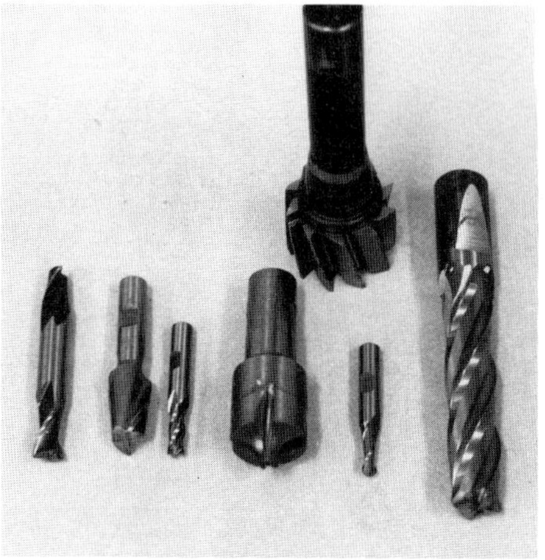

Figure 53
A variety of end mills.

Figure 51
Profile milling cutters.

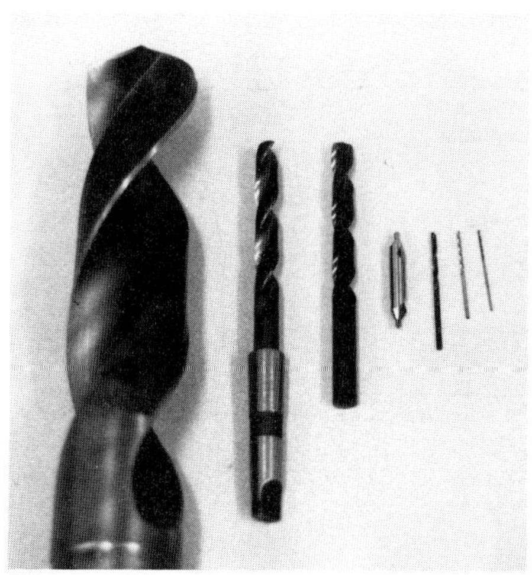

Figure 54
A variety of high-speed twist drills.

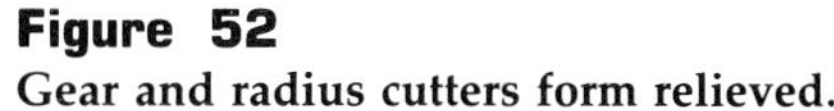

Figure 52
Gear and radius cutters form relieved.

Figure 55
Helix of twist drill provides low or high rake angle.

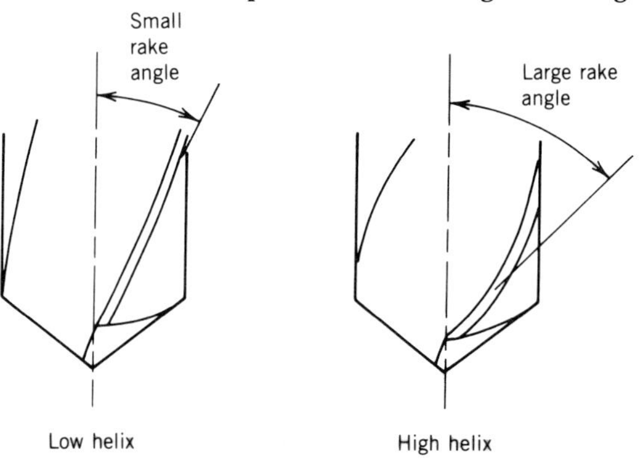

glasses and often hard hats are required. Proper shoes should also be worn; chips can penetrate soft-soled shoes.

Milling Cutters Milling cutters also must have correct relief and rake angles. Circular cutters (Figure 50) are used on horizontal milling machines for slabbing (surfacing) and slotting. These are called profile milling cutters (Figure 51). Form or cam relieved cutters are also used on horizontal spindle machines for gear cutting, producing radii, and making other shapes (Figure 52). Large vertical spindle milling machines use face mills for making large flat surfaces, but more often small diameter two-flute or multiple-flute end mills are used (Figure 53).

Drills, Taps, and Reamers High-speed steel twist drills (Figure 54) are the most common type used for general drilling in metals. The twisted flutes provide for chip clearance and removal. The helix of the flute also provides rake (Figure 55) which is needed for such metals as steel. Straight flute drills (Figure 56) are used for some plastics, brass, and bronze which have the tendency to "grab" when drills have positive rake. Straight flute drills provide zero rake. However, this can also be provided on a standard twist drill by grinding a flat on its cutting edge (Figure 57). Spade drills (Figure 58) are made in sizes from very tiny to 12 in. diameter. Spade drills are widely used for drilling holes 1 in. and larger. They have the advantages of being very rigid in the shank enabling heavier cuts to be taken and the main body is often provided with holes to pump coolant through to the blade. Blades are easily sharpened and replaced with far less expense than that for replacing a twist drill. They are frequently provided with a step to provide a bevel or chamfer on the edge of the hole to facilitate tapping and assembly.

Gun drills (Figure 59) are capable of making very precise holes many feet deep. They are used with special machinery that guides the drill and pumps coolant through it. Centerdrills are used for making centers in shafts for turning in a lathe, supported by a 60 degree center in the tailstock. They are also used for spotting, that is, for providing a starting hole for other larger drills.

Figure 56
Straight flute drill.

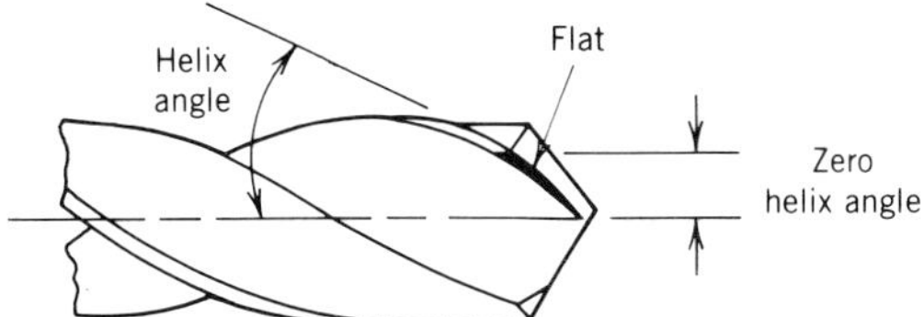

Figure 57
A flat ground on the cutting edge of the drill is often necessary when drilling brass, bronze, and plastics.

Figure 58
Taper shank spade drill (Courtesy of DoALL Company).

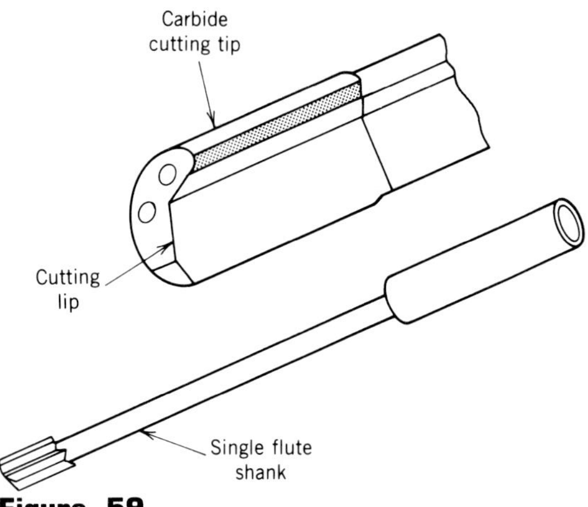

Figure 59
Gun drill.

Figure 62
Tapping with a spiral flute tap (Courtesy of DoALL Company).

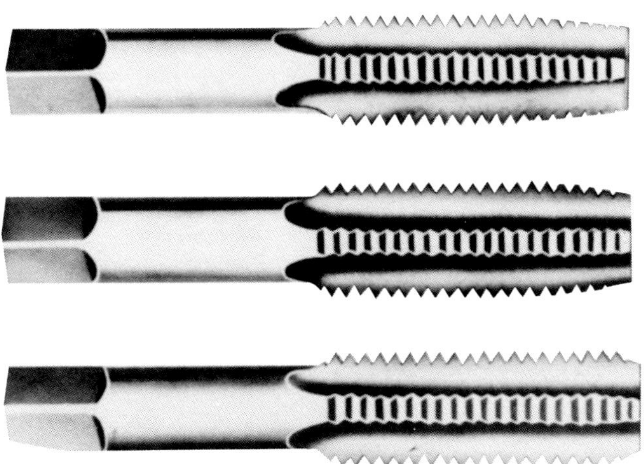

Figure 60
Standard set of three taps. Top to bottom: taper, plug, and bottoming.

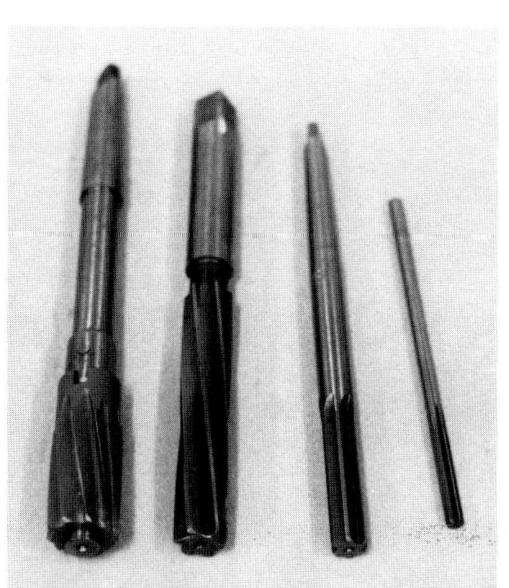

Figure 63
A variety of various reamers.

Figure 61
Spiral point tap (Courtesy of DoALL Company).

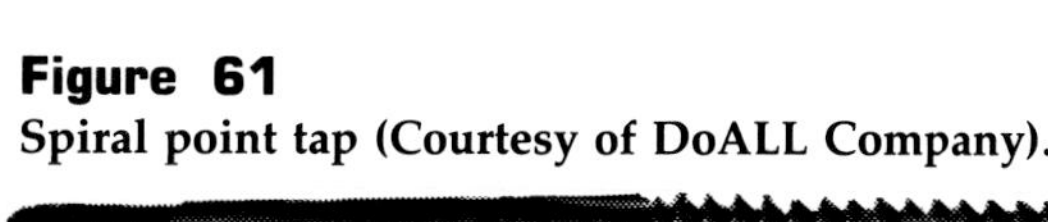

Figure 64
(*a*) Counterbore. (*b*) Countersink. (*c*) Spotfacing.

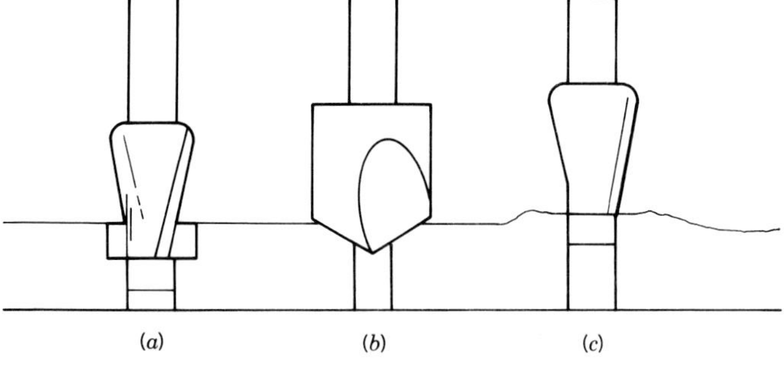

Taps are hardened, threaded cutting tools with flutes that contain or remove chips. Standard taps (Figure 60) come in sets of three: taper, plug, and bottoming. Threaded internal holes can be made by twisting the tap into a properly sized drilled hole by means of a tap handle. Sulfurized cutting fluid is often used when tapping. Machine tapping is a higher speed operation in which plug taps are sometimes used, but spiral point taps (Figure 61) are far more efficient and are less likely to break. Taps are brittle and breakage is a serious problem in manufacturing processes. A blind hole (not a through hole) that is tapped presents problems because of incomplete threads on the end of starting taps and the problem of chips collecting at the bottom of the hole. Bottoming taps are used in hand tapping. In machine tapping, spiral point taps push the chips forward to the bottom of the hole, sometimes causing jamming. Spiral flute taps (Figure 62), on the other hand, direct the chips outward along the flutes.

Threading dies for making external threads are used in a diestock, a handle that holds the die, for hand threading. Machine dies for high production threading are covered in Chapter 12.

Reamers (Figure 63) are used to enlarge an existing hole and give it a more accurate dimension with an acceptable finish. A drilled hole is usually oversized and somewhat rough. However, if the drilled hole is off center, rather than relocating it the reamer will follow the eccentric hole. Boring operations can restore an off-center hole to true position. Reamers are used on drill presses, lathes, and sometimes on vertical milling machines. Special tools called counterbores are used to enlarge existing holes (Figure 64). Countersinks provide a bevel to receive certain bolt heads and spotfacing is done to provide a machined seat for bolt heads or nuts.

Work Materials and Their Effects

Along with tool shape, work materials play an important part in chip formation. Tougher and greater strength materials require greater cutting forces which cause more tool deflection, greater heat generation, and therefore higher operating temperatures. Hard or abrasive materials can accelerate tool wear. Soft, ductile materials, such as soft pure aluminum or hot rolled (HR) low carbon steel, are somewhat "gummy" when cut on machines and tend to tear chips away causing roughness unless large back and side rake angles are ground on the tool (Figure 65). The roughness is caused by a built-up edge (BUE) (Figure 66). The large back rake helps to eliminate the BUE and produce a clean cutting action with a good finish, but the chip is formed as a coil or helix that is, as a continuous chip. Because the tool face is in constant contact with the chip, operating temperatures are higher than those developed with broken chips. When brittle metals such as some bronzes or cast iron are cut, fragmented chips are produced that are called segmented or discontinuous chips. Harder materials tend to cut cleaner and to coil the chip tighter with the same chipbreaker, when other factors are the same. Machinability (ease of cutting) of metals describes the ease or difficulty of metal cutting. Surface finish, horsepower required, and cutting tool life are factors to be considered. Machine steels have been rated on a scale based on annealed carbon steel containing .12 percent carbon (AISI B-1112) which is 100 on the scale (Table 2). Metals that are more easily

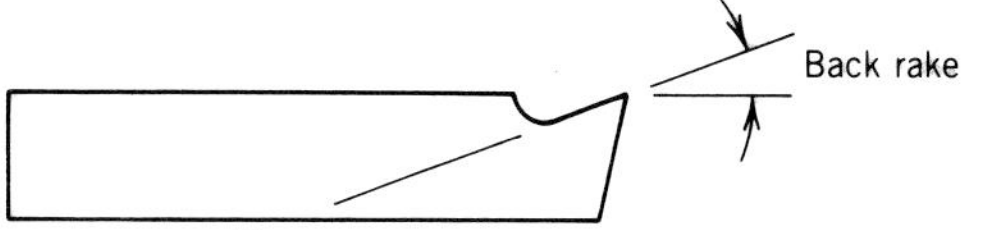

Figure 65
Lathe tool with large back rake angle.

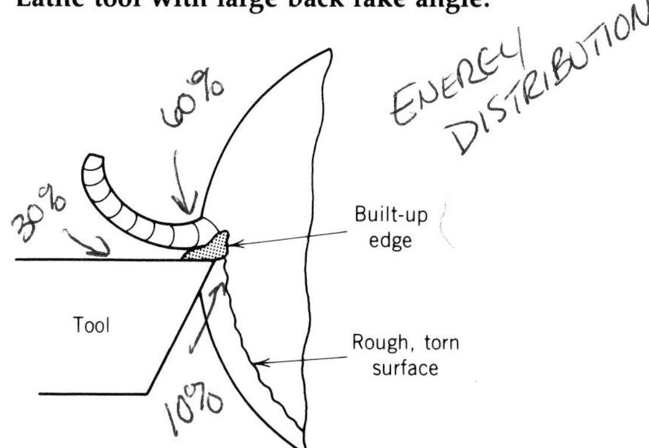

Figure 66
Tool with built-up edge.

Table 2
Machinability Ratings of Annealed Steels

AISI Classification	Machinability Rating	Approximate Hardness BHN (Brinell)
B1113	135	200
B1112	100	205
C1118	80	160
C1020	65	150
C1040	60	200
A8620	50	220
A3140	55	200
A5120	50	200
C4140	50	200
Cast iron	40 to 80	160 to 220
302 stainless	25	190

SOURCE John E. Neely, *Practical Metallurgy and Materials of Industry*, Second Edition, Wiley, New York, © 1984.

Table 3
Cutting Fluid and Speed Table for Machining Various Metals with High-Speed Tools

Material	Cutting Fluid	Speed (ft/min)
Aluminum and alloys	Soluble oil, kerosene, light oil	200 to 300
Brass	Dry, soluble oil, synthetic coolant solution, light mineral oil	150 to 300
Bronze	Dry, soluble oil, mineral oil, synthetic coolant solution	200 to 250
Cast iron, soft	Dry, air, synthetic coolant solution	100 to 150
Cast iron, medium	Dry, air, synthetic coolant solution	70 to 120
Cast iron, hard	Dry, air, synthetic coolant solution	30 to 100
Magnesium and alloys	60-sec mineral oil	300 to 600
Steel, alloy	Soluble oil, synthetic coolant solution	30 to 90
Steel, low carbon	Soluble oil, synthetic coolant solution	90 to 120

machined will have a higher number than 100 and materials more difficult to machine are numbered lower than 100.

Speeds and Feeds

Speed in machining is the number of surface feet per minute (SFM or FPM) that work material moves against the tool or that the tool cutting edge moves against the work. Feed rate is the motion of the tool being fed into the work or the work being moved into the cutting tool. On lathes and drill presses, this tool movement is in inches per revolution (IPR) and on milling machines it is table feed in inches per minute (IPM). Table 3 gives some cutting speeds for these machines for cutting several metals using high-speed tools. Remember that with carbide tools, these speeds can be increased three or four times.

Table 4 gives approximate feeds. As a rule, the greater the feed is, the closer the chip curls; small feeds tend to produce stringy chips. Since speeds are three to four times faster for carbides than for high-speed tools, better finishes are often obtained with carbide tools. Slow speeds tend to tear away metal and produce rough finishes and to distort the surface grains which can cause early failure of a machined part (Figures 67 and 68). Speeds for manufacturing should be as high as can be maintained without excessive tool breakdown. Feeds for roughing (rapid metal removal) should be as coarse as the tool material can withstand without fracture and

Table 4
Lathe Feeds, Inches per Revolution (IPR)

Material	Roughing Feed	Finishing Feed
Cast iron	.010 to .020	.003 to .006
Low carbon steel	.010 to .020	.003 to .005
High carbon steel, annealed	.008 to .020	.003 to .005
Alloy steel, normalized	.005 to .020	.003 to .005
Aluminum alloys	.015 to .030	.005 to .010
Bronze and brass	.010 to .020	.003 to .010

Figure 67
This micrograph shows the surface of the specimen that was turned at 100 SFM. The surface is irregular and torn, and the grains are distorted to a depth of approximately .005 to .006 in. (250 ×).

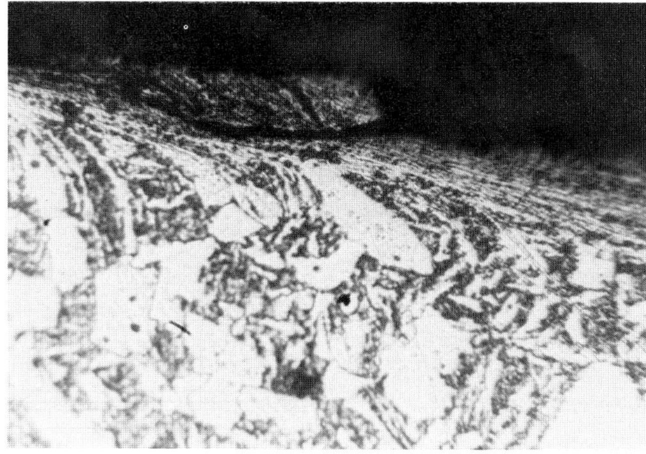

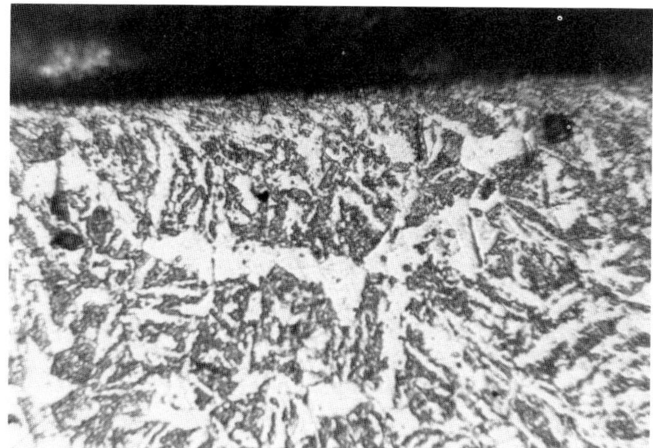

Figure 68
At 400 SFM this micrograph reveals that the surface is fairly smooth and the grains are only slightly distorted to a depth of .001 in (250 ×).

the machine can handle. Finishing feeds should be fine enough to produce the desired finish on the workpiece. Minimal machining time must be balanced against cost and breakdown time of tool material and the limits of the machinery.

Cutting Fluids

Cutting fluids are usually liquid, but air is sometimes used to avoid contamination on certain metals. In general, cutting fluids serve to dissipate heat, to lubricate the tool–work interface to lower friction, and to carry away chips from the cutting area. Some cutting operations, such as cutting threads by tapping or using a die, have considerable friction and need a fluid with good lubricating qualities; whereas high-speed cutting operations, as on a milling machine where the tool work interface is small, require less lubricity and more cooling qualities in a cutting fluid. There are essentially two types of cutting fluids, cutting oils and coolants. Cutting oils can be either animal fats or petroleum oils. Where extreme pressure lubricants are needed, as in threading operations, chlorinated or sulfurized oils can be added to cutting fluids. Coolants are usually water-based soluble oils that have excellent heat transfer qualities. Petroleum oils or waxes are treated so that they emulsify when mixed with water to produce a milky-white solution. Other chemicals such as wetting agents, rust inhibitors, antibacterial agents, and polarizing agents are added to improve the coolant. Many synthetic coolants have been developed for use as emulsions or cutting oils for specific and general uses. Some materials, such as cast iron, brass, and plastics, can be machined dry, but new cutting fluids are now used on these materials to greatly enhance the cutting speeds and finishes.

Some lathe operations on all materials including steel are done without coolant. Tool room or engine lathes used in machine shops are quite often operated without coolant, but all production machines are fitted with coolant pumps and tanks. Also all milling machines normally require coolant to avoid burning cutting tools. There are some exceptions; for example, in an intermittent cutting setup with a carbide insert tool where there is alternate heating and cooling of the carbide, coolants would only add to the thermal shock which can crack the carbide insert.

REVIEW QUESTIONS

1. In terms of interchangeability of manufactured parts, what characteristic of machining processes is virtually unique and most important?
2. Define the process of machining metal.
3. If you had only an inch-reading micrometer and obtained a measurement on a part of .315, you might suspect that it is a metric measurement since most inch dimensions are usually found in decimals of fractions; the nearest fraction is $\frac{5}{16}$ which equals .3125 in. Using the conversion factor 1 in. = 25.4 mm, what would the metric equivalent of your inch measurement be?
4. Is a machinist's steel rule considered to be a precision measuring tool, or a semiprecision measuring tool?
5. How can a precision measurement be transferred from an inside diameter of a bore to an outside direct reading instrument such as a micrometer caliper?
6. If an inch micrometer can discriminate to .0001 in. and a metric micrometer to .01 mm, which instrument can measure to the smallest or finest increment? Use the conversion factor 1 in. = 25.4 mm or 1 mm = .03937 in.
7. Since angular measure is in degrees, minutes, and seconds, which instrument, the bevel protractor or the combination set protractor, has the most precision?
8. How can surface roughness be measured?
9. When sizes between two tolerance dimensions need to be checked quickly by a machine operator at a production machine, what type of measuring tool is used?
10. Gage blocks can have an accuracy of (*a*) ten thou-

sandths, (*b*) hundred thousandths, or (*c*) millionths of an inch.

11. Describe the process by which metal is removed by a cutting tool in the machining process. Is the metal removed in chips by splitting the metal ahead of the tool, or by plastic deformation?

12. The most common cutting tools used in machining operations of all kinds are high-speed steel, ceramic inserts, diamond tools, and varieties of carbide tools. Of these, which type of tool is by far the most commonly used in production machining operations?

13. Stringy, wiry chip formations are hazardous and cannot be easily handled or disposed of. How can this kind of chip be avoided? What is a more desirable chip form?

14. Holes made by twist drills are usually rough and often oversized. How can a drilled hole be made to have a more accurate dimension with a smoother finish?

15. How can a built-up edge cause a reamer or other cutting tool to produce a rough finish and inaccurate dimensions?

16. Cutting speeds (SFM) for manufacturing should be as high as possible without burning the tool or workpiece. If the speed for a lathe operation has been determined to be 300 RPM for high-speed tools, how fast should the machine turn for carbide tools?

17. Cutting fluids come in two types. Name each and describe its' special qualities.

18. When are lathe operations sometimes performed without cutting fluid?

CASE PROBLEMS

Case 1
Chip Control Problem

High-speed steel cutting tools were used for production turning stainless steel fittings on an automatic screw machine. The setup person originally sharpened the cutting tools by offhand grinding on a pedestal grinder. A problem developed in which wiry chips entangled the mechanism causing lost machining time in shutdowns for cleaning out the entangled chips. What would you recommend to correct this problem?

Case 2
Translating Shop Terminology into Precise Dimensions

A millwright telephones your shop to request a number of roller chain sprockets, some to be bored to $2\frac{7}{8}$ in. and some to $1\frac{7}{16}$ in., "for a sliding fit." To do what the millwright wishes, you would have to make the bore $1\frac{1}{2}$ to 2 thousandths of an inch over these standard shafting sizes. Convert this crude data into precision decimal inch dimensions first and then to metric, because your shop is now on metric measure.

CHAPTER 12 MACHINE TOOL OPERATIONS

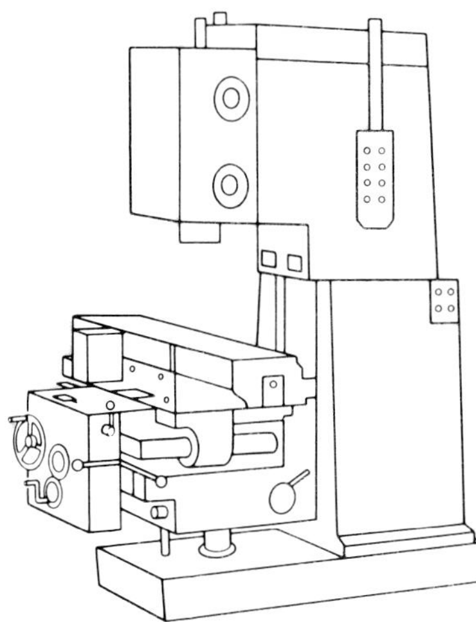

Machining processes that involve the removal of metal in the form of chips are used not only in the field of product manufacturing, but also in what could be called "conventional machining." The principle purpose of production machining of such items as gears, housing, and other mechanical devices is to reduce manufacturing costs by means of high production rates. Costs can be further reduced in some operations by use of automatic or computer controlled machining. These high production machines require set-up machinists, operators, and technicians. By far, most jobs in machining manufacturing are in these categories.

Much machining is still performed in single or one-of-a-kind operations that require the constant attention of a machine operator or general machinist. This conventional machining involves the use of nonspecialized, versatile machines such as drill presses, engine lathes, and milling machines. General machinists require much more training than production machine operators because of the many skills needed to produce precision machine parts from mechanical drawings. An even more highly skilled trade is that of a tool and die maker who is a specialized general machinist, trained in toolmaking. Usually the most highly regarded and highest paid of all machinists in the trade, these tool room machinists are always in demand to make prototypes of new mechanisms (Figure 1). When these new prototypes have been sufficiently developed in the tool room machine shop, a manufacturing plant can then tool up or adjust machines, as in flexible manufacturing systems (FMS), to mass produce the parts. Besides making new tooling, these machinists often repair existing tools and dies.

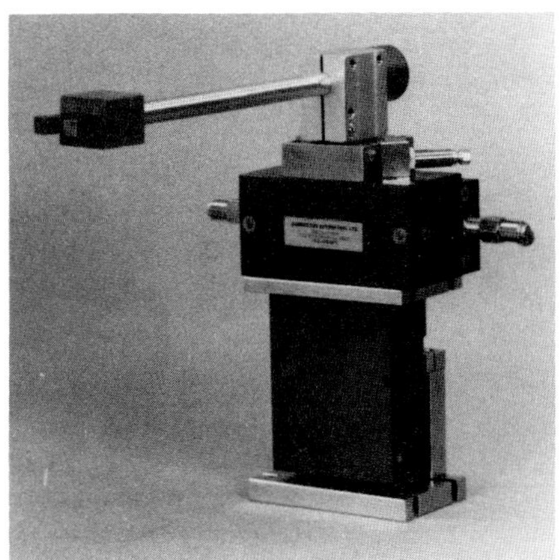

Figure 1
Robot positioner. Robot arms and components such as this one must first be made by conventional machining processes before being put into production (Barrington Automation Ltd.).

Small privately owned job shops can be found in virtually every city. They utilize basic machine tools to produce short runs of products not being manufactured elsewhere. Of course they could not compete with a high production manufacturer. These machine shops also repair machine parts for local industries. Some large manufacturing plants have their own machine shops for machinery repair.

OBJECTIVES

This chapter should enable you to:

1. Decide when to use tool room machinery and when to use production machinery for making a machined part.
2. Decide which type of drilling or sawing machine is needed for certain drilling or sawing operations.
3. Explain the special uses of vertical and horizontal spindle turning, milling, and boring machines.
4. State how machine threads are cut and how they are designated.
5. Describe the processes of making many kinds of internal and external gears and splines.
6. State the principles and uses of abrasive machining.
7. Determine how grinding machines and processes are used in machine shops and in manufacturing.

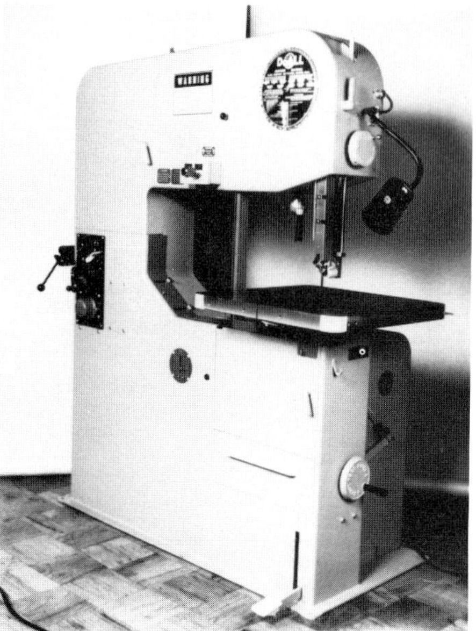

Figure 2
Vertical band saw (Courtesy of DoALL Company).

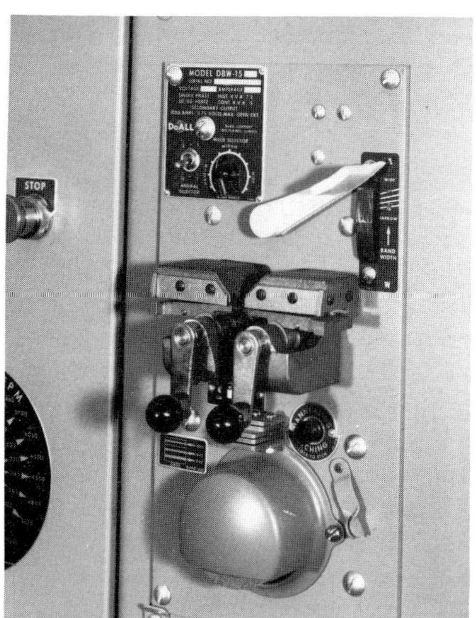

Figure 3
Flush-mounted blade and grinder enable the operator to weld saw blades for external and internal sawing (Courtesy of DoALL Company).

BASIC MACHINE TOOLS

Basic machine tools such as metal cutting saws, drill presses, lathes, and milling machines are used in conventional machining. They are not normally used for production purposes unless attachments such as jigs and fixtures are used with them to guide tools and hold workpieces.

Metal Sawing Operations

Conventional machine operations are designed to remove all unwanted metal on a solid piece of metal to produce a finished part. The vertical band saw (Figure 2) removes unwanted metal without turning all of it into chips. For some operations the saw cut can be made to the layout line and no subsequent machining is required. In other cases, finishing operations are required after the part is sawed, but considerable time is saved in the sawing operation. Band saw blades are made of carbon steel and on some bands high-speed steel teeth are attached to the cutting edge. Bands are cut to length and welded to form endless bands on a resistance butt welder (Figures 3 and 4), sometimes called "flash" welders. Internal areas can be cut by inserting the blade through a drilled hole in the material and then welding the blade together. Contours, small or large radii, and many other shapes can be cut with vertical band saws.

Cutoff Saws Metal bar stock for manufacturing is usually shipped in 20- or 40-foot lengths and range from $\frac{1}{4}$ in. diameter round bars to those of 12 in. diameter and over. Universal milled flat bar stock is rolled in sizes up to 14 in. wide and there is a great variety of round, square, and rectangular bar stock and a wide variety of round, square, and rectangular tubing, plus many other shapes. All of these must be cut to the

Figure 4
Ends of band being welded in a resistance "flash" butt welder.

Figure 5
Horizontal band saw (Courtesy of DoALL Company).

length needed for specific part manufacturing. The most common type of cutoff saw used for this purpose is the horizontal band saw (Figure 5). This type is often used by machinists or welders in machine or welding shops. Automatic band saws, often controlled by microprocessers (Figure 6), cut off long bars to programmed lengths, feeding in the bar after each cut. Most of these saws use coolant to prolong saw blade

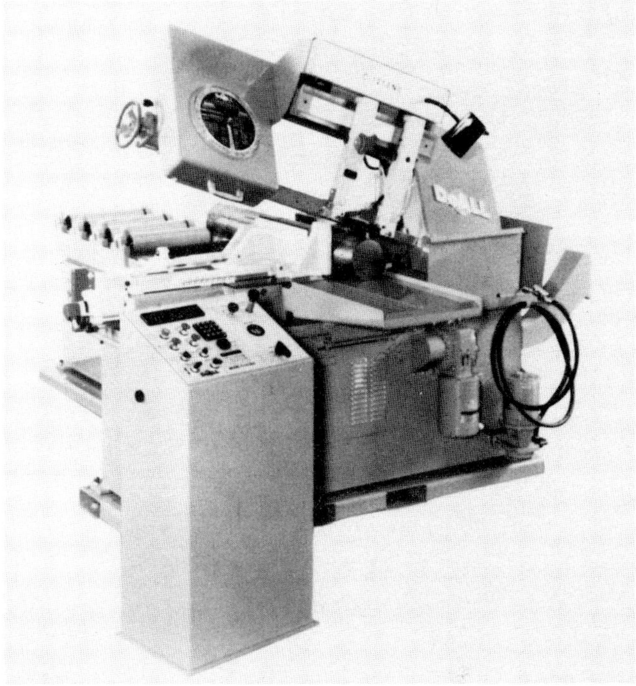

Figure 6
Automatic band cut-off saw (Courtesy of DoALL Company).

life. These machines are mainly used for production.

The reciprocating cutoff saw uses a straight blade and is often called a power hacksaw since its action resembles the operation of a hand hacksaw with its back-and-forth motion. It is an older machine tool than the band saw, having a somewhat slower cutting action; it is still used in many shops because it can cut harder material than the band saw with less blade breakdown. It is useful for cutting off tough, hard metals such as titanium, zirconium, stainless steel, and tool steels.

Drilling Machines

Drilling machines are designed primarily for making holes in metal and other materials; however, other operations such as counterboring, spotfacing, reaming, and tapping are often performed on them. Since hole drilling operations range from a few thousandths of an inch diameter (size of a hair) to over a foot in diameter, it follows that several kinds of drilling machines are needed. Also some drilling machines are geared for shop work and others are for manufacturing. The three basic types of drill presses are sensitive, upright heavy duty, and radial arm. Other variations used are multiple spindle drill presses, turret types, and deep hole drilling machines.

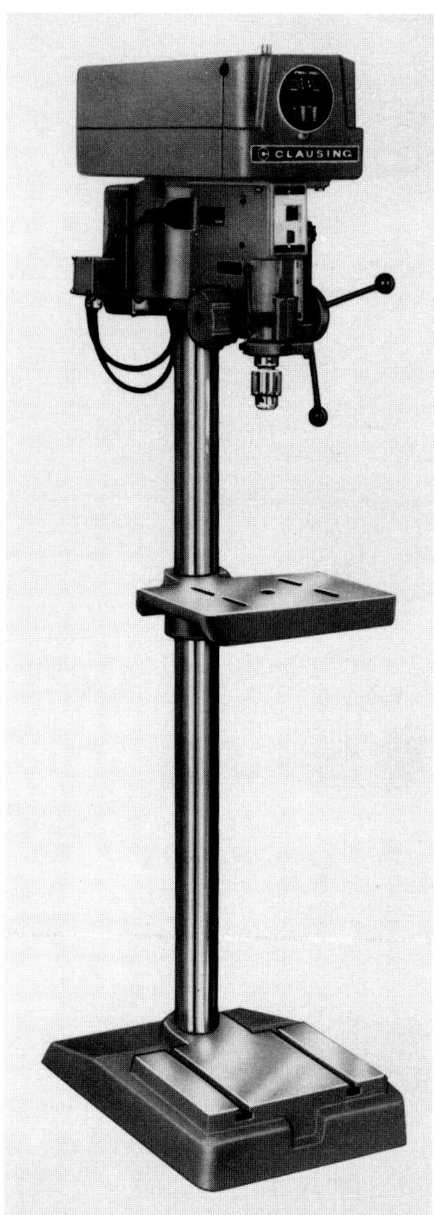

Figure 7
Sensitive drill presses are used for light duty drilling (Clausing Machine Tools).

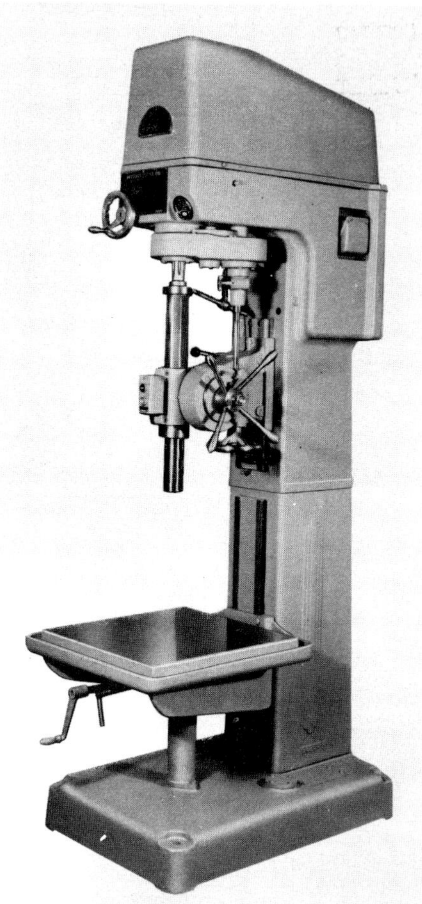

Figure 8
Upright heavy duty drill press (Buffalo Forge Company).

Figure 9
Radial arm drill press (Clausing Machine Tools).

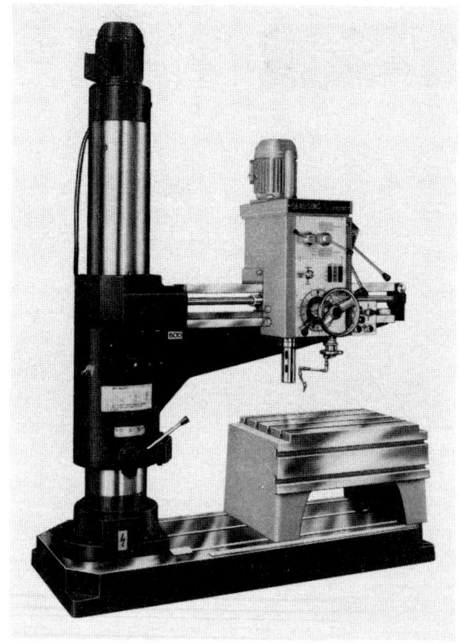

The sensitive drill press (Figure 7) is so named because the operator "senses" or "feels" the cutting action of the drill while holding the handle that feeds the drill into the work. These machines have no power feed and are usually limited to drills of up to $\frac{1}{2}$ in. diameter. Sensitive and upright heavy duty drill presses are measured by the largest diameter of a circular piece that can be drilled in the center. Thus if the distance from the column to the center of the spindle is 10 in., it is a 20-in. drill press. Ordinary sensitive drill presses used in machine shops can drill holes no smaller than about $\frac{1}{16}$ in. diameter, possibly only as small as .30 in. in some cases. Holes smaller than this must be drilled on a microdrill press. The smaller the drill size, the higher the RPM must be to have the same surface speed on a drill. For example, using the simplified formula $\text{RPM} = \frac{CS \times 4}{D}$, where CS is the cutting speed and D is the diameter, with the cutting speed of 100 SFM, a 10 in. diameter drill would need to turn 40 RPM, which would be realistic for this diameter drill. If the drill were only .010 in. diameter the RPM would have to be 40,000. However, it has been found that this formula is not realistic at very small diameters and high rotational speeds. Even at 10,000 RPM, which would be more reasonable for a .010 in. drill, a fine, powdery chip formation often develops that can quickly dull the drill. Most microdrilling takes place at speeds between 5,000 and 12,000 RPM.

Heavy duty, upright drilling machines (Figure 8) are provided with power feeds and reverse capability. On some machines tapping accessories are provided that feed the tap in with the correct thread pitch and with rapid reversal to remove it from the work. A coolant pump is often provided with these heavy duty drill presses. A wide range of holes can be drilled on these machines, from $\frac{1}{4}$ to $3\frac{1}{2}$ in., which is the largest twist drill size. Larger holes are normally made on radial arm drilling machines with flat or spade drills.

Radial arm drill presses (Figure 9) are made in a variety of sizes. They are measured by the diameter of the column and the length of the arm as measured from the center of the spindle to the outer edge of the column. Small workpieces can be shifted around by the operator until the desired hole location is directly under the twist drill, as is done on sensitive drill presses. However, when the workpiece is a large casting that weighs many tons, it is not convenient to move it for each drilled hole. But the drilling tool of the radial arm drill press is easily moved and located over the workpiece. When it is located at the drilling position, the radial arm can be clamped to the column and the drill head can be clamped to the arm by moving a lever or pressing a switch. Like the upright machine, the radial arm drill press has a power feed mechanism and a hand-feed lever. It also has a coolant tank, usually located in the base, with a pump.

Gang drills (Figure 10) are used where a series of drilling and reaming operations must be performed in a sequence. Each drill head has its own special tooling and a fixture for holding the workpiece. Time-consuming tool changing and the making of setups with hold-down clamps are eliminated by using preset gang drills. Drilling machines with multiple spindles (Figure 11)

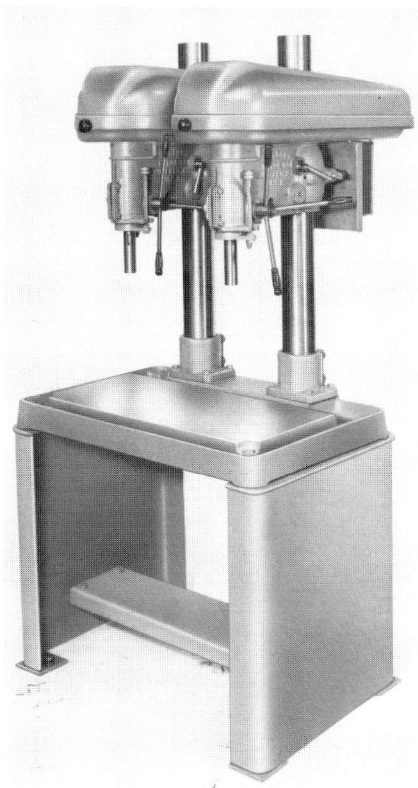

Figure 10
Gang drills (Buffalo Forge Company).

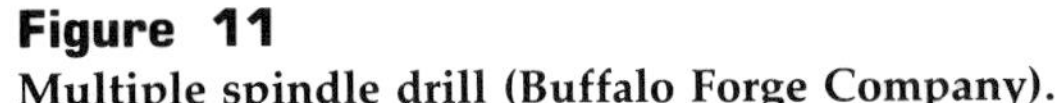

Figure 11
Multiple spindle drill (Buffalo Forge Company).

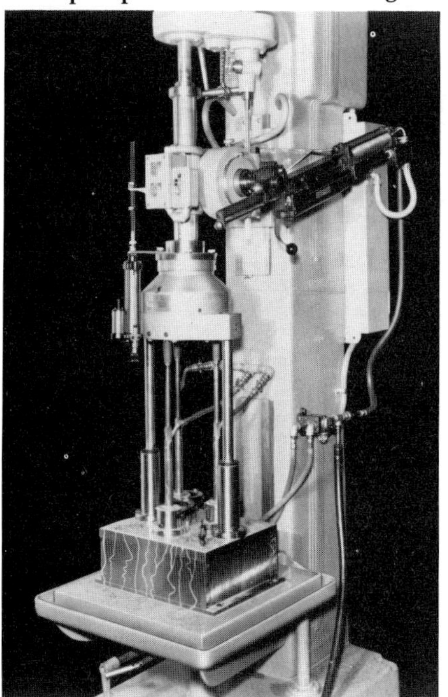

Figure 12
Machining centers such as this one perform many different cutting operations automatically (Cincinnati Milacron, Inc.).

Figure 13
NC turning lathe (Cincinnati Milacron, Inc.).

Figure 14
Curved and tapered turning operation on a shaft in an NC lathe (Cincinnati Milacron, Inc.).

Figure 15
A 15-in. swing engine lathe. Swing (largest diameter that can be turned) is one measurement of lathe size. (Clausing Machine Tools).

Figure 16
Vertical boring mill.

Figure 17
Vertical turret lathe (Ductr Mfg.).

allow many drilled holes to be made at one pass. Drill jigs that guide the drills to exact locations are often used with multiple spindle drills. Modern machining centers that are controlled by computers (Figure 12) are often used for precision drilling and reaming operations. They also perform many other machining operations which will be discussed later in this chapter.

TURNING MACHINES

Turning machines are those that rotate a workpiece against a tool. The tool is fastened to a sliding member that gradually moves the tool along the work, producing an accurate cylindrical or tapered (conical) surface. Tracing attachments make possible almost unlimited curved shapes on the turned cylinder. Numerical control (NC) and computer numerical control (CNC) operated turning machines (Figure 13) can produce a great variety of cylindrical, curved, and tapered shapes (Figure 14). These NC machines are fully automatic and require an operator only to set up the workpiece. Turning machines include the engine lathe with its horizontal spindle (Figure 15), the vertical spindle machines, the vertical boring mill (Figure 16), and the vertical turret lathe (Figure 17).

The Engine Lathe

This machine got its name from the steam engine that first powered it, an improvement over the earlier foot or hand powered lathes. It has been the most important tool in machine shops and manufacturing plants for over 150 years. Until production machines were developed, cylindrical parts were made with new tooling and measuring setups each time a part was made by machinists on engine lathes. These valuable machines are still very much in use in tool and die shops, job machine shops, and machine repair shops, although they are no longer used for production purposes. These turning machines range in size from tiny jeweler's lathes to huge machines that turn massive forgings (Figures 18 and 19).

The principal function of the lathe is to remove unwanted metal to form cylindrical and conical shapes. This is done with a single-point tool (as contrasted with a multiple point tool such as a milling cutter). However, other tool shapes are used to make grooves, to cut off round stock, and to make various kinds of threads by chasing, that is, taking a series of cuts in the same groove until a full thread is made.

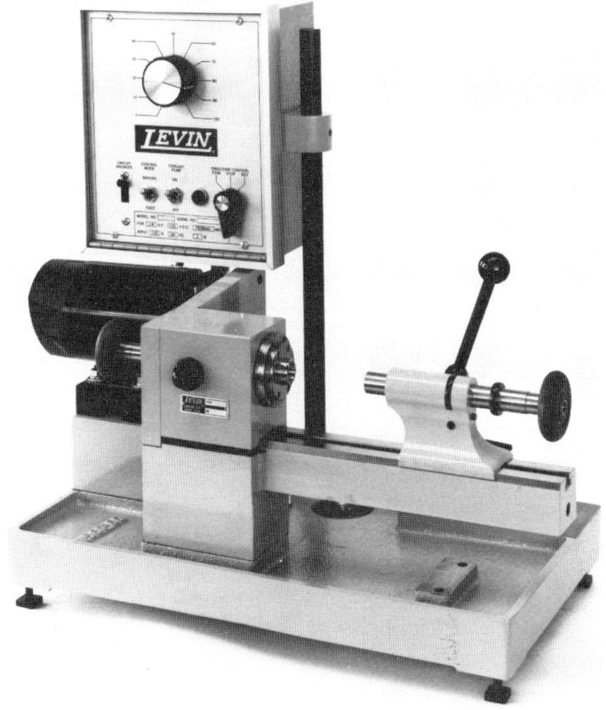

Figure 18
Jeweler's lathes are used to make very tiny precision parts. Many attachments for toolholding can be used with these machines (Photograph courtesy of Louis Levin & Son, Inc.).

Figure 19
A large heavy duty engine lathe (Clausing Machine Tools).

Thread Cutting

Although cutting threads is one of the important functions of the engine lathe (Figure 20), most threads are cut on production machines. There are a number of thread forms that have been standardized. Some of the more commonly used ones can be seen in Figures 21 through 32. Screw thread sizes are expressed in terms of major (outside) and minor (inside) diameter, pitch diameter, number of threads per inch, and pitch (distance between threads) (Figure 33).

Figure 20
Single-point threading on a lathe.

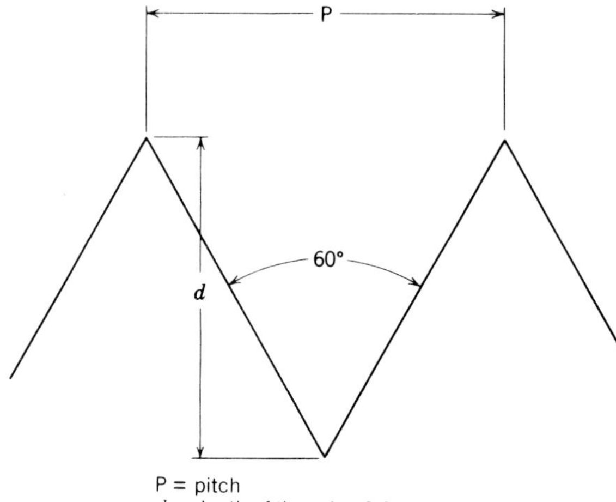

P = pitch
d = depth of thread = 0.866P

Figure 21
The 60-degree sharp V-thread form (White, Neely, Kibbe, Meyer, *Machine Tools and Machining Practices*, Vol. I, © 1977 John Wiley & Sons, Inc.).

Figure 22
The American National form of thread (White, Neely, Kibbe, Meyer, *Machine Tools and Machining Practices*, Vol. I, © 1977 John Wiley & Sons, Inc.).

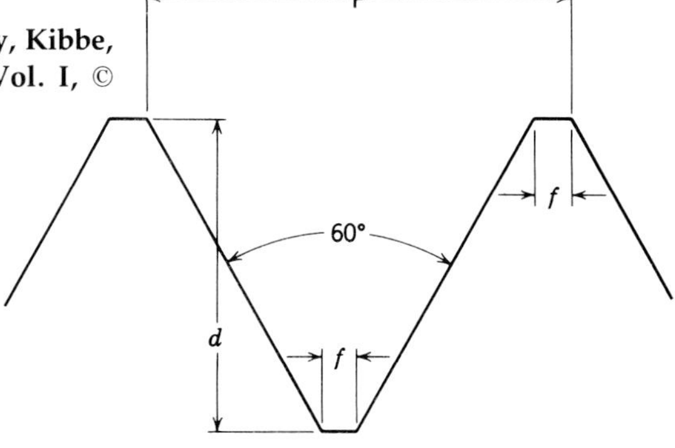

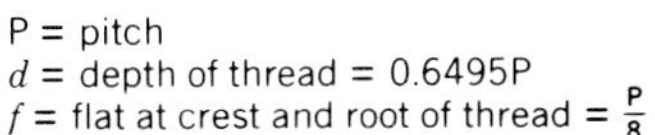
P = pitch
d = depth of thread = 0.6495P
f = flat at crest and root of thread = $\frac{P}{8}$

Figure 23
Unified thread forms.

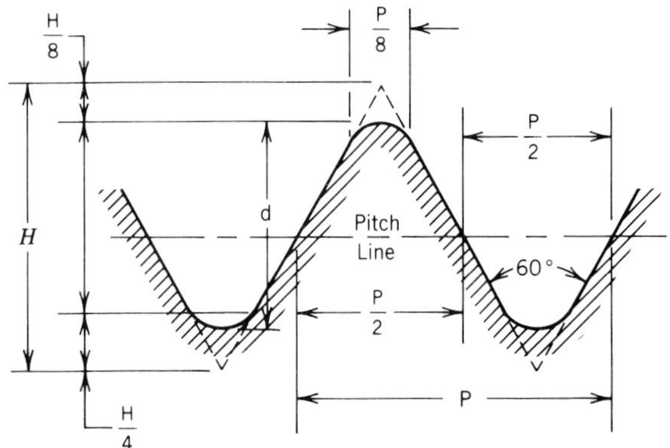

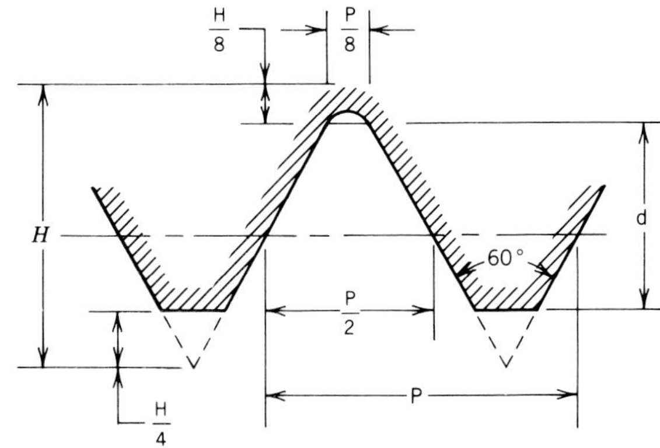

External unified threads.

P = pitch = $\frac{1}{n}$

d = depth of thread = 0.61343 × P
= $17H \div 24$

Flat at crest = 0.125 × P

Rounded root optional or resulting from tool wear.

Rounded or flat crest.

Unified threads.

Internal unified threads.

P = pitch = $\frac{1}{n}$

d = depth of thread = 0.54127 × P
= $5H \div 8$

Flat at crest = 0.125 × P
Flat at root = 0.125 × P

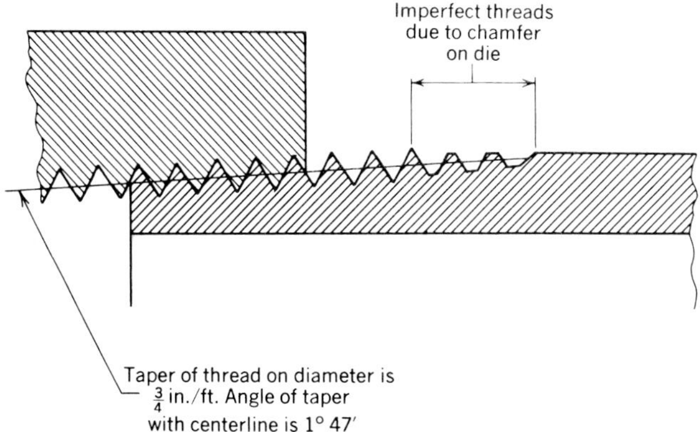

Figure 24
American National standard taper pipe thread form (White, Neely, Kibbe, Meyer, *Machine Tools and Machining Practices*, © 1977 John Wiley & Sons, Inc.).

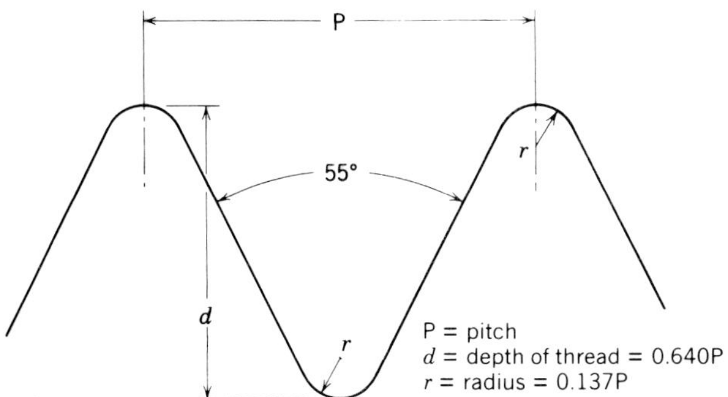

Figure 25
Whitworth thread form (White, Neely, Kibbe, Meyer, *Machine Tools and Machining Practices*, Vol. I, © 1977 John Wiley & Sons, Inc.).

Figure 26
The SI metric thread form (White, Neely, Kibbe, Meyer, *Machine Tools and Machining Practices*, Vol. I, © 1977 John Wiley & Sons, Inc.).

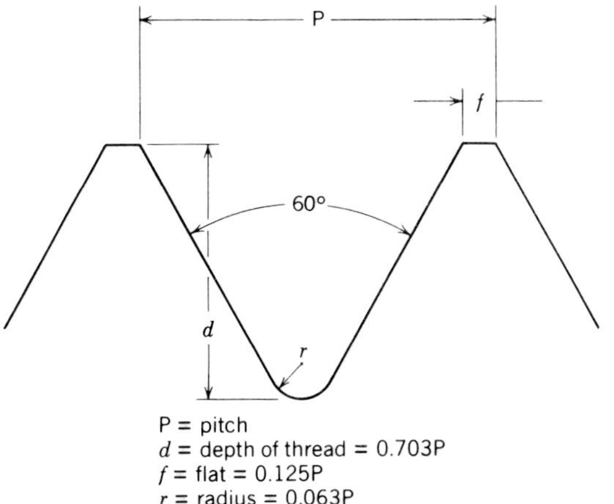

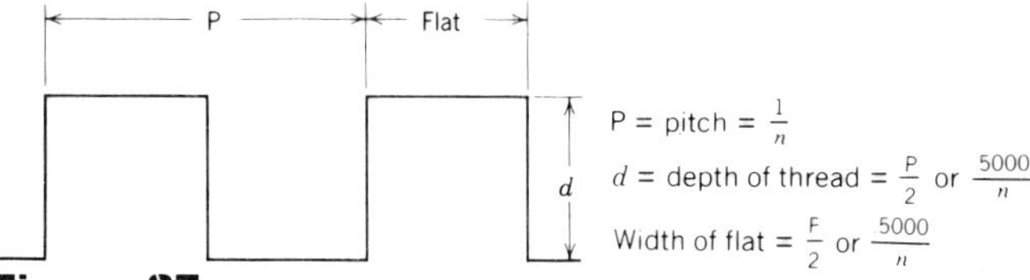

Figure 27
The square thread form (White, Neely, Kibbe, Meyer, *Machine Tools and Machining Practices*, Vol. I, © 1977 John Wiley & Sons, Inc.).

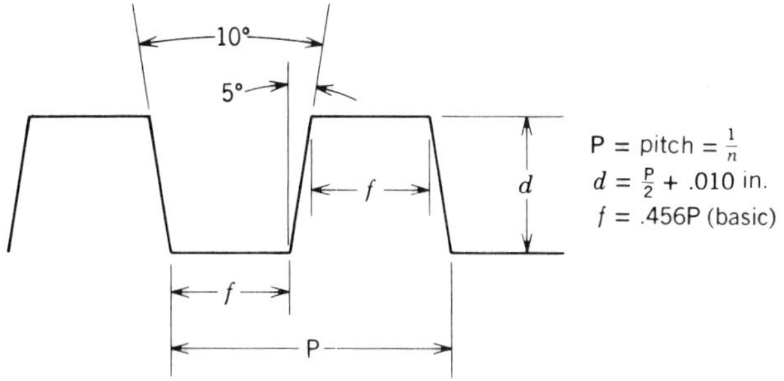

Figure 28
Modified square thread form (White, Neely, Kibbe, Meyer, *Machine Tools, and Machining Practices*, Vol. I, © 1977 John Wiley & Sons, Inc.).

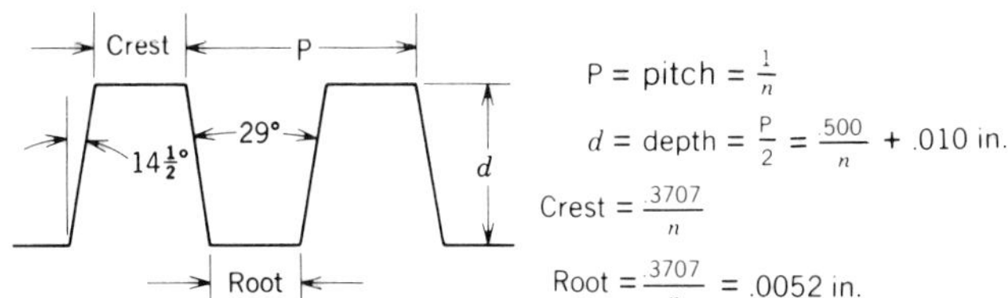

Figure 29
Acme thread form (White, Neely, Kibbe, Meyer, *Machine Tools and Machining Practices*, Vol. I, © 1977 John Wiley & Sons, Inc.).

Figure 30
American Standard Stub Acme thread form (White, Neely, Kibbe, Meyer, *Machine Tools and Machining Practices*, Vol. I. © 1977 John Wiley & Sons, Inc.).

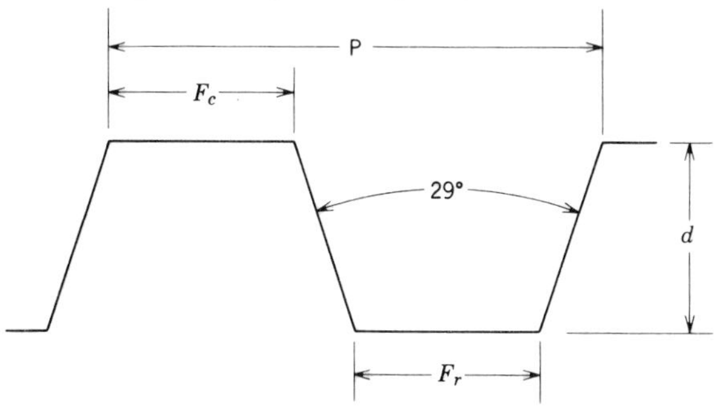

P = Pitch = $\frac{1}{n}$
d = Depth = .3P
F_c = Basic flat at crest = .4224P
F_r = Basic flat root = .4224P − .0052

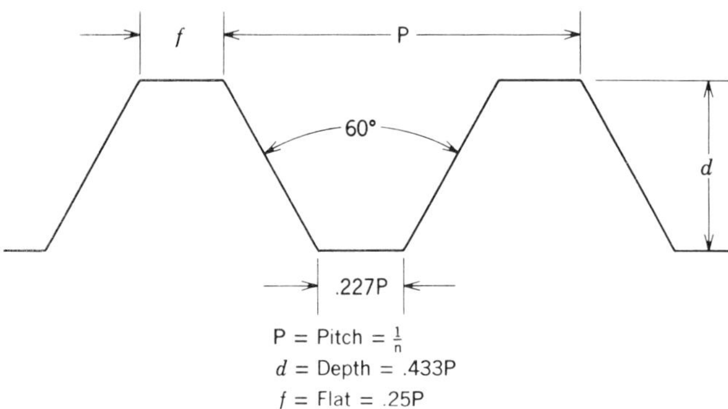

Figure 31
Stub Acme thread form (White, Neely, Kibbe, Meyer, *Machine Tools and Machining Practices*, Vol. I, © 1977 John Wiley & Sons, Inc.).

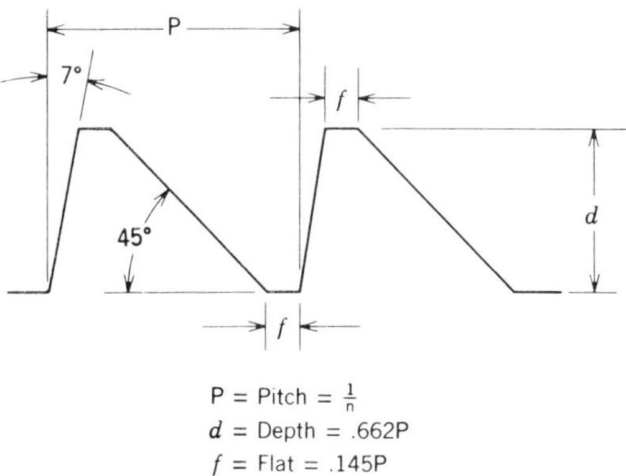

Figure 32
Buttress thread form (White, Neely, Kibbe, Meyer, *Machine Tools and Machining Practices*, Vol. I, © 1977 John Wiley & Sons, Inc.).

Even though the American National Form as a thread standard with its common thread series National Coarse (NC) and National Fine (NF) is obsolete, it is still in use in many machine shops. Even some tap drill charts are still being produced based on this popular system. The American standard for Unified threads is the official thread standard at the present time with its basic Unified coarse thread (UNC) and Unified fine thread (UNF). American and Unified standards will interchange for the most part, fits and tolerances make up the principal difference.

Screw thread manufacture in the United States is either in the Unified or Metric series. Both have similar thread forms but with different pitch, pitch diameter, and major diameter. Metric thread forms in different countries often have different pitches and pitch diameters and are not interchangeable. Besides the British ISO, there is the Systéme International (SI), the Löwenherz metric thread used in Germany, SAE spark plug threads, the British standard for spark plugs, and numerous differences between manufacturers worldwide.

In the Unified, American National standard, and the ISO metric thread systems there is a coarse and a fine series, the pitch varying with the diameter. The angle of the threads is 60 degrees on all three series. Threads are designated in the Unified standard with the nominal size first, followed by the number of threads per inch, the thread series, and class of thread. If the letters *LH* are added, the thread is left-hand. An example of this is $\frac{3}{4}$—10 UNC—2A. The symbols *1A, 2A,* and *3A* indicate an external thread tolerance and are right-hand. 1A is a loose fit and 3A is a tight fit. Three other tol-

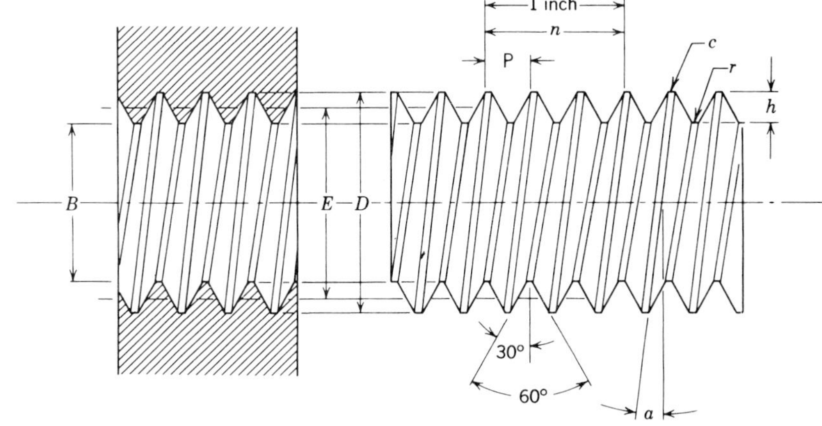

Figure 33
Thread nomenclature and dimensions (White, Neely, Kibbe, Meyer, *Machine Tools and Machining Practices*, Vol. I, © 1977 John Wiley & Sons, Inc.).

erances are given in the internal thread, 1B, 2B, and 3B. These can be adjusted to get a desired tolerance; that is, a 2A bolt can be fitted to a 3B nut. Metric threads are likewise designated with the major diameter first, followed by the pitch, and then the tolerance grade. An example is M20 × 2 6H/5g6g. Internal threads that are tapped must be drilled to a certain diameter to produce a correct percentage of thread for ease of tapping in a given material. For example, a 75-percent thread is correct for soft machine steel, but a 60-percent thread (larger tap drill) may be better for 301 stainless steel. The percent of thread and tap drill size or **actual minor diameter** can be determined with the following formula:

$$\text{percentage of thread} = \frac{\text{major diameter} - \text{actual minor diameter}}{2 \times \text{basic thread height}} \times 100$$

or,

$$\text{actual minor diameter} = \text{major diameter} - 2 \times \text{basic thread height} \times \text{percentage of thread}$$

Basic thread height or single depth for Unified threads equals the pitch × .541. See Appendix tables for Unified and Metric thread sizes and corresponding tap drill sizes. Tapered threads such as those used to join pipe are also made on an engine lathe by using a taper attachment. The symbol for American standard taper pipe thread is *NPT*.

Workholding Devices on Lathes These include independent (usually four-jaw) chucks (Figure 34), universal (two, three, and six-jaw) chucks (Figure 35), and collets (Figure 36). Independent chucks have jaws that must each be adjusted separately to grip and locate the workpiece. Square, rectangular, and odd-shaped parts are usually held in four-jaw chucks. Most universal chucks have three jaws that move together or apart the same amount when a chuck wrench is turned. Round and hexagonal parts or bar stock can be quickly gripped and centered with these chucks. Six-jaw universal chucks are able to grip round materials more securely than three-jaw chucks. Three-jaw chucks for engine lathes often have two interchangeable sets of jaws, one set to grip hollow parts, such as rings, internally, and one set to hold workpieces externally. However, most chucks now used on engine lathes or on manufacturing machines have removable jaws (called top jaws) that are either bolted on or have a quick change feature so they can be quickly removed and turned over for internal

Figure 34
Four-jaw independent chuck (Buck Chuck Company).

Figure 35
Three- and six-jaw chucks. These standard chucks require an extra set of jaws for gripping large diameters (Buck Chuck Company).

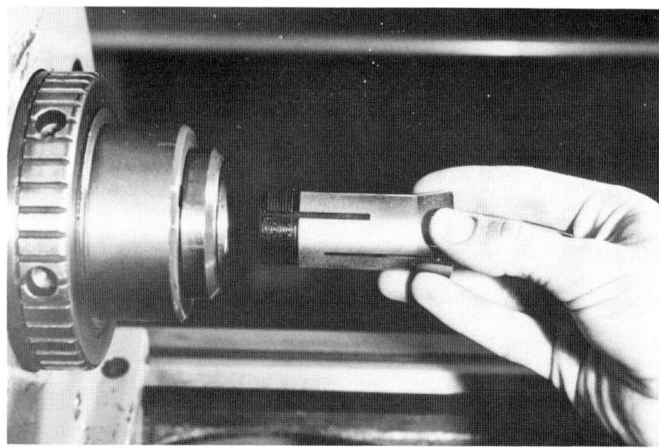

Figure 36
Collets are held in the spindle nose with an adaptor and a draw bar.

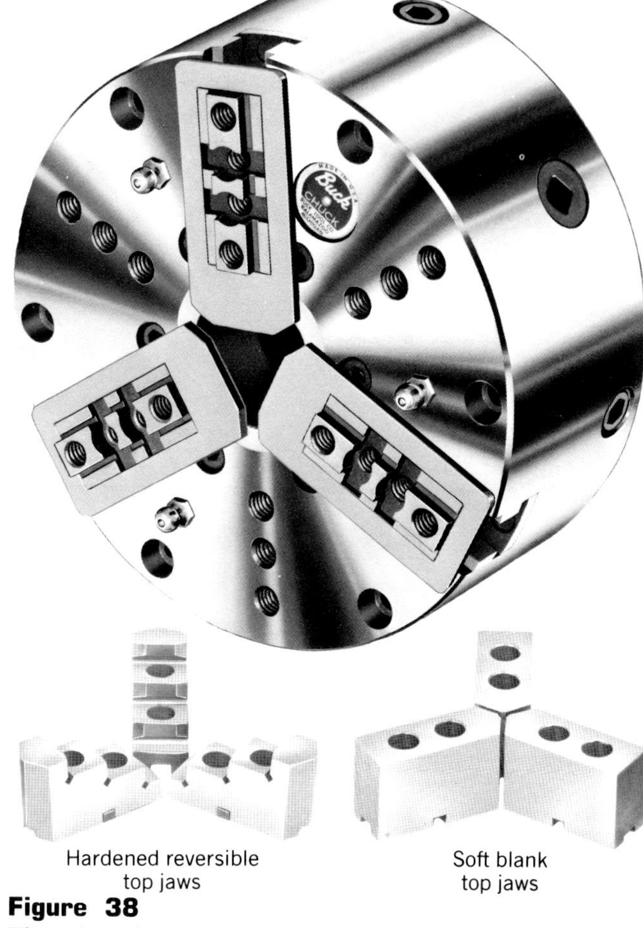

Figure 38
Three-jaw chuck showing hardened reversible top jaws and soft blank top jaws (Buck Chuck Company).

Figure 37
Three- and six-jaw chucks with top jaws (Buck Chuck Company).

gripping (Figures 37 and 38). Also soft jaws are provided that can be machined while in place to fit a particular odd-shaped part.

Power chucks (Figure 39) are used on turret lathes and other production machines. Usually the chuck is opened and closed by means of an air cylinder that is fastened to and rotates with the machine spindle. Most three-jaw chucks, when new, will hold round parts within less than .001 in. runout (eccentricity). However, when they become worn, runout may exceed .010 in. Some chucks have adjustment screws to take up wear; these can be adjusted to within .0005 in. (Figure 40).

Collets are capable of consistently holding small precision workpieces to less than .001 in. runout. Steel collets cannot hold parts that vary from the nominal size of the collet by more than a few thousandths of an inch. Rubber flex collets can hold parts with a range of $\frac{1}{8}$ inch. All of these workholding devices for lathes are fastened to the spindle in any one of several methods, flange nose, camlock, long taper key drive, and threaded nose (Figure 41).

Mandrels (Figure 42) are a form of workholding device for parts that have a hole or bore in their centers. The mandrel holds the work concentric to the bore, so any machining done is also concentric. Workpieces are

Figure 39
Power chucks are often used on production machinery (Buck Chuck Company).

Figure 40
Adjust-Tru® chuck. Chuck is adjusted at *G* to eliminate runout using a dial indicator. Jaws are tightened at *C* (Buck Chuck Company).

Am. St. flange nose

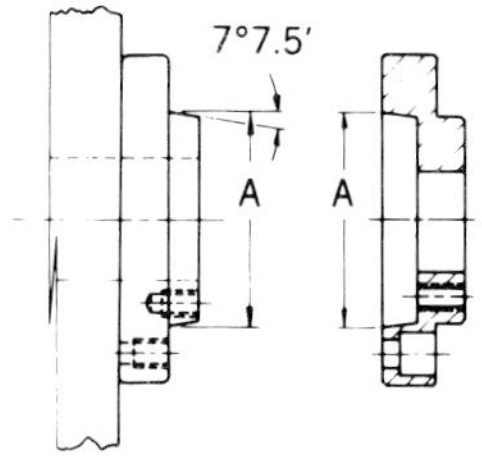

Camlock
Camlock stud length is adjustable to suit spindle.

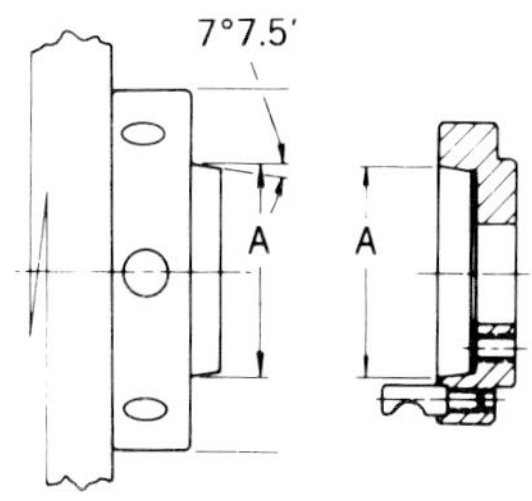

Long taper key drive

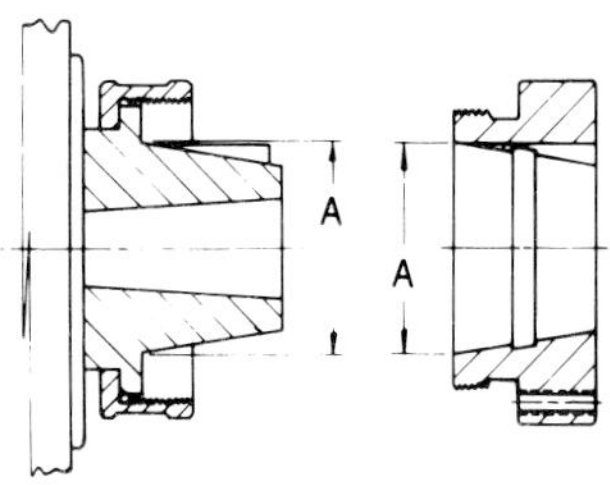

Threaded mounting plates

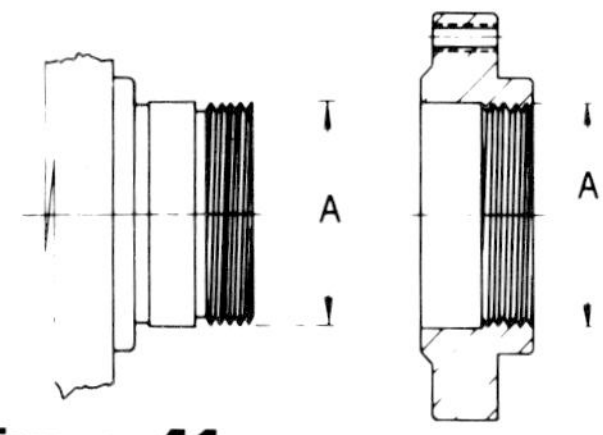

Figure 41
Four basic types of spindle nose arrangements for mounting chucks and other workholding devices (Buck Chuck Company).

held by friction, either by means of a slight taper (.006 in./ft) on a mandrel that is pressed into the bore, or by an expanding mandrel. Stub arbors, as they are sometimes called, are another form of mandrel that is held in a chuck and not between centers (Figure 43). The term *arbor* correctly refers to a rotating member that holds a cutting tool such as the arbor of a milling machine, and *mandrel* correctly refers to a rotating member that holds a workpiece. However, these terms are often interchanged in technical literature and in common shop terminology. For example, an arbor press is used to force mandrels in and out of workpieces and is never used to apply pressure to arbors.

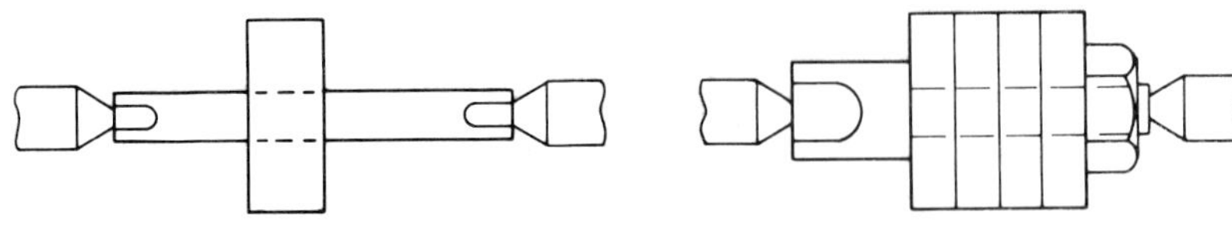

Figure 42
Left, tapered mandrel (.006 in./ft) holds part to be turned by friction in the bore. Right, gang mandrel holds several parts for machining on a straight threaded screw.

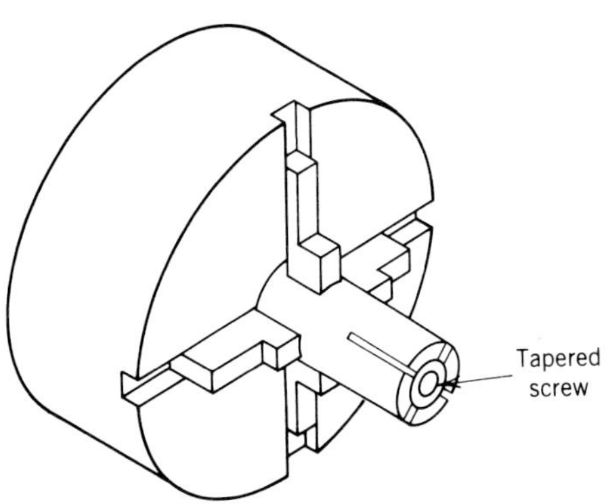

Figure 43
Stub mandrel. Part to be machined is gripped in the bore by expanding the mandrel with a tapered screw.

Figure 45
Short workpieces can be turned while supported in the chuck only.

Figure 44
Turning operation. Long pieces should be supported between the chuck and the center.

Figure 46
Drilling in the lathe. The drill is supported in the tailstock and is forced into the metal by means of a handwheel.

Turning Operations

Engine lathes are designed to remove metal to form cylindrical shapes. This can be done either by holding the work between centers, one in the spindle and one in the tail stock, or by simply holding the part in a chuck (Figures 44 and 45). This is called **turning**. When a cut is taken on the end of a piece by moving the cross slide instead of the carriage, it is called **facing**. If a bore is required in a solid workpiece, a hole is drilled by holding the drill in the tailstock and feeding it in with the handwheel (Figure 46). The hole can then be reamed (Figure 47) or bored with a single point tool on a boring bar that is fastened to the carriage (Figure 48).

Reaming is limited to the exact size of the reamer and, if the drilled hole is eccentric, the reamed hole will also be eccentric as the reamer follows the hole. Boring has the advantage of always making a hole concentric to the centerline of the spindle rotation, even if the drilled hole is off-center. Also any variation of size is possible using a boring bar and counterbores, internal grooves, and threads can be made. However, accurate inside dimensions are attainable only by experienced machinists, whereas any semiskilled machine operator can make a relatively accurate hole using a reamer. For the most part, only highly skilled machinists use engine lathes and semiskilled machine op-

Figure 47
A drilled hole can be finished to size by reaming.

Figure 49
Turret lathes are semiautomatic production machines (Clausing Machine Tools).

Figure 48
Accurate holes may be made by boring with single-point tools.

erators use turret lathes and some other production machines.

Turret Lathes

Turret lathes (Figure 49) are similar in many ways to engine lathes. The difference is that they are arranged for rapid production rates of a single item, often requiring several operations in quick succession. The work is fastened in a chuck, often air operated. A turret, usually six sided, contains different tools such as drills, taps, and turning and knurling tools. The cross-slide usually holds a parting or cutoff tool to sever the finished part from the bar stock. The operator simply

Figure 50
Radial die head (TRW Cutting Tools Division).

Figure 51
Tangent die head (TRW Cutting Tools Division).

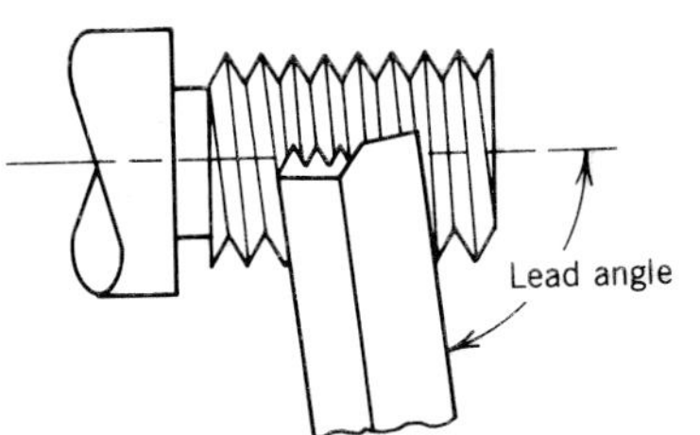

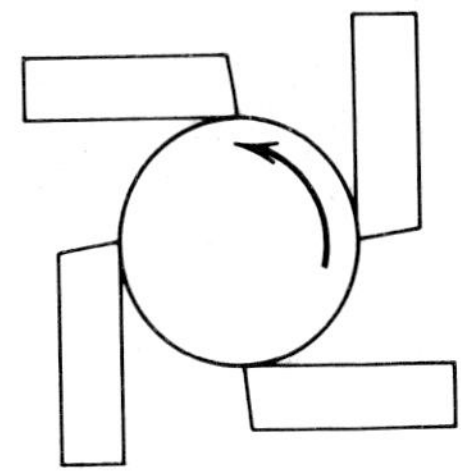

Figure 52
Tangential principle of thread cutting.

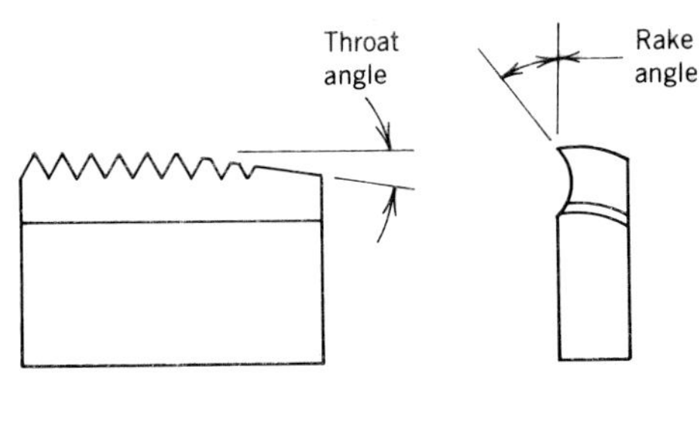

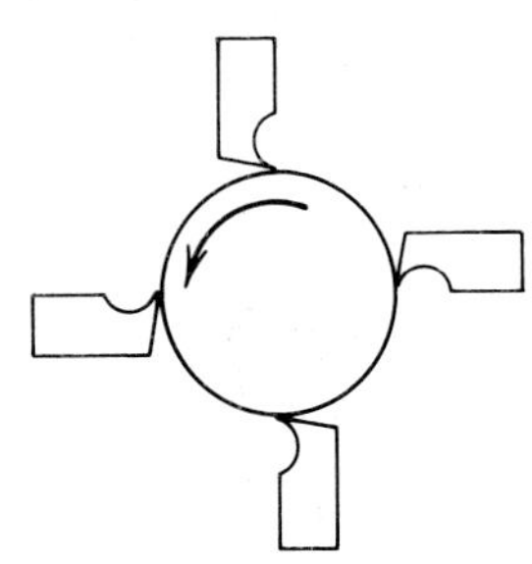

Figure 53
Principle of radial thread cutting.

moves levers in a proper sequence to make the part. Tooling is all preset by a set-up machinist and the operator only checks dimensions from time to time. Turret lathes require an operator at all times. Automatic screw machines bear some resemblance to turret lathes but the operator controls the operation of several machines. These machines make small machined parts from bar stock that is automatically fed into the machine.

External threading is done on turret lathes by means of die heads. There are two basic types, radial (Figure 50) and tangent (Figure 51). Both types make the full depth thread in one pass and can be quickly opened to remove the die by means of a handle; some types automatically open when the thread is completed. The thread chasers on the radial type are arranged to cut on their side on the radius (Figure 52). Tangent chasers cut on their ends on the tangent (Figure 53). Tangent die chasers are easier to sharpen and last longer than radial chasers and are probably more economical in terms of upkeep. Internal threads are also quickly produced in turret lathes by means of collapsing taps (Figure 54). Ordinary taps, after making the thread, must be screwed out of the work by means of reversing the spindle rotation. Collapsing taps (Figure 55) simply reduce their diameter when the thread is completed and can be quickly removed. These quick-opening die heads and collapsing taps are not exclusively used in

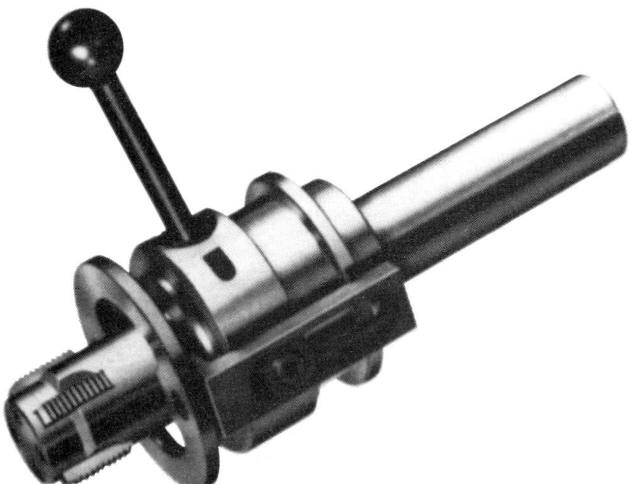

Figure 54
Collapsing taps make possible the removal of the tap without stopping or reversing the spindle rotation (TRW Cutting Tools Division).

Figure 55
Here a collapsing tap is mounted on the turret of a turret lathe so it can be quickly swung into position for making an internal thread (TRW Cutting Tools Division).

Figure 56
Automatic screw machine.

turret lathes. They can be adapted for use in engine lathes, drill presses, and for manufacturing machining centers.

Automatic Screw Machines

Automatic screw machines (Figure 56) are similar to turret lathes in some respects, in that they have preset, indexable tooling and follow a set sequence of operations to produce a part. Most screw machines are completely automatic, producing small parts at a rapid rate without the need of constant attention of an operator. A stock feed mechanism feeds metal bars into the machine as needed. Stock feeders are kept full by an operator who is in charge of a number of machines. In addition to small threaded fasteners, screw machines can produce a large variety of small cylindrical and tubular parts.

VERTICAL TURNING MACHINES

Vertical turret lathes (see Figure 17) have the advantage of supporting large heavy parts on a horizontal chuck surface that rotates on a vertical spindle. These machines are capable of taking heavy cuts in tough materials with a high stock removal rate.

Vertical boring mills (see Figure 16) are similar machines but are built on a much larger scale, some having rotating tables over 50 ft in diameter. Massive steel castings, weighing many tons, are placed on the table and clamped in place. These are then bored, counterbored, faced, or O.D. turned. Many of these machines are equipped with indexable turrets on the ram heads so various tooling arrangements can be quickly brought into use.

HORIZONTAL BORING MILLS

Horizontal boring mills (Figure 57) make use of a single point tool for making precision bores, especially in large or awkward parts. Like the lathe, they have a horizontal spindle to which boring bars may be attached, or they have long, line-type bars with an outboard bearing support. The workpiece is mounted on a sliding table that can be moved in two horizontal axes. The spindle and outboard bearing may be raised or lowered. These useful machines are rapidly being replaced by machining centers with modular tooling systems.

Figure 57
Horizontal boring mill (Ductr Mfg.).

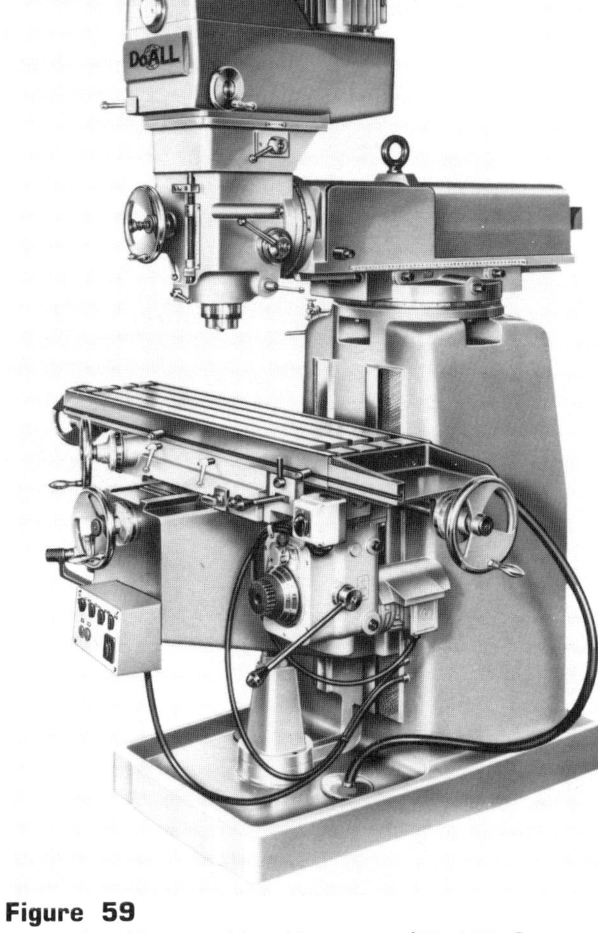

Figure 59
Vertical milling machine (Courtesy of DoALL Company).

Figure 58
Horizontal milling machine (Cincinnati Milacron, Inc.).

MILLING OPERATIONS AND EQUIPMENT

Milling machines make use of a rotating cutter with two or more teeth to remove metal from a workpiece. There are basically two types of milling machines, those with horizontal spindles (Figure 58) and those with vertical spindles (Figure 59). There are many combinations of these two types; some machines incorporate both principles. Sizes range from small bench-top models to huge planer-type mills that can surface castings weighing many tons. Large profiling mills (Figure 60) are extensively used in the aircraft industry for machining aluminum alloy spars and other structural parts.

Conventional horizontal spindle milling machines are used in machine shops and tool rooms. These machines make use of a circular cutter with many cutting teeth. These cutters are either made of high-speed steel or they have inserted carbide cutting edges. Milling cutters are classified into two groups, **profile** and **form relieved**. Profile cutters are sharpened on their outer edge and cutting clearance or relief is provided by the

Figure 60
Profiling mill. Aircraft plants use these NC machines for milling parts (Lockheed Corporation).

sharpening or grinding operation. Form-relieved cutters have a set form such as a radius or gear shape and must be sharpened on the tooth face instead of on the periphery of the cutter. Many types of milling cutters are used on horizontal milling machines: slabbing mills for making flat surfaces, side milling or staggered-tooth for deep slots, saws or slitting cutters for grooves, and face mills for surfacing. Horizontal mills are sometimes used in machine shops for making splines and gears (Figure 61).

Vertical spindle milling machines are the most useful machines in tool rooms and machine shops. Tooling ranges from end mill cutters to face mills and fly cutters. These versatile machines can be used for drilling, reaming, and boring operations as well as for milling operations. End mills cut on the sides (periphery) and on the end, so vertical plunge cuts as in drilling can be made. This can be followed by horizontal milling cuts. Thus a blind keyseat (Figure 62) can be easily made in a shaft or an internal pocket can be milled on

Figure 61
Horizontal milling machine, gear cutting operation.

Figure 62
Blind keyseat.

Figure 63
Shaper.

a part with this machine. Vertical mills are sometimes operated by numerical control (NC) for hole drilling patterns or milling operations.

SHAPERS AND PLANERS

Shapers (Figure 63) are designed to cut metal by forcing a single point tool across a workpiece by a reciprocating action. These machines are now virtually obsolete as machine tools; however, some are still in use in machine shops. The principle of shaping is used in gear manufacture where it is very useful for producing certain types of gears.

Like shapers, planers (Figure 64) are virtually obsolete as a machine tool, but they are still useful for machining some extremely large machine parts. On the shaper, the ram moves back and forth, holding the tool that cuts the stationary workpiece. In contrast, on the planer, the workpiece is clamped to the table and moves back and forth past a stationary tool. Some large pla-

Figure 64
Planer shop floor. Planers are useful for machining extremely large heavy parts, such as castings for presses (Illustration courtesy of National Machinery Company).

Figure 65
Twin-ram with in-line transfer progressive broaching machine that cuts blind external splines in two different parts. One part is loaded while the other is broached (Apex Broach & Machine Co.).

ners have been retrofitted with a milling head and the table speed is altered to convert them into planer mills which are much faster and more versatile machine tools for some purposes.

BROACHING

Broaching is the precision cutting of a material by passing over, or through it, a series of progressively stepped teeth. Flat surfaces, contours, internal splines, and external splines are some of the shapes produced by broaching. Broaching is usually a one-stroke application in which roughing, semifinish, and finish cut is all done in a single machine stroke. However, when excessive stock prevents a one-stroke application, multiple strokes can be used, either using several different broaches or by adjusting the position of a single broach.

Broaching is ideally suited for high production machining (Figures 65 to 67). The forged connecting rods that were shown in Chapter 8 would require a ma-

Figure 66
Part being loaded under the front ram while rear ram is stroking down (Apex Broach & Machine Co.).

Figure 67
Progressive tooling for two different parts (Apex Broach & Machine Co.).

Figure 68
Broaching machine making connecting rods and caps (Apex Broach & Machine Co.).

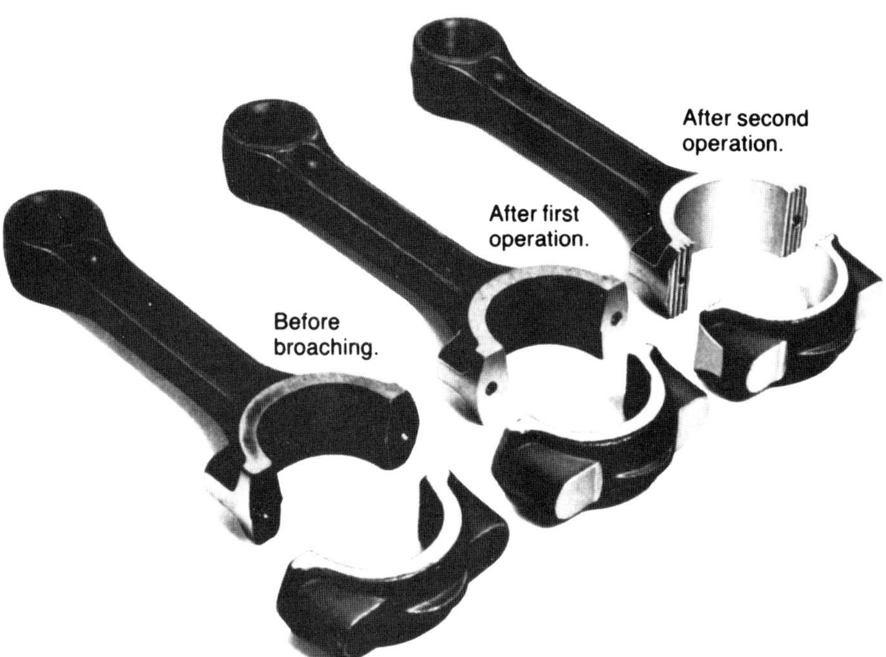

Figure 69
Connecting rods and caps (Apex Broach & Machine Co.).

chining operation to make them ready for use (Figures 68 and 69). These diesel engine connecting rods are finish-machined by broaching at the rate of 112/hr. These rods have 12 surfaces that are flat, round, and serrated. Another high-production broaching operation on a vertical machine is seen in Figures 70 and 71. Where quite long broaches are needed, horizontal broaching machines (Figures 72 and 73) are used. External broaching of large flat surfaces is an economical means of machining when a large number of piece parts are involved.

GEARS

One of the most efficient methods of transferring rotary motion from one shaft to another is that of gearing.

Figure 70
Blind spline broaching machine. Inset shows blank and internal broached part. These automatic transmission converter-clutch hubs are splined at the rate of 240 hubs per hour, 12 times faster than conventional gear shaping methods (Apex Broach & Machine Co.).

Figure 71
Closeup of turntable on broaching machine. Two parts are broached at the same time under a single ram. Automated pick-and-place arms unload the machine in just 10 sec (Apex Broach & Machine Co.).

Figure 72
Horizontal broaching machine (Apex Broach & Machine Co.).

Figure 73
Surface-to-air missile casings. Internal grooves are made with the horizontal broach in Figure 72 (Apex Broach & Machine Co.).

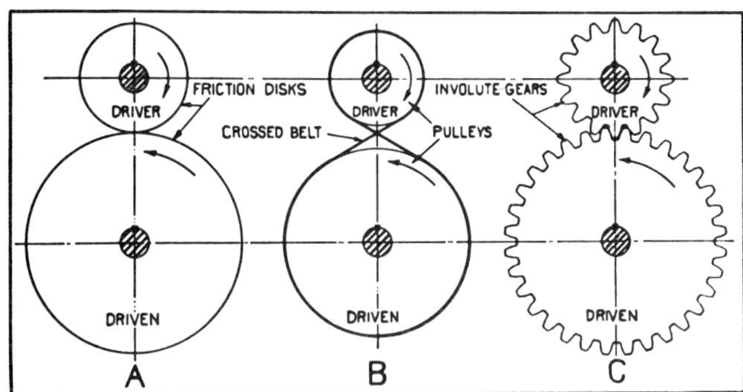

Figure 74
Three means for transmitting rotary motion from one shaft to another (Courtesy Fellows Corporation).

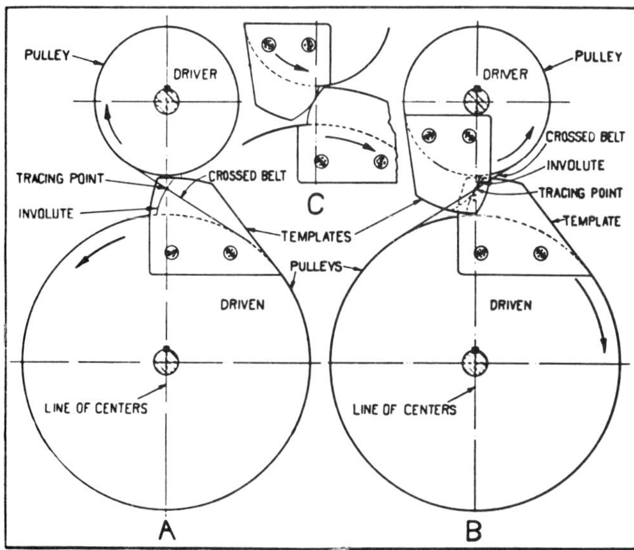

Figure 75
Diagram illustrating that the involute is that curve traced by a point on a crossed belt as it moves from one pulley to another (Courtesy Fellows Corporation).

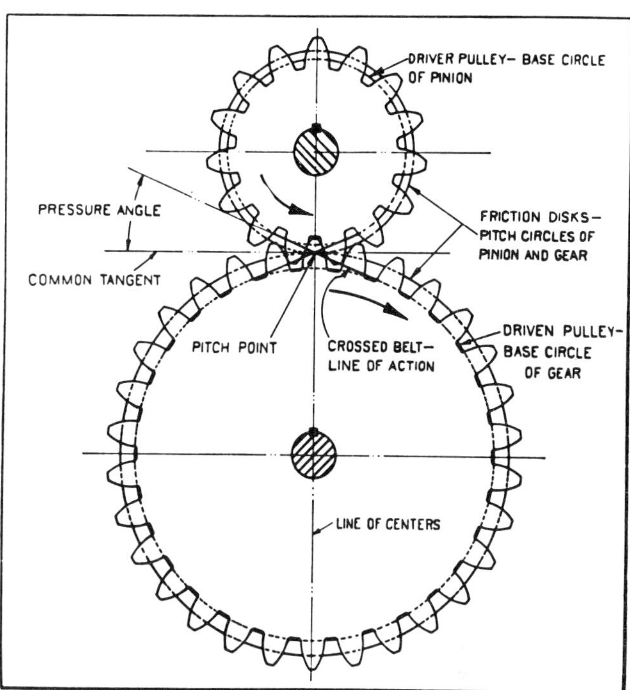

Figure 76
Diagram showing gears, pulleys, and crossed belt superimposed on friction disks, illustrating basic gear elements (Courtesy Fellows Corporation).

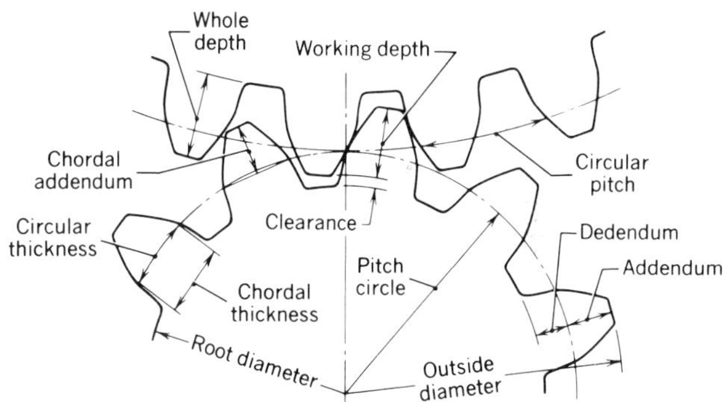

Figure 77
Spur gear terms (White, Neely, Kibbe, Meyer, *Machine Tools and Machining Practices*, Vol. II, © 1977 John Wiley & Sons, Inc.).

Large amounts of torque with a definite ratio can be transmitted in a small space by using gears. When they are properly made and kept lubricated, they are quiet, dependable, and long lasting.

Two friction disks or crossed belts can be used as examples to explain the operation of gears (Figure 74). The involute curve has been proven by long experience to be the most satisfactory profile for spur and helical gear teeth. This gear tooth form transmits motion smoothly and uniformly. The involute can be described as a curve traced by a point on a crossed belt as it moves from one pulley to another without slippage (Figure 75). If the crossed belts are superimposed on the friction disks (Figure 76), it can be seen that the outer diameters of the disks represent the pitch circles of the gears; the pulleys are the base circles, and where the belt crosses the line of centers is the pitch or tangent point of the friction disks. Figure 77 shows the terms used to identify parts of spur gears.

Many types of gears are made for various purposes. *Spur gears* are used to transmit motion between parallel shafts where they turn in opposite directions. *Helical*

Figure 78
Hobbing machine producing gears (Illinois Gear, a Household International Company).

gears are used for the same purpose, but they are quieter running gears. *Bevel gears* connect intersecting shafts on the same plane but at an angle. *Miter gears* are bevel gears that connect shafts at a 90 degree angle. *Worm gears* connect shafts at 90 degree angles but not on the same plane. *Hypoid gears* such as those used in automotive differentials are similar to worm and bevel gears. They are quiet running like worm gears and have some of the efficient characteristics of bevel gears. *Crossed helical gears* also connect two shafts that are at an angle but not on the same plane. Internal gears can be helical or spur types. *Herringbone gears* consist of two sets of helical teeth with opposite angles on the same gear. Their main advantage is that they eliminate the end thrust characteristic of helical gears. *Gear racks* are designed to convert rotary motion to linear motion using a rack and pinion.

Gear Cutting Operations

Several methods are used to manufacture gears. External gears are cut one tooth at a time on conventional horizontal milling machines with form cutters where high production is not a requirement. A single gear cutter mounted on an arbor of a milling machine removes material on the periphery of a gear blank to form the gear teeth. This is usually done in two cuts, first a heavy roughing cut at each tooth location all around the blank, and then a light finishing cut. These precise tooth divisions are made by means of a dividing (indexing) head. This is a relatively slow process of gear making and it is only useful when a special gear is needed for a prototype or a repair operation.

High production rates for external gears can be realized in the process of hobbing. Hobbing (Figures 78 and 79), like gear shaping, is a generating process in which a rotating cutter is slowly fed into a rotating blank. Extremely large gears can be generated by the hobbing process (Figures 80 and 81). Racks (Figure 82) can be produced by hobbing. The hobbing process is

Figure 79
G-Tech™ CNC gear hobbing machine (The Gleason Machine Division).

Figure 80
Extremely large gears can be produced by the hobbing process (Illinois Gear, a Household International Company).

Figure 81
Hobbing a large helical gear for a hot strip mill (Illinois Gear, a Household International Company).

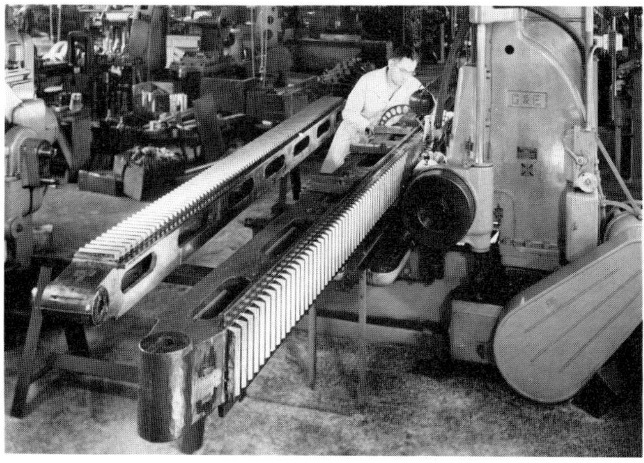

Figure 82
Cutting large heavy-duty racks for steel mill equipment (Illinois Gear, a Household International Company).

Figure 83
Herringbone gears are made on special machines (Illinois Gear, a Household International Company).

Figure 84
Large bevel gears are also produced on special machines (Illinois Gear, a Household International Company).

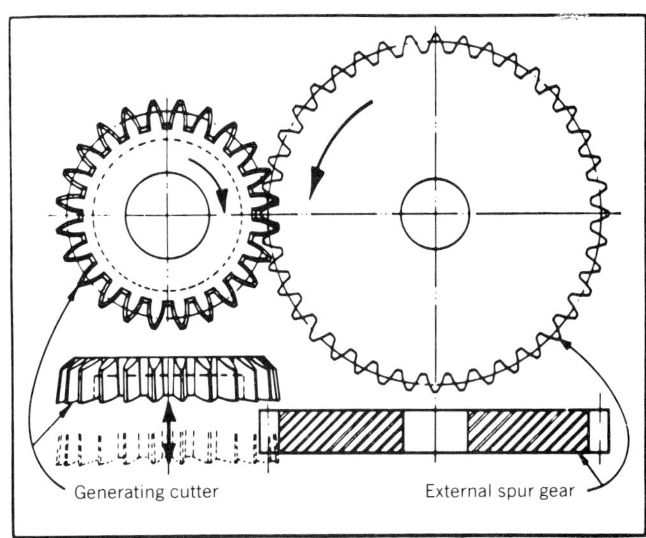

Figure 85
Illustration showing how the gear shaper employs the "molding-generating" process for cutting gears (Courtesy Fellows Corporation).

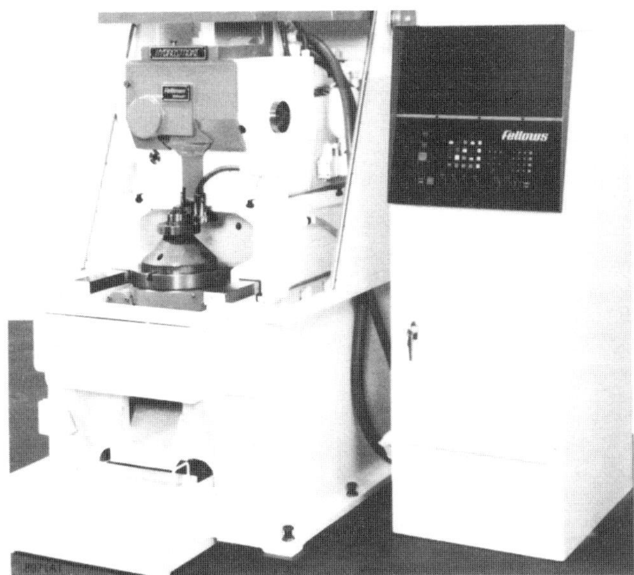

Figure 86
CNC gear shaper (Courtesy Fellows Corporation).

Figure 88
Internal gear shaping (Courtesy Fellows Corporation).

Figure 87
Schematic of CNC gear shaper (Courtesy Fellows Corporation).

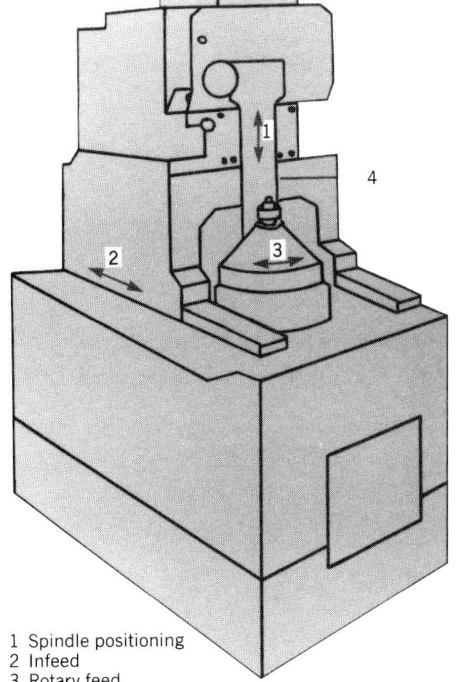

also used for making small gears, splines, and roller chain sprockets. Often, when the blanks are thin as with plate sprockets, many blanks are stacked together on a mandrel and cut in one operation. Large herringbone gears (Figure 83) and large bevel gears (Figure 84) can also be produced on special machines.

Where internal gears, helical teeth, and special gears are required at higher production rates, gear shaping is used. The principle of gear shaping is based on the fact that the teeth on two meshing gears rotating together will follow an involute curve. If one gear is hardened and provided with suitable clearances, it can be used as a generating cutter for a mating gear blank (Figure 85). The cutter is made to reciprocate like the tool in a vertical shaper, while the tool and blank are slowly rotating. The cutter is fed into the blank until the gear is full depth (Figures 86 and 87). These machines are often computer controlled (CNC). Both internal and external gears can be produced in these machines (Figures 88 to 91). Very large and very small gears as well as helical and special shapes can be produced on gear shapers. Internal splines and some external gears that are easily made on gear shapers cannot be made by conventional gear cutting processes such as hobbing.

Figure 89
External gear shaping. Here a crowned gear (higher in the center) is being generated on a gear shaper (Courtesy Fellows Corporation).

Figure 90
Quite large external gears may be produced by the gear-shaping process (Courtesy Fellows Corporation).

Figure 91
A large internal ring gear being generated by the gear-shaping process (Courtesy Fellows Corporation).

Figure 92
Spiral bevel and hypoid gear generator (Illinois Gear, a Household International Company).

Figure 93
Close-up of a hypoid gear being cut from a solid blank on a G-Plete® generator (The Gleason Machine Division).

Hypoid gears are generated on special machines (Figures 92 and 93). Like other automotive gears, they must be finished after they are machined by other processes (Figure 94).

Finishing operations include gear shaving, flame hardening, and gear grinding. A rotating cutter, shaped like a gear, is made to mesh and turn with a machine cut gear before it is heat treated. This eliminates most indexing errors and roughness. A flame or induction hardening process (Figure 95) is often performed on the teeth of a gear or splines of a shaft, followed by a finish grind to restore accuracy. This assures strong, shock and wear resistant gears that mesh together accurately and run quietly.

Figure 94
Lapping of intermediate size spiral bevel and hypoid gears (Illinois Gear, a Household International Company).

Figure 95
Flame hardening gear teeth with pyrometer controlled machine (Illinois Gear, a Household International Company).

ABRASIVE MACHINING

Grinding processes are actually a form of metal removal by means of abrasive grains. Abrasives can be silicon carbide for cast iron and nonferrous metals, aluminum oxide for ferrous metals, and boron nitride and diamond for grinding hard materials such as tungsten carbide. Since abrasives are harder than hardened metals, the grinding process is ideally suited to the removal of unwanted material on hardened workpieces. Also, since there are many small cutting edges traveling at a high surface speed, extremely good finishes can be obtained by abrasive machining. Accurate dimensions can be maintained on a production basis with these precision machines. Precision grinding operations often follow a conventional machining process or heat treatment process to provide the final dimension and finish to the workpiece.

Many variables exist for uniform precision production of parts by grinding technologies. The complexity in this area of machining is greater than that of single point tooling for metal cutting. Grinding wheels tend to break down and become smaller as they are used. This must be compensated for by adjusting downfeed or infeed. Grinding wheels tend to be self-sharpening. When the grains dull, they break away, exposing new sharp grains. Wheels must be trued occasionally to maintain this shape. This is usually done with a diamond dresser on production machines. Feed rate, wheel speed, sparkout time, and wheel dress can be set and adjusted on the grinding machine. Production factors of grinding machines include surface finish, surface integrity, cycle time, problems of burning the workpiece, and cost per piece. Furthermore, a large variety of grinding wheels are available, from soft to hard and from open to dense bond, and various grit sizes as well as other variables. Choosing the right wheel for the job requires extensive study of abrasive wheels and their uses.

Grinding Machines

Grinding machines range from shop pedestal grinders for sharpening small tools to large specialty machines

Figure 96
Surface grinders are used to produce flat surfaces (Clausing Machine Tools).

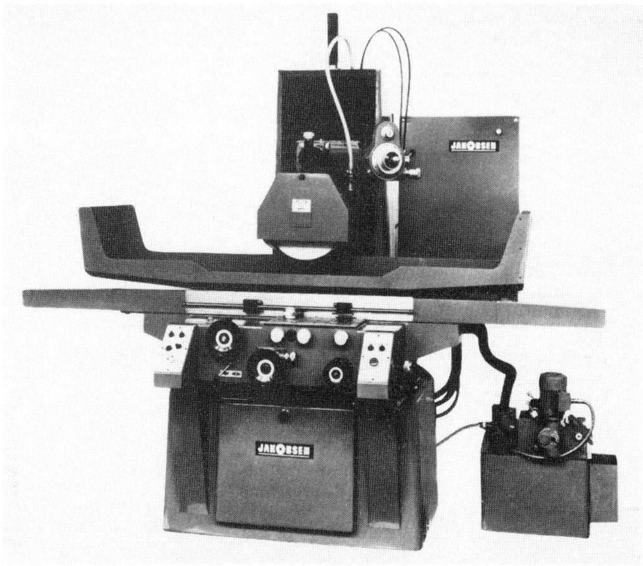

Figure 97
Cylindrical grinder.

such as crankshaft grinders. Vertical spindle machines are useful for grinding surfaces on many small parts at one time. Horizontal spindle machines such as small surface grinders (Figure 96) are useful to produce flat surfaces and other forms. Workpieces are usually held on a magnetic base or table if they are ferromagnetic. The table passes under the rotating abrasive wheel in a reciprocating motion while the table is fed crosswise. Cylindrical grinders (Figure 97) are designed to grind finish surfaces on rotating parts that are cylindrical in shape. Some of these machines are designed to be used in small shops where a machinist finish grinds one or only a few piece parts at a time. Many of these machines are equipped to do internal grinding of parts that are held in a chuck. Some production machines are designed for internal grinding.

Production Grinding Processes

Three methods of grinding diameters on cylindrical parts are used in manufacture: *plain grinding, plunge feed grinding,* and *centerless grinding.* Plain feed grinding in which the grinding wheel moves along the work axis (Figure 98) is usually used where long cylinders or shafts must be ground to a uniform diameter. This method is most often used in small machine shops.

Plunge grinding (Figure 99) is most often used in production grinding where large numbers of parts are manufactured to exacting specifications. Plunge grinding is most useful where short lengths are involved. In production grinding, automatic indexing on the fly (without stopping shaft rotation) greatly reduces cycle time. Wheels are dressed with diamond nibs or wheels at certain times by computer programming and wheel compensation is automatic. In-process gaging (Figure 100) allows workpiece diameters to be checked as the grinding process proceeds. With programmable machine controls (Figure 101) dimensions can be controlled and measured to .00004 in. or .001 mm.

Figure 98
Plain feed grinding in which the workpiece traverses across the wheel to a preset stop (Rodco Precision, Honing & EDM).

Figure 99
Straight plunge grinding on a shaft. In this operation the wheel is fed straight in (Photo Courtesy Landis Tool).

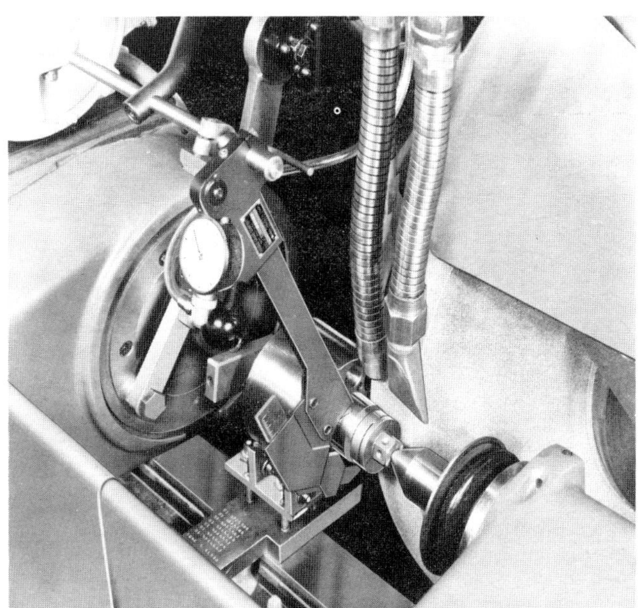

Figure 100
In process gaging. Dimensions can be checked while the machine is in operation with this setup (Photo Courtesy Landis Tool).

Figure 101
Small crankshaft grinder. Grinding process can be programmed on this machine for a precision automatic operation (Photo Courtesy Landis Tool).

Figure 102
Plunge grinding on a crankshaft (Photo Courtesy Landis Tool).

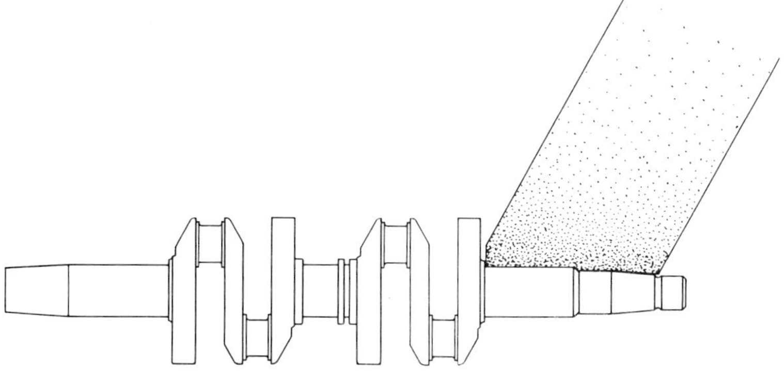

Figure 103
Angular plunge grinding on a crankshft. Two diameters, fillets, faces, radii, and a tapered diameter are made in one plunge grind (Photo Courtesy Landis Tool).

Cylindrical grinders often have swiveling grinding heads called angular center type grinders (Figures 102 and 103). In this plunge grinding operation, both diameters and shoulders can be finished in one operation. Grooves in wheels or rings can be plunge ground by fastening them on a mandrel (Figure 104). Many dimensions on one shaft can be finish ground in one operation with multiwheel grinders (Figures 105 and 106).

Centerless grinding is done by means of a large grinding wheel and a regulating wheel (Figures 107 and 108). A work rest supports the workpiece, allowing it to rotate. Centerless grinding is a high-production method of obtaining precision rod, shafting, and small

Figure 104
Grooves can be made by the plunge grinding process. Grinding the diameter of a large bearing race mounted on an arbor (Photo Courtesy Landis Tool).

Figure 106
Programmable machine control for the multispindle grinder allows the operator to control wheel feed, in-process gaging, wheel balance, and monitoring dimensions (Photo Courtesy Landis Tool).

Figure 107
Centerless grinding. Through feed grinding of bearing rings (Lidkopings Mekaniska Verkstads AB, Sweden).

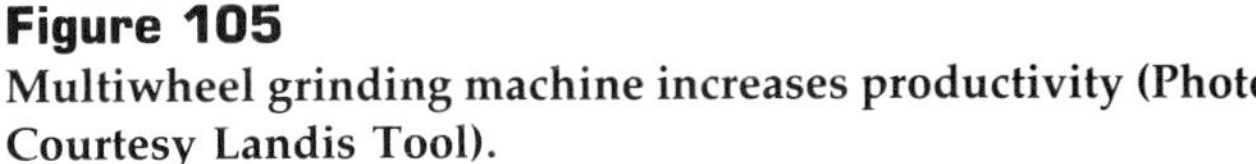

Figure 105
Multiwheel grinding machine increases productivity (Photo Courtesy Landis Tool).

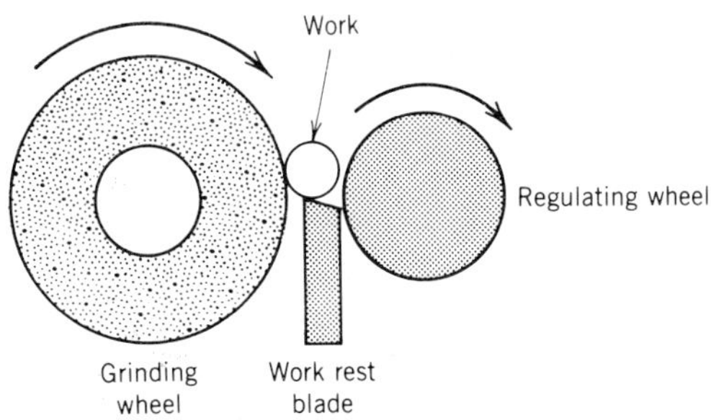

Figure 108
Principle of centerless grinding.

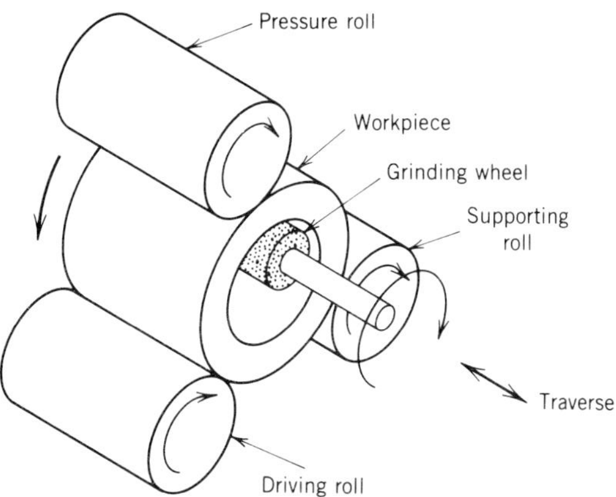

Figure 109
Internal centerless grinding.

Figure 110
Honing machine (Rodco Precision, Honing & EDM).

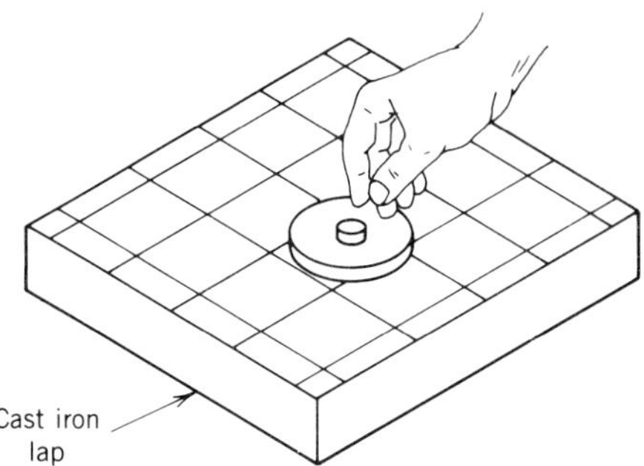

Figure 111
Lapping by hand. Grooves hold the abrasive. A circular motion is used to ensure flatness.

Figure 112
Lapping machine (Courtesy of DoALL Company).

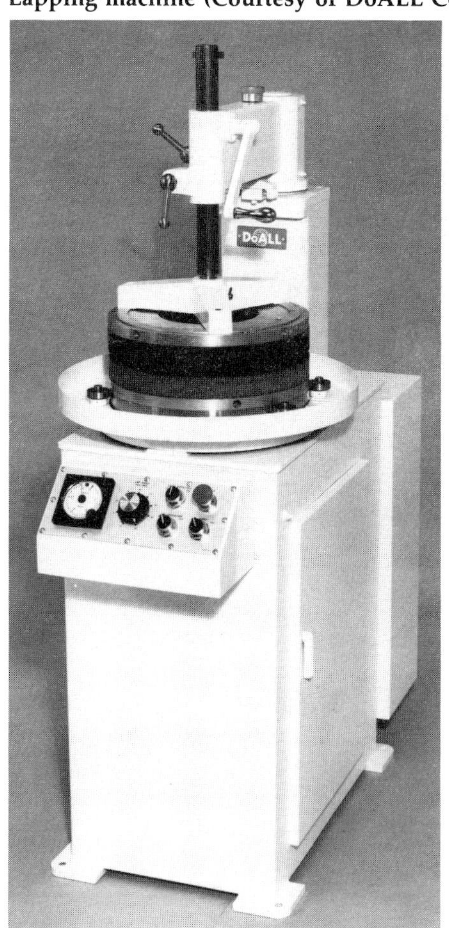

parts. Commercial ground and polished (G and P) shafting is produced on centerless grinders. Centerless internal grinding is accomplished with three rolls supporting and rotating the workpiece while the inside surface is ground (Figure 109).

HONING, LAPPING, AND SUPERFINISHING

When smooth internal finishes are required, such as in tubes for industrial hydraulic cylinders, a preferred method of finishing is honing (Figure 110). The honing head follows the existing hole or work surface and rotates while it oscillates in the bore. If the hole needs to be realigned, a boring operation must first be performed before honing. Multiple spindle honing machines are used in manufacturing. An example is honing engine blocks in which all cylinder bores in a block are honed at one time.

Lapping is an abrasive superfinishing process in which fine abrasives are charged (embedded) into a soft material called a lap. Lapping can be done by hand (Figure 111) on a cast iron surface or it can be done on a special machine (Figures 112 to 114).

Superfinishing (Figure 115) is similar to honing and it can be done on a flat or a cylindrical surface. In superfinishing, a lubricant is used to maintain a separation of a honing stone and the workpiece. Tiny projections that cause roughness on the workpiece are cut off by the honing stone because they extend through the lubricant. By maintaining a controlled pressure, a desired degree of smoothness or finish is obtained and the controlled-viscosity lubricant prevents further cutting action.

Figure 114
Typical workholders are custom designed for each individual flat or cylindrical workpiece application (Courtesy of DoALL Company).

Figure 113
The planetary lapping attachment provides for complete coverage of laps during operation (Courtesy of DoALL Company).

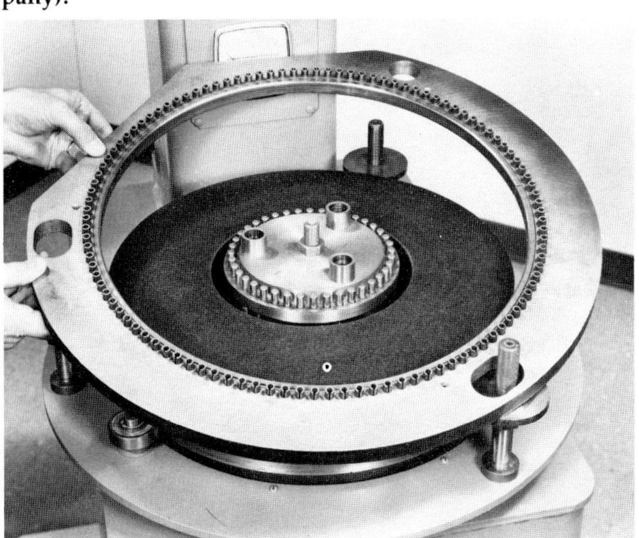

Figure 115
Superfinishing.

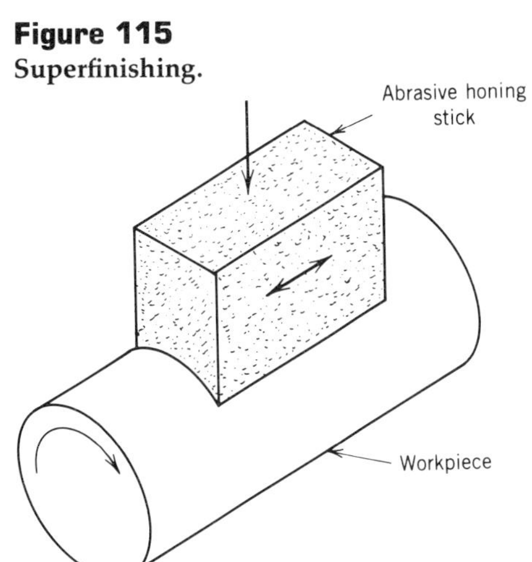

REVIEW QUESTIONS

1. Saw blades (bands) for vertical and horizontal band saws are made in long strips that are coiled for packaging and shipment. How are these strips made into endless bands?
2. What is the greatest advantage of vertical band sawing over other machining processes?
3. Two drilled holes are required on a part, one .010 in. diameter and one .100 in. diameter. Which drilling machine or machines should you use to drill these holes?
4. Using the formula $\text{RPM} = \frac{CS \times 4}{D}$, how fast should the drill press be turning for a $\frac{1}{2}$ in. diameter drill if the cutting speed is 90?
5. A massive steel casting weighing 3 tons requires that several 3-in. diameter holes be drilled into it. Which type of drilling machine would be best suited for this job?
6. An NC lathe produces accurate shafts having several diameters without the need for an operator to be in constant attendance. Would you call this an engine lathe or a production machine? Explain.
7. Threads are often designated omitting the tolerance symbols. Explain the meaning of the following: $\frac{3}{8}$ − 16 UNC, and M16 × 2.
8. Internal and external threads can be made with hand taps, dies, or on an engine lathe with a single point tool. How are threads cut at rapid production rates on turning machines?
9. How can a round or hexagonal bar be securely and accurately gripped for turning operations in a lathe? How can a square or rectangular piece be gripped for turning?
10. On which type of turning machines are air-operated chucks commonly used?
11. What principal advantage does a turret lathe have over an engine lathe?
12. A blind keyseat is a short groove that begins and ends anywhere along a shaft but not on the very ends. Which machine, a vertical or a horizontal spindle milling machine, would be most appropriate for cutting a blind keyseat?
13. Name two methods by which internal splines can be cut.
14. Hobbing is used almost exclusively for what machining operation?
15. How can hardened workpieces that cannot be machined with ordinary cutting tools be finish machined to a precise dimension with a very smooth finish?
16. How are steel workpieces usually clamped to the reciprocating table on a small surface grinder?
17. What is the difference between plunge grinding and plain feed grinding on cylindrical grinders?
18. A precision part must be held to a tolerance of ±.0001 (±.0025 mm). Would you attempt to finish such a part on a turning machine? Explain your answer.
19. Fifty thousand hardened capscrews are made a few thousandths of an inch oversize and must be ground to a diameter of ±.0001 in. on a cylindrical surface. Which grinding process and machine should be used to get the job done most economically? (*a*) center-type cylindrical grinder, (*b*) multiwheel grinder, or (*c*) centerless grinder.
20. Where is honing most likely to be performed, on flat surfaces, outside surfaces of cylinders, or in bores? What is the principal purpose of honing?

CASE PROBLEMS

Case 1
The Manufacture of a Transmission Gear and Shaft

A manufacturer is asked to design and build a small tractor with a four-speed manual transmission. You are asked to decide upon the means of manufacture, on a cost effective production basis, of a transmission shaft (Figure 116). One end of the shaft has a spline and the

Figure 116

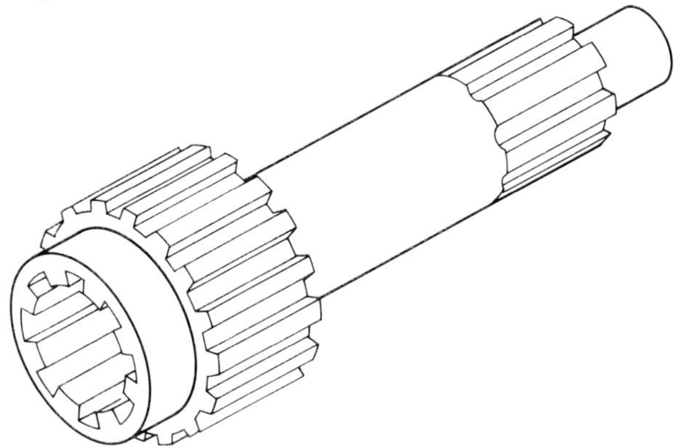

other end has a spur gear. The gear end is hollow and has a blind end internal spline. The gear, internal and external splines, and the two bearing surfaces must be hardened. The gear must also be a precision one with close tolerances. The blank will be forged from a high chromium carbon steel that can be hardened. What processes would you use to produce and finish this geared transmission shaft?

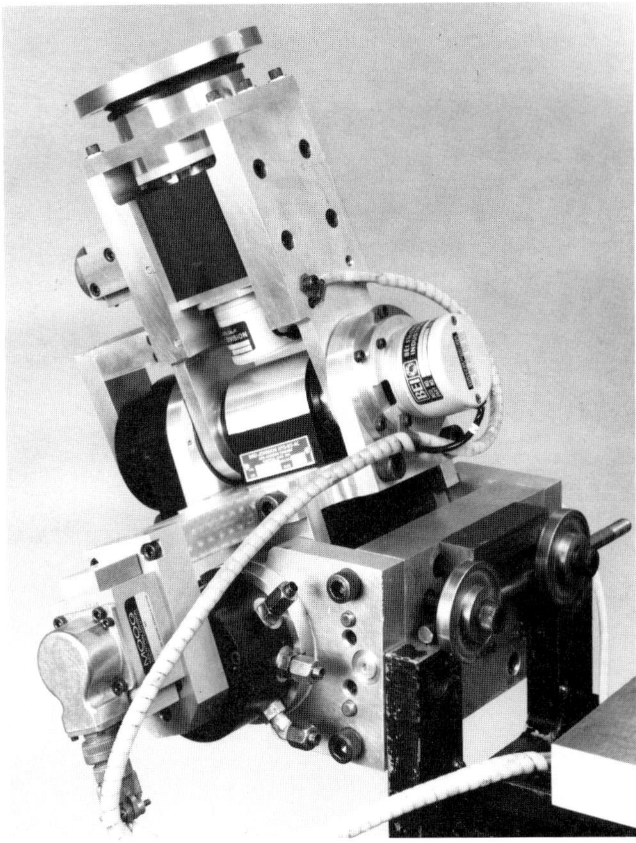

Figure 117
(Bird-Johnson Company.)

Case 2
The Robotic Wrist Component

A prototype (one only) of a robotic wrist component similar to the one shown in Figure 117 is required before it is put into production. It must be tested and perhaps revised and retested before it is ready to be manufactured in quantity. The prototype is mostly made of aluminum alloy. Where and how should it be made? Should the parts be:

1. Cast in a foundry?
2. Machined from aluminum alloy bar stock on production machines?
3. Machined from aluminum alloy bar stock by broaching the holes and surfaces?
4. Machined from aluminum alloy bar stock in a tool room (machine shop) one piece at a time?

Explain your reason for your choice of answer.

Case 3
RPM for Drilling Operations

A production drilling machine needs to be adjusted to the correct RPM for several drill sizes, $\frac{1}{4}$, $\frac{1}{2}$, and $1\frac{1}{2}$ in. diameter. The drilling operations are all in steel which requires a cutting speed of 90. Using the simplified formula

$$\text{RPM} = \frac{CS \times 4}{D}$$

where

RPM = Revolutions per minute
CS = Cutting speed
D = Diameter of the drill

what would be the correct RPM settings for each of the drill sizes?

CHAPTER 13 NONTRADITIONAL MACHINING PROCESSES

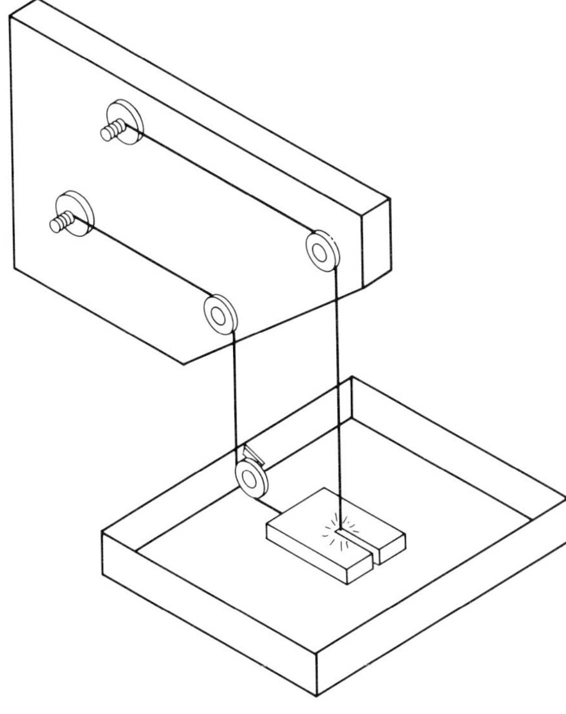

The previous chapters described machining processes in which workpiece material is removed in the form of chips, by direct contact of a cutting tool. These processes are often called traditional machining processes and account for the majority of materials processing done by this method.

The design and manufacturing engineer, in the continuing search for improvements in designs, both stimulates development of alternate materials processing methods and takes advantage of new processes that evolve naturally throughout manufacturing industries.

Machining processes making use of electricity, laser light **electrochemical** methods, and extremely high temperature **gas plasmas** for cutting tools, have developed in more recent years. These processes have become well established and because of their unique capabilities, have opened many new avenues of design to the engineer. The processes described in this chapter are often referred to as **nontraditional machining processes,** only because they do not involve the chip-shaving process of conventional machining. They are nonetheless well established as standard manufacturing processes.

Nontraditional machining processes include the following:

1. Electrodischarge machining (EDM)
2. Electrochemical machining (ECM and ECDB)
3. Electrolytic grinding (ELG)
4. Laser machining
5. Ultrasonic machining
6. Hydrojet machining
7. Electron beam machining (EBM)
8. Plasma beam machining

OBJECTIVES

This chapter should enable you to:

1. List common nontraditional machining processes.
2. Describe in general terms how these processes work.
3. Describe applications of the processes.

ELECTRODISCHARGE MACHINING

Electrodischarge machining, commonly known as **EDM,** removes workpiece material by an electrical **spark erosion** process. This process is accomplished by establishing a large potential (voltage) difference between the workpiece to be machined and an electrode. A large burst (spark) of electrons travels from the electrode to the workpiece. When the electrons impinge on the workpiece, some of the workpiece material is eroded away. Thus the machining is accomplished by electrical spark erosion.

The EDM process (Figure 1) takes place in a **dielectric** (nonconducting) oil bath. The dielectric bath concentrates the spark and also flushes away the spark-eroded workpiece material. A typical EDM system consists of a power supply, dielectric reservoir, electrode, and workpiece. The EDM machine tool has many of the same features as its conventional counterpart. These include worktable positioning mechanisms and measurement devices. Many EDM machine tools (Figure 2) are equipped with computer numerical control systems. Thus the versatility of CNC for workpiece positioning and tool control functions can be effectively used in the process.

EDM Electrodes

In the EDM process, the shape of the electrode controls the shape of the machined feature on the workpiece. EDM electrodes may be made from **metal** or **carbon (graphite)** and shaped by molding or machining to the desired geometry. Erosion of the electrode takes place as well during the EDM process. In time it will become unusable. To circumvent this problem, a roughing electrode may be used to generally shape the workpiece and then a finish electrode may be applied to complete the process and establish final dimensions and geometry.

Figure 1
Basic EDM system components.

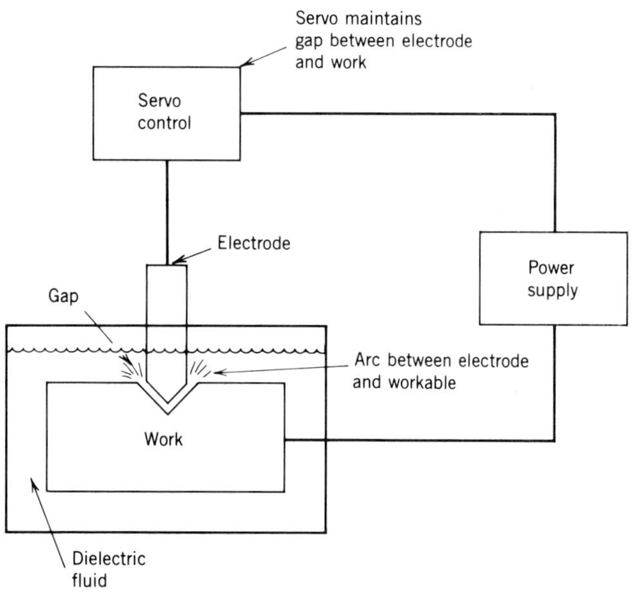

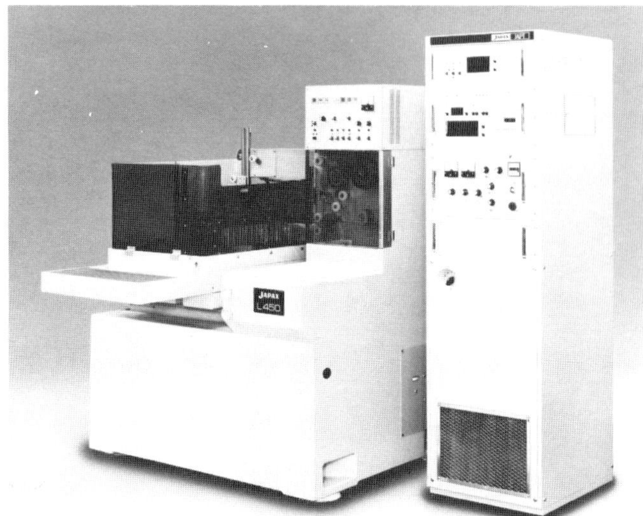

Figure 2
CNC wire cut EDM machining center (Japax Inc.).

Wirecut EDM

Wirecut EDM uses a **slender wire** as an **electrode** and is extremely useful for making narrow slots and detailed internal features in the workpiece (Figure 3) and for detailed tool and die making. Wire-cut EDM is similar in some ways to band sawing in conventional machining. Whereas the saw teeth do the cutting in band sawing, in wire-cut EDM a spark occurs between the wire electrode and the workpiece. Wire feed is continuous during the process and the wire may not be reused.

Figure 3
Blanking die for watch parts made with wire cut EDM (Japax Inc.).

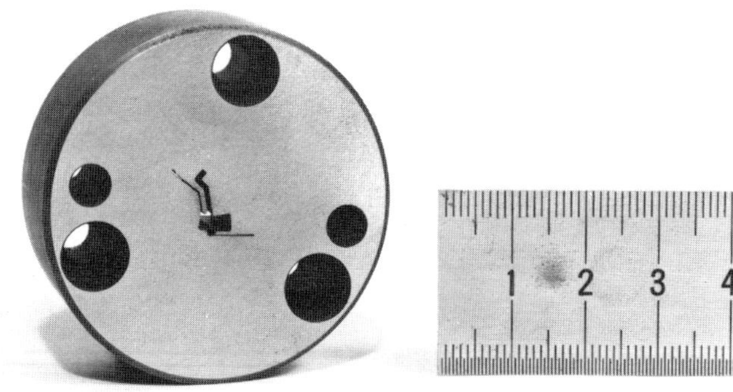

Figure 4
Plastic mold machined with ram-type EDM machine (Japax Inc.).

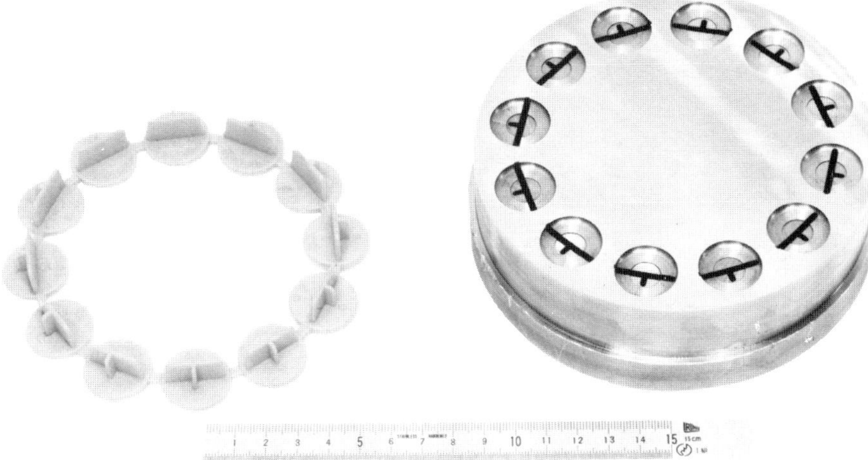

Different metals are used for wire electrodes, brass is one popular material. EDM wire can be of very small diameter, on the order of .005 in. Sapphire guides maintain the alignment of the wire as it passes through the work. Wire tension is maintained by weighted pulleys or other mechanical tension schemes.

Wirecut EDM using such small diameter wires permits extremely narrow slots to be machined in the workpiece. The kerf is only slightly wider than the wire diameter. This process is highly suited to cutting intricate slot features in tools and dies.

Advantages and Applications of EDM

EDM processes can accomplish machining that would be impossible by the traditional methods. Any odd electrode shape will be reproduced in the workpiece. Fine details can be made, such as those needed in tool, die, and mold work (Figures 4 and 5). When EDM is coupled to numerical control, an excellent machining system for the tool and diemaker is available.

Since machining is accomplished by spark erosion, the electrode does not actually touch the workpiece and is therefore not dulled by the hard workpiece materials that would dull conventional cutters. Thus, the EDM process can spark erode very hard metals and it has found wide application in the removal of broken taps without having to destroy an expensive workpiece in the process.

Disadvantages of EDM

The EDM process is much slower, from a metal removal standpoint, as compared to conventional machining processes involving direct cutter contact. There is also a possibility of over cutting and local area heat treating of the parts being machined.

However, EDM can accomplish many machining tasks that could never be done by conventional machining. Thus, the EDM process has become well established as a very versatile manufacturing process in modern industrial applications.

ELECTROCHEMICAL MACHINING AND ELECTRICAL CHEMICAL DEBURRING (ELG AND ECDB)

Electrochemical machining (ECM) is essentially a reverse metal-plating process (Figure 6). The process takes place in a conducting fluid or electrolyte that is pumped under pressure between electrode and workpiece. As workpiece material is deplated, it is flushed away by the flow of electrolyte. Workpiece material is removed from the electrolyte by a filtration system.

Advantages and Applications of ECM

As in EDM, ECM can accomplish machining of intricate shapes in hard-to-machine material. The process is also burr free, and does not subject the workpiece to distortion and stress as do conventional machining processes. This makes it useful for work on thin or fragile

Figure 5
Intricate internal geometry machined with wire cut EDM (Japax Inc.).

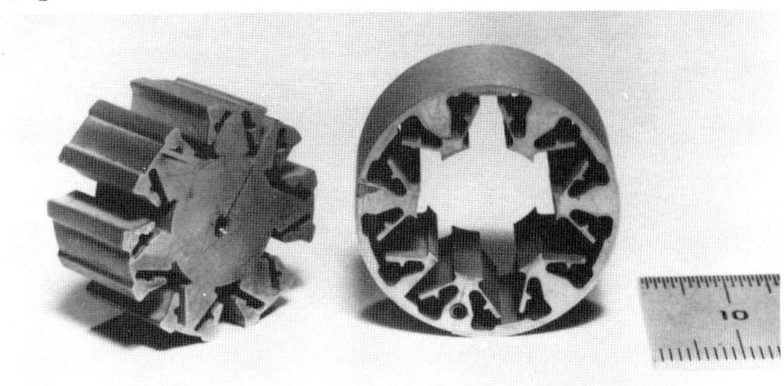

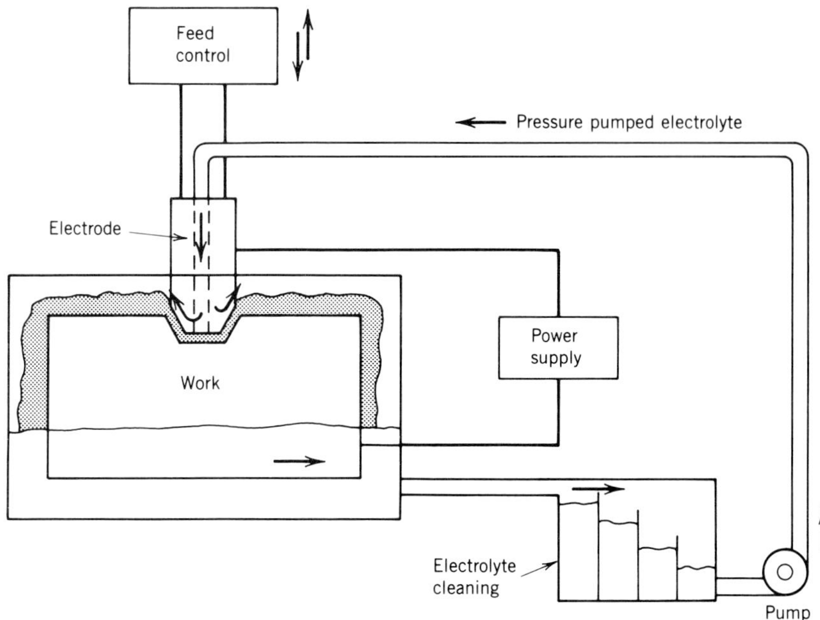

Figure 6
Basic ECM system components.

workpieces. ECM is also used for part deburring in the process called ECDB or **electrochemical deburring.** This process is very useful for deburring internal workpiece features that are inaccessible to traditional mechanical deburring processing.

ELECTROLYTIC GRINDING (ELG)

In the process of **electrolytic grinding (ELG)** an abrasive wheel much like a standard grinding wheel is used. The abrasive wheel bond is metal, thus making it a conducting medium. The abrasive grains in the grinding wheel are nonconducting and aid in removing oxides from the workpiece while helping to maintain the gap between wheel and work. ELG, like ECM, is a deplating process and workpiece material is carried away by the circulating electrolyte.

The ELG System

The basic ELG system (Figure 7) consists of the appropriate power supply, the electrode (metal bonded grinding wheel), workholding equipment, and the electrolyte supply and filtration system. Workpiece material is deplated and goes into the electrolyte solution.

Advantages and Applications

Since this process is primarily electrochemical and not mechanical as is conventional grinding, the abrasive wheel in ELG wears little in the process. ELG is burr

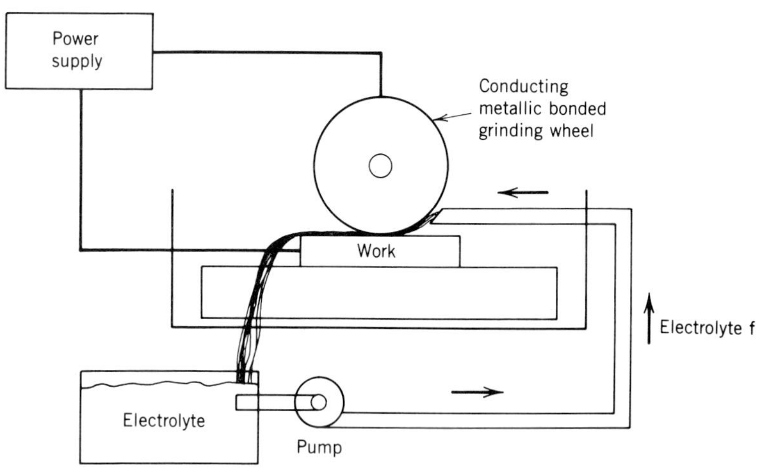

Figure 7
Basic ELG grinding system.

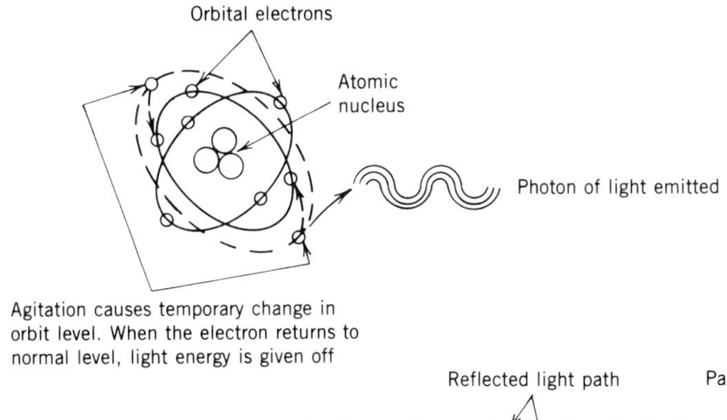

Figure 8
Photon emissions and laser generation.

free and will not distort or overheat the workpiece. The process is therefore very useful for small precision parts and thin or fragile workpieces.

LASERS AND LASER MACHINING

Lasers

LASER is an acronym for **Light Amplification by Stimulated Emission of Radiation.** By electrically stimulating the atoms of certain materials, such as various crystals and certain gasses, the electrons of these atoms can be temporarily displaced to higher electron shell energy level positions within the atomic structure. When the electrons fall back to their originally stable levels, photons of light energy are released (Figure 8). This light energy can be enhanced and focused into a coherent beam and then used in many manufacturing, medical, measurement, and other useful applications. The applications of lasers in manufacturing are widespread. Laser energy finds many applications other than using its energy for a cutting tool.

Figure 9
CNC laser machining center (Photo courtesy of Coherent, Inc, Palo Alto, California).

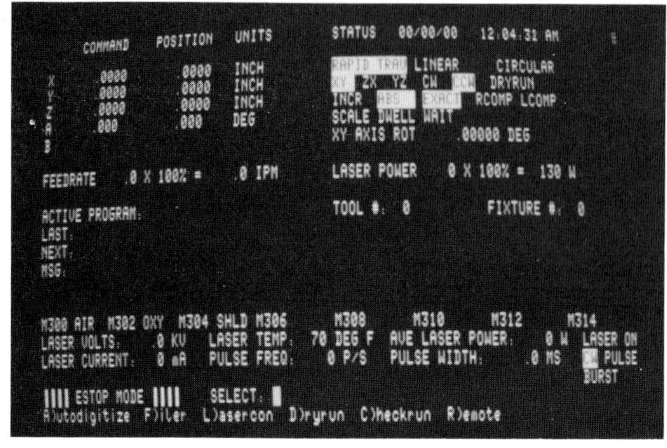

Figure 10
CRT on CNC laser machining center showing process status during production (Laserdyne).

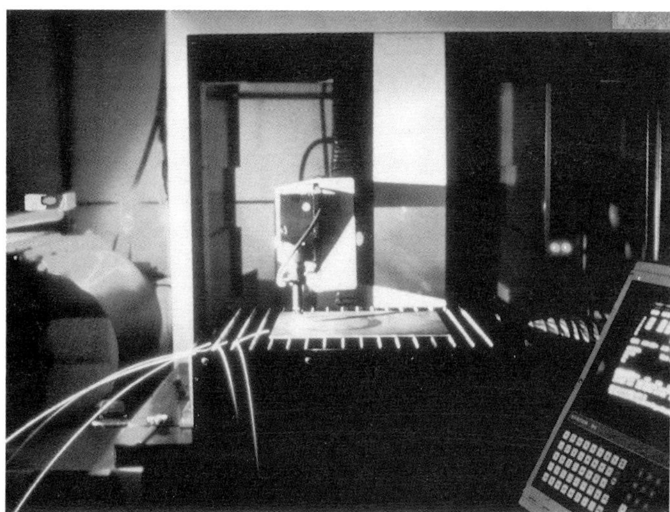

Figure 11
Laser plate cutting (Laserdyne).

Figure 12
CNC laser cutting of circular saw blades (Coherent).

Laser Machining

In laser machining, the **coherent laser light** becomes the cutting tool. When the laser is integrated with CNC machine tool positioning and CRT process data display (Figures 9 and 10), an extremely versatile system with many capabilities is available.

Applications of laser machining include cutting plate (Figure 11) for shapes such as saw blades (Figure 12), slotting stainless tubing (Figure 13), and marking systems (Figure 14). Lasers are also effectively applied in welding processes. These are discussed in Chapter 14.

Figure 13
Stainless steel tube slotting (Lasedyne).

Figure 14
Laser marking and other laser machined products (Laser Fare).

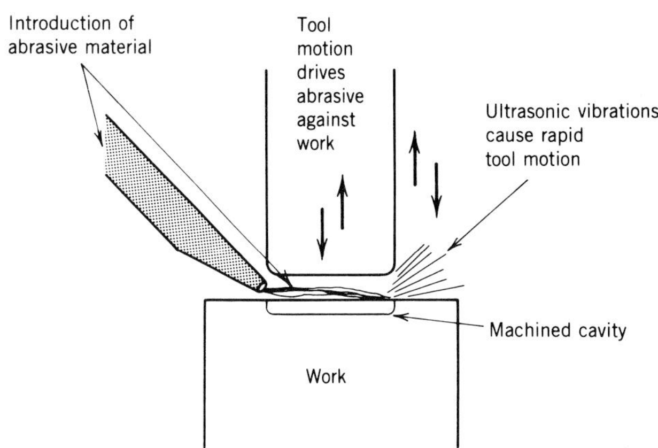

Figure 15
Ultrasonic machining.

ULTRASONIC MACHINING

Ultrasonic machining is similar to an abrasive process like sand blasting. High frequency sound (Figure 15) is used as the motive force with which to propel abrasive particles against the workpiece. Advantages of this process include the ability to machine very hard material with little distortion. Good surface finishes may be obtained and part features of many different shapes can be machined.

HYDROJET MACHINING

The cutting tool in this process is an **extremely high pressure water jet.** Pressures of several thousand pounds per square inch are used to pump water through a small nozzle, thus directing a narrow stream at the workpiece. The efficiency of the process is enhanced by adding an **abrasive material** to the water. Thus both pressure and mechanical abrasion join forces to accomplish cutting and machining tasks. The process may be used for hard metallic materials. However, there are limitations as to the thickness of the materials being machined. In the fabric industry, the process has efficiently been applied to fabric cutting applications.

ELECTRON BEAM MACHINING

Electron beam machining (EBM) is in some ways related to EDM and to electron beam welding. In EDM, however, a burst of electrons (spark) impinges on the workpiece, whereas electrons in a continuous beam are used in the EBM process. The workpiece material is heated and vaporized by an intense electron beam. The process like that of electron beam welding must be

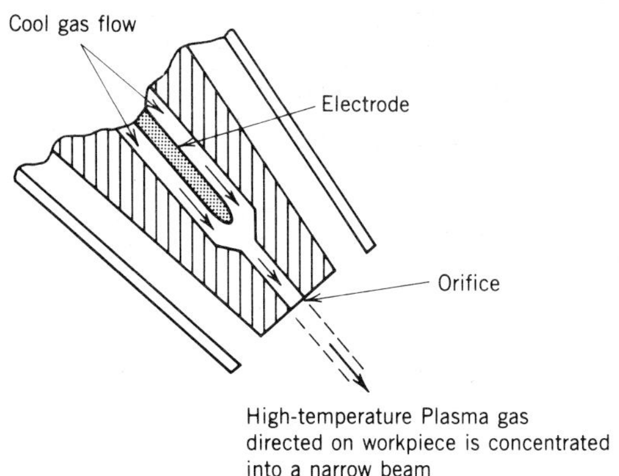

Figure 16
Plasma arc.

carried on in a vacuum chamber and appropriate shielding must be employed to protect personnel from X-ray radiation.

PLASMA TECHNOLOGY

A great deal of interest and development of **plasma technology** has occurred in recent years. Plasma is created by passing a gas through an electric arc. The gas is ionized by the arc and an extremely high temperature results. Temperatures in plasma arcs can exceed 40,000° F., many times hotter than any arc or flame temperature produced by other processes (Figure 16).

Plasma arcs of such high temperatures have a multitude of uses, not only in machining, but for other applications such as ore smelting and waste metal recovery operations. There is ongoing research on the applications of high temperature plasmas for incinerating industrial wastes.

The process is useful for cutting, welding, and machining nonferrous metals and stainless steel. Plasma torch machining is extremely fast and will produce smooth surface finishes to close dimensions.

REVIEW QUESTIONS

1. Describe in general terms how the following processes work: EDM, ECM, ELG, laser machining, ultrasonic machining, hydrojet machining, and plasma technology.
2. Which processes might be used for deburring applications?
3. Which process would be used for detail die slotting?
4. What temperatures can be reached in a plasma system?
5. Which process is essentially deplating?
6. What is the primary function of the grinding wheel in ELG?
7. What provides the energy in ultrasonic machining and what does the cutting?
8. Which process is accomplished in a dielectric fluid?
9. Which process uses a conducting fluid?
10. Describe in general terms how a laser functions.

CASE PROBLEMS

Case 1
Punch and Die Tooling

A punch and die set is required as shown in Figure 17. Of the processes discussed in this chapter which one would be most applicable to the manufacture of this tooling?

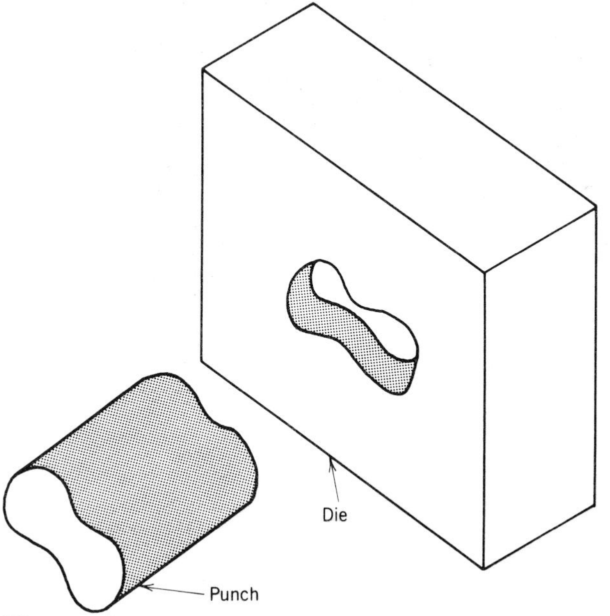

Figure 17
Punch and die set.

Case 2
Internal Deburring

A complex part with several intersecting drilled holes must be as burr free as possible. Since many of the internal features are inaccessible for mechanical deburring, which process discussed in this chapter might be used to accomplish the task?

CHAPTER 14 JOINING PROCESSES

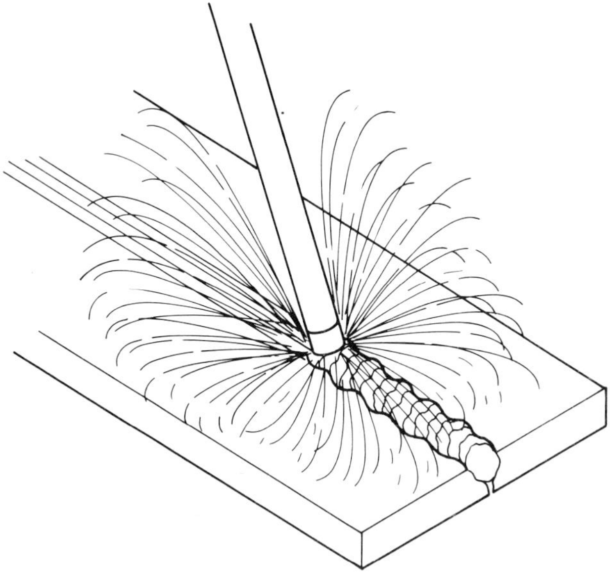

Any manufacturing process will yield either raw materials, further-processed raw materials, or many different materials assembled into a final product. When manufacturing any type of assembly requiring several different pieces, or an assembly involving several different materials, it will almost always be necessary to *join the materials* at some point in the process. Plastics and wood material must be glued or bonded, fabrics must be stitched, and metal components must be welded, bolted, bonded, or riveted. It is often necessary to join the same types of material within a given assembly, and it is often necessary to join many different types of materials within the same assembly or product. The purpose of this chapter is to survey the major methods of joining materials in manufacturing.

Joining processes can be divided into the following general categories.

1. Mechanical fasteners.
2. Adhesive bonding.
3. Welding.
4. Soldering and brazing.

OBJECTIVES

This chapter should enable you to:

1. List common methods of joining materials and cite advantages and disadvantages of them.
2. List common welding processes and describe how they work and what their general applications are.

MECHANICAL FASTENERS

With mechanical fasteners, there are many separate mechanical fastening systems and components that join materials. These include **threaded fasteners, nails** and **stapling, riveting, stitching** and **tying, pins** and **retaining rings, pressing, crimping,** and **specialty fastening** systems.

Threaded Fasteners

Threaded fasteners include **screws, bolts,** and **nuts.** These fasteners are very common in all mechanical assemblies. All threaded fasteners make use of the inclined-plane principle from mechanics to put pressure on the parts that are to be assembled. They will withstand large loads, are available in a wide variety of sizes and materials, and can be designed to meet many specialty fastening applications.

Bolts and nuts are found in almost every mechanical assembly. The bolt is designed to work in conjunction with a nut, with the material to be joined compressed between the nut and bolt head (Figure 1). Bolts with threads all the way to the head, called cap bolts or cap screws, are designed to screw into tapped holes, capturing the parts to be joined between bolt head and base material.

Figure 1
Common bolt and nut assembly.

Machine screws are similar to bolts, but are usually in smaller sizes (Figure 2). They have many of the same fastening applications as bolts and nuts. Sheet metal screws are made with coarse threads and are designed for joining thin sheet materials. All threaded fasteners have features that permit them to be secured with various tools such as wrenches and screw drivers.

The advantages of threaded fasteners joining include a disassembly capability and they come in a wide variety of types and sizes. This makes the system extremely versatile and likely to always remain in wide use.

Disadvantages include labor intensive assembly, requirements to stock a large number of types and sizes, and the possible loss of the hardware during assembly, disassembly, or service. Easily damaged threads on fasteners can make assembly or disassembly difficult or impossible. Any threaded fastener will require a predrilled or punched hole in the workpiece before assembly or joining of the materials can occur.

Figure 2
Machine screw assembly.

Nails and Staples

The foremost application of fastening by nails is in **wood building construction.** Although threaded fasteners may also be used in some wood construction applications, the additional time required to drill holes for bolts and to make up assemblies delegates use of threaded fasteners to large wood beam constructions that are designed to support large loads.

A typical nail consists of a stiff wire with a head on one end for driving and a point on the other end to penetrate the material to be joined. Nails come in a variety of sizes, styles, and materials including steel, brass, and aluminum (Figure 3). As the nail is driven into wood it pushes aside the grain of the material. The frictional forces exerted by the wood recompressing holds the fastener in place. Nailing at an angle to the forces trying to separate the material can also add some additional mechanical holding capacity. More holding capacity can be obtained by using fast helix nails. These have some of the characteristics of threaded fasteners and are used in crate and pallet manufacturing.

Nails are extremely suitable for static assemblies such as houses. They do require the expending of considerable energy for driving and this is often done by hand. However, nailing by machine and by explosive charges can be done to expedite this joining process.

Nailing is not suitable for dynamic loads in an assembly. The fastener will pull out if the structure is sufficiently distorted. A material assembled with nails

Figure 3
Common types of nails.

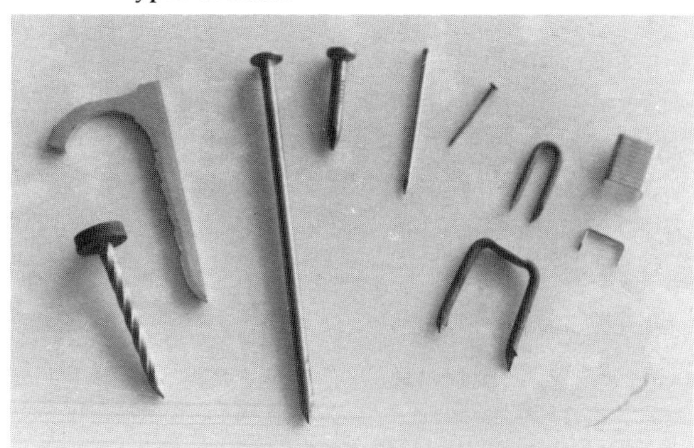

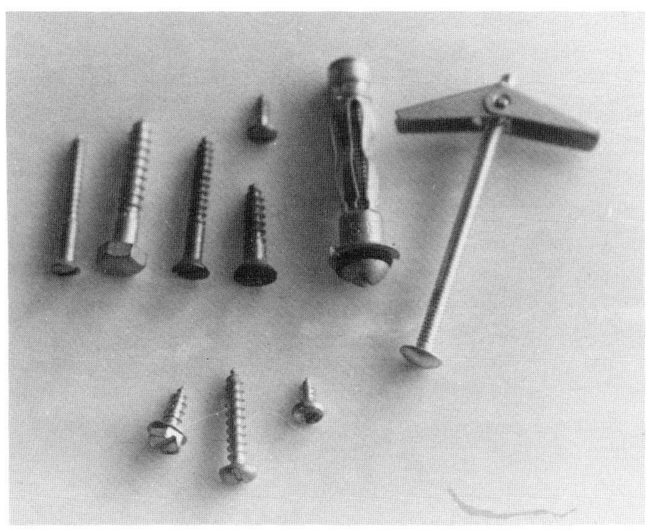

Figure 4
Wood and sheet metal screws and special hollow wall fasteners.

must also be compressive by nature so that the fastener will not fall out.

Stapling is related to nailing in that a staple is basically a two-pointed nail. Stapling is a versatile mechanical fastening system, highly adaptable to high production of furniture, paper boxes, and installation of many materials such as insulation and ceiling tiles. Stapling is best used where loads on the structure are moderate. It is not suitable for heavy loads because of the usually short penetration of the fastener into the material being joined. Stapling can be most unsatisfactory in this respect and this system should be selected only after a careful engineering evaluation of the loads the assembly will encounter.

Threaded fasteners including sheet metal screws, wood screws, and many types of specialty hollow wall fasteners are available to the building industry (Figure 4).

Rivets

Rivets are a very common and extremely versatile mechanical joining system. Use of rivets for joining materials range from heavy steel structurals to light gage

Figure 5
Riveting flush with surface is used extensively in aircraft cowl and skin applications in order to reduce air flow turbulence (Lockheed—California Company).

Figure 6
Riveting internal aircraft structures makes them very strong but also very light.

sheet metal, plastics, and composites. They are widely used in the **aircraft industry** where it is desirable to have as little projection of the fastener head as possible above the surface of the materials being joined. This greatly reduces air friction on skin and cowl surfaces subjected to air flow during flight (Figure 5). But since supersonic aircraft require almost totally smooth skins, riveting is giving way to adhesive bonding in this application.

Riveted structures are used extensively in internal aircraft structure (Figure 6). Extremely strong but very light ribbed structures can be fabricated by this method (Figures 7 and 8). Machine riveting is widely used in aircraft production. Since rivets are retained on the back side as are many threaded fasteners, they are well suited to dynamic loads and will not come loose through distortion of the assembly.

Much of the riveting is accomplished with power machinery. The equipment used may be portable or may be located on the bench where the work can be brought to the machine. The equipment illustrated in Figure 9 swages rivets using a kneading action from the center of the rivet outward.

Multipoint riveting machines can swage several riv-

Figure 7
Strong but light-weight internal riveted rib structures in aircraft (Lockheed—California Company).

Figure 8
Riveting major aircraft components (Lockheed—California Company).

ets at one time (Figure 10). Many riveting tools and rivet styles are available to meet a broad range of riveting applications (Figure 11).

Advantages of riveting are strength with a minimum-sized fastener and the ability to join material by insertion and attachment of the fastener from only one side of the assembly. An example of this is the popular pop rivet used in many common fastening applications.

Disadvantages include labor intensive installations, requirements for predrilled or punched holes in the material, and difficulty in disassembly. Most rivets must be drilled out in order to take an assembly apart. Reassembly requires replacement of the fastener with new hardware.

Stitching, Tying, and Snaps

Stitching and tying are found primarily in the fabric industry. Machine lock stitch is an essential process in the making of clothing. The modern sewing machine has perfected this method of material joining. Other applications include sack sewing with an easily unraveling chain stitch. Tying with rope, string, or cord has many temporary applications. Mechanical snaps are

Figure 9
Bench rivet machine forms rivets with a center-outward (radial) action causing little distortion of the parts being joined (Bracker Corporation/Leland J. Balber Associates).

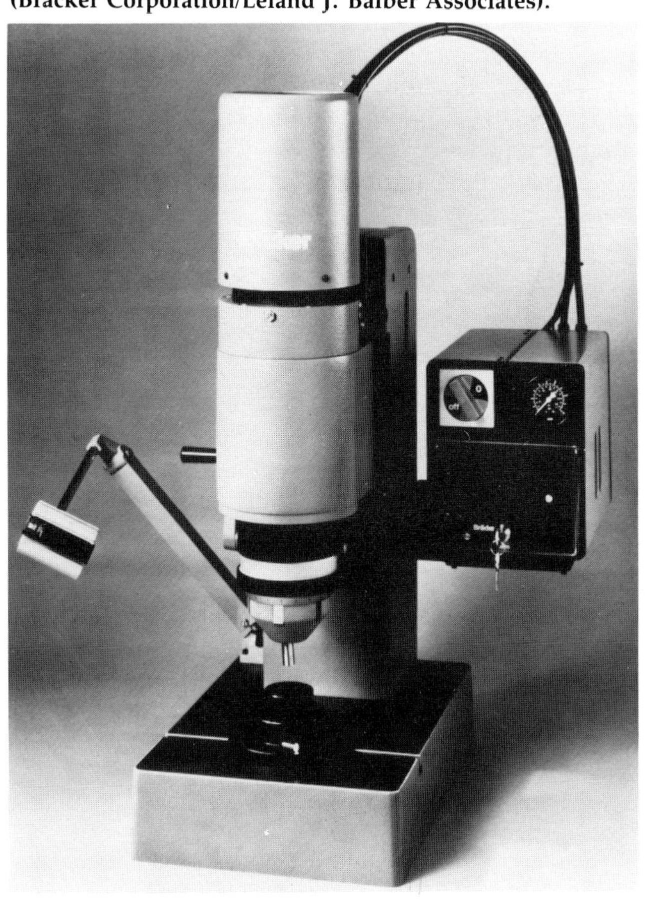

Figure 10
Multipoint radial riveting machine for die cast metals and plastics (Bracker Corporation/Leland J. Balber Associates).

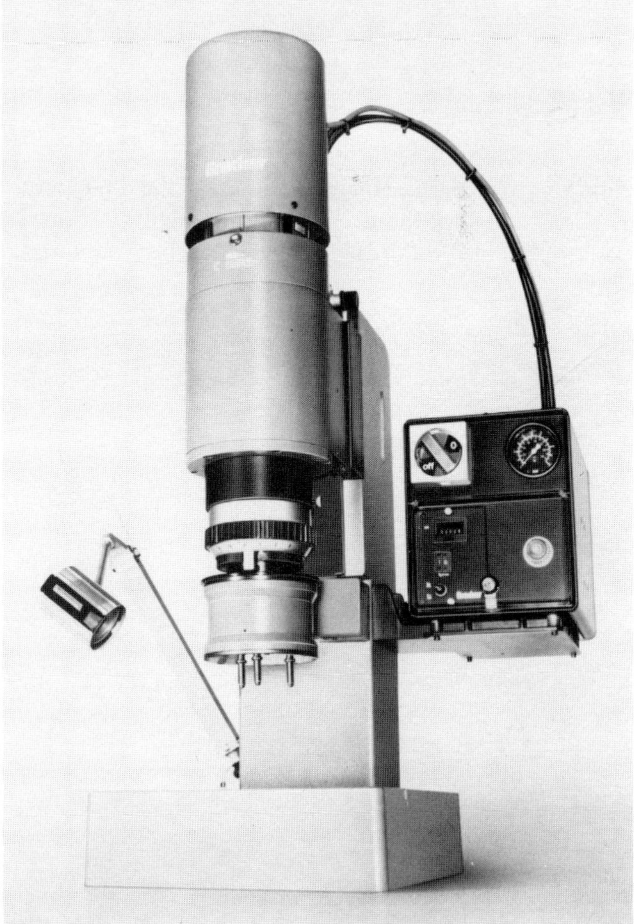

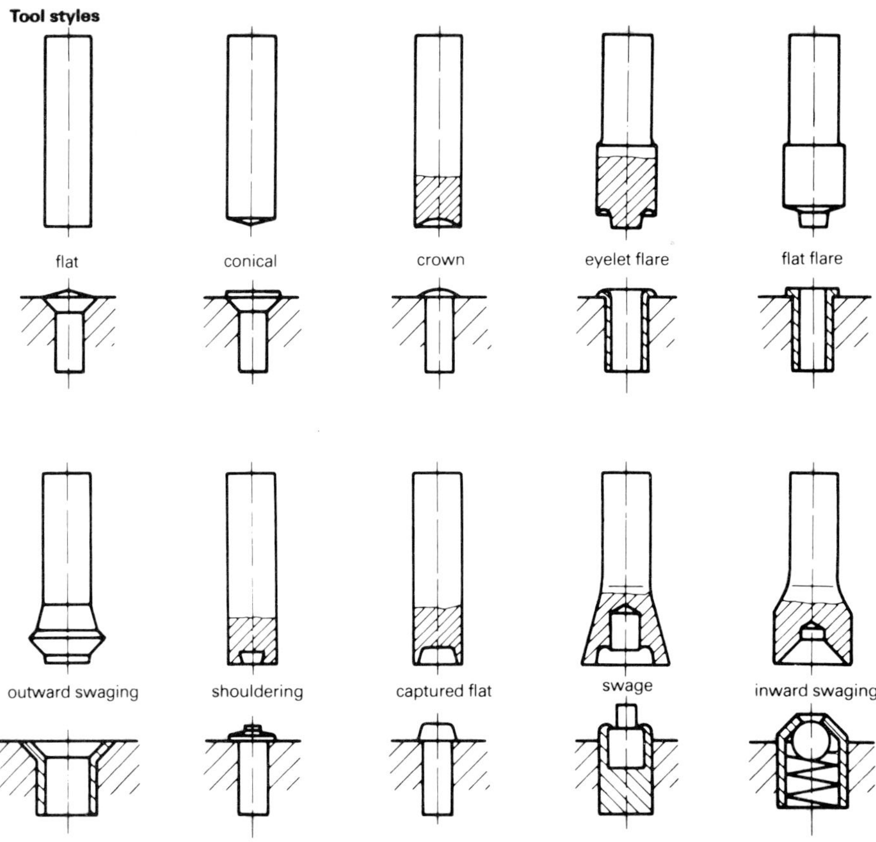

Figure 11
Rivet types and applications (Bracker Corporation/Leland J. Balber Associates).

frequently used for temporary joining of fabrics and canvas.

Pins and Retaining Rings, Pressing, and Crimping

Assembly of mechanical components will often make use of **pins** and **retaining rings.** These fasteners make disassembly possible, but precision machining of the parts is often required. Retaining rings may be used to hold mechanical parts together (Figure 12). Pins are often found where parts must be joined and disassembled frequently (Figure 13) or to retain assembly of routine parts.

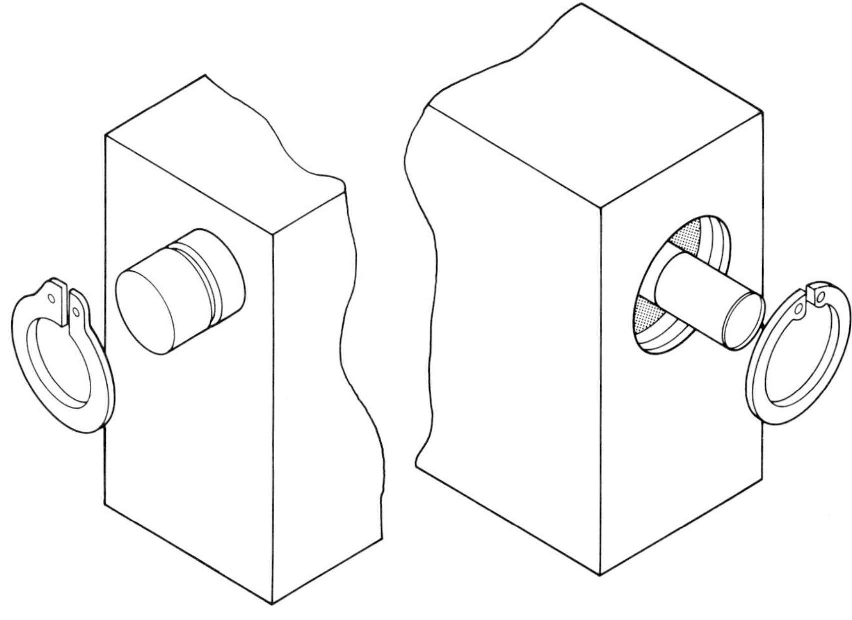

Figure 12
Mechanical assembly using retaining rings.

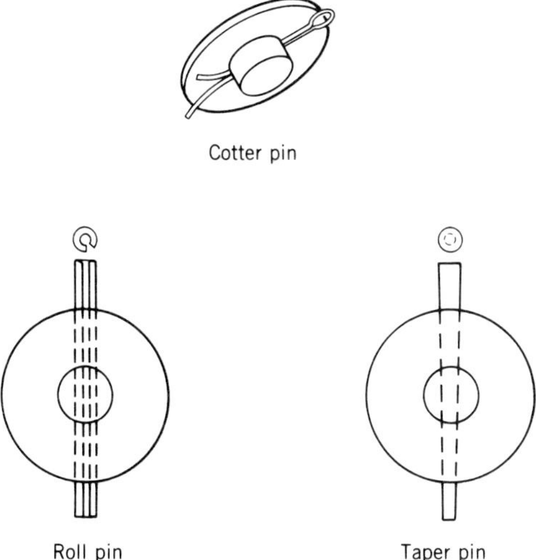

Figure 13
Mechanical assembly using pins.

Figure 14
Mechanical assembly using press fitting.

Many parts in mechanical assemblies can be joined by **pressing.** In this method, the mating part dimensions overlap slightly and the parts are forced together under mechanical pressure. The method has many applications, especially in assemblies where bearings and bushings are to be retained without the need for additional mechanical hardware (Figure 14).

Related to mechanical pressing, are **shrink** and **expansion fitting** where parts in an assembly are joined by heating or shrinking one or the other, fitting the parts, and then returning them to normal temperature. This process makes use of the normal expansion and contraction of material as temperature is varied.

Crimping is another method of material joining, where mechanical force is applied to compress or expand one piece of material against another piece to join the parts. Generally, crimping requires a groove or other feature to provide a place where the material of the mating part can grip (Figure 15).

Seaming is related to crimping. This joining process is widely used in the sheet metal industry for closing duct work and other sheet metal fabrications. Common sheet metal seams include those illustrated in Figure 16. The industry makes use of a number of hand and power operated machinery to roll, form, and crimp the various types of joint seams.

Figure 15
Mechanical assembly with crimping.

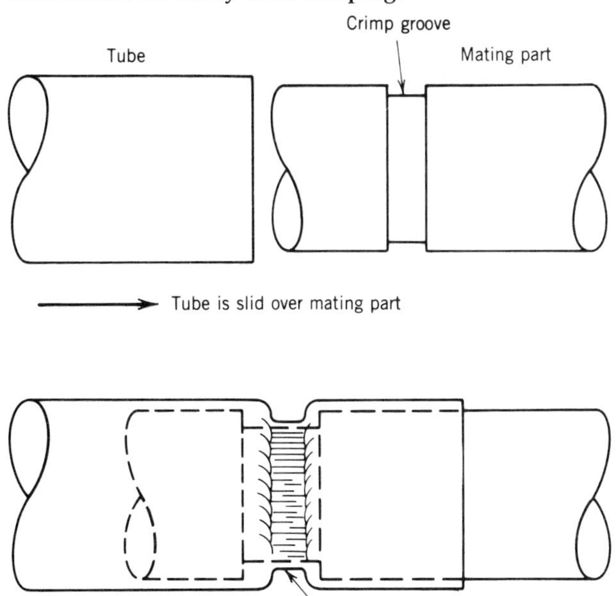

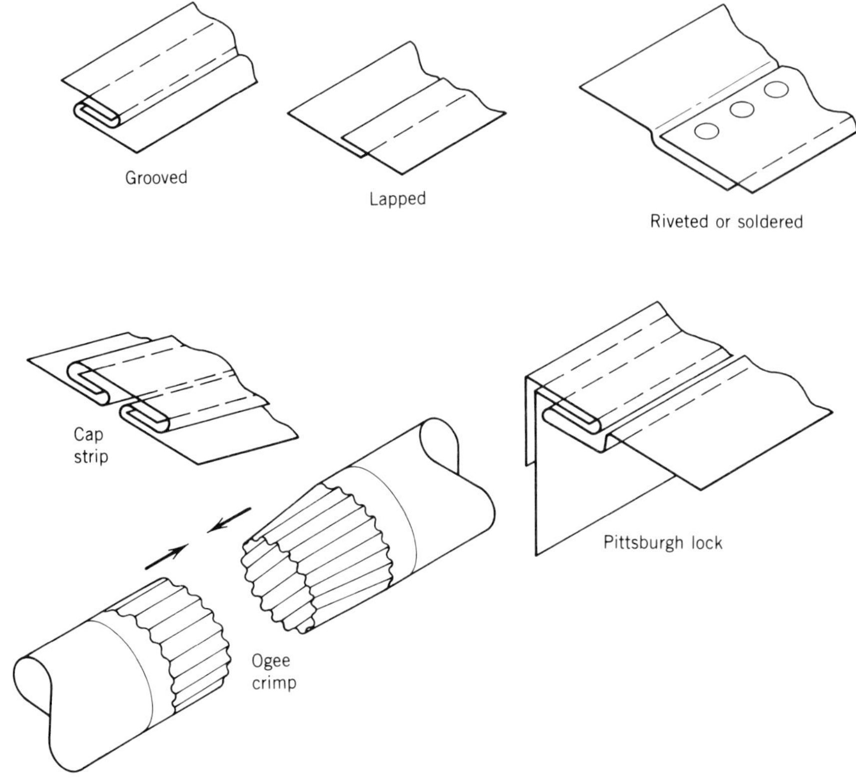

Figure 16
Joining sheet metal by various seams.

Specialty Fastening Systems

Specialty fastening systems have found wide applications. One example, **plastic barbs,** consists of a pad of plastic barbs designed to grip into mating materials (Figure 17). This fastening system has many uses where temporary joining of material is desirable. Plastic barbs are not suitable for joining materials that will be subjected to heavy pull-apart loads. However, this system is extremely useful for temporary joining of fabrics and other light-duty holding applications.

ADHESIVE BONDING

It is naturally desirable to join material without the need of labor intensive preparation such as hole drilling or punching and the insertion of mechanical hardware. Processes involving adhesives have developed rapidly in recent years and are now in wide use. Research and development into adhesive joining of metals, plastics, and composite materials is under continual development.

Advantages of Adhesive Bonding

Advantages gained by adhesive bonding include:

1. Dissimilar materials can be joined that cannot be joined by other processes.
2. Mechanical preparation such as hole drilling is eliminated.
3. Adhesive bonding joins materials with no gaps as there would be between bolts or rivets in a me-

Figure 17
Plastic barbs, a unique material-joining method.

chanical assembly. Overall strength of the joint can therefore be enhanced.

4. Many adhesives are not subject to deterioration from service environments and therefore can be used in a wide range of temperature and corrosive environments.
5. Joined materials are not subjected to mechanical or heat distortion as they would be if joined with mechanical or welding processes. Hence, precision parts can be joined without risk of damage or distortion.

Disadvantages of Adhesive Bonding

Disadvantages of adhesive bonding include:

1. Specific joint preparation may be required.
2. Adhesive material may be expensive and require application in controlled environments.
3. Curing time may be long and may require a special environment.
4. Mechanical strength of the adhesive-bonded joint may depend greatly on how the joint is loaded. Shear loading is best followed by tension. Where loading of the joint tends to peel materials apart, adhesive bonding may prove unsatisfactory (Figure 18).

Types of Adhesive Bonding Materials

Adhesives are divided into two major groups:

1. Natural adhesives.
2. Plastic resin adhesives.

Natural Adhesives

Natural adhesives include the following:

1. Vegetable and animal glues and casein.
2. Sodium silicate.
3. Natural gums.

Vegetable and Animal Glues Vegetable glues are used in the paper industry and they have applications in woodworking. Examples include wallpaper paste made from starch and the gum arabic used on envelopes and postage stamps. Animal glues made from bone and hide are adequate for some applications, but they tend to deteriorate in moist environments. Casein, made from milk, is a water-based adhesive.

Sodium Silicate Sodium silicate shows good temperature characteristics and it can be used as a binder to make some types of composite material.

Natural Gums Natural gums have long been used as adhesive materials. These include natural rubbers, sealing waxes, and asphalt.

Plastic Resins Synthetic resins (plastics) used for adhesives have undergone much development in re-

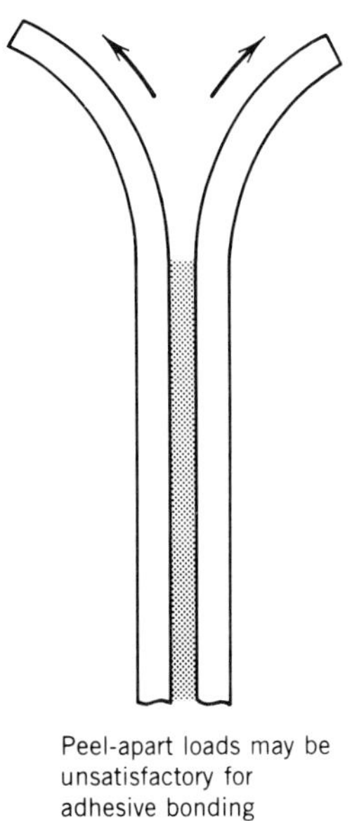

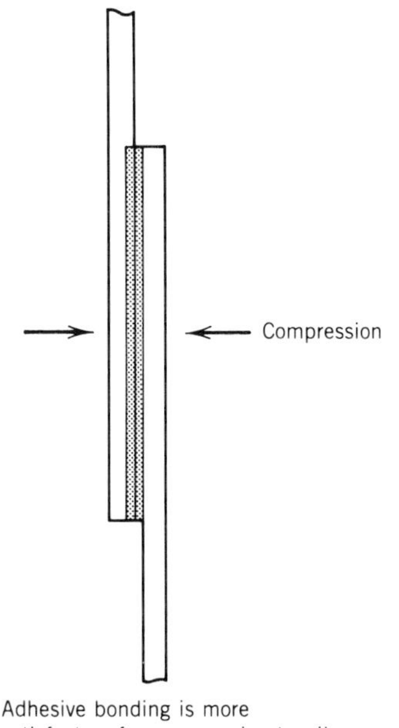

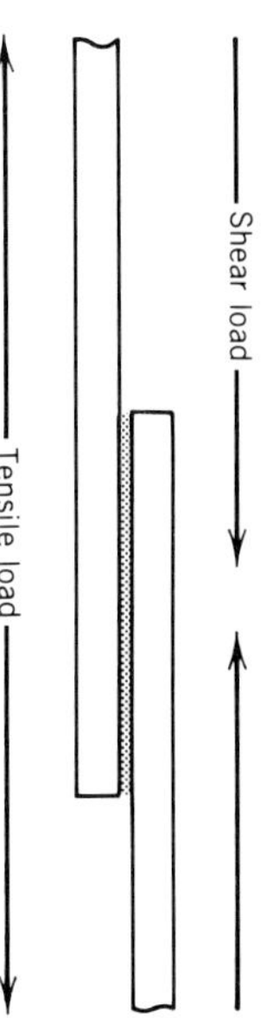

Figure 18
Types of loads applied to adhesive bonded joints.

Figure 19
Bonded joints are resistant to corrosion.

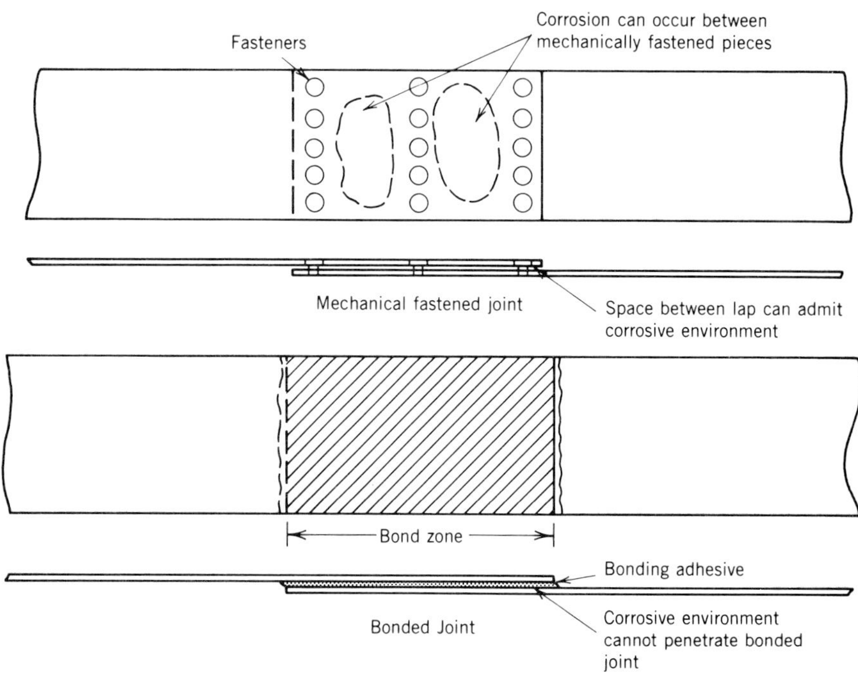

cent years and have found an important place in material joining processes. In many cases plastic resin adhesives have proved to be superior to mechanical joining systems.

Types of Plastic Resin Adhesives Resins used for adhesives come from both major plastic classifications, of thermosetting and thermoplastics. The thermosetting plastics undergo fundamental chemical change while hardening and cannot be reused or recycled. Thermoplastics, on the other hand, can be heated, resoftened, and reused.

Thermoplastic resins used for adhesives include cellulose nitrate, acetate, acrylic, cyanoacrylate, and vinyls. Thermosetting resins used for adhesives include polyesters, epoxies, phenolics, ureas, and silicones.

Adhesive Bonding in Aircraft Production

Adhesive bonding in aircraft assembly has undergone much development in recent years. The ability to bond

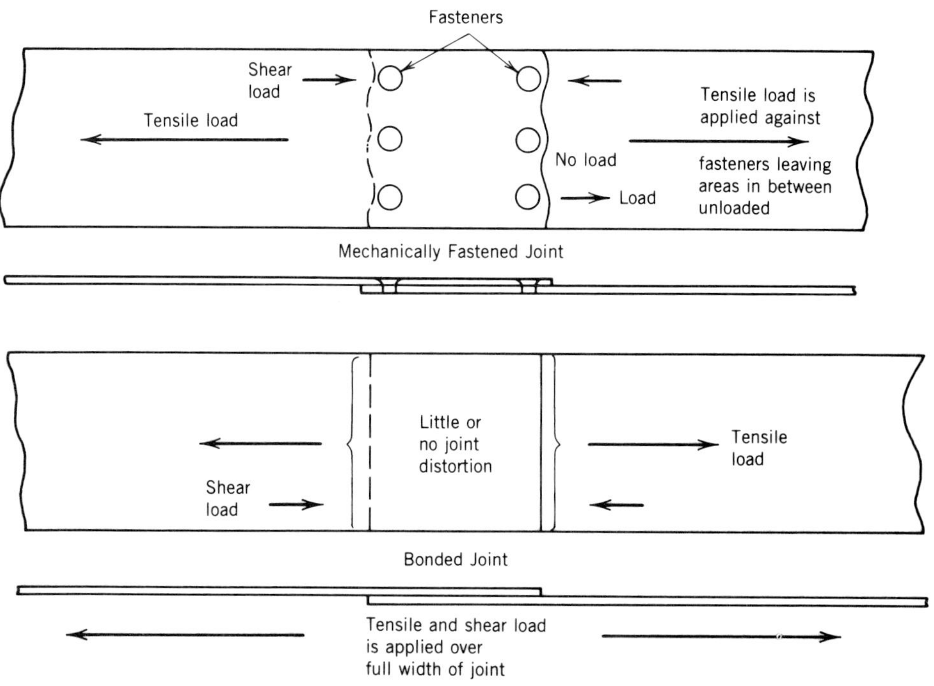

Figure 20
Bonded joints distribute load better than mechanical joints.

aircraft skin panels has numerous advantages over joining with screws or rivets. Advantages include:

1. Resistance to corrosion (Figure 19).
2. As strong or stronger than riveting joints.
3. Joint loading is more uniform (Figure 20).
4. Joint fatigue and crack development are reduced using bonded joints.
5. Bonded joints are more effective in sound and vibration dampening in aircraft structures.
6. Bonded joints eliminate the need for many fasteners and thus effects a substantial weight saving, essential to modern fuel-efficient aircraft.

The requirements of aircraft skin panel bonding are stringent. The adhesive used must have high resistance to peel apart and shear loads. The bond joint must be able to resist penetration to corrosive fluids and must also be able to withstand high temperatures in its service environment.

Other requirements of bonded joints include consistency of the bond throughout the joint area and the adhesive material must contain a corrosion inhibitor and must not crack or crumble when drilled for attachment to any mechanically fastened components. The most versatile adhesives, demonstrating these properties that are used for aircraft bonding requirements, come from the epoxy group.

Aircraft Bonding Technology Skin panels are first sprayed with a protective coating. And then they are formed by pressing to the contour shape of the aircraft fuselage. The protective coating previously applied is not affected by the forming process. The formed skin panels are then cleaned (Figure 21). Cleaning, which is essential to a proper bond, is done with an alkaline solution and water rinse. This is followed by an acid rinse used to slightly etch the surface of the skin panel. Acid etching provides an effective bonding surface.

The panel then goes to a temperature and humidity controlled adhesive bonding room where the epoxy-based adhesive primer is applied. Air spraying is used, with the spray air having been carefully cleaned of all oil and moisture. A full and uniform coating, required for proper bonding, is applied at this stage in the process. Primed panels are then cured and stored in a clean room environment.

Bonding occurs in a clean-room environment where the atmosphere is maintained at a positive pressure. The epoxy-based adhesive is applied in a layer .010 in. thick. Parts are joined and held in position by nylon tape (Figure 22). The joined assemblies are then placed

Figure 21
Aircraft skin panels being cleaned prior to bonding (Lockheed—California Company).

Figure 22
Panels are held in position with nylon tape prior to bonding (Lockheed—California Company).

on bonding fixtures in plastic bags and a vacuum is drawn creating a downward pressure on the parts.

The next process phase is bonding and curing. This is accomplished in a large capacity **autoclave** (Figures 23 and 24). The autoclave is closed, heated, and purged with an inert gas under pressure. After a suitable period, pressure and temperature are reduced and the bonded assembly is removed from the autoclave.

Inspecting Bonded Panels. After completion of the bonding process, assemblies are cleaned and prepared for inspection. Samples of the bonded assemblies undergo peel and shear testing. Bonded joints undergo ultrasonic inspections (Figure 25) to determine joint integrity. Water is sprayed on the joint and an ultrasonic reading is graphed. In this way, joint quality can be determined. Final phase is the painting and assembly of the skin panel on the aircraft (Figure 26).

Large aircraft skin panels are precision components and are subject to damage unless handled carefully. It is interesting to note the methods by which this is accomplished. An array of vacuum cups are used to pick up and handle the skin panel. The methods are

Figure 23
Bonded panels enter the large autoclave for curing (Lockheed—California Company).

Figure 24
Special transporters are used to handle the large panels (Lockheed—California Company).

Figure 25
Ultrasonic testing of panel bond joints (Lockheed—California Company).

Figure 26
Suction cups hold the panel while it is positioned for installation (Lockheed—California Company).

very secure but, at the same time, will not damage the precision panel component (Figure 27).

WELDING PROCESSES

It would be difficult to overstate the importance of welding in modern material joining processes. These processes make fabrication of complex metal assemblies possible. In welding, the materials to be joined are heated to the point of fusion and then joined and allowed to cool. Filler stock, generally of the same material as the parent metal, may be added to fill the weld joint during the process. Common industrial welding processes include **oxy-acetylene gas welding** and **electric welding.**

Oxy-acetylene Welding

The oxy-acetylene welding process uses an oxygen supported acetylene flame to heat the parts to welding temperature. This process, often known as **gas welding,** is mostly used for repair welds on light to medium applications (Figure 28). Gas welding requires considerable technique on the part of the operator and fuel supplies are expensive. However, gas welding does find application in repair welding where it might be difficult to join material by other processes, or where electric welding might not be available.

Oxy-acetylene flame processes are also used extensively for **flame cutting** applications (Figure 29). Although this is not welding, it plays an important part in manufacturing. Oxy-acetylene cutting torches can

Figure 27
Bonding technology requires special equipment handling as well (Lockheed—California Company).

Figure 28
Oxy-acetylene welding and cutting equipment.

Figure 29
Flame cutting with an oxy-acetylene cutting torch is an efficient, fast, versatile, and widely used process (RRK Machine and Manufacturing).

cut through metal several inches thick, making them indispensable for manufacturing of structural fabrications such as bridges, ships, and oil field and farming equipment. As with most other manufacturing processes, flame cutting has been computer automated.

In the flame cutting process, fuel gas flow (acetylene, methane, hydrogen, or propane) is fed under regulated pressure to the cutting torch. Pure oxygen is also fed to the torch under regulated pressure. The fuel oxygen mixture is ignited as it exits the torch nozzle and the operator adjusts the mix by valves on the torch assembly. The flame mixture may be adjusted to be neutral (equal fuel and oxygen mixture), oxidizing, or reducing. The oxidizing flame has an excess of oxygen and the carburizing or reducing flame has an excess of fuel gas.

The particular requirement of the torch application will determine which flame is to be used. For example, the carburizing flame can be used to inject carbon into the surface of the metal being heated, thus effecting the process of carburizing for surface heat treatment. Neutral flames will be used for cutting. During this process, preheated jets in the torch tip heat the metal to a red-hot state and maintain this temperature. When the proper cutting temperature is reached, a pure stream of oxygen is directed through a hole in the center of the torch tip. When the oxygen makes contact with the red-hot preheated metal, a high degree of rapid oxidation takes place almost instantly and the metal is cut at that point. An operator skilled in handling the torch, using the right gas pressure, can result in a very smooth torch-cut surface. Flame cutting has also been computer automated and is used where duplicate workpieces must be produced. Cutting torches that are solidly mounted and have their movement controlled by mechanical feeding mechanisms can produce very smooth flame-cut surfaces that may require little or no machining operations for finishing. Mechanical components such as gears may be manufactured directly by this process. Various torch tips are used depending on the flame cutting applications. These include torch tips for cutting, welding, brazing, and heating.

Electric Welding

Electric welding processes are mainstays in joining metals during manufacturing processes. The bulk of steel structural fabrication and construction is done by electric welding processes (Figure 30).

Although the individual processes differ to some extent, they all convert the flow of electric current to heat so that the material to be welded may be heated to proper fusion temperatures. The welding system consists of a power supply capable of sustaining large

Figure 30
Large structural steel fabrication joined by electric arc welding processes (Miller Electric Manufacturing Company).

Figure 31
Electric arc welding system components.

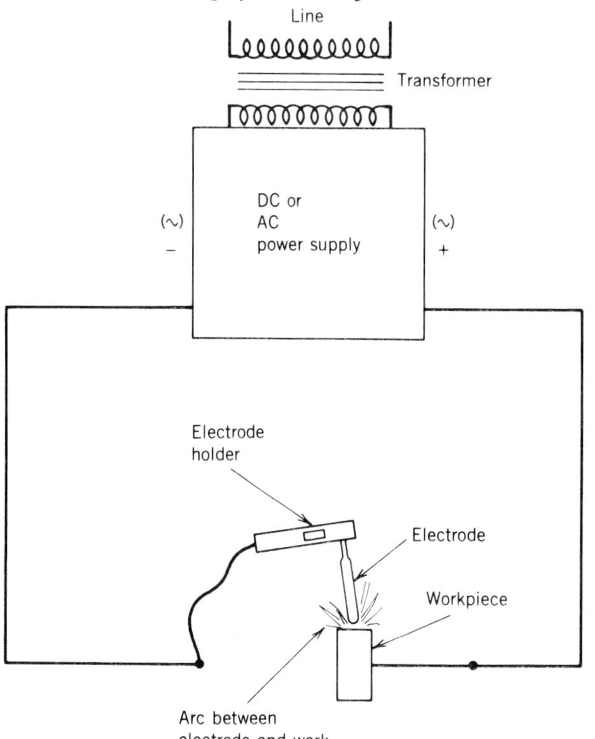

currents at relatively low voltages and appropriate cables connected to the workpiece, thus completing the electrical circuit (Figure 31).

There are several electric welding processes in common use. They have wide capability and can be used on a variety of materials ranging from heavy sheet, structural, and plate to very thin sheet metal. Electric welding is also applied to many different metals. Common electric welding processes include **rod electrode arc welding, inert shield gas arc welding, resistance welding, submerged arc welding,** and **stud welding.**

Rod Electrode Arc Welding In this process, current flows from the power supply through copper wire cables to the electrode holder which is usually held in the operator's hand. The welding power supply may be powered from the utility line or it may be engine driven. Engine-driven power supplies permit electric welding to be accomplished at locations where no line power is available. This makes the welding system extremely useful for field repairs such as those often made on farm equipment.

A wire electrode is clamped in the electrode holder (Figure 32). The electrode, often called a welding rod, is about one foot long and is usually coated with a hard dry fluxing material. The rod flux coating burns away during the welding process and provides a fluxing or cleaning action in the immediate area of the weld. The flux also serves to shield the weld zone from oxygen

Figure 32
The engine-driven power supply for electric welding is indispensable equipment for farm machinery repairs. The electrode holder and rod are seen in the welder's hand (Miller Electric Manufacturing Company).

Figure 33
Electric arc weld.

and nitrogen in the air. Thus, contamination of the weld zone is reduced, resulting in a cleaner yet stronger weld and better quality joining of the materials.

The wire core of the welding rod acts as the filler metal filling in the joint between the material being welded. As the operator touches the rod to the work, current flows across the junction and an electric arc (short) is created. Rapid heat develops because of the large current flow heating the workpiece while at the same instant the filler rod melts and is deposited into the weld joint (Figure 33).

Considerable skill is required on the part of the operator to properly maintain the arc length and to make correct movements so as to deposit filler metal neatly into the welded joint. Heat is controlled by adjusting the current flow from the power supply. Current flow is based on the type of rod used and the thickness of the metals being joined. The operator (welder) must be protected from the intense heat and bright light of the welding arc by wearing appropriate clothing and a face hood containing a darkened-glass view port. This enables the welder to observe the welding being done while eyes are protected from the intense light emitted by the welding arc.

Rod electrode arc welding is widely used in all types of welding manufacturing processes. It is especially popular in fabrication industries such as ship, bridge, pipeline, and tank manufacturing and in many other industries that fabricate products from structurals, plates, and heavy-gage sheet steel materials.

Inert Shield Gas Arc Welding In the previously discussed process, the gasses resulting from burning the flux coating on the welding rod provide a certain degree of protection from external air contamination of the weld zone. Since the weld takes place at elevated temperatures, contamination from the air surrounding the weld zone is a significant problem. On quite thin steel material such as sheet metal, the heat from the welding process cannot be dissipated fast enough before complete oxidation of the material being welded occurs. Also, nonferrous metals with an even higher affinity for oxygen than steel, will be consumed before they can be joined by the welding process.

An excellent method of controlling this problem is inert gas shield arc welding. In this type of welding, the entire weld zone is surrounded by an inert gas such as argon, helium, or carbon dioxide (Figure 34) or by various mixtures of these gasses. The inert gas displaces the air in the immediate vicinity of the weld zone

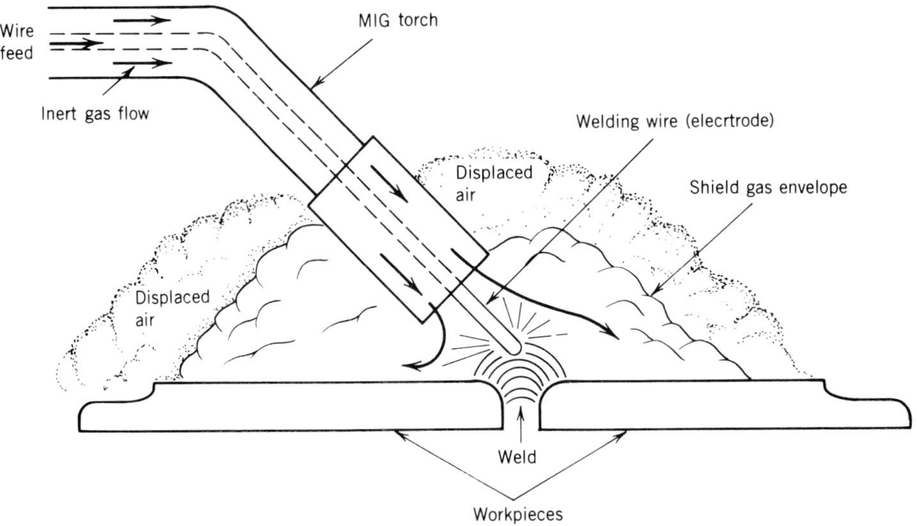

Figure 34
GMAW or MIG welding shields weld zone with inert gas.

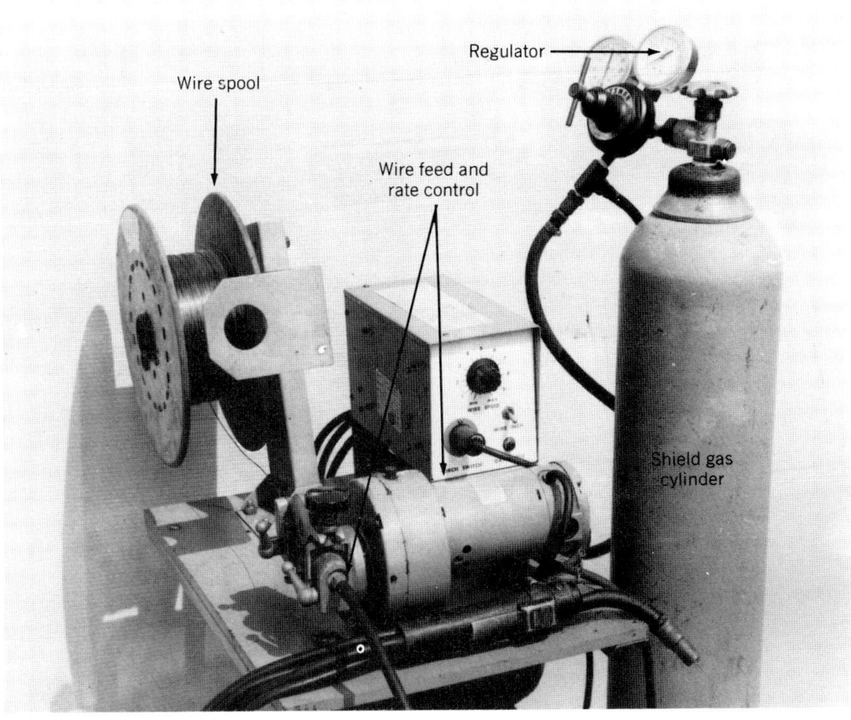

Figure 35
MIG welding equipment includes wire feeder and shield gas cylinder.

Figure 36
MIG electrode and shield gas is fed to the weld in a continuous flow (Miller Electric Manufacturing Company).

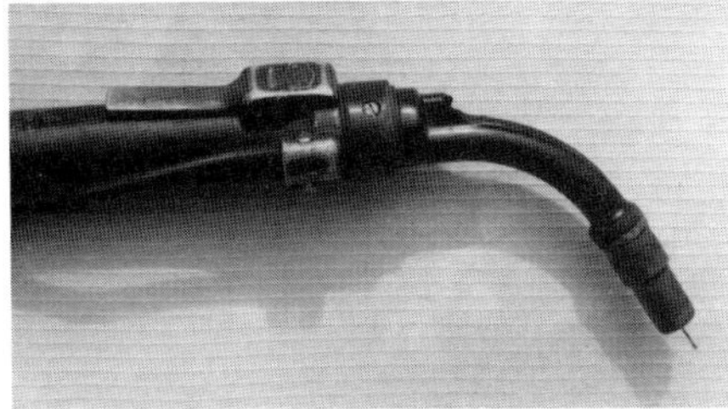

Figure 37
The MIG torch.

and the weld can take place in what amounts to a totally uncontaminated environment. The major shield gas welding processes include the following:

1. **Gas metal arc welding (GMAW or MIG)**
2. **Gas tungsten arc welding (GTAW or TIG).**

MIG Welding Gas metal arc welding (GMAW), also known as MIG, is a widely used and efficient process that can be applied over a wide range of metal thicknesses and materials. MIG is extremely well suited to thin metals and the process works well when welding steel and aluminum on a production basis.

The MIG system consists of a variable current power supply and wire feed unit (Figure 35) and a shield gas cylinder with pressure and flow regulation. The MIG electrode is a continuous wire of the same material that is being welded. Electrode wire is fed continuously from a reel while the welding process is taking place (Figure 36). At the same time shield gas flows from the electrode gun (Figure 37) surrounding the wire and displacing air from the weld zone area. Welding current, shield gas flow, and electrode feed rate may be varied to meet the exact requirements of the welding job. The process is not unlike rod electrode welding, except that the wire electrode is available on a continuous feed basis.

Figure 38
Lathe mounted workpiece undergoing weld buildup by the MIG process (Miller Electric Manufacturing Company).

Figure 39
Weld built up by the MIG process is virtually free of slag pockets and inclusions. After weld buildup is completed, the part may be remachined to restore original dimension. (Miller Electric Manufacturing Company).

Variations of the MIG process include the use of flux cored or magnetic flux electrode wires. The fluxing agent may adhere to the outside of the electrode wire by magnetism or it may be contained inside a hollow electrode wire (flux core). Fluxed electrode MIG welding may be done with or without shield gasses.

Because the process takes place in the inert gas environment, weld quality of the MIG process is unsurpassed. It is therefore very suitable to rehabilitating worn parts by weld buildup (Figure 38). In the illustrations, a shaft has been mounted in a lathe where it can be rotated. The MIG bead can be evenly and accurately laid down on the workpiece (Figure 39). When the welding is complete, the part may be machined to restore its original dimensions.

Figure 40
The operator can easily reach the work using an efficient portable MIG welder with overhead suspended wire feeder (Miller Electric Manufacturing Company).

Figure 41
Automated MIG welding equipment (Miller Electric Manufacturing Company).

With the convenience of the wire-fed electrode and the inert gas shield, the MIG process is highly suited to production welding of nonferrous metals such as aluminum, steel, and stainless steel (Figure 40). The MIG process is highly suited to both semiautomation (Figure 41) and full automation using industrial robots. Figure 42 shows automated robotic welding. The production welding of the assembly was accomplished in a half hour using the robot, whereas the same job took one and one half hours by hand methods (Figure 43).

TIG Welding In MIG welding, the continuously fed electrode wire is consumed during the process and becomes the filler material deposited into the weld joint. The **GTAW** or **TIG** process uses a tungsten metal electrode that is not consumed during the welding process. Filler material is fed externally into the weld joint during welding. The process is somewhat like the oxyacetylene process in this respect, the difference being that the heat is supplied by an electric arc established between the tungsten electrode and the workpiece. TIG also uses a shield gas to surround the weld zone to displace air and other contaminants.

The TIG system (Figure 44) consists of a power supply, cables, and the TIG torch that is usually held in

Figure 42
The MIG welding process is highly suited to industrial robot automation (ESAB Robotic Welding Division).

one hand by the operator. With the other hand the operator may feed the filler rod into the weld joint (Figure 45). Shield gas flow and welding current are controlled by a foot pedal. Some TIG torches are water cooled. This helps dissipate heat, thus preserving the electrode and making the torch more comfortable to hold.

Figure 43
Robotic MIG welding can be done with great speed and quality (ESAB Robotic Welding Division).

Figure 44
GTAW or TIG welding system.

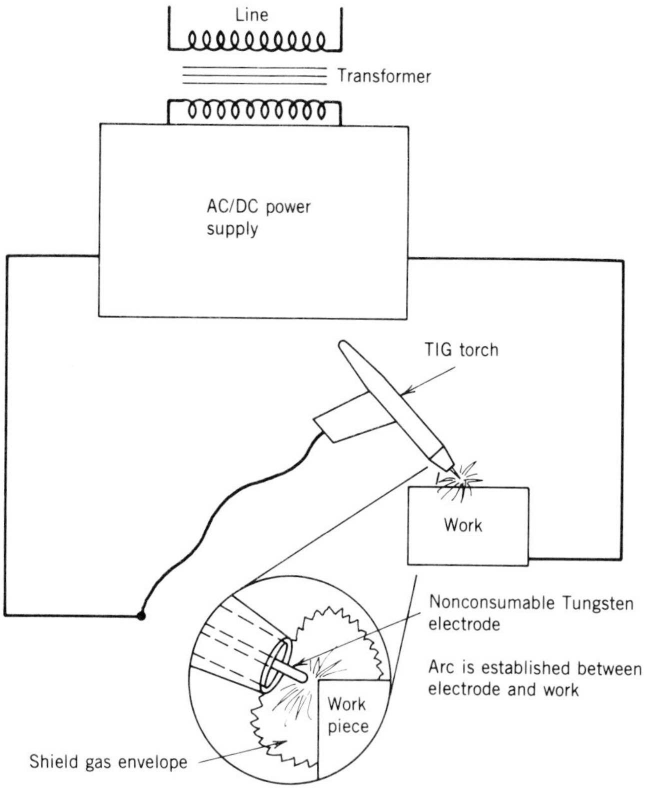

Line
Transformer
AC/DC power supply
TIG torch
Work
Nonconsumable Tungsten electrode
Arc is established between electrode and work
Work piece
Shield gas envelope

Figure 45
TIG welding on a turbine rotor. The TIG torch and filler rod are visible in the weldor's hands (Miller Electric Manufacturing Company).

The TIG process is extremely useful for small welding applications where detail welding is required. TIG can be used to weld metal a few thousandths of an inch thick. Since the process takes place cleanly in the inert gas envelope, it is possible to achieve superior quality welds. For this reason, TIG may be used where a weld of outstanding quality is required. An example would be the root pass in joining nuclear system piping. After the TIG root pass is made at the bottom of the weld joint, the remaining weld filling of the joint may be done by other processes such as rod or MIG. TIG is also superior on nonferrous metals, especially aluminum.

Resistance Welding Most metals, particularly steels, have some resistance to electrical current flow. This characteristic can be used to heat the metal to welding temperatures so that the parts to be joined can be forced together by mechanical pressure completing the weld. No externally supplied filler material is required and the welding process does not require any shield gas. One popular and versatile resistance process is spot welding (Figure 46). Workpieces are clamped between two small-diameter electrodes. The current flow heats the metal to a fusion temperature and the mechanical pressure of the electrodes forges the material together.

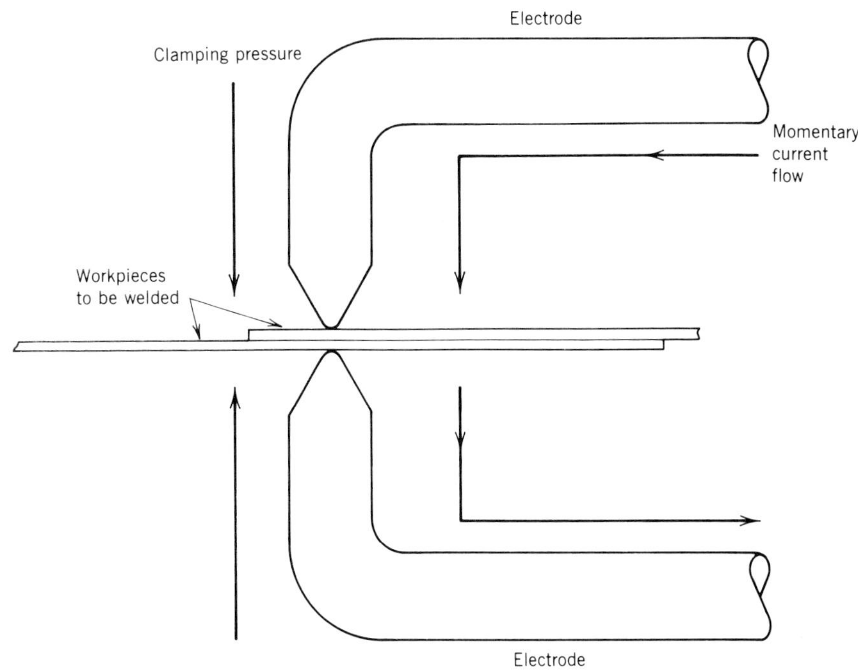

Figure 46
Spot welding.

Figure 47
Industrial robot spot welders on each side of the production line efficiently weld auto bodies with great speed and reliability (GM—Pontiac Division).

Since the process takes place in a small-diameter area, usually several spot welds are necessary to ensure adequate strength in large fabrications. Water cooling is used to cool spot welding electrodes.

The **spot welding process** is almost instantaneous and is therefore highly suited to volume production. Spot welding is widely used in the automobile industry and the nature of the process is such that it has been highly automated using industrial robots (Figure 47).

Microprocessor controls are used on spot welding equipment to control current and the length of time that the welding electrode is in contact with the workpiece.

Resistance welding is also used to join the ends of bandsaw blades, in the process called **resistance flash welding.** The process is similar to spot welding in that the heated material is mechanically pressed together after being heated to fusion temperature (Figure 48).

Figure 48
Flash butt welding a band saw blade.

Figure 49
Roller electrode resistance welding.

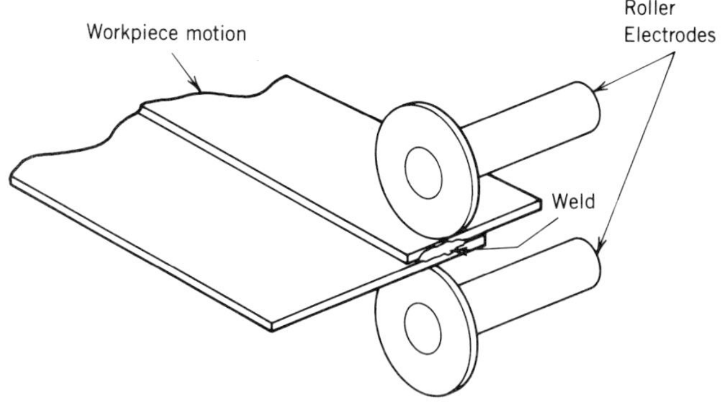

A bandsaw weld will require annealing (softening) after the welding process so that the band will have the flexibility required in service. Bars and rods may be welded in a similar fashion in the process of butt flash welding.

Resistance welding can be applied to seams and long joints by using roller electrodes. The process called **resistance seam welding** is used in the manufacture of pipe and tube (Figure 49).

Other Welding Processes

Other common welding processes include the following:

1. Thermit.
2. Forge.
3. Electron beam.
4. Laser.
5. Ultrasonic.
6. Induction.
8. Submerged arc.
9. Plasma arc.
10. Stud welding.

Thermit Welding In the thermit process, parts to be joined are submerged in a bath of molten metal that is sufficiently hot to heat the parts to be welded to a fusion temperature. The thermit mixture consists of metallic oxide of the metal to be welded and another metal that will achieve a very high temperature when ignited. Aluminum or magnesium is often used in the thermit process because of its ability to burn at a high temperature.

The thermit weld must take place in a mold so that the molten metal will be retained as it solidifies. In this respect thermit welding is somewhat like casting the superheated metal around the parts to be welded (Figure 50). Sand may be used for a thermit mold. Parts to be welded must be cleaned and correctly aligned. Preheating is often required. An example of a thermit welding application is the joining of railway rail ends so that the rail becomes continuous. This makes for a much smoother roadbed.

Forge Welding Forge welding (Figure 51) is probably best known from the days of the village blacksmith. Parts to be welded are heated in a forge consisting of charcoal or coal which is ignited. The fire is forced-air fed by a bellows or a motor driven fan under the coal bed to increase the temperature. Parts to be welded are placed in the fire and are heated to a point slightly below melting. The parts are then removed from the fire and placed in contact. Pressure is applied by hammering, in the case of the blacksmith, or by mechanical pressing or rolling in more modern techniques. Assuming correct heating, a weld occurs.

Electron Beam Electron beam welding (Figure 52) is a high-technology process that has found wide application in modern industry. Heat for the process is achieved by allowing an intense beam of electrons generated by large potential differences to impinge on the

Figure 50
Thermit welding.

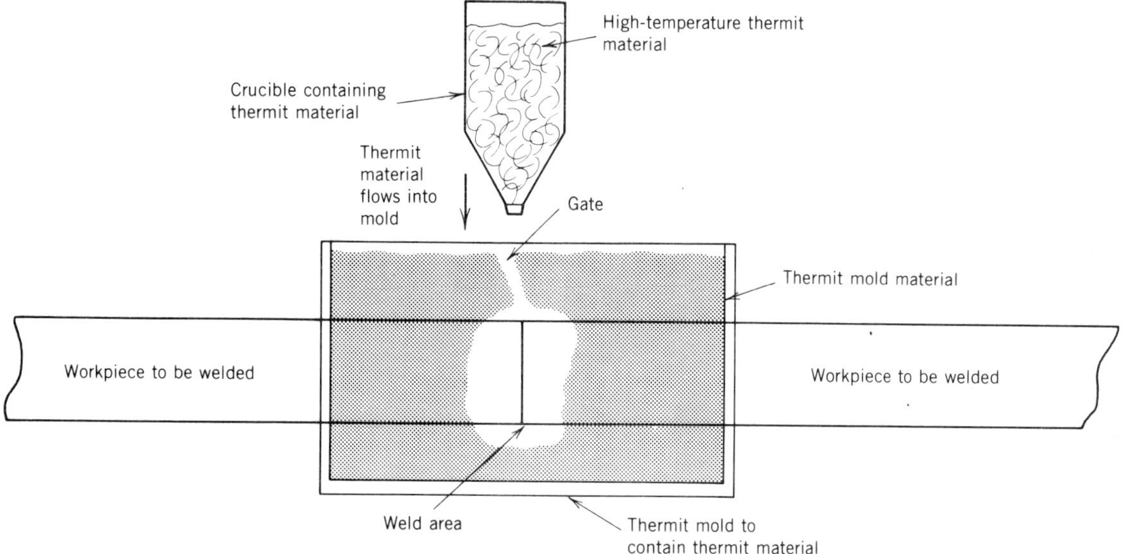

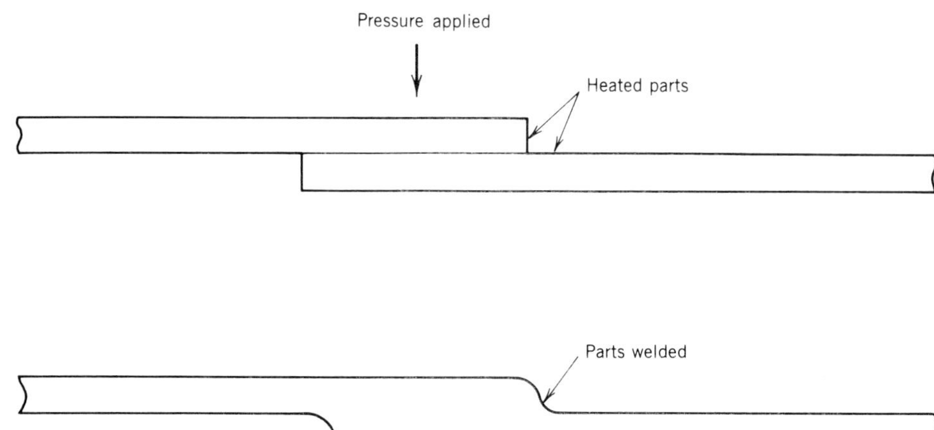

Figure 51
Forge welding.

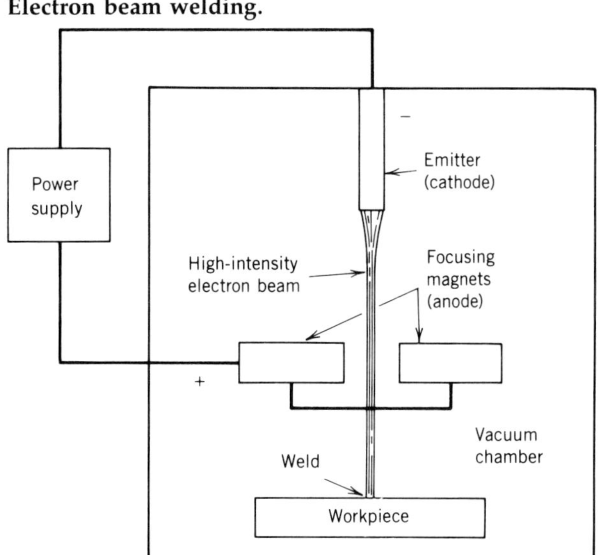

Figure 52
Electron beam welding.

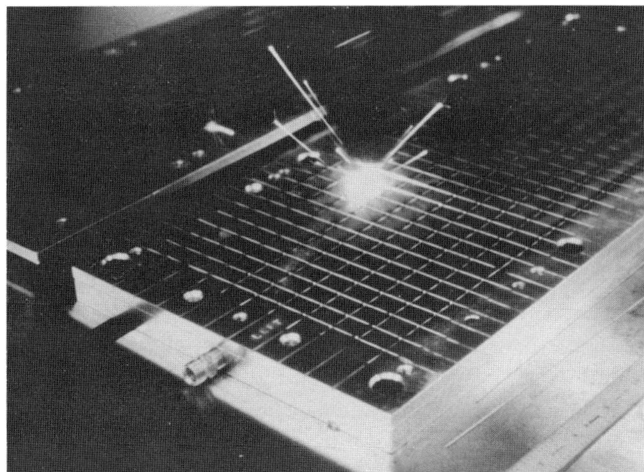

Figure 53
Laser welding (Laserdyne).

parts to be welded. Electron beam welding is accomplished in a vacuum. The process is highly suited to precision welding on parts that have already been final machined. Since the weld depth and location can be precisely controlled, this process has wide application in joining precision assemblies with a minimum of heat distortion. The electron beam process is also used for hardfacing teeth on bandsaw blades, thus permitting the manufacture of superior cutting bands for sawing difficult-to-cut materials.

Laser Welding The laser, discussed in the previous chapter, is an intense beam of coherent light. It has not only found application in joining processes, but has found many other applications in modern industrial processes as well. **LASER** is an acronym for **Light Amplification by Stimulated Emission of Radiation.** Laser light is generated by particular agitation of atoms or molecules in other gasses such as carbon dioxide and neon or crystals such as rubies. The light emitted from the gas or crystal laser is single color and in an extremely tight or coherent beam. The coherent beam of a high-power laser can be extremely intense. This property along with the coherent property is what makes the laser so useful for industrial applications.

In laser welding (Figure 53), heat is generated by directing the laser beam onto the parts to welded. The beam is concentrated and can also be applied in areas of extremely small diameter. Precision welds (Figure 54) of excellent penetration are possible on very thin materials (.001 in. thick).

Related in many ways to welding of metals, the ability of the laser to seal other materials such as human tissue has led to its many amazing applications in med-

Figure 54
Laser welding results in excellent penetration and little distortion of surrounding material (Photo courtesy of Coherent Inc., Palo Alto, California).

icine. The lasers can also be used to cut (Figure 55), drill, plate, heat for selective hard facing, and etch markings on parts.

Ultrasonic Welding Heat for ultrasonic welding (Figure 56) is generated by subjecting the parts to be welded to low-end radio frequency energy in the ultrasonic range. The ultrasonic energy causes the parts to heat as they are forced together by mechanical pressure. The process has applications for precision welding in a variety of materials.

Figure 55
Laser cutting (Photo courtesy of Coherent Inc., Palo Alto, California).

Figure 56
Ultrasonic welding.

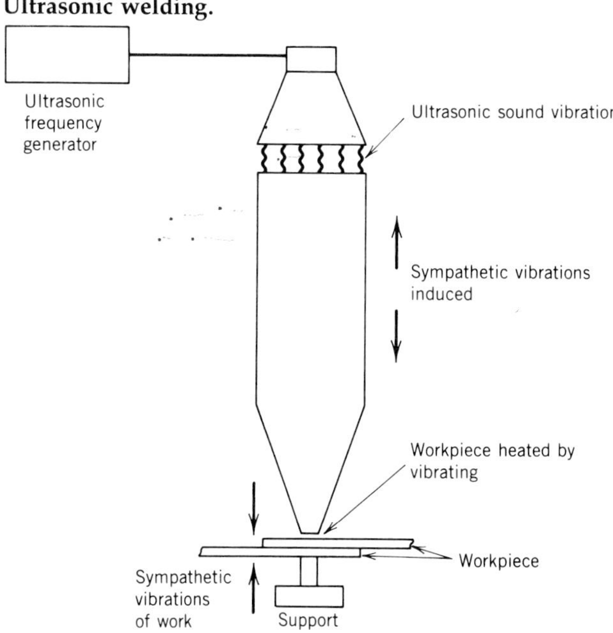

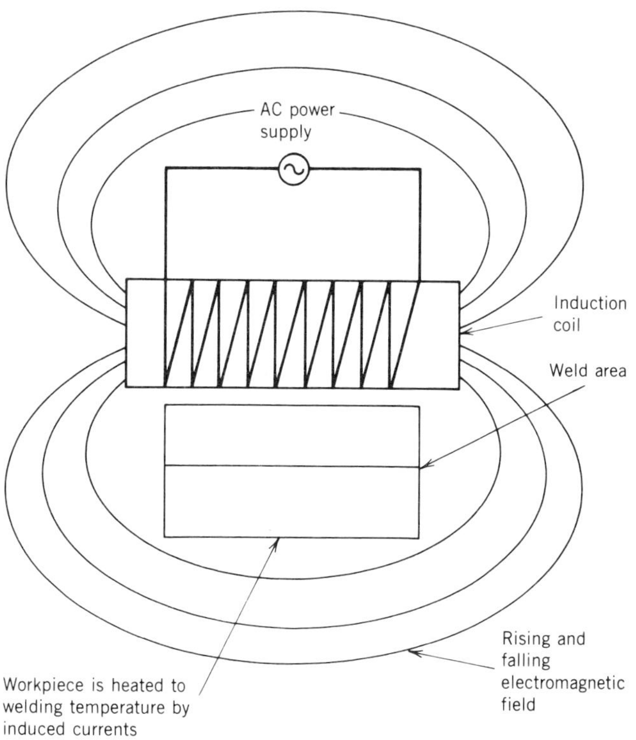

Figure 57
Induction welding.

Induction Welding Heat generated for induction welding (Figure 57) is accomplished by inducing a rapidly oscillating electromagnetic field in the parts to be welded. Parts may be placed in an induction coil field and induced current flowing in the workpiece as the field rises and falls in the induction coil causes rapid heating. Induction heating for welding is useful for production work and it is used as a heating method for brazing.

Submerged Arc Welding Submerged arc welding (Figure 58) is a shielded electric arc process. Instead of using a shield gas as in MIG or TIG, the welding process takes place buried under a fluxing agent. The fluxing agent acts in much the same way as a shield gas, but it is more efficient, resulting in a better quality weld (Figure 59). Submerged arc welding equipment consists of the wire electrode feed mechanisms, the flux tank, and the power supply. The flux tank and wire feeder may be portable, and in the field they can be connected to the power supply by cables (Figure 60).

The flux tank is air pressurized to provide the flux feed. The process is suited to automatic welding of major structurals and pipe, but it is limited to flat position welding so that the fluxing agent will remain in the weld area.

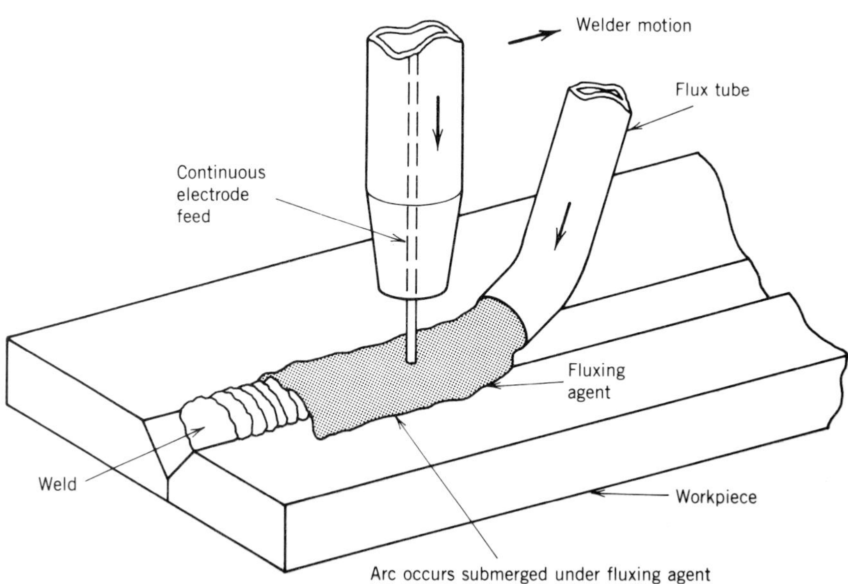

Figure 58
Submerged arc welding.

Figure 59
Submerged arc results in a weld of exceptional quality (Miller Electric Manufacturing Company).

Figure 60
Submerged arc welding equipment includes the wire feeder and the flux tank (Miller Electric Manufacturing Company).

Plasma Arc Heat for the plasma arc process (Figure 61) is derived by ionizing (creating an imbalanced electric charge) a gas by an electric arc. The resulting concentrated plasma arc can deliver a large amount of heat to the weld zone. Plasma arc welding is fast and clean and its arc concentration properties permit quality welds with minimal effect of the surrounding area. The process is suitable for welding many materials and may be used for cutting. Plasma arcs can generate temperatures of over 40,000° F, much higher than can be achieved by any fuel combustion process. Additional uses of extremely high temperature plasma arcs are presently being investigated and applied in many other areas of industrial processing.

Stud Welding A stud is a mechanical threaded fastener, usually with threads on both ends. Studs used in stud welding have threads only on one end with the other end being welded to the mounting surface. Stud welding is a very convenient method to attach a threaded fastener to a flat surface. Nonthreaded fasteners may also be installed by stud welding. These can be used as hooks or clips to which something may be attached.

In the stud welding process, an arc is established between the stud and the mounting surface. The stud

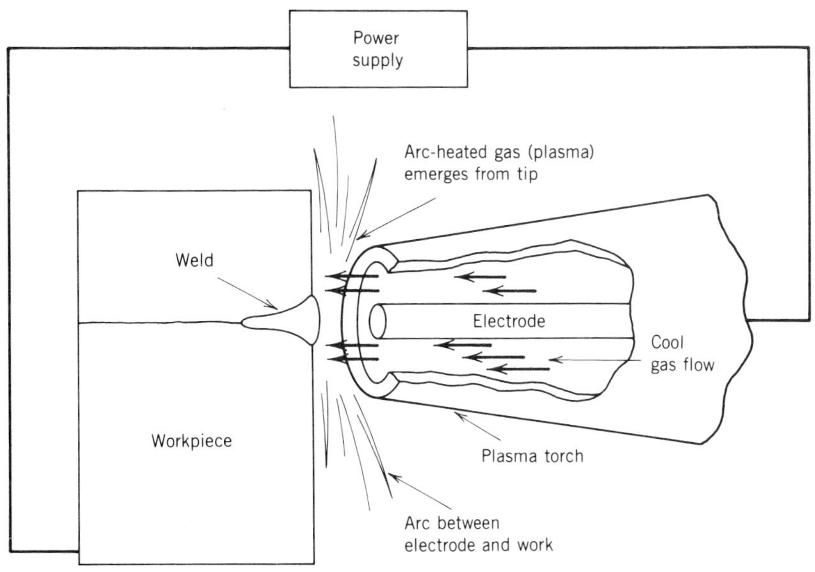

Figure 61
Plasma arc welding.

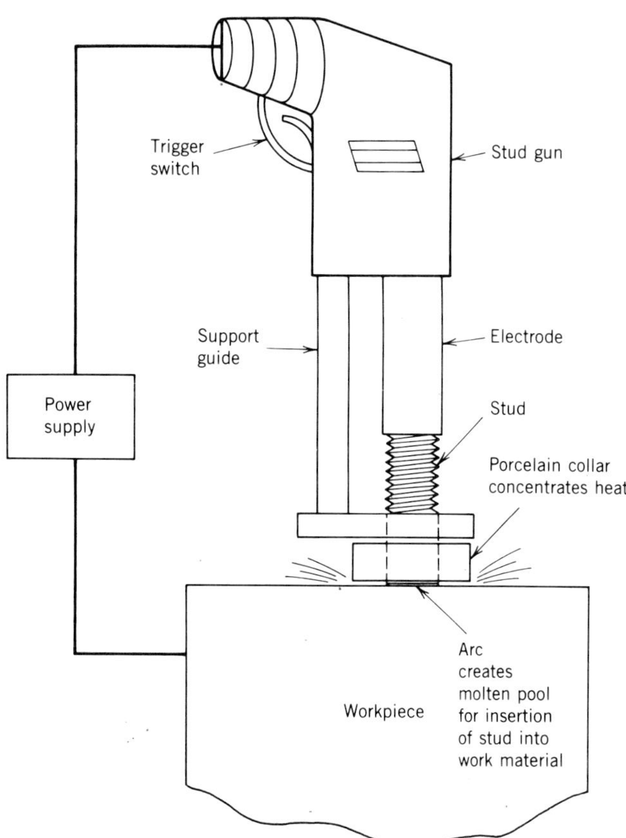

Figure 62
Stud welding.

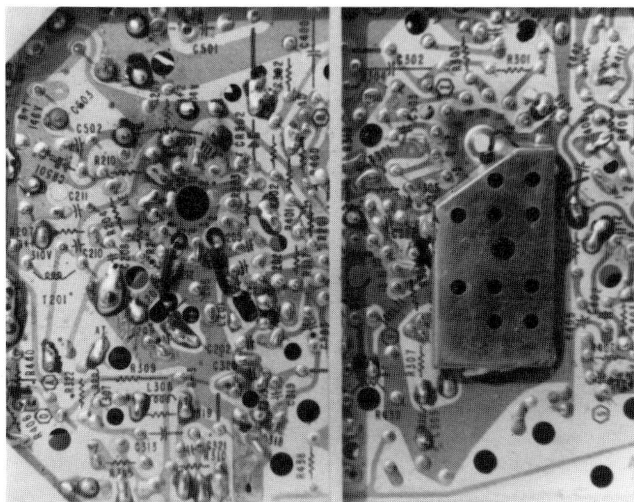

Figure 63
Electronic printed circuit card soldering using soft lead–tin alloy solders.

is secured in a stud gun (Figure 62) and a ceramic ring surrounds the base where the weld will occur. This concentrates the heat of the weld and will be broken off after the weld is completed. When the stud gun is energized, arc heats the base metal and mechanical pressure forces the stud in the molten pool of metal, accomplishing the weld. The process takes place almost instantly. Studs of many different sizes and styles may be installed by this fast and efficient process.

SOLDERING AND BRAZING

In the previously discussed welding processes, it is necessary to heat the parts to be joined to a temperature where they will fuse into a common mass across the weld zone. If filler material is used, it too becomes part of the common mass. In the processes of brazing and soldering, the filler metal used melts at a much lower temperature. It does not actually become part of the common material in the weld. The joining of the material is purely mechanical in that the filler metal grips the parent metal purely by mechanical attraction.

Soldering and brazing do not provide as strong a mechanical joint as do the fusion welding processes. Nonetheless they have wide application as efficient material joining processes in manufacturing. It is important that these processes be applied only where soldering or brazing is a suitable joining method. They must not be used in applications where strength requirements necessitate the complete fusion bond accomplished in the welding processes. When properly done, solder joints can actually be stronger than the parent metal when a tension load is applied.

Typical soft solders are made from nonferrous metals. Lead and tin are very popular metals for soft solders. These will melt at temperatures below 800° F. Soft solders are used in joining sheet metal requiring low to medium strength but leak-proof joints. Soft solders are used extensively in the electronics industry for PC card soldering (Figure 63). The soft solder alloys have excellent electrical conductivity and lend themselves to automated soldering applications. These solders can be melted with electric soldering irons or low-temperature air–fuel gas torches.

Hard solders, those made from silver–tin alloys, are used where a high degree of mechanical strength is required. These solders are considerably more expensive and require more heat for applications. They may be done with oxy-fuel gas torches or assorted furnace equipment with the proper heating capability. Hard silver solders are acceptable for joint soldering in high pressure piping systems. These solders will melt above 800° F.

Both soft and hard solders require proper joint preparation and material cleaning to ensure properly soldered joints. Fluxes are almost always used to aid in cleaning during the soldering process. Soldering can be applied to many materials; however, there are some metals for which soldering is unsatisfactory.

Brazing and Braze Welding

These joining processes use higher temperature filler metals. The same joint preparation and fluxing that is required in soldering is required in brazing and braze welding. Heat sources are usually torches and in some cases furnace brazing may be done, especially on a production basis. Joint strength is much higher with the brazing and braze welding process than with soft soldering processes.

Figure 64
Flame spraying.

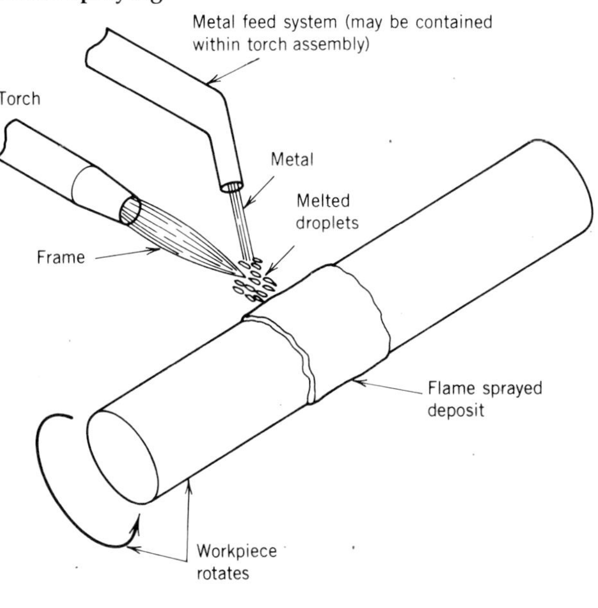

Hard Facing, Flame Spraying, and Metallizing

Flame spraying or metallizing is a process of building a layer of filler metal onto the surface of a cylindrical part. The process is extremely useful in repair applications where the wear surface or bearing surface can be rebuilt and then remachined, thus saving complete replacement of the component.

Another application of these processes is to provide a hard wear surface that will withstand abrasion. Examples include hard facing of plow and disk teeth on farm equipment to reduce the wear caused by rock and dirt abrasion. Several of the previously discussed welding processes can be used to hard surface or hard face parts. Several electric welding electrodes have been developed for this application.

In the processes of metallizing and flame spraying (Figure 64), small droplets of molten metal are carried by the flow of the flame and deposited on the workpiece, where they adhere. The plasma arc may also be used in this process. A layer of filler material may be built up by this process. If the part is rotated during the flame spraying process, buildup may occur evenly on a cylindrical object. Mechanical bonding at the interface between the part and the sprayed-metal buildup is increased by machining a rough surface on the part to be built up. This makes the process very useful for rehabilitating a worn bearing surface. The part may be remachined after cooling to restore its dimensional specifications.

PLASTIC WELDING

Plastics may be welded by heat, ultrasonic sound, and solvents. Heat welding may be accomplished by contacting the plastic material with a heated tool. The material melts and runs together, fusing as it cools. Heated gasses may also be used. Ultrasonic sound may be applied in plastic welding much as it is in metal welding, where the heat is derived from ultrasonic vibrations (Figure 65).

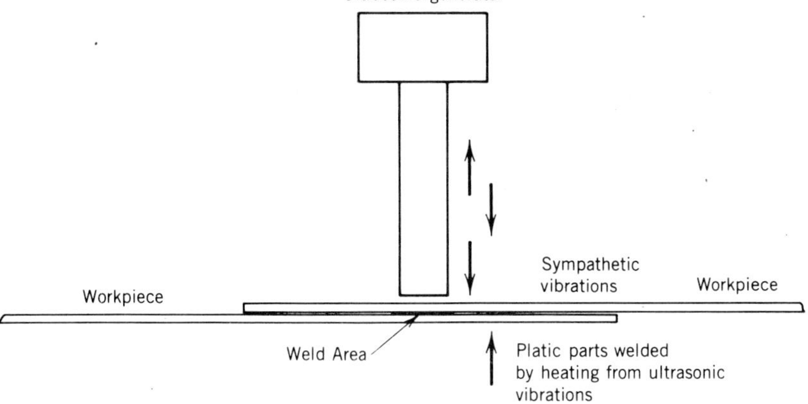

Figure 65
Welding plastic with ultrasonic sound.

Solvents that dissolve the plastic material may be used in plastic joining. The plastic surface in the joint to be welded is rendered into a liquid state by the application of appropriate solvents. After the parts are joined and the solvents have evaporated, the joint becomes homogeneous as the parent material.

REVIEW QUESTIONS

1. Name three mechanical fastening methods.
2. Discuss the relative advantages and disadvantages of common mechanical fasteners.
3. An assembly must be taken apart often for maintenance. Which mechanical fastening system would be the best for this application?
4. Where would you most likely find that stitching and tying and snaps used as a material joining process?
5. What are the advantages of adhesive bonding?
6. What are the disadvantages of adhesive bonding?
7. In what industry would you likely find adhesive bonding and why?
8. List three plastic resins used in adhesive bonding.
9. Describe the inspection method used in the aircraft industry to check for defects in bonded joints.
10. List four common electric welding processes.
11. What is GMAW or MIG?
12. What are the advantages of GMAW?
13. What is GTAW or TIG?
14. How does GTAW differ from MIG?
15. Describe rod electrode welding and discuss its applications.
16. What is stud welding?
17. What is the function of the shield gas in shielded arc welding?
18. In the MIG process, how is the electrode delivered to the weld area?
19. What material is used for TIG electrodes?
20. If welding were needed where no access to utility company electric service was available, what welding processes could be used?
21. What is the resistance in resistance welding and what does it have to do with the process?
22. Describe laser welding and discuss its advantages.
23. Describe the thermit welding process.
24. What conditions are necessary for electron beam welding?
25. Describe the submerged arc process.
26. How does soldering and brazing differ from welding?
27. What are the constituents of soft solders?
28. What are the constituents of hard solders?
29. Discuss where flame spraying, metallizing, and hard facing might be used.
30. How can plastic be welded?

CASE PROBLEMS

Case 1
Welding Thin Materials

A new business begins to manufacture patio furniture made from thin-wall steel tubing. Rod arc welding is initially employed, but metal burn through occurs immediately. The production supervisor, an experienced welder, recommends that rod arc welding be replaced with MIG. The vice president of manufacturing suggests that this would be expensive and the problem could be just as easily solved by further training of the production welders in order to improve their welding skills. In your opinion, who is right and what would you recommend in this case?

Case 2
Collapsing Furniture

The Super Cut Rate Wood Furniture Manufacturing Co. is receiving customer complaints that some of its furniture is falling apart soon after purchase. Super Cut Rate sends its director of quality control to check on the problem. The QC director finds that on the furniture assembled by high production stapling, the fasteners are pulling out when the furniture is subjected to normal-use loads. A new method of fastening must be used. Considering the primary factors of high production of a low cost product, screws, bolts and nuts, and gluing are ruled out as being too expensive and labor intensive. What other methods might be employed to solve this production problem?

CHAPTER 15 PLASTICS AND COMPOSITES PROCESSING

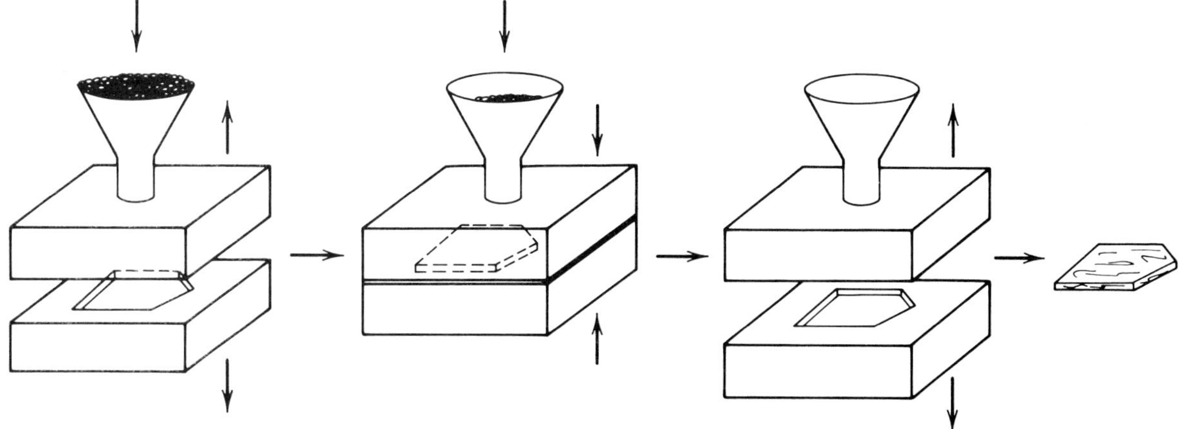

Product performance and manufacturing costs are major factors in motivating the designer to seek new product materials and new manufacturing methods to process them. For example, fuel efficiency in both automobiles and aircraft has become an important consideration in recent years. To meet these requirements, the designer has turned to nonmetallic materials that demonstrate both light weight and strength. Numerous other advantages in using nonmetallic material such as plastics and advanced composites include resistance to rust and other corrosion, denting, chipping, changes in temperature, and ease of manufacturing.

The purpose of this chapter is to survey the major methods of processing these materials.

OBJECTIVES

This chapter should enable you to:

1. Identify common methods of processing plastics and composites.
2. Describe in general terms how the processes work.

THE WORLD OF PLASTIC AND COMPOSITE PRODUCTS

It would be difficult to envision the modern world without plastics and composites. The product line of these materials is vast and expanding every day. Although metals will always remain in widespread use, plastics and composites are fast replacing metals in a vast array of manufactured products.

Reasons for this include the often high cost of metals and the large energy requirements in metal processing, the ease by which plastics and composites may be processed compared to metals, and the high strength and light weight of plastic and composite materials permitting their use in applications where previous designs required metals.

In recent years, engineering development into plastic and composite materials has yielded products that are equal or better than their metal counterparts in every way. One significant factor is that of weight. Plastic and composite structures are generally lighter than their equivalent metal counterparts; they have become very popular in product applications where weight is a significant factor. Lighter weight vehicles both in the automotive and aircraft fields have generally increased fuel efficiency. Plastic and composite materials have therefore generated much interest as materials for these applications.

Composite engineering has yielded structures that meet design strength requirements, thus permitting extensive use of these materials in modern jet aircraft (Figure 1). The illustration shows a prototype aircraft in which the fuselage and other major structural components are completely made from composite materials. When this aircraft is commercially produced, and powered by the efficient turbojet engine pusher propellers (Figure 2), it will have the speed and altitude capability of the pure jet, while at the same time it will also demonstrate very high fuel efficiency because of its light weight.

Advanced composite as well as plastic materials have also been used for major structural components on full sized commercial and military jet aircraft (Figure 3). The advanced composite vertical stabilizer fin in the

Figure 1
Gates–Piaggio truboprop aircraft is an example of the use of composite material for structural aircraft components (Gates Learjet).

illustration weighs 25 percent less than its aluminum counterpart. Several aircraft manufacturers are building entire major structural components such as fuselages and wing structures from these advanced materials. The weight saving in aircraft and auto structures will play an important part in the fuel efficiency ratings of these products.

Another common example is the use of plastic material such as laminated products as a very durable substitute for wood. These materials are easily fabricated, easily cleaned and cared for, and can be made very thin and therefore very light while still retaining the beauty of wood grain designs. Figure 4 shows plastic laminate material as a wood substitute in aircraft cabinetry.

Since plastic and composites are generally in a liquid

Figure 2
High strength but light weight plastic and composite materials greatly enhance fuel efficiency in aircraft (Gates Learjet).

Figure 3
The vertical tail fin of the Lockheed L1011, a full-sized commercial jet aircraft. This component is 25 percent lighter than its aluminum counterpart (Lockheed–California Company).

or softened state prior to their formation into products, these materials are highly suited to manufacturing one-piece complex shapes such as automotive body components. The truck and bus hood and fender components seen in Figure 5 are made from reinforced plastics and weigh less than 100 lbs. They are also resistant to corrosion, impact damage, and they are unaffected by variations in temperature.

Figure 4
Plastic laminates simulate the look of wood grain while taking advantage of the properties of plastics (Lockheed–California Company).

Figure 5
Plastic and composite materials are highly suited to the manufacturing of one-piece complex-shaped products (Goodyear Tire and Rubber Company).

PLASTICS AND COMPOSITES PROCESSING METHODS

So far in this text, you have studied many processes in which metals are processed into useful products. Many of the same manufacturing process principles are applied in the processing of plastic and composite materials. These materials may be cast, molded, extruded, machined, fabricated, laminated, and rolled. The equipment used for processing plastic and composite material is specifically designed to process these materials.

Three major methods make up a large portion of plastic and composite processing techniques. These include blow molding, injection molding, and extrusion.

Blow Molding

In blow molding, air is used to force a mass of molten plastic against the sides of a mold shaped in the form of the desired end product (Figure 6). The sequence of illustrations (Figure 7 *a,b,c,* and *d*) shows a common and versatile application of the blow molding process for the production of refrigerator liners. This process is particularly well suited to the forming of large fabrications, such as the example shown. After cooling, the mold is opened and the finished part is removed.

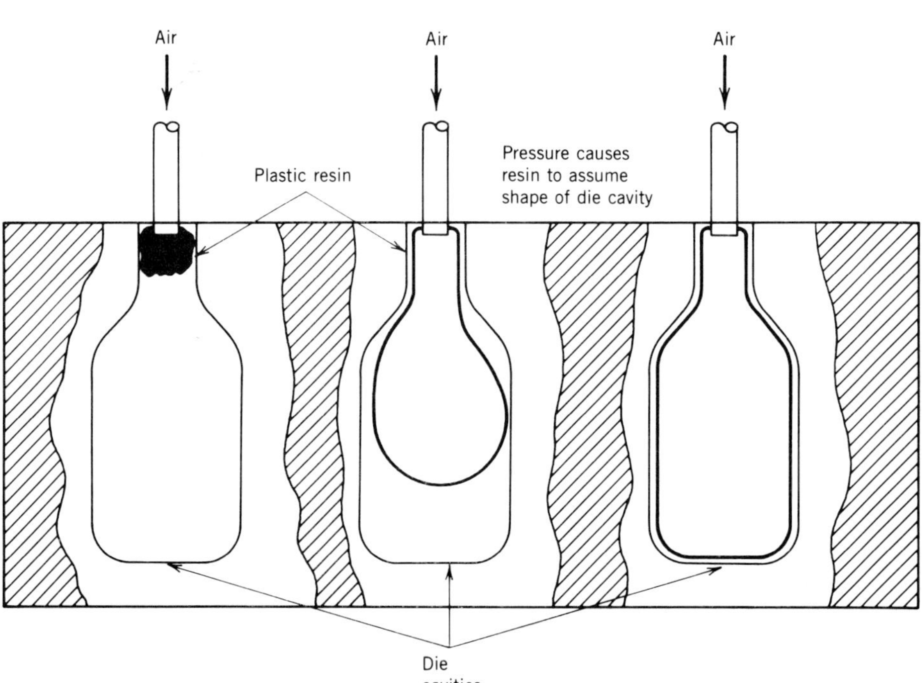

Figure 6
Blow molding process.

Figure 7
(a) Partially blow-molded refrigerator liner. The blow molding die is shown in the open position. *(b)* The liner begins to assume the shape of the die. *(c)* The completed liner just prior to removal from the blow molding die. *(d)* Completed refrigerator liners after removal from the blow molding die (all photos courtesy of Whirlpool Corporation).

(a)

(b)

(c)

(d)

This method can also be used for sheet forming where a sheet of plastic is clamped between a form and die. Air pressure forces the heated plastic against the die where it assumes the die shape.

Blow molding may be used when the end product must have an enclosed hollow internal shape. An example of this application is containers such as the plastic jugs illustrated in Figure 8. Machinery used in this process will produce units at a high rate of production. The dies used to shape the container are seen in Figure 9.

Related to this process is vacuum forming where a vacuum is drawn on one side of the material and air pressure on the opposite side forces the material against the mold or form.

Injection Molding

Injection molding is another versatile and widely used process. Examples of injection molded products are found everywhere. In this process, molten plastic is forced into a metal die cavity that has been machined to the shape of the desired end product. When the

Figure 8
Continuous production blow molding machine for plastic containers (Plastic Machinery Incorporated).

Figure 9
Blow molding machine for plastic containers showing die cavities (Plastics Machinery Inc.).

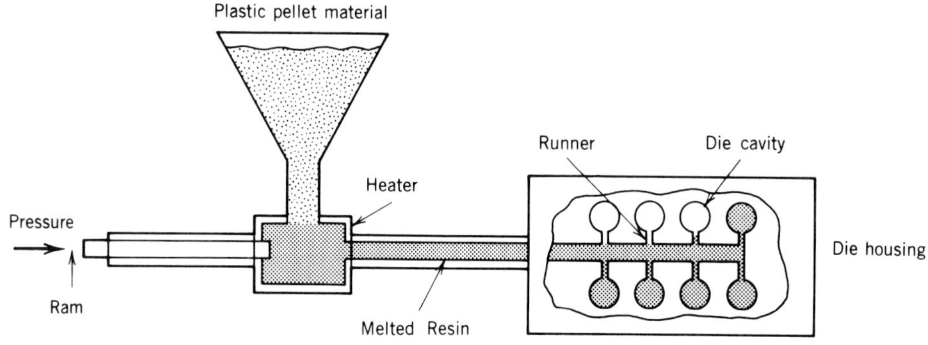

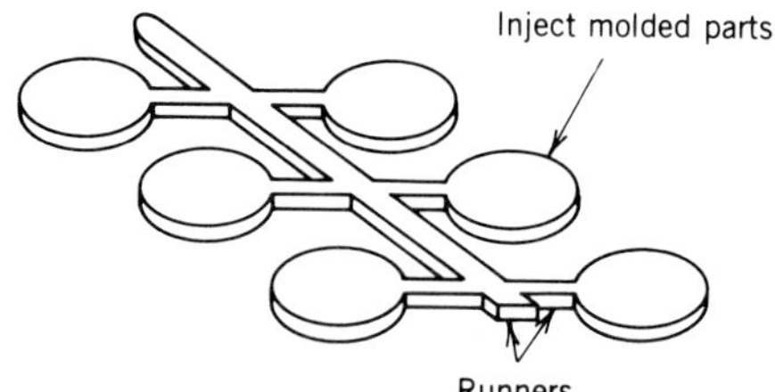

Figure 10
Injection molding process.

Figure 11
Cutaway view of an injection molding machine (Cincinnati Milacron Plastics Machinery Division).

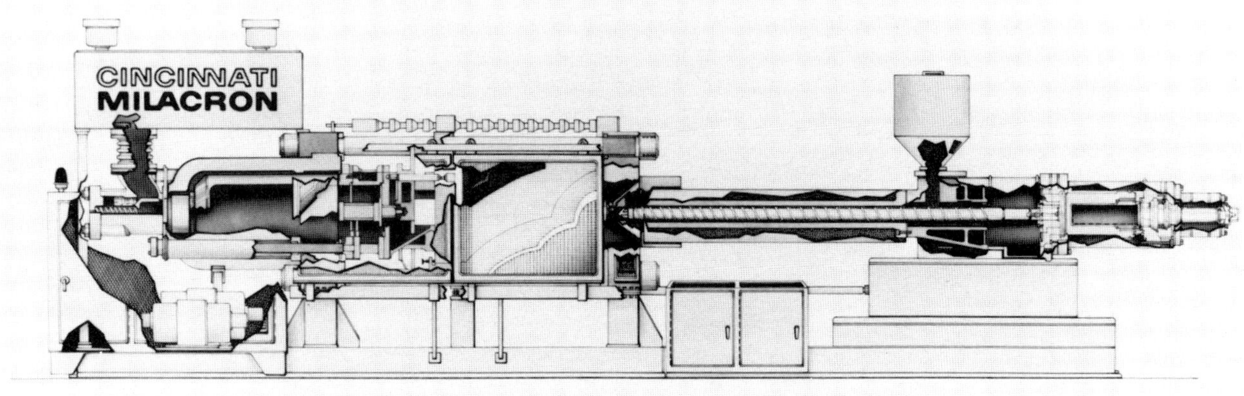

Figure 12
Plastic pellets, the raw material for injection molding and other plastic manufacturing processes (Whirlpool Corporation).

plastic material has solidified sufficiently, the die is opened and the part removed. Figure 10 shows a cutaway view of a plastic injection molding machine (Figure 11). Raw plastic material in small pellets (Figure 12) is placed in the machine hopper. The raw material then enters the heater where it is melted. Molten plastic is then pushed into the die cavity by hydraulic or direct mechanical pressure (Figure 13).

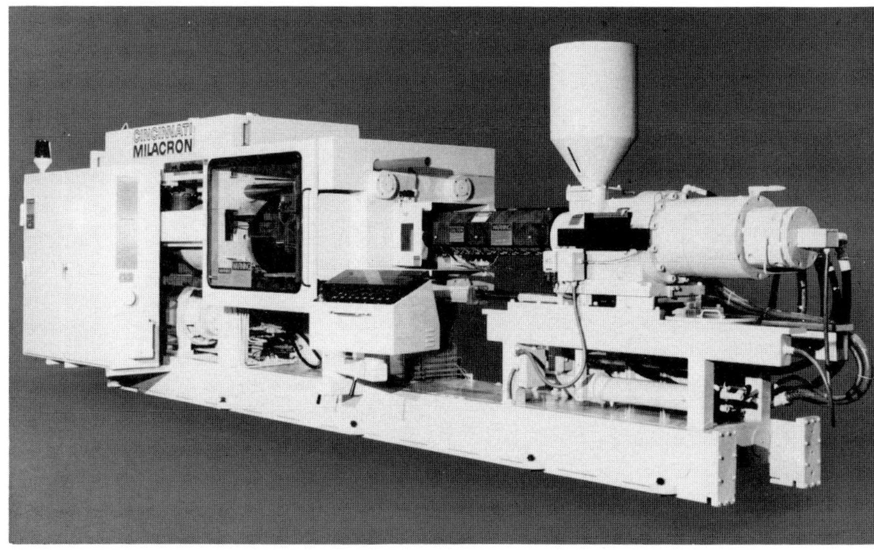

Figure 13
Injection molding machine, 250 ton capacity, with computer microprocessor controls (Cincinnati Milacron Plastics Machinery Division).

Large-capacity injection molding machines may exert several hundred tons of pressure and can be used to fabricate large one-piece plastic parts. Examples include automotive body components including fender/hood assemblies, bumpers, and grilles (Figures 14 and 15).

The injection molding process is extremely versatile for making parts with fine detail and complex shapes (Figure 16). Examples include high production of small precision detailed parts such as those for plastic models.

Reaction Injection Molding Reaction injection molding (**RIM**) is the latest injection molding technique presently in use. In this process, two base resins (monomers) are mixed together just as they enter the mold (Figure 17). A chemical reaction occurring at low heat takes place and the plastic material (polymer) of the end product is formed at that instant. The RIM process does not require the plastic materials to be heated before molding. Thus, the process is both energy efficient and very fast. Large parts weighing many pounds can be made in a few seconds and the process minimizes scrap.

Figure 14
One-piece plastic fender and cowel components made by injection molding (Goodyear Tire and Rubber Company).

Figure 15
Injection molding can produce intricate parts with extremely smooth and uniform surfaces such as these auto grille components (Goodyear Tire and Rubber Company).

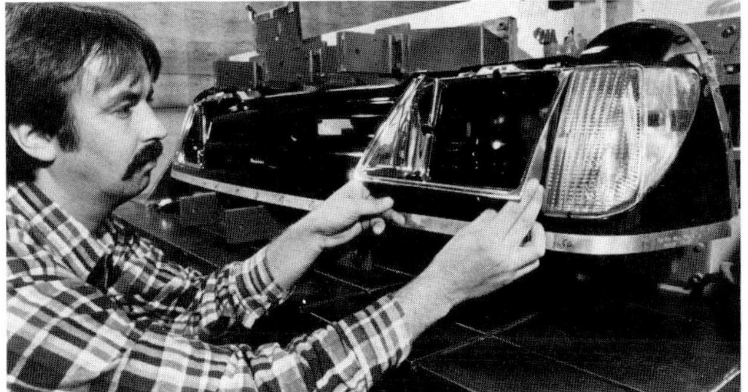

Figure 16
Complex shapes are easily produced by injection molding processes (Goodyear Tire and Rubber Company).

Figure 17
Reaction injection molding process (RIM) (Cincinnati Milacron Plastics Machinery Division).

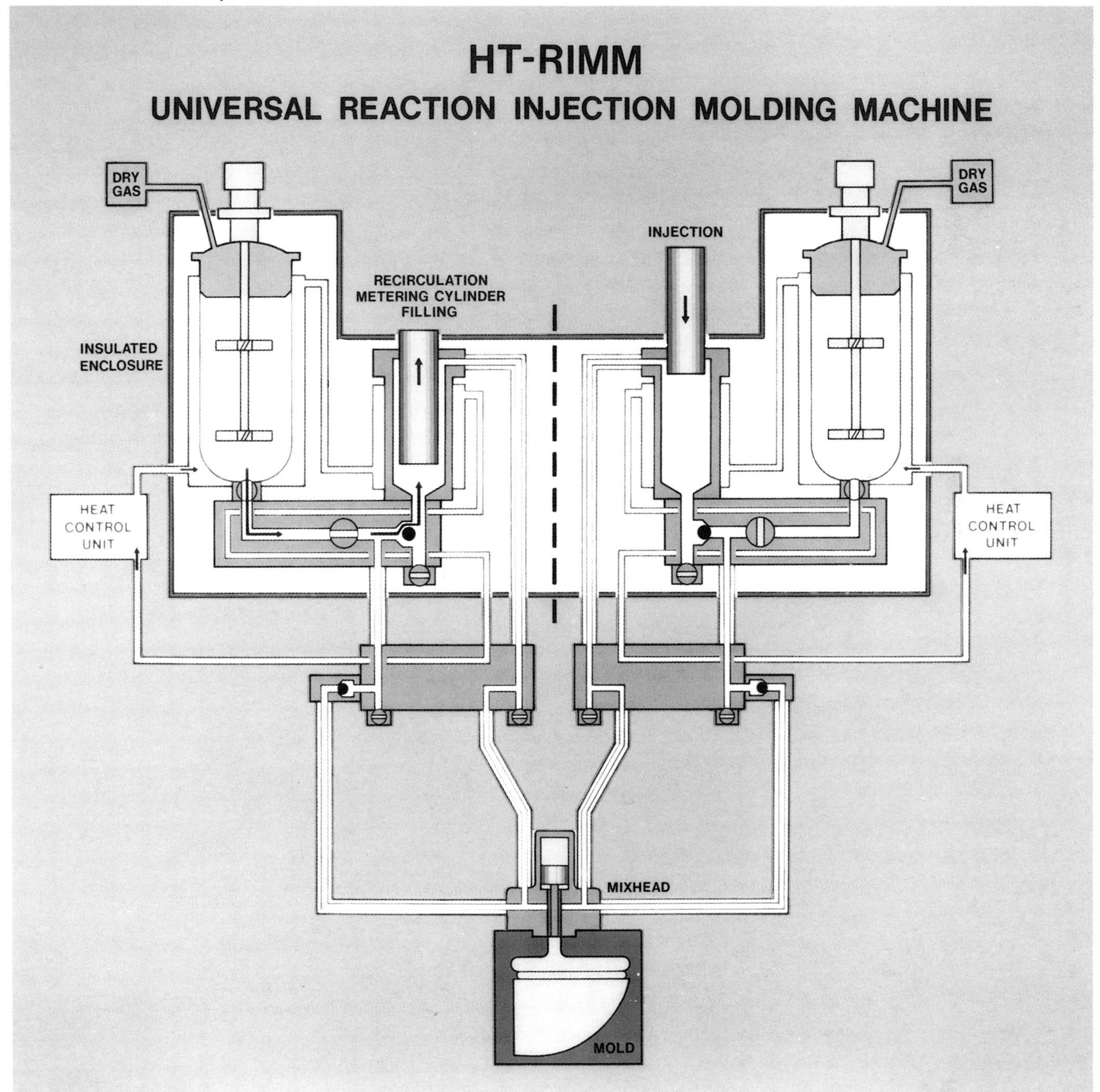

Figure 18
Platen type reaction injection molding machine (RIM) the latest technique in injection molding processes. This illustration shows the machine with essential safety guards removed so as to better show machine parts (Cincinnati Milacron Plastics Machinery Division).

Reaction injection molding machines (RIMM) may be of the platen design (Figure 18). The machine illustrated is seen with essential safety guards removed so that details of construction may be seen. This machine is highly suited for production of large parts such as auto bumpers and fenders.

Extrusion

Extrusion is another versatile and widely used plastic and composite processing method. In this process, plastic material is forced through an extrusion die that forms the end product (Figure 19). The major component of a plastic extruder is the extruder screw (Figure 20). The extruder may be of the twin screw design where two screws with variable lead sections are in mesh. Parallel screw extruders are also used. This system provides the mechanical force to push the plastic material through the extrusion die. Extrusion is the popular manufacturing process for products requiring long dimensions in one axis. Examples are plastic pipe and plastic molding.

The illustration (Figure 21) shows a plastic extruder for manufacturing plastic pipe. The specific tooling for this product is seen attached to the machine on the left side.

Other Molding Processes

Other processes that make use of molding principles include compression molding, transfer molding, rotational molding, and solvent molding.

Compression Molding This process is similar to injection molding, but it is primarily used with ther-

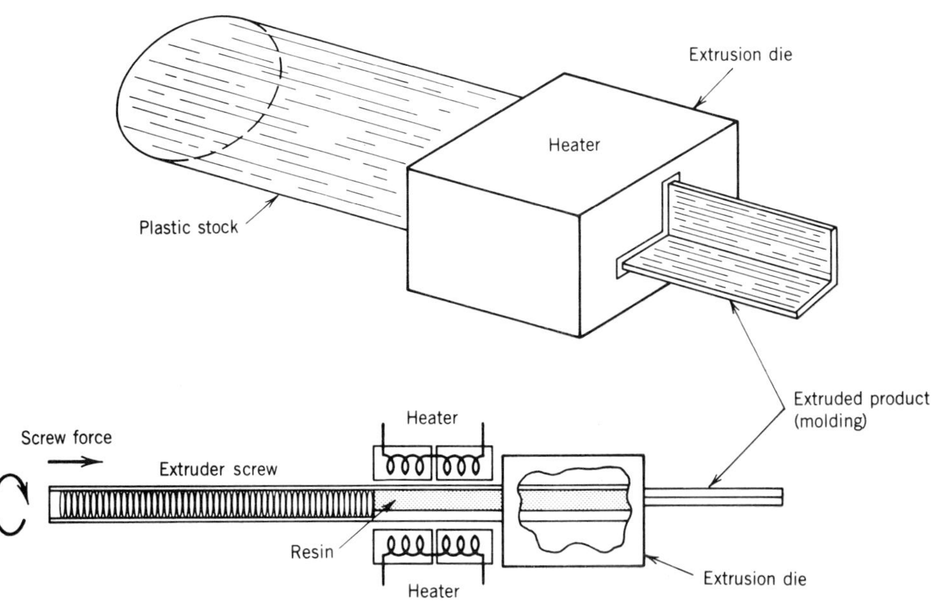

Figure 19
Extrusion process.

Figure 20
Twin-tapered extruder screws, uniquely engineered major components of a plastic extruder (Cincinnati Milacron Plastics Machinery Division).

mosetting resins. In compression molding, the resin is in pellet or dry powder form and may be mixed with a binder agent to provide strength. In this respect it can become a reinforced plastic material. The dry resin is mechanically compacted into a mold shaped to produce the desired end product. Heat and pressure are then applied to melt the resin and cause it to flow throughout the mold. A thermosetting plastic material undergoes a fundamental chemical change in this process and a permanently hardened part is produced. The process is used to make handles and knobs for kitchen utensils since thermosetting resins can be subjected to heat without becoming soft.

Transfer Molding Transfer molding is similar to compression molding except that the resin is heated to a liquid state before being forced into the mold. This process may be used where intricate-shaped parts are required or where metal inserts are molded into plastic. Metal parts that are positioned in the mold will remain in location while the liquid resin flows around them. When the plastic has solidified, inserted metal parts are securely held in place.

Rotational Molding In rotational molding, the mold cavity is spun and centrifugal force causes the resin to flow into the mold. Heating is required to place the resin in a liquid state. The process is suitable for large

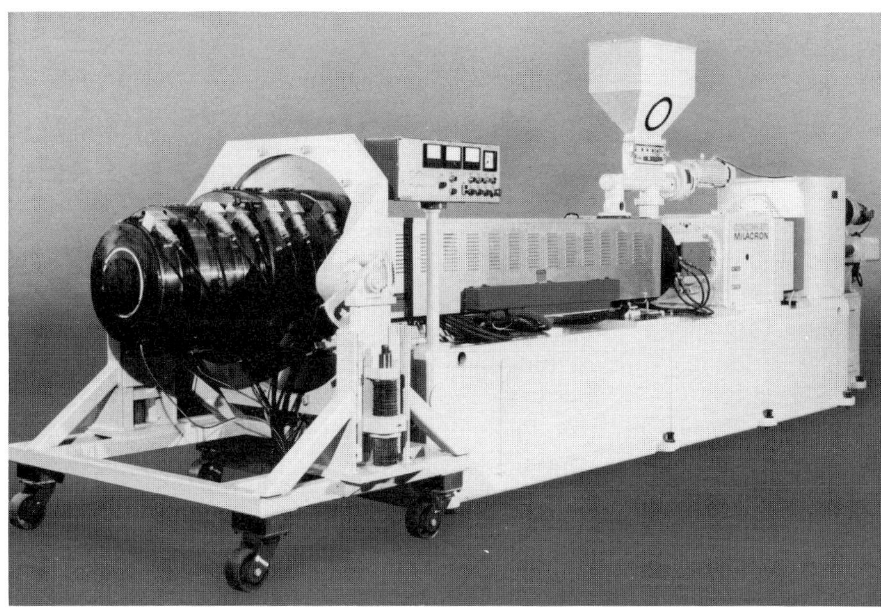

Figure 21
Plastic extrusion machine for plastic pipe. Tooling is seen attached to the machine on the left side (Cincinnati Milacron Plastics Machinery Division).

parts and will yield products having little internal stress and with even-wall thicknesses.

Solvent Molding In solvent molding, a mold of the desired end product shape is dipped into or filled with a liquid resin material. When the mold is withdrawn or emptied out, a thin layer or coating of resin adheres to the sides. When the solvent has evaporated, the formed part may be removed. Additional layers may be built up to increase the thickness of the end product and the process may also be used as a method to plastic-coat another part.

Other Plastic Processing Methods

Other plastic processing methods include casting where liquid resins are poured into molds. The resin contains a catalyst that causes the resin to harden after a short period of time. Resin casting is popular in the home-craft area and will yield many decorative products. The process can also be used to encase an object in a transparent block of resin. Color may be added to clear casting resins. The process must be done quickly or the catalyst will harden the resin before it can be poured. Casting can be used to make sheets, rods, tubes, special shapes, film, and sheeting from both thermoplastic and thermosetting resins.

Rigid plastic and composite materials can be effectively shaped by most of the common machining processes. The material can be milled, turned, sanded and ground, broached, reamed, drilled, and tapped. A number of plastics are readily available in bar stock form for processing by machining methods. Plastics suitable for machined products include nylon, acetyl (delrin), fluorocarbon (Teflon), ABS, and PVC. Common machined plastic products include gears, bearings, threaded fasteners, and bushings.

Processes used to join pieces of plastic in assemblies include welding by heat and ultrasound, gluing with solvents and other resin adhesives, sewing (sheet and film), and mechanical fasteners such as bolts and nuts, screws, and rivets.

Reinforced Plastic Materials

The mechanical properties of plastics can be greatly enhanced by adding a reinforcing agent to the resin. Plastic resins may be reinforced with **cloth, paper, glass fibers,** and with other fibers such as **graphite,** thus forming advanced **composite materials.** Reinforced plastics differ from composites in the way in which the reinforcing fibers are established in the base resin material. Fibers used in reinforced plastic are generally in short pieces and are randomly oriented within the resin. Composite fibers, on the other hand, **run unbroken** through the resin and provide the major structural component for the material.

Figure 22
Reinforcing resin with glass fibers in the process of fiber glassing, yields extremely durable structures such as boat hulls (Wellcraft Marine Corporation).

Figure 23
Glass fiber application (Wellcraft Marine Corporation).

Although reinforced plastics may not have quite the same strength and load carrying capability as their true composite counterparts, they are nonetheless excellent structural material for a variety of products.

A popular process and one with wide application is **fiber glassing,** in which alternate layers of glass fiber fabric and resin are coated over a mold or form built in the shape of the desired end product. Short pieces of glass fiber may also be mixed with the resin while it is still in a liquid state. A popular application of this

Figure 24
Smoothing fiber-glass reinforced resin in boat manufacturing (Wellcraft Marine Corporation).

Figure 25
Finishing a fiber-glass reinforced boat hull (Wellcraft Marine Corporation).

method is in manufacturing hulls for boats (Figure 22) and other large hollow fabrications such as swimming pools.

The process is accomplished by building up layers of fiber strands or sheet fabric (Figure 23) and resin. Pieces of glass fiber are laid over the desired form and the resin is applied by spray application or hand spreading (Figure 24). The liquid resin and fiber mixture may be troweled and smoothed while it is still in a liquid state. After hardening, the fiber glass fabrication may be refinished by mechanical abrasive processes such as sanding and buffing (Figure 25).

The end product of reinforced plastic is extremely strong, durable, and lightweight. It is also impact resistant and not subject to corrosion, and almost any shape can be fabricated given the proper tooling.

The process of **calendering** or **rolling** plastic material is used to form sheets and film. Plastic is considered to be film if its thickness is less than .010 in.; over this thickness, the material is considered to be sheet. Sheet and film are formed by pressing the material between heated rollers. Thickness is determined by the spacing between the calendering rolls. Plastic coatings are also applied to cloth by this method, forming the many useful plastic-coated fabrics appearing in a large number of consumer and industrial products.

Plastic **coatings** may also be brushed or sprayed on metal or wood to protect them from corrosion. Clear plastic coatings are used as wood and metal finishing mediums. Plastics are also used in paints, where color as well as durability properties of the plastic can make a superior coating medium. Examples include epoxy and polyurethane-based paints.

In the process of **high-pressure lamination,** which generally makes use of thermoplastic resins, sheets of the material are placed between heated steel platens which are then compressed with hydraulic pressure. The process bonds the layers of the laminate into a rigid sheet. Laminated sheets can also be formed around corners and then bonded to wood bases. The products resulting from this process are used extensively for kitchen counter tops. Very attractive products that simulate woods, but have the advantages of plastics, are created by this process.

MANUFACTURING AND PROCESSING COMPOSITE MATERIALS

Composite materials are resins that have been reinforced with organic or inorganic materials. Composites differ from reinforced plastics in that the fiber structure is **continuous** throughout the structure. It is this particular structural design that gives the composite its superior mechanical properties. An example of a composite material is a plastic such as **epoxy resin** reinforced with **graphite (carbon) fibers.** These materials have many unique properties, especially their strength-to-weight characteristics, resistance to corrosion, resistance to impact, and their ability to be formed into complex shapes by many of the previously discussed processes. Composite materials are presently undergoing extensive development and their use in automotive and aerospace applications is already widespread and will be expanded in the future. Both thermoplastics and thermosetting plastics are used in advanced composite materials. Research has shown that the thermosetting plastic will probably be more popular for use in advanced composites. The reason for this is that many applications of advanced composites required the product to be subjected to considerable heat. The thermosetting plastics are more suitable to these applications.

Advanced composites consist of the plastic material reinforced with some type of stranded fiber material. The fiber reinforcement material usually makes up about half of the total material weight. Examples of the fiber materials used in composites are **graphite** and **glass.** The fiber functions as a structural component of the composite and is designed to take the load stresses applied to the composite structure. Were it not for the fiber portion of the composite material, only the resin portion of the material would be subjected to the applied loads. This structure alone would not be able to withstand the forces applied and would fail in service (Figure 26).

Composite Manufacturing Methods

Three major techniques of manufacturing composite material are in common use. The first process is a reverse extrusion process or **pulltrusion** (Figure 27) in which the fiber portion is pulled or drawn through the liquid resin and then through a heated die that forms the desired shape. The process is much like extrusion except for the pulling rather than pushing of the material through the extrusion die. Composite products manufactured by pulltrusion methods include structural members and tube.

In the technique of **filament winding** (Figure 28), the fiber is wound back and forth on a cylindrical form. This method is used to produce cylindrically shaped products such as tanks or other pressure vessels. After curing, the form is removed leaving the hollow composite product.

Laminating alternate layers of resin containing the structural fiber is the third major method of composite manufacturing (Figure 29). This process is similar to the fiber-glassing technique previously discussed. However, the composite fibers are continuous throughout the material, whereas in fiber glass short pieces of glass fiber are randomly situated in the resin structure.

COMPOSITE APPLICATIONS

As composite materials undergo further development, their uses will extend to a larger product line. In the beginning of this chapter, the applications in aircraft were briefly mentioned. In this application, the design engineer seeks materials with favorable strength-to-weight ratios. Strength coupled with light weight can result in products that are less heavy and thus require

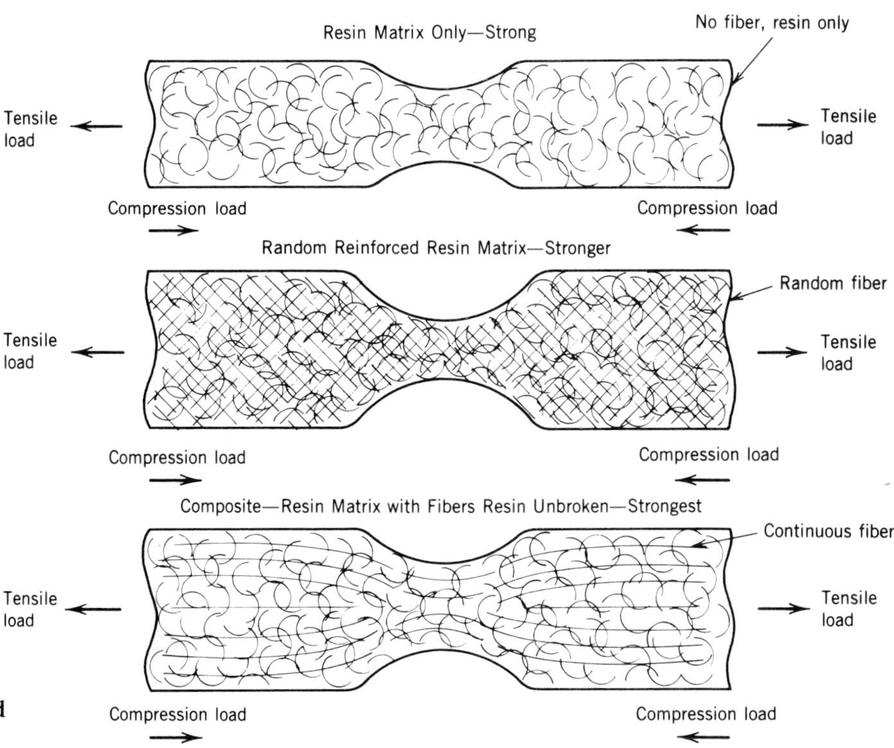

Figure 26
Load applications to resins, reinforced plastics, and composites.

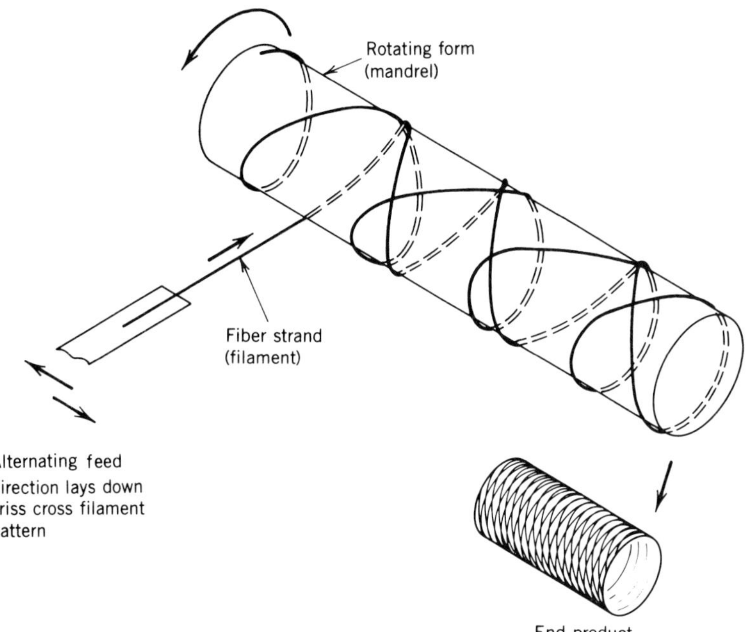

Figure 27
Manufacturing composites by the pulltrusion process.

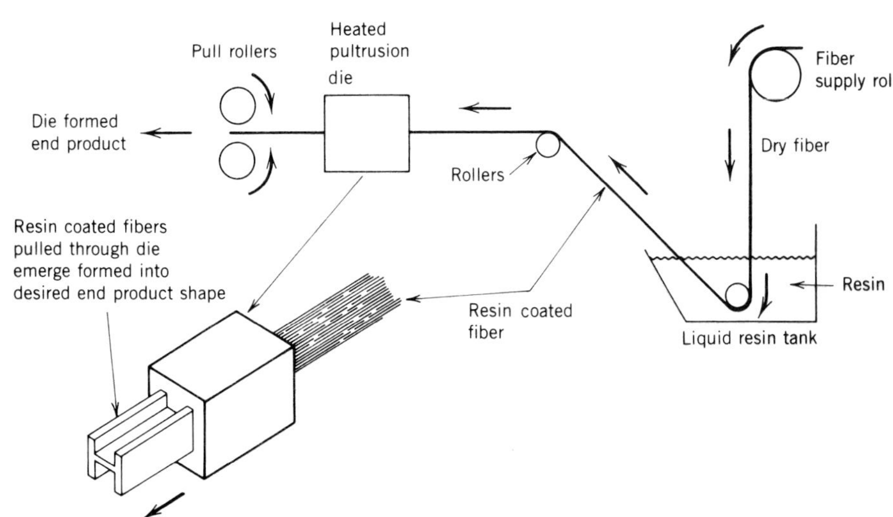

Figure 28
Manufacturing composite by the filament winding process.

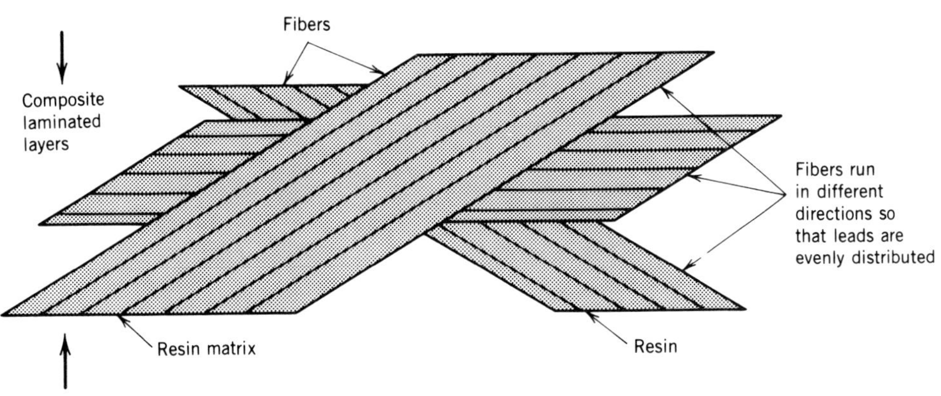

Figure 29
Manufacturing composites by the lamination process.

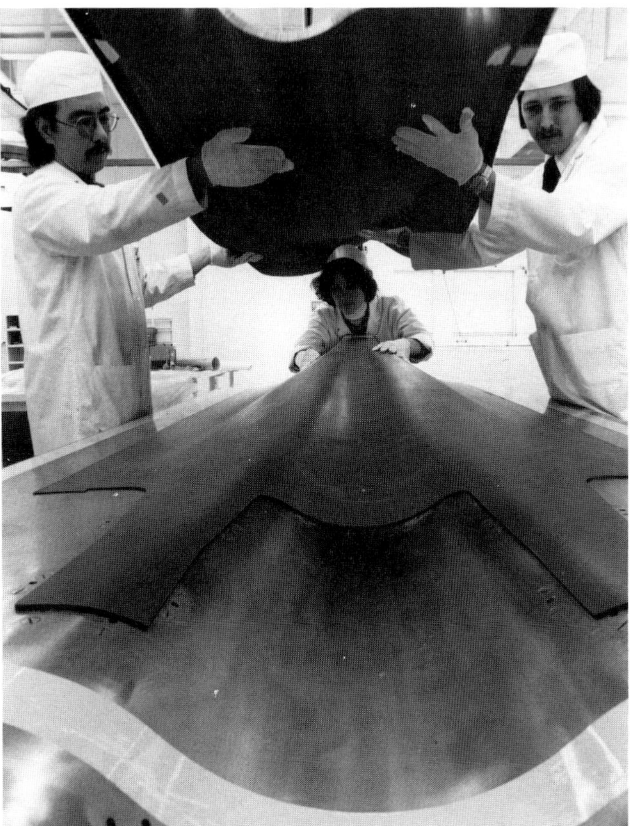

Figure 30
Aluminum honeycomb sandwiched and bonded between graphite composite skins makes an extremely strong and lightweight structure for the F-18 fighter aircraft (Northrop Corporation).

Figure 31
Commercial jet liner lightweight transcowling made from graphite composite material (Martin Marietta Corporation).

less fuel to propel them. This consideration is extremely important in the production of aircraft, spacecraft, and automobiles of today and tomorrow.

The honeycomb composite structure creates a product that demonstrates these highly desirable characteristics of light weight and strength. In Figure 30 aerospace technicians lay an aluminum honeycomb core on a graphite composite skin. The metal honeycomb, that is in itself a high strength light weight material, will be sandwiched between two layers of the graphite composite. The entire assembly will then be cured by heat and pressure, thus making a durable aircraft structural component.

Large components can be made from composite materials easier than with metals. Figure 31 shows a large graphite composite component for a commercial jet liner. Use of honeycomb composite structure is seen in Figure 32. This large fabrication will become part of an aircraft jet fan reverser, once again an application where high strength and light weight are desirable engineering design specifications.

TOOL AND DIE MAKING FOR PLASTIC AND COMPOSITE PROCESSING

As is the case in processing most raw materials into useful products, special tooling is required in the plastic and composite processing industry. Here the raw material is most often in a liquid or flexible state prior to its formation into the end product. Therefore, considerable tooling is required to hold or form the material until steps are taken in the process that alter the state of the material to make it rigid.

In processing plastic and composites, much of the tooling appears in the form of **molds** or **forms.** Most

Figure 32
Aircraft fan reverser component showing composite fiber honeycomb structure (Martin Marietta Corporation).

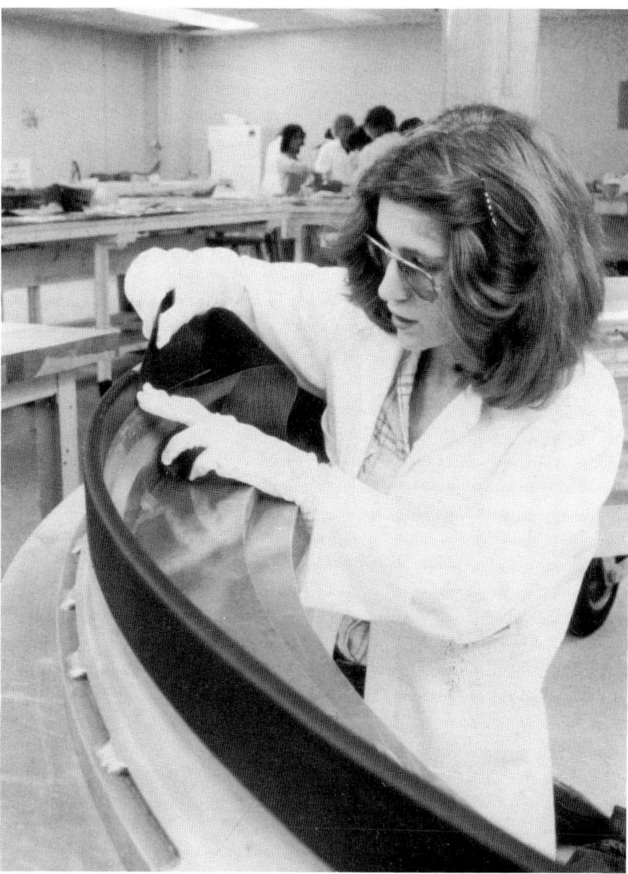

Figure 33
Tooling is all important to the formation of plastic and composite products. A technician lays up graphite composite material on a large precision bonding fixture (Martin Marietta Corporation).

of this tooling is made from metal and is manufactured to exact dimensional specifications. Figure 33 shows a large form tool and bonding fixture designed for shaping graphite composites.

Tool and die making is therefore an integral and indispensable part of any manufacturing industry. The tool and die maker, who represents the upper end of the skilled machinist trade, will always be needed to build the tooling that makes mass production possible. The toolmakers of today and tomorrow have at their disposal all of the modern improvements that an advancing technology makes possible. Among these are Computer Aided Design (CAD) and the versatility of computer numerical control machining (CNC) for making the high precision tooling required by space-age manufacturing.

Modern methods of high production plastic and composite part production will yield parts at very low cost. However, the tool and die cost must be included in the product price. Each plastic part produced includes a portion of the tooling price. The first part produced represents the full cost of tooling plus the material consumed in making the part. As each successive part is produced, the cost of tooling is spread further and further out. Eventually enough parts are produced to underwrite the tooling cost and from then on, the only cost is for material and production overhead expenses. If a large number of production parts are further produced, a good return is received on the tooling investment.

Production tooling, once manufactured, can be stored and reused at a later date. If a manufacturer receives an additional order for the same product, it is not necessary to remanufacture the tooling and the customer may be able to purchase the manufactured product at a more reasonable price.

REVIEW QUESTIONS

1. What factors are involved in selecting a plastic or composite material over metals?
2. What are the advantages of using these materials rather than metals?
3. List three major methods of processing plastics and describe how they work.
4. Give an example of a blow molded product.
5. Cite an example of an injected molded product.
6. Cite an example of an extruded product.
7. What is RIM and how does it differ from conventional injection molding?
8. Briefly describe molding processes other than RIM.
9. What types of plastic products can be made by machining?
10. Describe reinforced plastic material and cite a common example.
11. What is the purpose of reinforcing plastic and what materials may be used in this process?
12. Cite a use of high-pressure laminates.
13. How does a composite material differ from a reinforced plastic?
14. What medium provides the structural component in a composite?
15. List and describe three major methods of composite manufacturing.
16. What process would be best for manufacturing a composite water tank?
17. What process would be best suited to manufacture a composite structural beam?
18. What process would be best suited to manufacture an aircraft composite deck walkway?
19. What is tool and die making and where does it fit into the plastics and composite manufacturing industry?

CASE PROBLEMS

Case 1
Choosing Different Material

To conserve weight in an aircraft structure, an analysis is made of the aluminum structural beams that support the internal platform decks. Since these components are not flight-critical items, their strength-to-weight ratios can be maximized. In other words, they need only be strong enough to support required loads, but need not have the safety factors required of such flight-critical components as wing and tail spars.

One proposal advanced by structural engineering is to cut lightening holes in all deck beams to conserve weight. However, CAD analysis suggests that if enough material were cut from the beams to make a significant weight saving, the beams would be borderline as to their load carrying capabilities. What would you suggest to solve this problem?

CHAPTER 16 CORROSION AND PROTECTION FOR MATERIALS

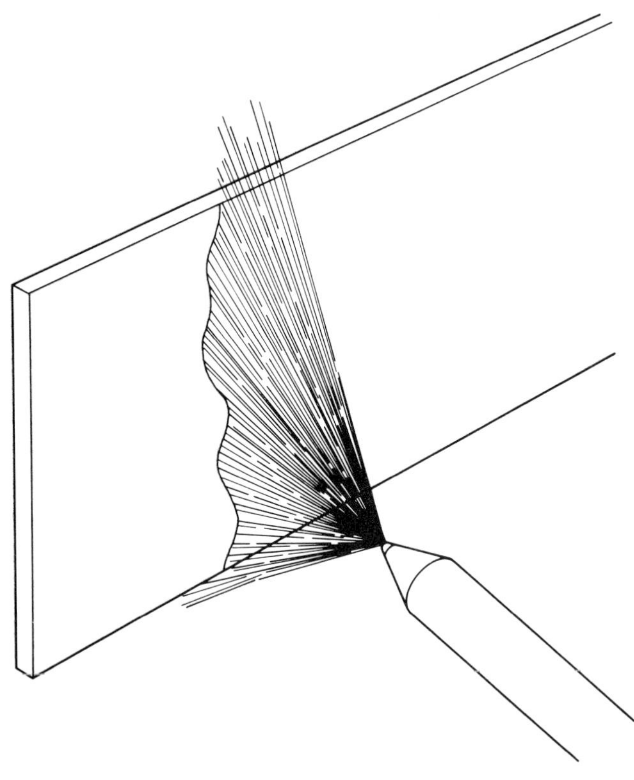

Since many materials, especially metals, react chemically with their service environments, it is often necessary to provide protective coatings and coverings for them so that surface deterioration will be prevented or slowed. Failure to provide surface protection can result in rapid destruction of many types of materials thus requiring expensive replacement. Other material, such as wood and plastics, subject to physical deterioration from weather, sunlight, and water must also be protected from the environment. The purpose of this chapter is to survey the various means by which materials may be protected from environmental degradation.

OBJECTIVES

This chapter should enable you to:

1. Discuss the causes of corrosion.
2. Describe the various ways in which corrosion may be slowed or prevented.

CORROSION IN METALS

Most common metals exist in nature chemically combined with other elements thus forming the various **oxides (ores).** Examples include iron ores from which metallic iron is extracted and bauxite from which aluminum is derived. These ores are mined and refined by various processes to extract the metallic components that in turn become the common familiar metals. There is a natural tendency for many refined metals to revert to their natural state after the metallic component has been extracted. Since both air and water are common service environments for many metals and oxygen is a large percentage of both, many metals will try to reunite with the oxygen present as they attempt to revert to their original oxide (ore) state.

Iron is a good example. Existing in nature as an oxide (iron ore), the metallic element iron is chemically combined with the oxygen in the air. When the metallic portion of the ore is separated from the oxygen during the iron ore refining process, the iron will tend to reunite with any oxygen present and revert to its original more stable state of iron oxide. This process is a slow oxidation of the metallic iron and is known to most of us as rusting. If the temperature of the iron is raised to a high level, the process of oxidation is much more rapid. It should be noted that not all metals oxidize in this manner. Gold, for example, does not readily unite with oxygen and is therefore found in nature in its pure form.

Corrosion of metals can be classified according to the two major processes by which it occurs. One is the process just described, oxidation, and is called **direct oxidation corrosion;** the second process is **galvanic corrosion,** in which corrosion takes place at normal temperatures in a moist or electrolytic service environment.

The first process, direct oxidation corrosion, occurs slowly, as in normal rusting, or quickly, as in the case of steel, where heating to high temperatures for the purpose of hot rolling or heat treating leaves a layer of black oxide on the material. This is called mill scale and is noticeable as the black coloring seen on most structural steel.

Galvanic corrosion gets it name from the galvanic effect. This is the electrical phenomenon of current flow resulting from the presence of a potential (voltage) causing a current to flow through an electrolytic (conducting) medium. Galvanic corrosion is greatly enhanced if the metal being corroded is surrounded by an electrolyte. An electrolyte is any solution that can conduct an electric current. Fresh and salt water are examples of electrolyte environments where it is often necessary to have metals. Under these circumstances, the materials are subject to galvanic corrosion.

For galvanic corrosion to occur, the electrolyte must form a path for an electric current to flow through and a situation must exist where a potential (voltage) difference can exist. These conditions are easily met by placing two different metals in an electrolytic environment or by a situation in which two different areas on the same piece of metal act as the cathode and anode in the electric circuit. During the process of galvanic corrosion a small current flows in the circuit established between the cathode and anode. The electrolyte provides the path for current flow. In this corrosion process, the metal is actually deplated. Galvanic corrosion can be greatly accelerated if the electrolyte is efficient and the potential difference between cathode and anode is great. But the process can also take place where the electrolyte is barely identifiable. A finger print, for example, acting as the path for current flow, can cause galvanic corrosion to occur on a metal surface. For this reason some manufacturing of metal products subject to corrosion may be accomplished in clean-room environments where air moisture is carefully controlled. Parts are never touched directly.

PROTECTION METHODS

There are a number of means by which a given material may be protected from influences of the environment. The particular method used will depend on several factors such as cost, how corrosive the environment is, and how the need for protection of the material is directly tied to structural design requirements.

Fundamentally this can be done by:

1. **Shielding the material** from its service environment by coating it with another material. Examples include painting, plating, cladding, and treating with various oxides. In this manner, service environments work to degrade the coating material and as long as its integrity is maintained, the base material will be preserved.
2. **Establishing the chemical make up** of the parent material such that service environments do not cause it to degrade. An example here would be the alloying of metals with anticorrosive alloys thus making them less subject to corrosive forces. Although this is frequently done for many manufactured products, it is generally much more expensive than coating a less expensive material with a surface protection medium.
3. **Allowing the corrosive activity** to continue in a normal way, but **to be concentrated on** other **nonstructural parts** made from materials that are designed to accelerate corrosion and thus concentrate the corrosive activity. This is to say that nonstructural parts of the assembly are sacrificed in order to concentrate and thus reduce the corrosive activity on other essential structural parts.

An example of this third method is the **sacrificial anode** method. This technique is used quite extensively where steel and water must coexist. This service environment is highly favorable to galvanic corrosion. To slow the process as it reacts against the base material, a sacrificial anode may be used. The anode is made from a different metal, one that has a high susceptibility to galvanic corrosion. Thus, the process tends to concentrate on the anodes and they are sacrificed in order to preserve the base material. The process is effective; however many tons of sacrificial anodes may be required for large fabrications, and they must ultimately be replaced in order to maintain the concentration of electrolytic activity on the nonstructural members of a fabrication.

Cladding

Cladding is a thick layer of one metal applied to another metal for the purposes of corrosion protection, of improving mechanical characteristics such as hard facing, or in some cases, for appearance. Cladding may be done by weld build up of a layer of metal, or by adhesive bonding one metal to another.

In the clad welding process, weld beads are laid down on the base metal side by side and layer by layer. The process is tedious if done by hand and usually requires a shield gas welding process so that slag pockets and other inclusions in the cladding are eliminated. After clad welding, parts may be machined to achieve desired dimensional characteristics.

Alloying and Oxidizing

Alloying Alloying is an effective but often expensive method of material protection. Certain metals, such as chromium, have excellent anticorrosion properties. Plain carbon steel, on the other hand, has an affinity for the oxygen in air and will rust (oxidize) if exposed to normal outdoors environment. It is also subject to galvanic corrosion. If, however, the material is alloyed with chromium, creating the material called **stainless steel,** corrosion by rusting may be reduced or totally eliminated. Figure 1 shows pieces of stainless and plain carbon steel that have been exposed to normal outdoor environments for the same period of time. The plain carbon steel shows considerable rust, whereas the stainless steel shows very little evidence of corrosion. However, even stainless steel will corrode in some environments, so it is important to select anticorrosive alloys when using this material.

Other metals that will provide anticorrosion characteristics include **copper–nickel alloys (Monel), molybdenum, lead, nickel,** and **pewter.** Lead, used in lead acid batteries for plates, serves in an extremely corrosive acid environment. **Gold** is also an excellent anticorrosive metal and is often used to plate other metals for corrosion protection.

Another possible advantage of alloying is that the anticorrosion properties exist throughout the thickness of the metal. Thus, if the metal surface is damaged by abrasion, denting, or scraping, the anticorrosive properties will be preserved. This is not the case with plated, painted, or otherwise coated materials. Since the anticorrosion medium is only on the surface and most often is extremely thin, scratch or scrape damage can expose base material to the environment permitting corrosion to begin. Figure 2 shows painted steel on which the integrity of the paint has been breeched thus exposing the bare metal to the environment. This exposes a place where corrosion can begin.

Figure 1
Unprotected carbon steel (top) is heavily rusted. Stainless steel (bottom) shows little corrosion.

Figure 2
Damaged surface of painted steel exposes base metal and permits corrosion to occur.

Oxidizing A metallic material that tends to react with oxygen and oxidize may be protected by accomplishing this very process only on an outside layer of the material. Common oxidizing processes include black oxidizing and anodizing.

In black oxidizing parts are heated in a carbon or oxidizing salt environment and then quenched. Black oxidized parts take on a black color appearance and are thus surface protected from rusting.

Anodizing is a surface protection process that is popular, especially for aluminum. The process is electrically done and results in a thin layer of oxide on the surface of the part. This precludes further oxidation by the service environment (Figure 3). Once again, the anodizing layer must be maintained intact. This layer which is less than .0005 in. thick can be damaged to the point where the base metal is exposed to the corrosive environment.

Plating

The various plating processes make up a large portion of methods used to protect metals from their service environments. The theory behind plating is that if a thin layer of anticorrosive material can be plated to a base metal, then the material can be protected from corrosion. Figure 4 clearly demonstrates the application and advantage of plating as an anticorrosion measure. Note how the unprotected steel on the bracket is heav-

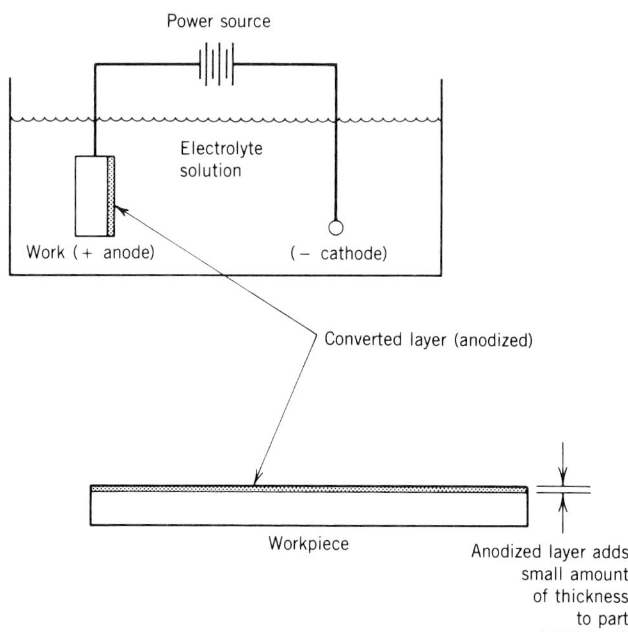

Figure 3
Anodizing process.

Figure 4
Unprotected steel is heavily rusted while plated fastener shows no evidence of corrosion.

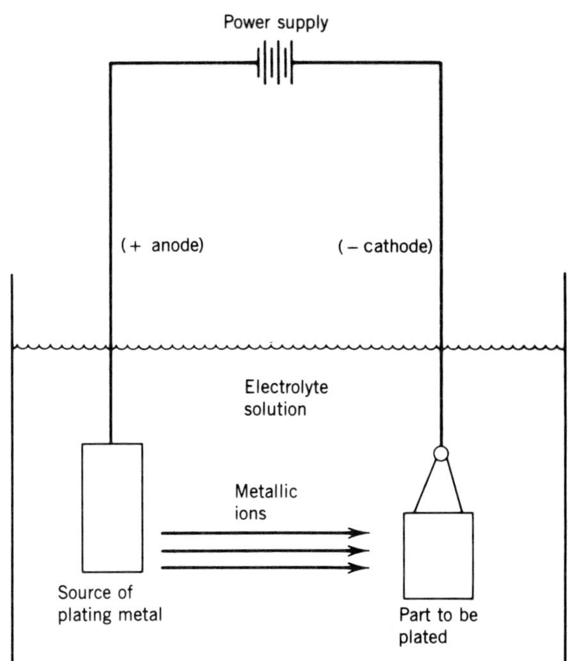

Figure 5
Electroplating process.

Figure 6
Examples of common galvanized products.

ily rusted. However the fastener which is zinc plated shows little or no degradation from corrosion.

Plating processes are widely used and have numerous advantages. For example, sheet steel is widely used for a vast number of products. It is easily formed, can be effectively assembled by high production methods such as spot welding, and is relatively inexpensive. Its primary drawback is that it is subject to rapid rusting in typical service environments. One solution is to use stainless sheet steel. However this product is much more expensive and less easily fabricated.

Common low-carbon sheet steel is one product that can be effectively protected from rusting by plating it with an anticorrosive metal (zinc) in the process of **galvanizing.** Thus, the useful production properties of sheet steel are maintained while preserving the material from corrosion from its service environment.

Electroplating Processes Many metals are electroplated onto other base metals by electroplating processes. Parts to be plated are placed in a tank containing a salt solution (electrolyte) of the plating metal. A direct current (DC) voltage is applied across the part to be plated (− cathode) and the source metal (+ anode). The current flow causes plating metal ions (imbalance charged atoms) to move from the source to the parts where they are deposited as plated metal (Figure 5). Galvanizing is one common process where a zinc coating is electroplated onto steel (Figure 6).

Other metals that are electroplated include **tin, cad-**

Figure 7
Silver plating (left) and brass plating (right) add both beauty and resistance to corrosion.

mium, nickel, chrome, brass, silver, and **copper.** Gold has many industrial plating applications in electronics because of its superior electrical conductivity properties. Silver and brass plate adds beauty and serves to resist corrosion (Figure 7).

Electroless Plating Processes Plating without electricity is also a widely applied process. Nickel is an example. The process of electroless nickel plating yields a more even thickness on complex parts than will electroplating processes.

Electropolishing

In the **electropolishing** process, a small amount of surface material is deplated from the part. The process is accomplished by placing the parts in an acidic electrolyte where a DC potential exists between parts (anode) and cathode (the other side of the power supply).

The process leaves an extremely smooth surface while not removing enough workpiece material to affect precision dimensional characteristics. It is also useful to smooth our part surfaces on irregularly shaped objects or where part surfaces are not visible or accessible.

Painting

Painting is probably the most widely used process for surface protection of metals and other materials. Its primary advantages include **ease of application, wide color variety,** and **reasonable cost.** Its primary disadvantage is that because it is a surface treatment, damage can expose the base material to the corrosive environment. However, the process is in wide use and will probably remain a mainstay of surface protection.

There are many types of paints and paint technology undergoes constant development. Lead-based oil paints have been used for many years but are giving way to water-based paints. Paints containing tin compounds are used as antifouling paints for marine applications such as ship hulls. These paints are toxic to the marine life that tends to attach itself to a ship's hull.

Many modern paints are plastic resin based, such as those from the acrylic group. Water is frequently the medium for carrying the pigment (solid paint particles with color). These paints are fast drying and do not cause the pollution problems that disposing of petroleum solvent fumes and liquids causes.

Paint Applications Paint is typically applied by **dipping, brushing, rolling,** and **spraying.** Spray technology is most popular in mass production applications and is the only way to achieve the truly smooth coat necessary in most manufactured product applications, such as on automobiles.

Spray painting can be done by air-driven sprayers (Figure 8), but overspray must be captured and retained in order to meet most current air pollution requirements especially if nonwater solvents are used. Personnel must also be protected during spray painting operations.

Electrostatic spray painting eliminates overspray and the resulting pollution. This process is extremely effective from both a coverage of product and a con-

Figure 8
Spray application of paint (Lockheed–California Company).

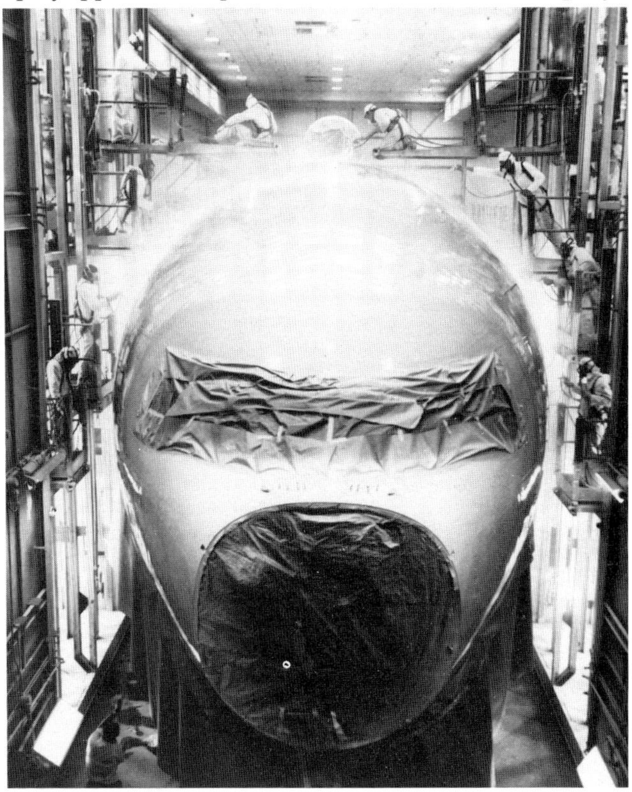

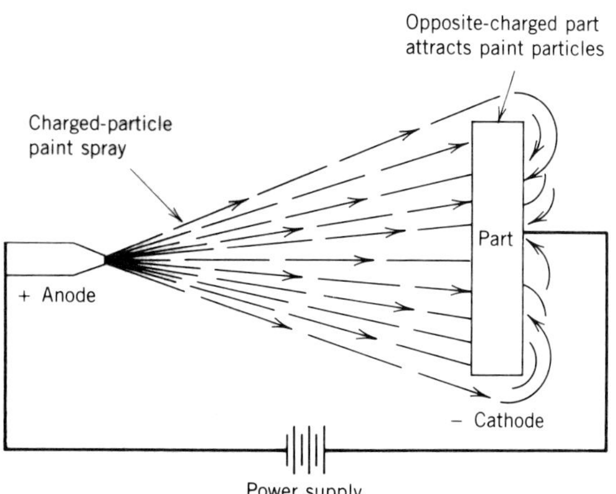

Figure 9
Electrostatic spray-painting process.

sumption of paint material standpoint. In the electrostatic process, high DC voltage difference is maintained between the paint sprayer and the part to be painted. Paint particles flowing from the sprayer acquire the charge opposite to that of the part and are therefore attracted to it. The process is extremely fast and since the part has an equal charge on all sides, the paint is attracted to the back as well as the front. Thus, parts are literally painted all over at once (Figure 9).

High-pressure paint pump sprayers have become very popular in recent years. In this equipment, the paint is atomized by pumping it through a small nozzle at very high pressure. High pressure pumping is very fast in application and the technology has been made available to the homeowner.

Paint spraying, regardless of the application equipment, requires considerable skill on the part of the operator in order to achieve a smooth and even coat on the workpiece. Like many other industrial processes, spray painting can be effectively accomplished with the computer driven industrial robot. Figure 10 shows some unique developments in this area. The auto bodies are positioned diagonally on the conveyer so that the spray-paint robots may access them for the automated production painting operation. Since painting must be accomplished on underside surfaces and overhead locations, the jointed-wrist type industrial robot is popular for this application.

Paints must dry after application. This can be a time consuming operation and can be speeded up by the application of heat and circulating air. Parts are conveyed to a heated drying facility where heaters expedite the drying process.

MATERIAL PREPARATION FOR SURFACE PROTECTION

Before material can be surface protected, it must be properly prepared. Failure to do this will result in a poor application of the particular surface protection process. Several methods are used for material preparation.

Mechanical Methods

Mechanical methods physically abrade the material in order to remove mill scale, dirt, and corrosion. Commom methods used include **sand** or **shot blasting** where high velocity abrasive particles are propelled against the workpiece, and the use of abrasives in the form of sand papers or abrasive wheels. Wire wheels (brushes) may also be used.

Tumble cleaning and deburring is another popular method for both cleaning and removing burrs from

Figure 10
Production spray painting with industrial robots (Chrysler Corporation).

Figure 11
160 ft³ capacity vibratory deburring machine showing compartments, workpieces, and abrasive pellets (Automated Finishing Inc.).

machined workpieces. In this process, parts to be cleaned or deburred are placed in a vat or vibratory bowl with small pellets of abrasive material (Figure 11). The bowl is vibrated rapidly causing the parts to be tumbled along with the abrasive material. Equipment used for this process ranges in size from small, with a few cubic feet of capacity (Figure 12), to quite large, with a high cubic foot capacity (Figure 13). Vibratory cleaning and deburring machinery can also be equipped with material washing, drying, and handling conveyer systems (Figure 14).

These methods are satisfactory but may be labor intensive, dirty, and sometimes dangerous. It is also difficult to clean all surfaces on a complex part by brushing or grinding. Sand blasting is a very abrasive process and will erode metal and wood surfaces rapidly. Using glass beads for the abrasive eliminates most of this effect, but this process must be done in a confined area where the fine dust can be collected and removed from circulation. Sand blasting must also be done under controlled conditions and the personnel must be protected from the flying sand.

Ultrasonic cleaning of parts is accomplished in a container with a cleaning medium. The bath is vibrated by ultrasonic sound waves. The process is a combination of both chemical and mechanical cleaning.

In the process of **buffing,** parts are brought into contact with a rapidly rotating wheel usually made from layers of cloth sewn together. Various grades of abrasive materials in cake form are hand applied to the wheel. These materials include various metallic oxides and rouge. The buffing process provides highly polished mirrorlike finishes on metals. However, it is labor intensive and often requires that the part be previously abraded with coarser materials so that final buffing will yield a desirable result. This process is popular in art metal and jewelry metal work.

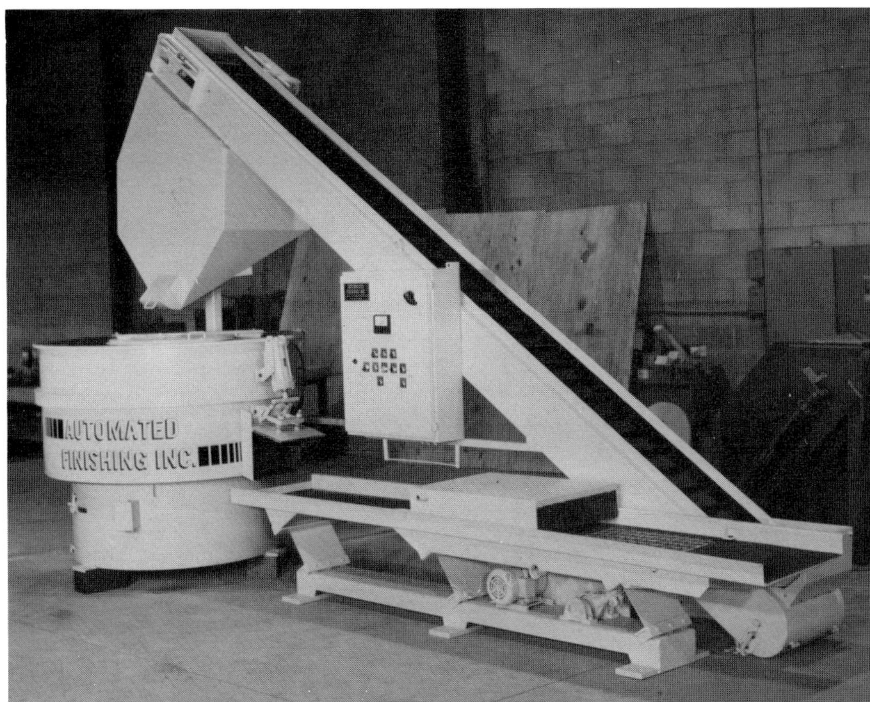

Figure 12
14 ft³ capacity vibratory deburring machine with material handling conveyers (Automated Finishing Inc.).

Figure 13
Large 160 ft^3 capacity vibratory cleaning and deburring equipment (Automated Finishing Inc.).

Figure 14
Through-flow vibratory machines equipped with part washers, dryers, and material handling conveyers. (Automated Finishing Inc.).

Chemical Processes for Cleaning and Material Preparation

Chemical preparations are usually necessary prior to painting. In **pickling,** material is dipped in an acid bath to remove oxides. This process must be followed by a rinse to neutralize the pickling acid.

Solvents can be used to remove grease and oil. The parts can be placed in the vapor of the solvent. When vapor condenses it washes the parts. This process is called **vapor degreasing.** Alkaline solutions are also used as a washing agent.

To provide for a better bond between paint and material, two phosphate coating processes are used. In **bonderizing** the parts are dipped in a phosphating solution. The resultant coating permits paint to adhere more readily. The phosphate process of **parkerizing** also provides a corrosion resistant primer coating under a painted surface.

REVIEW QUESTIONS

1. What are the primary causes of corrosion?
2. Why not manufacture everything out of solid anticorrosive materials?
3. Describe plating processes and discuss their advantages and disadvantages.
4. What is anodizing?
5. What is galvanizing?
6. How can paint coatings be applied?
7. Are there any disadvantages to painting as a corrosion prevention measure?
8. Describe electrostatic painting and discuss its advantages over air spraying.
9. How can materials be mechanically cleaned prior to painting or plating?
10. What chemical preparations are used to prepare materials for plating or painting?

CASE PROBLEMS

Case 1
Abrasion Resistance

An escalator side plate is subjected to light but constant abrasive forces. In your opinion, what would be the best method of protecting this material with an emphasis toward minimum maintenance and long-term preservation of the product?

- Plating.
- Painting.
- Anodizing.
- Making from solid material.
- Galvanizing.

Case 2
Rust Prevention

Large steel fabrications are designed to be used in a brackish estuary. The products are painted with the best available anticorrosive marine paints. What other methods may be used to further prevent corrosion of these items in their service environments?

CHAPTER 17 PROCESSING OTHER INDUSTRIAL MATERIALS

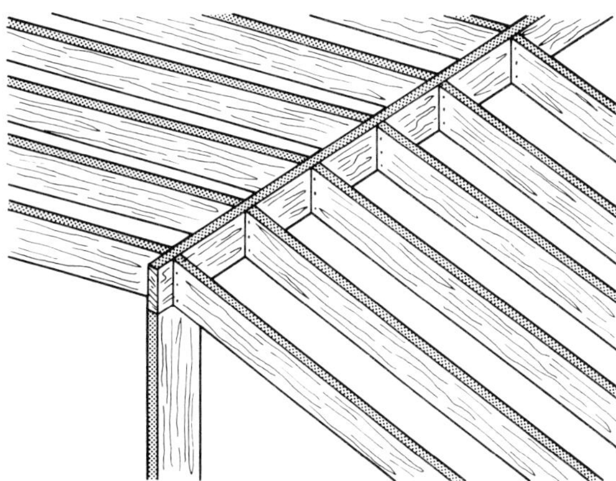

Although metals and metal working make up a large portion of materials and manufacturing processes, there are many other materials and processes that are equally important in manufacturing. Ceramics and wood, for example, have long been in use and in the age of high technology, these materials are undergoing new developments and many new applications are being found for them in the manufacture of industrial products.

OBJECTIVES

This chapter should enable you to:

1. Be generally familiar with common industrial materials other than metal and plastic.
2. Be familiar with their manufacturing processes.

Many of these materials are used in conjunction with metals in all types of products. In many cases some of these materials, because they are cheaper, stronger, lighter, and easier to process, are replacing metals. Foremost among these are plastics and composites, as previously discussed. Other important industrial materials include:

1. **Glass.**
2. **Ceramics.**
3. **Wood and paper.**
4. **Fabrics.**
5. **Rubber.**
6. **Natural materials.**
7. **Construction materials.**

GLASS

Glass consists primarily of fused silica SiO_2, a major constituent of beach sand. Its major uses are widespread and include many common products such as **windows, containers,** and **lenses.** The glass window wall commercial building has become very popular in modern building construction (Figure 1).

The applications of glass for containers are unsurpassed. The material will withstand **chemical attack** and can also withstand **large variations in temperature.** Lead crystal (Figure 2) is noted for its many beautiful designs in table ware. Lead is added to glass for windows that provide shielding from ionizing radiation while permitting radioactive equipment to be observed safely.

Another outstanding property of glass is its ability to be fashioned into precision optical lenses for eye glasses, telescopes, microscopes, and many other optical instruments.

A relatively new and developing technology in the uses of glass is **fiber optics.** Here, the glass is drawn into a thin strand, much like a glass wire. The stands are grouped into a fiber-optic bundle and then sheathed in plastic for durability and protection.

Fiber-optic technology is a rapidly developing field for the transmission of information in **telecommunications** and **computers.** Figure 3 shows a fiber-optic bundle designed to transmit digitized data in an aircraft communications system. Note the difference in size between the fiber-optic bundle as compared to the standard aircraft cable in the lower right side of the illustration. The fiber bundle is much smaller and lighter and it can transmit more channels of data than its wire cable counterpart.

Sheet glass is manufactured on a production basis by the latest computer controlled and automated pro-

Figure 1
Plate glass used in building construction turns entire walls into windows.

Figure 3
Glass fiber optics, a developing technology, uses strands of glass to transmit light signals in communications and computer systems. Note the difference in size of the glass fiber as compared to wire cabling that would carry equivalent electrical signals (Lockheed–California Company).

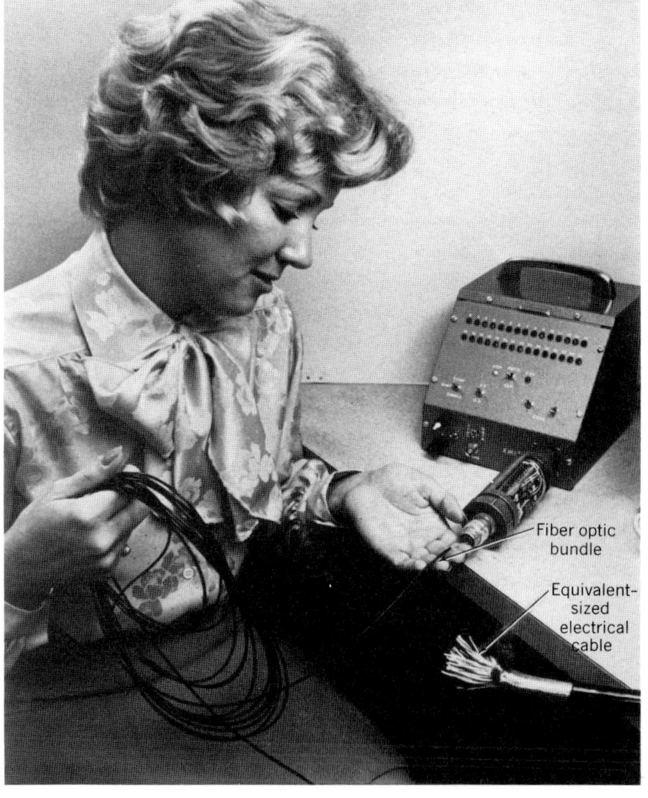

Figure 2
Lead crystal makes highly artistic and beautiful glass products.

Figure 4
Float glass manufacturing produces a continuous ribbon of glass (Libbey–Owens–Ford Company).

Figure 6
Curing automotive safety glass products in a large autoclave (Globe–Amerada Glass Company).

Figure 5
Assembly of laminated safety glass windshields involves placing a interlayer of plastic between two glass layers that have been cut and bent to the windshield shape (Libbey–Owens–Ford Company).

duction methods. The **float method** (Figure 4) produces a continuous ribbon of glass. This product is the material for sheet glass products such as windows, mirrors, and automotive items.

Glass for automotive windows is made by a **lamination process** in which a plastic interlayer is sandwiched between two layers of glass. This makes the material shatter proof. After assembly (Figure 5), the product is placed in an autoclave (Figure 6) where a bonding between glass and plastic interlayer occurs.

Other glass product manufacturing processes include **pressing, blowing,** and combination press and blow operations (Figure 7). These processes are done on production equipment in a highly automated manner (Figure 8) and are primarily used in glass container production.

Hand glass blowing is a long established process for producing containers and other glass art objects. Although the process is mostly automated for production applications, it is still done to some extent on an art and craft basis. In it a quantity of molten glass is retrieved from the furnace on the end of a blow pipe. Breath pressure is applied to the opposite end of the pipe causing the glass to inflate thus assuming a hollow shape. Rotating the blow pipe preserves the desired round shape of the product (Figure 9). Several types of tools and trimmers are used to further shape the glass before it cools (Figure 10). The process lends itself well to the production of many beautiful and artistic glass products.

Other glass manufacturing processes include **cutting** sheets into various shapes, **edge beveling, grinding,** and **hole drilling** so that glass pieces may be assembled by using mechanical fasteners.

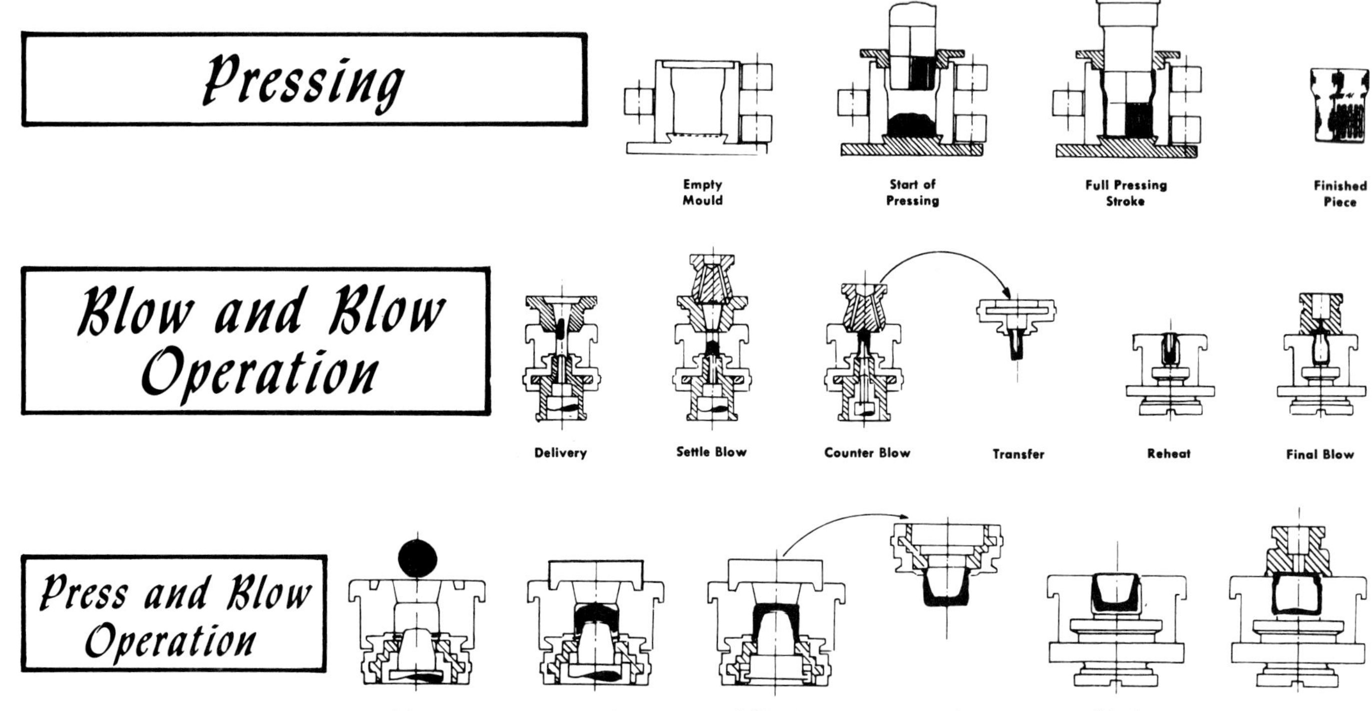

Figure 7
Glass manufacturing process for containers (Wheaton Glass Division, Wheaton Industries).

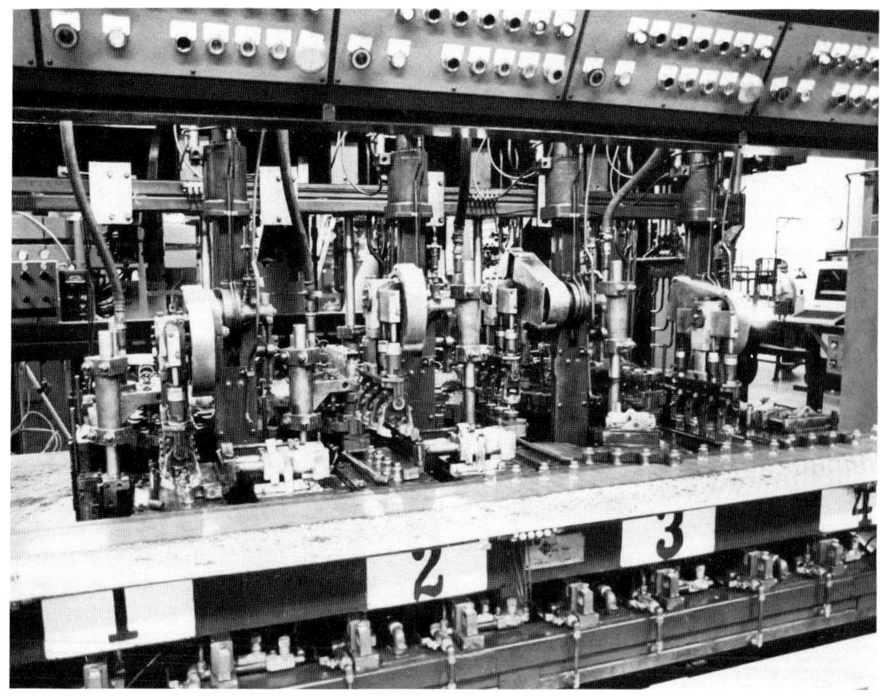

Figure 8
Automated glass container manufacturing (Wheaton Glass Division, Wheaton Industries).

Figure 9
Glass blowing and shaping by traditional methods (Wheaton Glass Division, Wheaton Industries).

Figure 10
Shaping blown glass using traditional hand tools (Wheaton Glass Division, Wheaton Industries).

Tempering turns standard glass into **plate glass.** The glass is heated in a large furnace to a semiplastic state. It is then removed from the furnace and quickly air quenched. This process sets up closely spaced swirl shaped stress patterns in the glass. Tempering is done to provide glass with mechanical strength for such applications as refrigerator shelves, store fixtures, doors, and windows. Should plate glass be broken, many small relatively harmless pieces result, as opposed to large very hazardous slivers. Thus, plate glass is specified for storefront windows, doors, and floor fixtures that might be subject to breakage by weather or impacts. Tempered glass also exhibits superior mechanical strength, a requirement for many glass applications.

Glass in its multitude of product forms is an extremely useful and versatile industrial material. Annual glass production in the United States amounts to several million tons per year.

CERAMICS

High technology product design has motivated great interest in and development of ceramic engineering materials in recent years. **Ceramics** are made from the **metallic oxides** of such metals as silicon, aluminum, magnesium, and others. Clay-based ceramic materials are also in wide use and are extremely popular in arts and crafts.

Materials used for industrial ceramic products include **carbides, borides, nitrides,** and **silicides.** Ceramic materials have some unique properties that make their applications for industrial products unique. These characteristics include **high melting points** (up to 7,000° F), **chemically inert, refractory** (reflect heat) and will withstand large **compressive (crush or pressure) loads.** Certain ceramics such as those from the silicon nitride group are as **hard as diamond** and as **light as aluminum.** Possible disadvantages of ceramics material include brittleness and a tendency toward weakness when subjected to tension (pull-apart) loads.

From the viewpoint of the industrial product designer, the ceramic property of **resistance to high heat levels** is one of the most interesting. Ceramic materials will withstand temperatures of several thousand degrees and still maintain their shape and strength. It is this characteristic that makes them useful for making parts that will be subjected to extreme temperatures. Examples include jet engine turbine components, reciprocating engine valves, and exhaust gas driven turbines in automobiles.

Manufacturing processes for ceramics include **extrusion, pressing, casting, machining, injection molding, flame sprayed coatings** and **sintering.** After the product is shaped using raw ceramic materials, it must then be furnace fired to achieve its final strength and form characteristics. The plasma arc with its extremely high temperature characteristics has been used to produce high quality ceramic powders, the raw material from which ceramic products are produced.

Uses of ceramics include **cutting tools** for machining, **abrasives for grinding,** and **coatings on cutting tools** to improve their performance characteristics. Industrial products include turbine rotors, engine valves, electrical insulators, bricks, tiles, dishes, and containers. Research is presently being done to develop mechanical ceramic components such as ball and roller bearings.

WOOD, WOOD PRODUCTS, AND PAPER

Throughout the history of manufacturing, wood has been an important and useful material with which to make a multitude of products. Since it is a natural and often readily available material, it is logical that its uses would be widespread. Wood is also a renewable resource and modern forest conservation practices should ensure that we have a continuing supply. With the increasing cost of metal and the energy expended in metal processing, the interest in wood as a structural building material is undergoing new interest.

It is interesting to note that wood was a fundamental fuel source well into this century. In fact, one factor that contributed to the industrial revolution, after which manufacturing shifted toward coal and petroleum-based fuel supplies, was a shortage of fire wood accompanied by high wood prices. Today there is a renewed interest in wood as a fuel source for domestic heating and even for industrial and power generation purposes. However, the return to wood fuel on a large scale is unlikely because of cost, availability, air pollution control, and its primary value as a construction material.

Wood and Wood Product Processing

Many types of woods are used in constructing wood products. The primary use of wood is in **structural lumber** for homes and small building construction (Figure 11). Trees that grow straight and tall are most useful for lumber. These include pine, fir, spruce, and redwood. Major lumber supply areas include the western forests of North America. Trees are selected, felled, and transported to high production saw mill operations where standard structural lumber is produced. The modern saw mill makes use of modern manufacturing automation and standard lumber is produced at a high rate.

Waste from sawing which includes saw dust, slabs, and trimmings was at one time burned at the mill site. Today, sawdust and wood chips are too valuable to be burned and are used to produce many other wood products that include plywood and chip wood products. These materials play an important part in supplementing structural lumber as a building material.

Wood Products In additional to structural lumber, wood products include **plywoods** in which thin layers of wood are glued into panels. The grain of each layer runs perpendicular to the next (Figure 12). Plywood is an indispensable building material in modern wood building construction. Plywood panels are extremely strong and are available in standard sized sheets in several thicknesses.

Other wood products include **wood chip products** such as **flake board** (Figure 13). This material allows wood waste to be used in a profitable manner. The material is quite suitable for a variety of products such as cabinets and furniture.

Wood should not be discounted as a structural build-

Figure 11
Structural lumber is a versatile building material for house construction.

Figure 12
Plywood consisting of veneer layers glued together in sheets is an exceptional building material.

ing material. **Glue laminated beams** (Figure 14) are well suited to significant structural building requirements. They will support large loads and are often lighter and cheaper than their metal counterparts.

Hardwoods The list of hardwoods used for furniture and flooring is long and distinguished. Common hardwoods used in furniture manufacturing include mahogany, rosewood, oak, teak, birch, maple, hickory, and walnut. Several other woods that are found in tropical areas are also used in furniture manufacturing.

Furniture manufacturing is a large industry employing many of the same manufacturing techniques and fastening systems used in metalworking. Wood must be cut and shaped and smoothed. Woodworking machine tools and processes resemble their metalworking counterparts in many ways. Assembly with modern glues and adhesives as well as mechanical fasteners including nails, staples, bolts, and screws is the most popular. Typical wood manufacturing equipment includes lathes, shapers, sanding machines, and various types of saws. Great effort is placed on surface finishing and artistic design in the wood products (Figure 15). Woodworking is one area of manufacturing in which

Figure 13
Flake board, a wood chip product, makes an effective building material and makes use of wood chips that would otherwise become waste products.

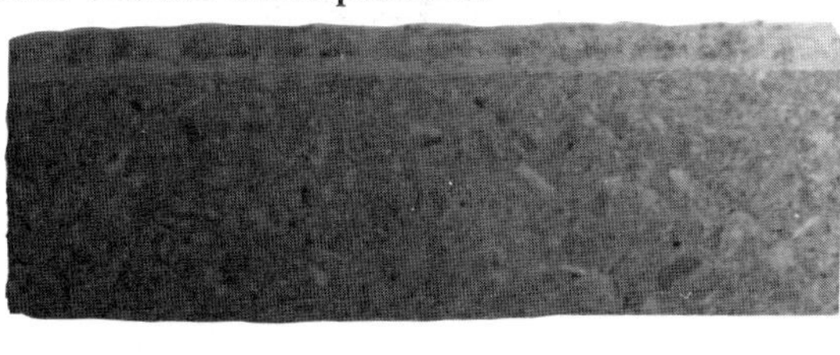

Figure 14
Glue laminated wood beams are an excellent structural building material for buildings of large size. This method demonstrates high strength-to-weight ratios while minimizing cost of the structural components.

Figure 15
The value of hardwoods for fine furniture construction will never be replaced (Ethan Allen).

Figure 17
Construction of fine wood furniture often involves delicate hand work during the manufacturing process (Photo courtesy of Ethan Allen).

hand crafting (Figure 16) and hand carving still have an important place. Artistic specialty areas are unparalleled (Figure 17).

Paper

The importance of paper as a modern industrial material should not be underestimated. Paper is an indispensable material for the support of the modern lifestyle. Major uses of paper include **printing** and **packaging.** Modern paper manufacturing makes available a large number of specialty paper types to meet the many specialized needs of the paper user. Heavy gage paper is used as an inexpensive construction material for many consumer products and paper has even been used for clothing. Rising energy costs and an emphasis on conservation of natural resources has prompted the growth of a large waste-paper recycling industry in recent years. Today, waste paper is a valuable commodity and can be effectively reused rather than dumped or burned.

Figure 16
The construction of fine furniture will always ensure employment of the artist and highly skilled craftsperson (Photo courtesy of Ethan Allen).

The primary raw material for paper is cellulosic fibers from wood or other plants. Various forest trees are used for wood pulp and, at present, much paper is made from recycled waste paper products. Various fabrics in rag form including felt, linen, and wool are also used in paper making to achieve desirable paper characteristics.

The paper-making process is somewhat complex and, like many other continuous industrial processes, is highly automated, carefully monitored, and computer controlled. In the paper-making machine, a liquid mass of wood pulp and rag pulp is fed onto a traveling wire screen conveyer. As the wire screen moves, water drains from the wet pulp fiber mass transported by the screen. Suction systems remove further water and the pulp mass begins to solidify as paper. Further operations

Figure 18
Air jet weaving machines produce quality woven fabrics at high production rates (Burlington Industries).

remove additional water by pressing and rolling. Final finishing is accomplished by calendering rolls exerting many tons of pressure on the paper and providing the desired thickness and surface finish.

FABRICS

Fabrics are important for **clothing** and for **upholstery** both in the furniture and the auto industry. Fabrics may be divided into two major classifications, **synthetic** and **natural.** Synthetic fabrics are primarily derived from **plastic** technology. Polyester is an example. Synthetic fabrics are in widespread use for clothing, carpeting and upholstery applications. Natural fabrics made from natural fibers include cotton, wool, and silk.

Many types of fabrics, both synthetic and natural, are made by weaving threads of the material into cloth of sufficient size for manufacturing fabric products. Modern fabric mills make use of highly automated and computer controlled water and air jet weaving machines (Figure 18).

Fabric processing involves cutting and joining primarily by stitching. All of the modern methods are applied including computer control of cutting and sewing. Mass production piece-part cutting of clothing pieces may be done by band sawing or with laser cutting systems. Modern fabric manufacturing is a continuous and highly automated process where computer process control (CAM) is widely used (Figure 19). Fabric mills have made use of the latest factory automation such as the wire guided computer controlled material handling system shown in Figure 20. Fabric finishing and dyeing are maintained at a high level of consistency using computer controlled processes (Figure 21).

Figure 19
State of the art fabric production making use of modern CAM results in products of high quality (Burlington Industries).

Figure 20
Automation in fabric manufacturing material handling involves driverless computer controlled transports (Burlington Industries).

Figure 21
Computer technology influences all phases of fabric manufacturing. Consistency, efficiency, and quality are computer monitored during the dyeing and finishing process (Burlington Industries Inc.).

RUBBER

Natural rubber or **latex** is derived from the natural gum of the rubber tree. Prior to World War II, this material was important in the manufacture of many products and sources of supply were considered to be valuable. The geography of the war placed sources of natural rubber out of reach and the needs of the allied military exceeded the available supply. The situation stimulated the development of synthetic rubber chemically derived from the basic chemical constituents of natural rubber. Once the chemistry of rubber was understood, a virtually unlimited supply of the material was then assured.

The applications of rubber are widespread. The material is used for coatings, tubing, hose, and tires. Rubber processing methods include many of the same processes used in plastics. Among these are various **molding processes** for products such as **tires. Extrusion** is used for products such as **tube, moldings,** and **hose. Calendering** (rolling) is used for sheet products and coating rubber on fabrics and metal.

NATURAL MATERIALS

Natural materials include **clay** and **adobe** from which **bricks** are made. These materials have long and distinguished histories as fundamental building materials. **Stone,** including granite, marble, slate, and sandstone have also been used as material for buildings and decorative facings. Although these materials are giving way to concrete, steel, glass, plastic, and aluminum in modern buildings, they are still used to some degree as construction materials and are still very much in evidence in distinguished architecture throughout the world.

Leather, derived from animal hides, is another natural material with a long history of use and it is still much in use today. Shoe manufacturing makes use of leather and although many modern shoes make use of plastic and other synthetic materials, leather still remains a superior material for this application.

Natural fabrics such as wool, cotton, linen, and silk still have wide appeal for clothing and other uses. Even though synthetic fabrics are in wide use, natural fabrics will always have a prominent place in fabric product manufacturing. They are processed by all the conventional manufacturing processes such as weaving, cutting, stitching, and bonding, and their product applications are widespread and well known.

CONSTRUCTION MATERIAL

Construction materials include all those from which structures are built, whereas **engineering materials** are those from which products are made. Construction materials include many natural materials such as stone and wood, or material made from natural components such as bricks and rammed earth (compacted dirt). All of these materials have long and distinguished histories and will probably always have some application for building certain types of structures.

In modern construction of large structures such as buildings, bridges, and highways, **steel and concrete** stand out as the most versatile and widely used materials. Structural steel is a totally essential material for modern large-scale construction. Modern steel production along with advanced metallurgical technology have developed high strength structural steel that has permitted the safe construction of extremely tall buildings. Plans are presently being drawn for skyscrapers that will exceed 2000 ft in height (200 stories) while occupying a minimum of space at their base. It is likely that future building construction in urban areas will favor high rise design in order to save ground space. Structural steels will play an ever-increasing part in this type of building construction.

When steel and concrete are used jointly, an unsurpassed construction material is created. **Concrete (portland cement)** is primarily a mixture of limestone and clay that has been pulverized to an extremely fine powder. This material is mixed with proper portions of sand, graded gravel (aggregate), and water, making a thick fluid mass. Large mixing machines that are often truck mounted are used to prepare the mix. Thus the concrete may be easily delivered to a job site.

In most cases where concrete is the primary construction material, it will be reinforced with **steel reinforcing rod,** often known as **re-bar.** Steel reinforcing allows much more weight to be supported by concrete structures. Thus, the material is exceptional for dams, bridges, runways, highways, and buildings. Steel reinforcement rod and bar is often arranged in a lattice work and laid in the form before the concrete is poured. The lattice may be welded or wired together to hold it in position. After pouring the concrete, the steel lattice is permanently encased and can then act to increase the load-bearing capacity of a concrete structure.

Reinforced concrete has many advantages as a construction material. It can be cast into almost any shape using simply constructed wood forms. It can bear large loads. **High strength concretes** containing polymers will withstand loads of **over 10,000 psi.**

Precasting of building wall components is a popular and versatile method of manufacturing building structural components (Figure 22). In this method, building walls are precast in molds, then removed and transported to the building site where they are bolted together. This method is very popular for commercial building and erection time is very fast. Cement walls for buildings can also be poured in place and when properly reinforced, are suitable for high rise buildings.

Figure 22
Buildings constructed from precast concrete walls are erected quickly in a large variety of sizes and floor plans.

REVIEW QUESTIONS

1. Name some advantages nonmetallic materials have over metallic materials.
2. What are ceramics and what are they used for?
3. What single property makes ceramic products so useful?
4. List some of the major manufacturing processes used to make ceramic products.
5. What are the major applications of wood and paper?
6. What processes are used in woodworking? Do these resemble machining processes in any way?
7. List some common natural materials.
8. What are the advantages, if any, to using natural material over synthetic materials? What are the disadvantages, if any?
9. What is concrete and what are its primary applications?
10. What can be done to concrete to enhance its strength characteristics?

MANUFACTURING SYSTEMS PART III

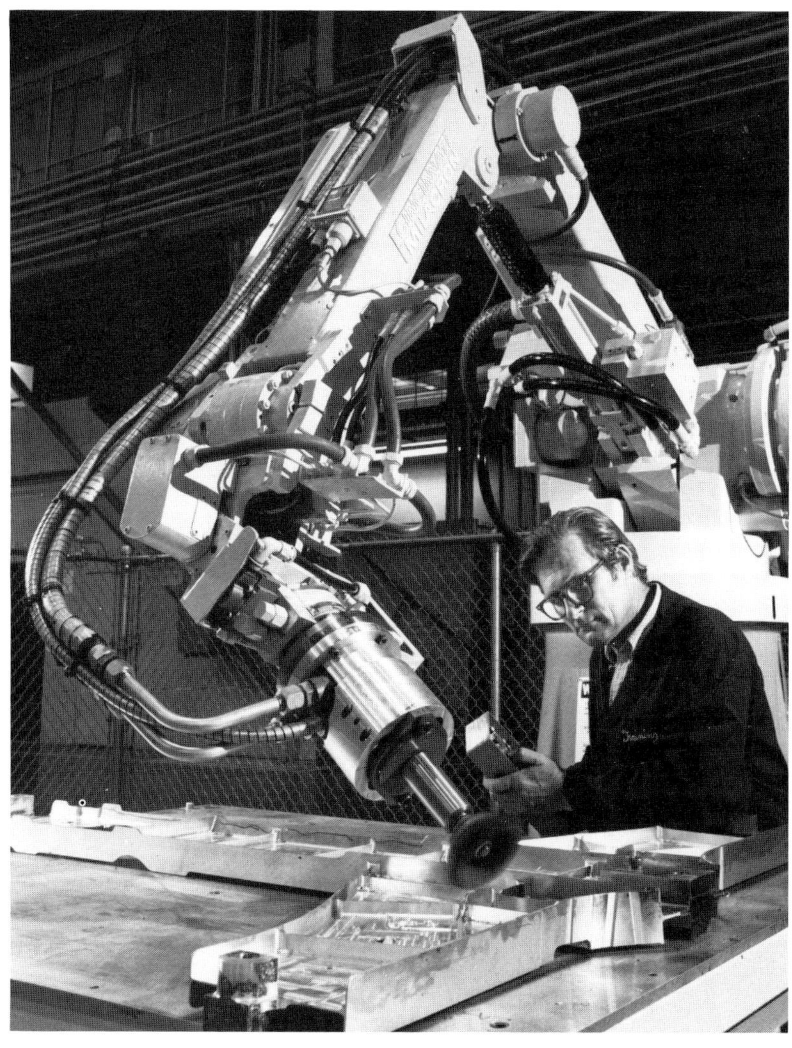

The computer controlled industrial robot, a major contributor to modern manufacturing automation, wire brush deburrs aluminium aircraft parts resulting in a 40 percent man-hour saving over hand methods (Northrop Corporation).

Materials and specific processes are, of course, essential to the manufacture of any product. However, the actual activities of manufacturing involve a great deal more than simply having material and processes available. In Part III you will see how the materials of Part I and the specific processes of Part II are brought together to form a complete manufacturing capability. This capability produces the vast variety of products, ranging from individual parts to complex assemblies such as autos, aircraft, and consumer products, that establish the lifestyle in a modern industrialized society.

Accomplishing the miracles of modern production manufacturing requires the fully integrated effort of all the manufacturing team members. Products must be **designed, prototyped,** and **tested. Production facilities** and **special tooling** must be created. Manufacturing must be **automated** so that large numbers of exact duplicate product units may be made quickly, efficiently, and at competitive prices. **Quality assurance and control** must be maintained at every step in the manufacturing process.

As the continuing research and development effort devises new materials and new methods, future manufacturing will be accomplished on an ever more automated and high-technology level. Individuals staffing manufacturing jobs in the age of high technology, will require a knowledge of **computers, industrial robotics,** and **fully automated factories.** Training in new methods will play an ever-increasing part in supplying the qualified manufacturing technicians and engineers of tomorrow.

CHAPTER 18 DESIGN, TOOLING, AND PRODUCTION LINES

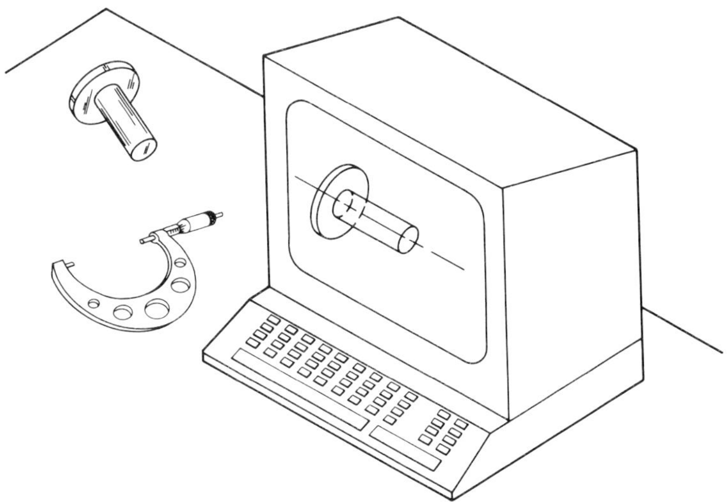

Products seldom just materialize. Most are available only because a great deal of time has been invested behind the scenes in product design work. **Design** is the process of **product creation on paper,** or in the mind of the design engineer. The purpose of this chapter is to survey the major phases of design and to examine some of the modern tools that a designer has to work with. This chapter also examines the development of the support tools and equipment necessary to support product production, and the production line system.

OBJECTIVES

This chapter should enable you to:

1. Explain the phases of the design process.
2. Describe the use of a CAD system and explain its function as a modern design tool.
3. Describe the functions of production tooling and describe common examples.
4. Describe the advantages and disadvantages of production lines.

DESIGN

The first step in product development is design. Design is motivated by the following factors:

1. The need for a specific product.
2. The level of available manufacturing technology.
3. Mass production capability and production cost.
4. Safety and reliability.
5. Marketability.

Need

Most products are designed and produced because a need has developed for them. With the tremendous capability of modern industrial manufacturing, needs for products are easily developed and, in fact, they are often created through such techniques as the advertising of features and options on products, thus inducing a prospective customer to buy a new and better model. This is especially true in the consumer goods area. Therefore product designers are engaged in a constant process of designing and redesigning products to fulfill a continuing need.

Capability of Available Manufacturing Technology

The capability of the existing manufacturing technology is an extremely important consideration in product design and development. The term **state of the art** is frequently heard in product advertising. This means that the product is designed and built to the latest design specifications, and from the latest materials available. State of the art is probably more relevant to high technology items, where products undergo significant design changes as new and better materials and techniques of manufacturing become available, than it is to continued production of standard products where the design is more or less fully developed. As technology develops new and better processes and materials, the designer naturally makes use of these to design and redesign products accordingly.

For example, during the vacuum tube age of electronics, hand-held calculators and small portable computers with low power requirements were not available. The advent of solid state electronics completely revolutionized this technology and permitted the electronic equipment designer to create an entirely new selection of products, heretofore unavailable.

The advent of integrated circuit electronics (ICs or chips) coupled with solid state devices, has profoundly altered the technology once again, not only in the selection of products available, but in the way in which they are and will be produced. This has opened many new avenues of design.

Production Capability and Costs

Most products that are made for the general public are only successful if they can be mass produced and marketed at affordable prices. The best product design is only as good as its ability to be competitively produced. Therefore a designer must consider these important questions when a product is being developed.

1. Is the production capability presently available?
2. If not, will sales of the product justify development of new production technology?
3. Can the product be manufactured and marketed at a cost that will return investment and profits?

These questions are complex and often occupy much of manufacturing and financial management's time. In some cases design of a product may also involve the parallel development of a new production capability necessary to its manufacture. This can be a costly and risky venture, but it can also be enormously profitable if successful. Manufacturing industries often dedicate large portions of their annual budgets to the research and development of new products and the manufacturing systems for them.

Safety and Reliability

Safety and reliability are critical factors in the design of many products, including aircraft, public works, ships, medical equipment, weaponry, automobiles and consumer products. In recent years product safety considerations have attained greater importance. This has been because of increased public awareness resulting from improved communications and more intense interest in consumer affairs. The designer's responsibility is equal where personal and public safety are concerned.

Reliability and safety are closely tied. Indeed product safety can be measured by reliability. For example a backup electrical generator for emergency power must start and run reliably as lives may depend on it. Aircraft control surface actuators must work reliably to ensure that the aircraft and passengers are safe, and automotive steering and braking systems must be extremely reliable. Safe products are generally reliable products, and reliable products can be designed for safety as well. These important considerations must govern the work of the designer in every design effort.

Marketability

If the manufacturing capability exists or can be developed, the next step involves marketing and distribution. It has often been stated that invention is one thing, but distribution is everything. This adage is very timely. If a product cannot be properly designed and manufactured in quantities to meet demand and then sold at an acceptable price, it will probably not be successful. The market environments of free enterprise economies are ruthless in selecting only those products that survive the tests of efficient cost-effective manufacturing.

Costs incurred in marketing include such things as advertising, national or international sales organizations, packaging, distributors and agents, distribution costs, and warranty services. The selling price to the end customer must reflect any or all of these expenses while being low enough to support sales and meet competition.

THE DESIGN PROCESS

The process of design generally begins with an idea that develops from a need determined in many cases by market research. Anyone can design; however a manufacturer of technical products will generally employ individuals with **specific engineering background** as designers. Design engineers have training in the selection and applications of materials and manufacturing processes. They must work closely with the marketing and distribution departments in a manufacturing company; thus, product development must be an integrated and coordinated effort.

Design Phases

Product design involves the following phases.

1. Sketches.
2. Production prototype.
3. Prototype testing.
4. Design review and evaluation.
5. Production drawings.

Sketch Phase

The designer will often begin work by freehand-draw-

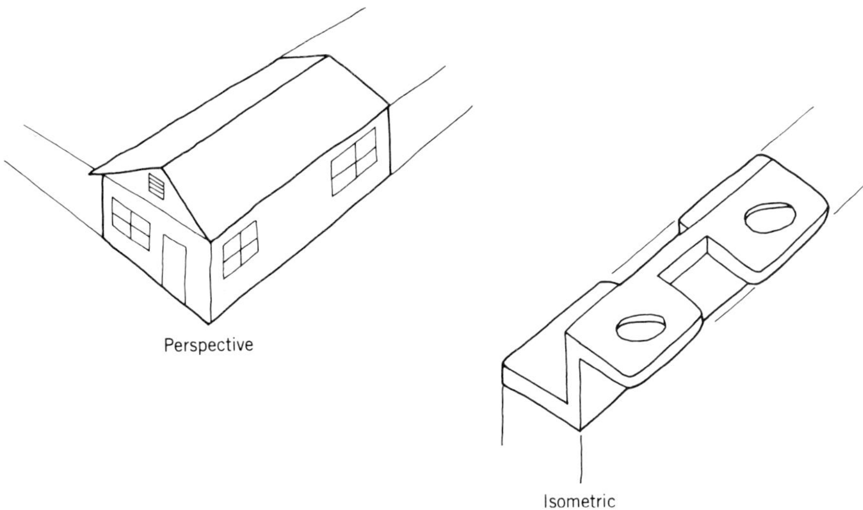

Figure 1
Freehand pictorial sketch formats.

ing a sketch of the item to be manufactured. Various sketch formats are used, often in pictorial styles or artist's conceptions (Figure 1). These enable the designer to see what the product will look like in its finished form.

Designing by Computer **Computer aided design and/or drafting (CAD)** can also be used to aid the engineer in the initial design phase. This developing technology is a spin-off from computer development and has rapidly become very popular. CAD systems are in fact revolutionizing traditional processes of design. The computer in the CAD system can create two- and three-dimensional graphics (Figure 2) that can be manipulated by the designer in order to make quick adjustments and changes to the design. CAD graphics may be transferred to paper, if required, on a plotter (Figure 3). An advanced CAD system can subject a computer model of the design to many variables that it could encounter in its service environment, before an actual prototype or model is produced. This capability of CAD makes the designer's work much more reliable while saving large amounts of money and material that would otherwise be required to produce and test prototypes. Building an actual prototype can be a very expensive and time-consuming process. In the case of large and complex products such as buidings, ships, dams, or bridges, prototyping would be impossible. A computer model can be created, and subjected to the environmental variables that the design will encounter in service. All this can be done quickly and at low cost. CAD is also extremely versatile in testing a new design without the need to risk money, personnel, or expensive materials.

Figure 2
Computer aided drafting design or CAD (Lockheed—California Company).

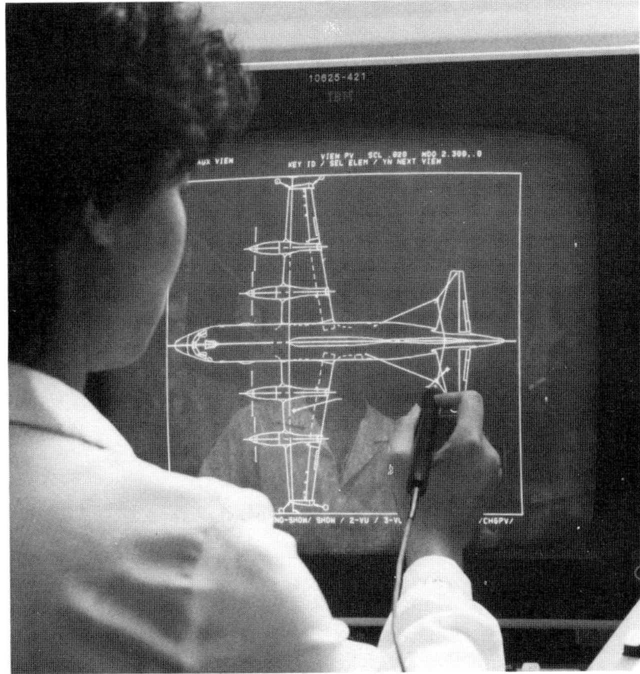

Figure 3
With the CAD plotter, hard copies of drawings are produced (Lockheed—California Company).

When using a CAD system, the designer can sketch ideas directly on the screen of a video display, use a digitizer pad (Figure 4), or enter data directly from the computer keyboard. The digitizer pad converts the fundamental elements of geometry such as points and lines into digital data (bits). The digitized data are understood by the system computer and the information appears simultaneously on the video display (CRT) as it is drawn on the digitizer pad. Some CAD systems employ a dual video screen and will produce screen images both in single color (monochrome) and in full colors.

Figure 4
Dual screen CAD terminal with digitizer (Intergraph Corporation).

Designing with a CAD system has many advantages. The computer can electronically generate three-dimensional drawings that may be rotated so as to view a part or assembly from different points. The operator may also zoom in and out on any portion of the drawing in order to reveal details. If the CAD system is used for schematic design, it may be done in layers of various color. Individual layers of the schematic may be electronically removed and replaced as required to study the interrelationship and position of overlapping systems in the design. This is particularly useful in integrated circuit (IC or chip) electronic design as well as in the layout of piping and electrical systems. Colors can also denote changes in stress or temperature in different areas of the design.

Another advantage of the CAD system is that many different part drawings may be permanently stored in the computer memory. These may be called up for review and revision at any time. Thus, the system speeds revised drawing information to the production shop floor.

The ultimate use of the computer as design tool is the ability to put designs to rapid test and evaluation prior to prototyping or manufacturing. This permits a design to be better evaluated prior to actual production. Considering the cost of modern production, the ability to evaluate designs without having to produce actual prototypes or parts is a distinct advantage.

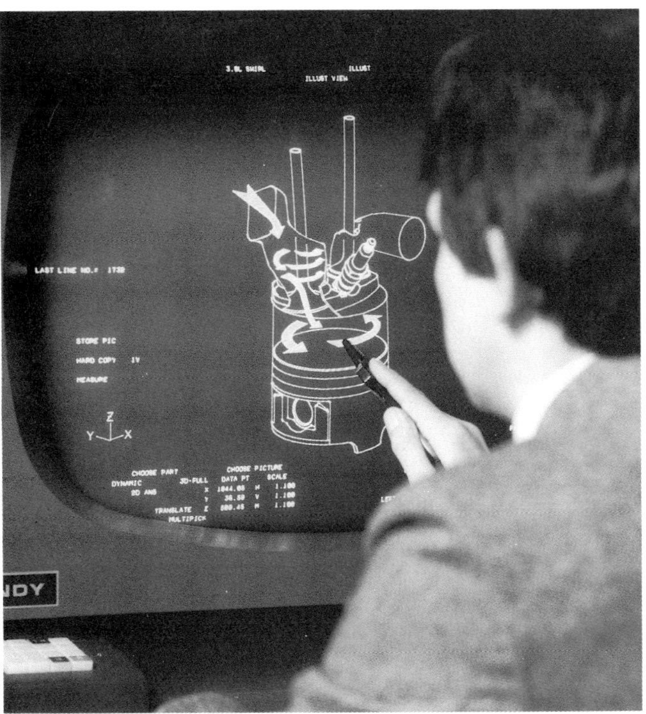

Figure 5
CAD used in evaluating fuel swirling in an internal combustion engine cylinder (Ford Motor Company).

Examples of design evaluation by computer include such diverse activities as studying the air–fuel mixing patterns in an internal combustion engine (Figure 5). Structural design may also be evaluated and weak points identified (Figure 6). Re-design may then be accomplished immediately to ensure that the design meets specifications. Also a computer may be programmed with a large data base representing a mathematical model of a prototype or production unit. By scanning the actual production model and recording measurements, an exact comparison of design dimensions and actual dimensions can be made by the computer (Figure 7). Through this process, the engineer can determine how accurately production equipment is producing product units in conformity with design specifications.

When CAD systems are coupled with their manufacturing counterparts, or **Computer Aided Manufacturing (CAM) systems,** a fully integrated CAD/CAM, CADAM, or CIM system may be developed. Computer generated information both for design and to control production may be fed directly to the production stations. In Figure 8, the computer locates and directs the robotic spot welding of the automobile body. Weld positions are shown on the computer display as the robot spot welder accomplishes its production welding task.

Figure 6
CAD for evaluating a structural design (Ford Motor Company).

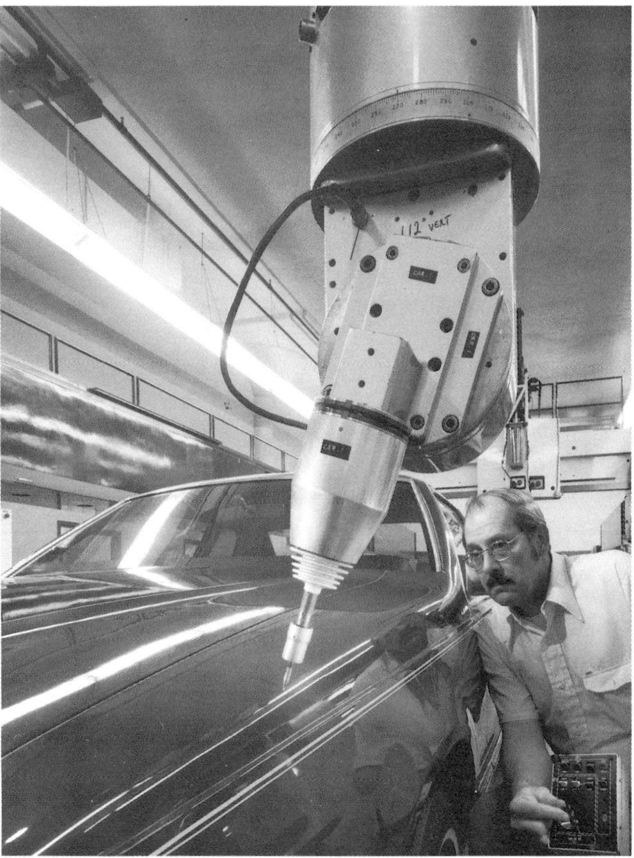

Figure 7
A scanner takes measurements on a production unit and compares them against a computer data base containing exact design specifications (Ford Motor Company).

Figure 8
Integrating CAD and CAM in production spot welding. This process is also known as CIM (Chrysler Corporation).

The integration of CAD and CAM has, in all its forms, become a major force in tying design to production in modern manufacturing. Integrated CAD/CAM systems and the steps toward complete factory automation will be discussed further in Chapter 19.

Prototyping Phase

A prototype is a model from which a product may ultimately be mass produced. Prototypes may be the same size as the final product or they may be scaled down (Figure 9). The purpose of prototyping is to build a working model of a newly designed product to determine if the design is acceptable. Although prototype designing can be done with the computer using a CAD system, sometimes there is no substitute for an actual working model, especially if physical product testing is to be done.

A manufacturing industry often employs people skilled in model making or prototyping; prototyping may also be jobbed out. However, costs can be high, especially if the design must be reworked. Nonetheless, some phase of prototyping is often an integral part of product design and development.

Prototyping in modern manufacturing can make use of some high technology techniques. For example, a master model of an automobile where parts are made from plastic materials may be used to permit reliability engineers to examine how parts fit in a complex auto body assembly (Figure 10). Each plastic part on the model can be removed and replaced with its production metal counterpart. The precision computer-linked scanning equipment shown in the illustration determines how each part fits compared to a perfect computer model of the desired design. Many thousands of mathematical data points representing design specifications are programmed into the scanner computer. This permits an extremely accurate comparison to be made between the prototype and its actual design specifications.

Design Review and Evaluation Phase

Test data from the prototype testing, or computer models if CAD is used, are carefully analyzed and evaluated (Figure 11). If there are problems, the design will be modified accordingly. This may involve further prototyping. Sketches and/or drawings used in the prototyping will be revised as needed.

Production Drawing Phase

When a decision is made to take a proven design into production, finished manufacturing drawings are made and copies are distributed to all departments responsible for product manufacturing. In fully integrated CAD/CAM systems, electronic drawings are stored in

Figure 9
Prototyping and modeling (Lockheed—California Company).

Figure 10
Plastic master cube used to check auto body parts against design specification (GM—Pontiac Division).

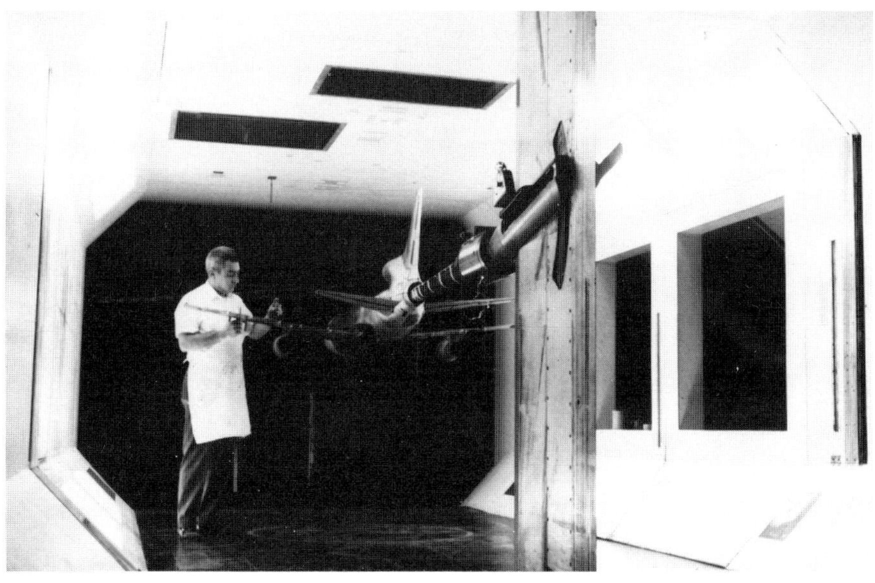

Figure 11
Wind tunnel testing of an aircraft prototype (Lockheed—California Company).

the computer and are called to the computer display terminals or plotted on paper if required. In conventional practices where design is done on paper, rough sketches and drawings used for constructing prototypes are converted into finished manufacturing drawings in the drafting department.

Typical manufacturing drawings are usually done in the standard **orthographic** format showing an object in any of several possible views (Figure 12). In addition, several types of **auxiliary** and **section** views may be used (Figure 13). Occasionally a pictorial format such as **isometric** may also be used on a manufacturing drawing (Figure 14). This is usually done to show relationships of parts in an assembly and is usually not dimensioned. Final-form drawings made from the designer's sketches are most often prepared in the company's engineering drafting department by professional drafters or generated by computer plotters.

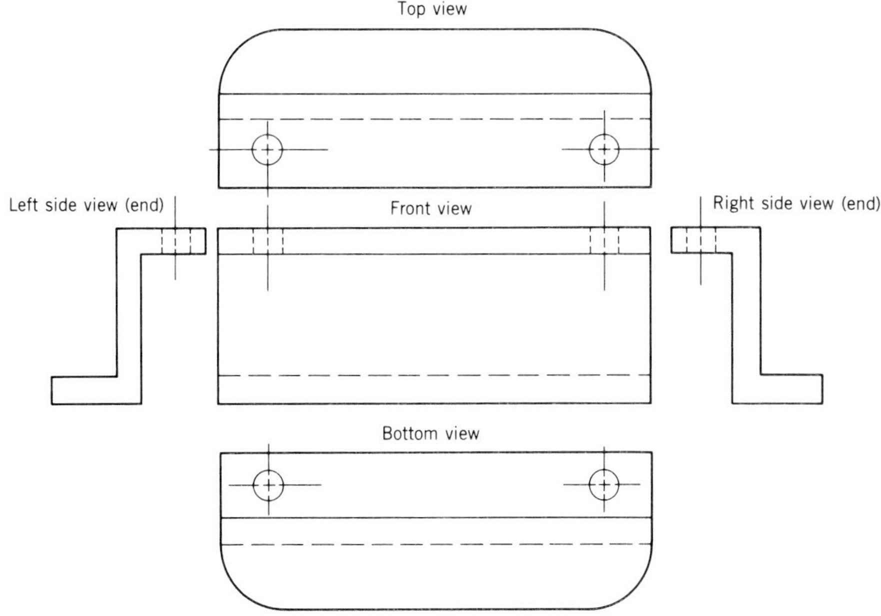

Figure 12
The standard orthographic format of manufacturing drawings.

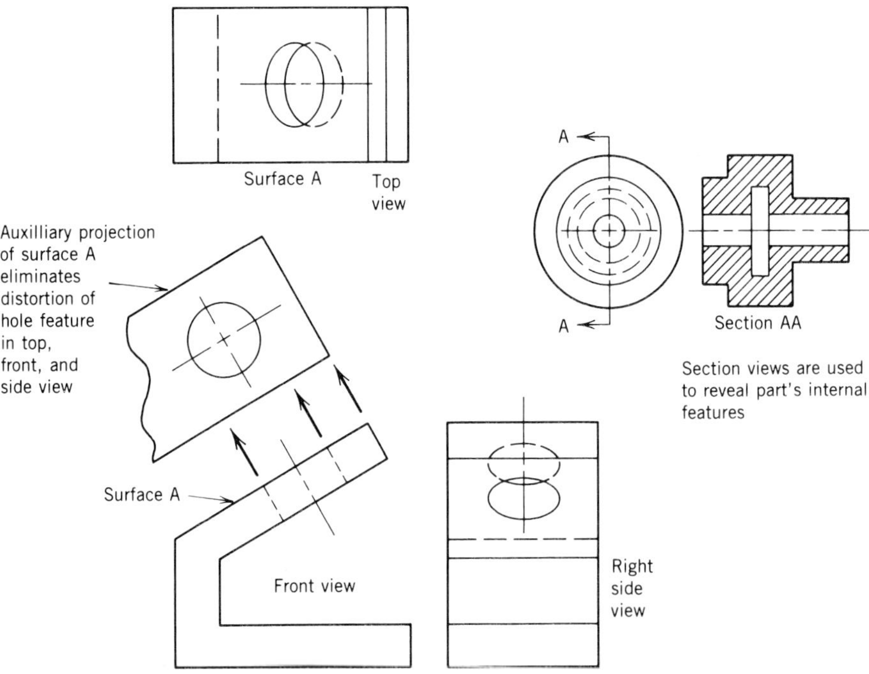

Figure 13
Auxiliary and section views on orthographic drawings.

Figure 14

Isometric pictorial of a complex assembly (Lockheed—California Company).

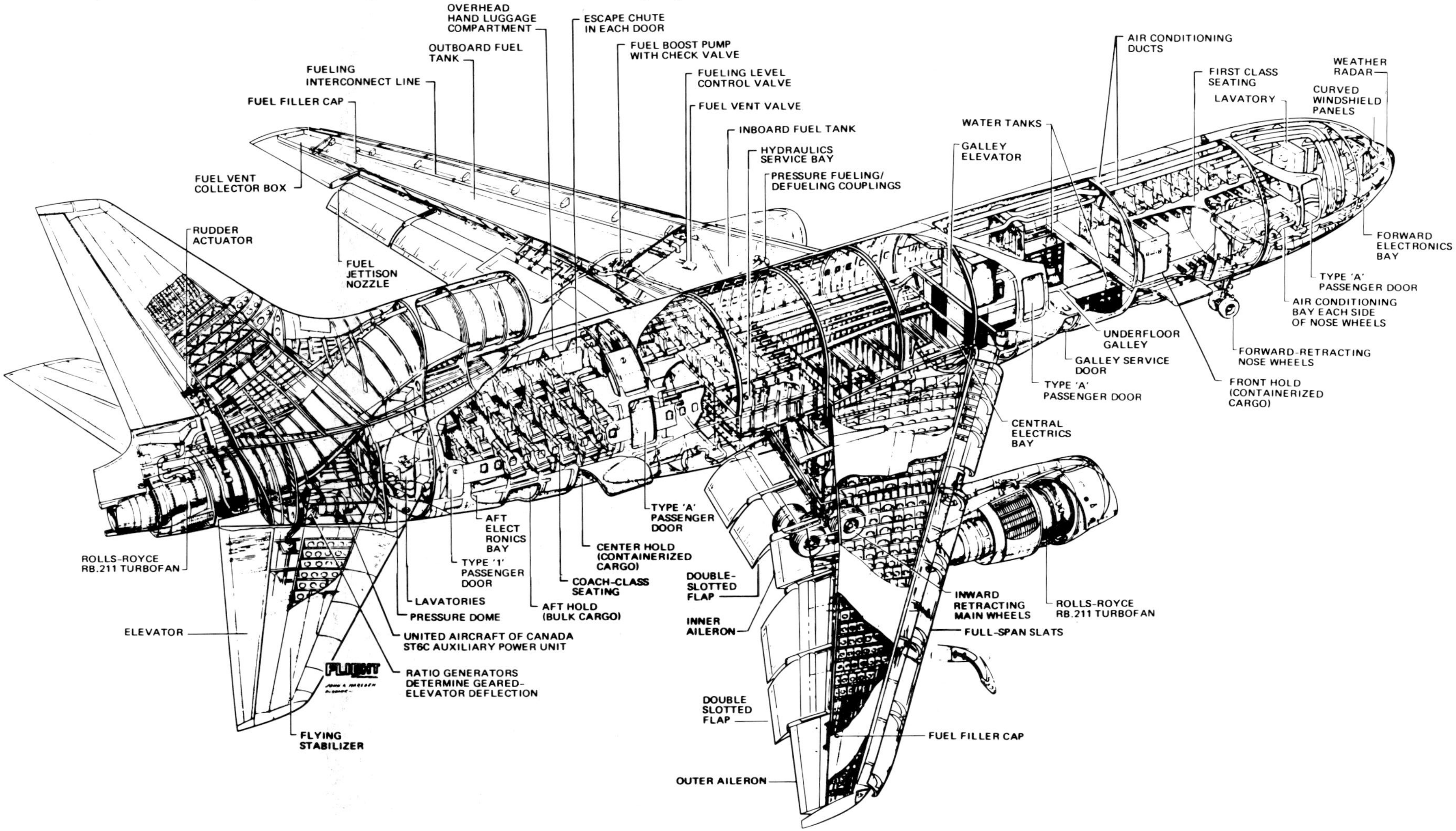

Drawings are done to exacting standards often on company-formatted drawing sheets. The completed drawings are then checked for accuracy and to see that no pertinent data have been omitted. They are then copied by various processes such as blueprinting and distributed to the various manufacturing departments or to jobbers that are involved in supporting a company's production requirements. The original drawings are carefully preserved and as production progresses, they are revised as necessary to meet changing design requirements, improve production flow, or solve production difficulties.

Methods engineering is generally the responsibility of industrial engineers, who select and copy portions of drawings and generate **operations sketches** (Op. Sketches), **manufacturing outlines** (MOs), or **assembly instructions.** These documents accompany parts as they flow through a production process. A typical operations sketch shows only a small portion of a production drawing. It emphasizes dimensional data and other pertinent information necessary to a specific manufacturing operation on that portion of the part.

Detail and Assembly Drawings Most product manufacturing uses two forms of drawings. First is the **detail drawing** (Figure 15), showing only one part and containing all information required for its manufacture. Important information contained on a detail drawing includes:

1. Part name.
2. Drawing number.
3. Material specifications.
4. Dimensions and locations of part and part features.

Figure 15
Detail drawing.

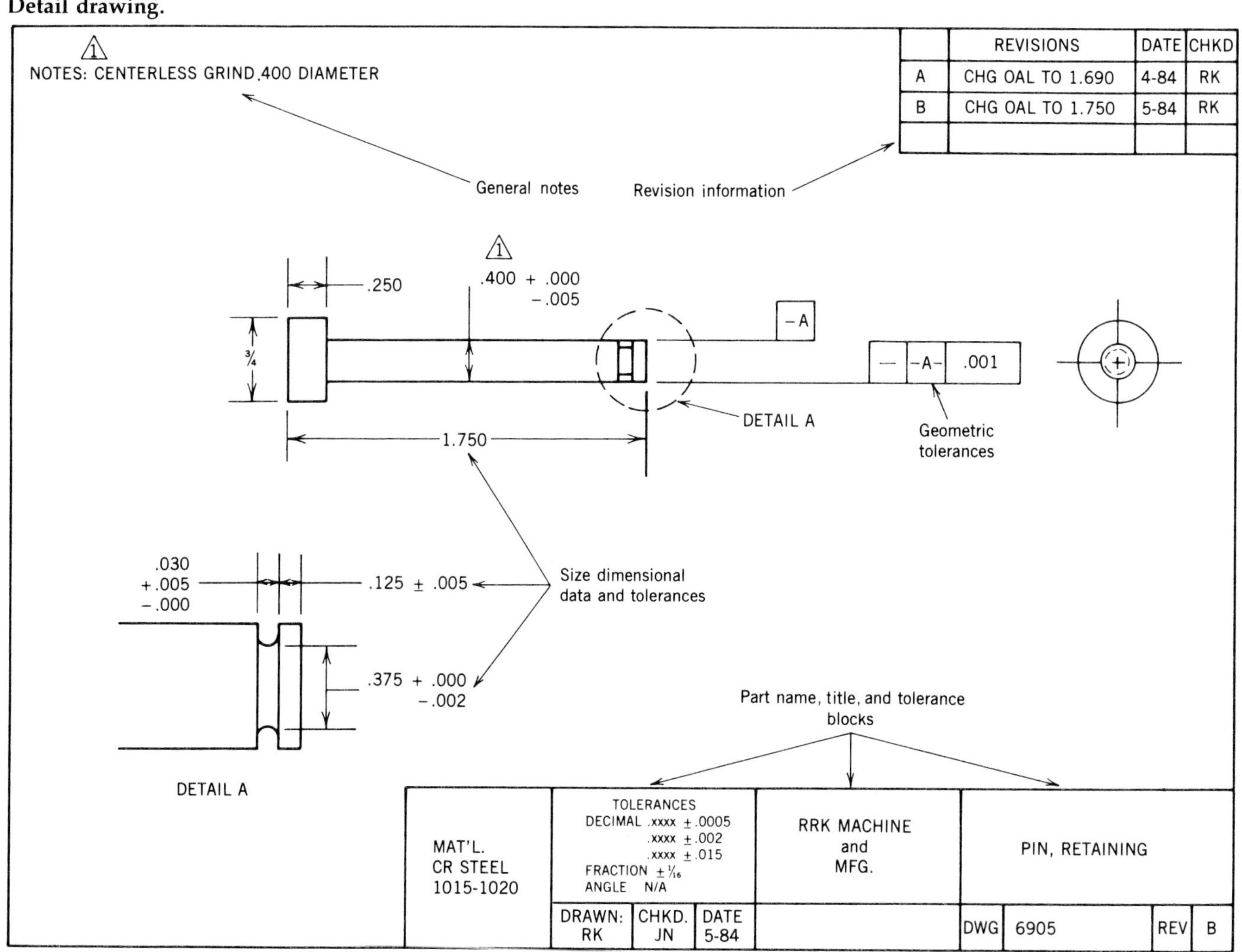

5. Tolerances for size and geometry.
6. Revision codes and dates of most current revisions.
7. Any special notes regarding special handling during manufacturing.

Second, is the **assembly drawing** (Figure 16), a drawing showing several parts as they would appear in a complete assembly or subassembly. Important information contained on an assembly drawing includes:

1. Assembly name.
2. Assembly drawing number.
3. Bill of materials, a list of individual parts with individual names and often cross-reference numbers to detail drawings.
4. Dimensions and tolerances pertaining to the assembly.
5. Revision codes and dates of last revisions.
6. Any special notes regarding the assembly.

DEVELOPING PRODUCTION TOOLING

Once a product design is complete, a prototype is built and tested, and the decision is made to take the product into production, the next phase is **production tooling.** Production tooling includes all of the machinery and equipment necessary to support product manufacturing. The specific product to be manufactured dictates the size and capability of the production line required. The estimated demand for the product dictates the required production line capacity.

Figure 16
Assembly drawing with bill of material.

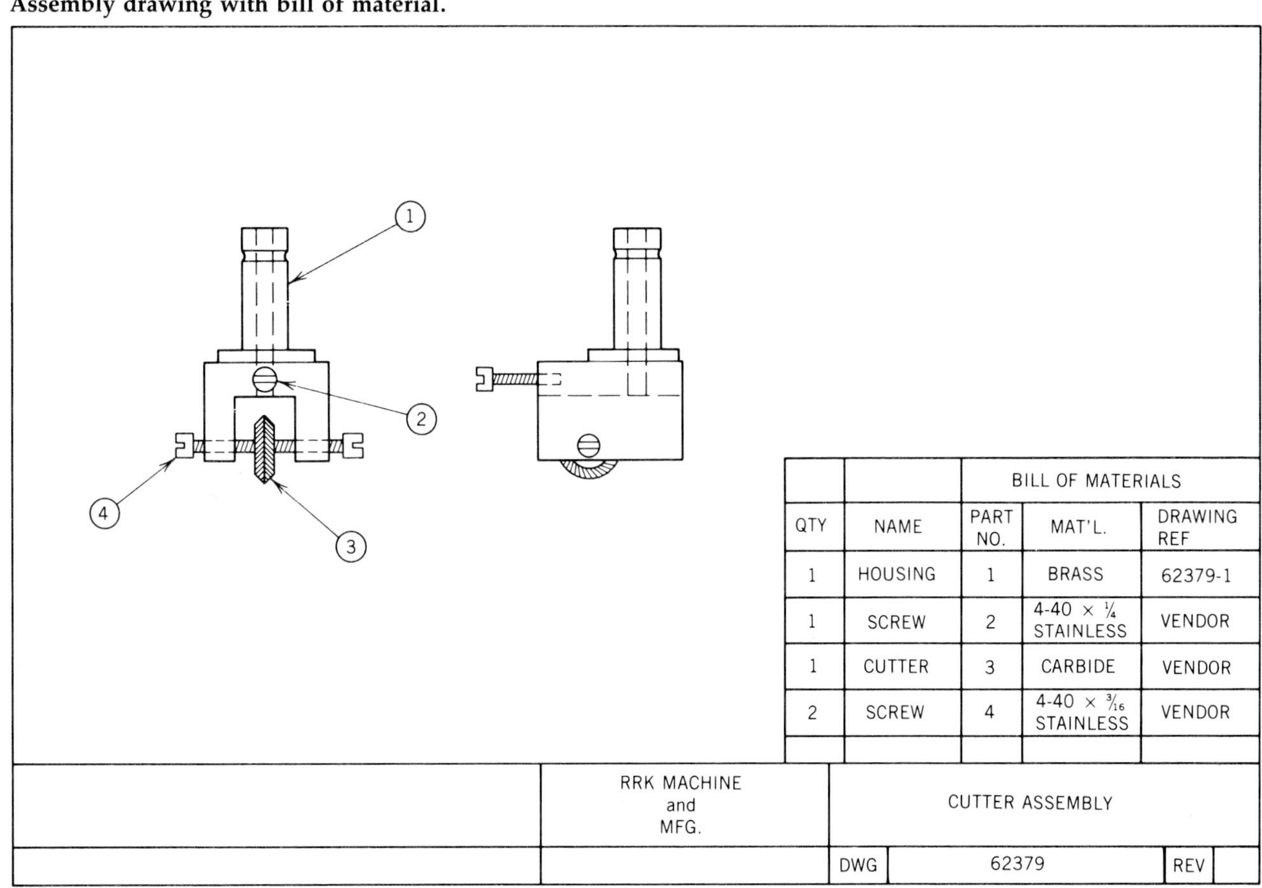

		BILL OF MATERIALS		
QTY	NAME	PART NO.	MAT'L.	DRAWING REF
1	HOUSING	1	BRASS	62379-1
1	SCREW	2	4-40 × $\frac{1}{4}$ STAINLESS	VENDOR
1	CUTTER	3	CARBIDE	VENDOR
2	SCREW	4	4-40 × $\frac{3}{16}$ STAINLESS	VENDOR

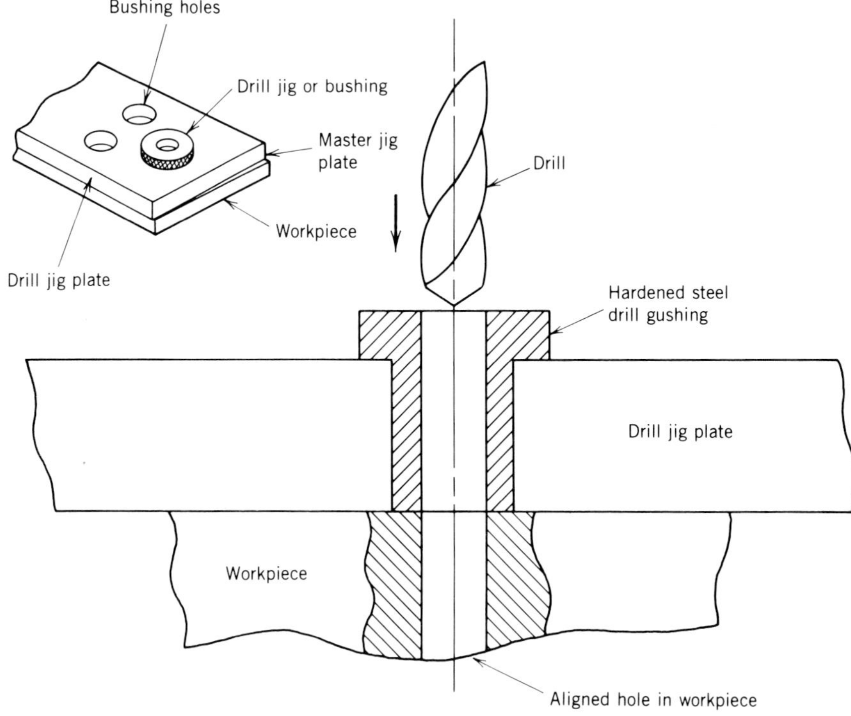

Figure 17
A drill jig is an example of simple tooling.

Any mass produced product will of course require applicable material processing equipment. This equipment may cover a wide range of standard machinery. An example would be machine tools. However, a given machine tool can usually accomplish many types of machining. A numerically controlled machining center, for example, can mill, drill, tap, ream, bore, index, and position a workpiece and perform other functions as well. If only hole drilling in a product were required then it would be more cost effective to use a machine that was limited to drilling only. On the other hand, a manufacturer involved with many different machined products may very well have use for the programmable multifunction use of the CNC machining center.

Even though a manufactured product may pass through many different processes during its manufacture, these individual processes are always repeated exactly on each part, making it necessary to hold and position the part while the specific process is being accomplished. Therefore, it is almost always necessary to develop special production tooling unique to the particular product being manufactured. One might say that this special tooling adapts the general multiuse material processing equipment to specific manufacturing requirements. Included in special production tooling are the following.

1. Jigs.
2. Fixtures.
3. Dies.
4. Molds.
5. Patterns.
6. Cutters.

Jigs

Jigs are special tools primarily designed to provide alignment of parts or tools during a manufacturing operation. A simple example of a jig is a **drill bushing**

Figure 18
Using a jig transit to align aircraft sections (Lockheed—California Company).

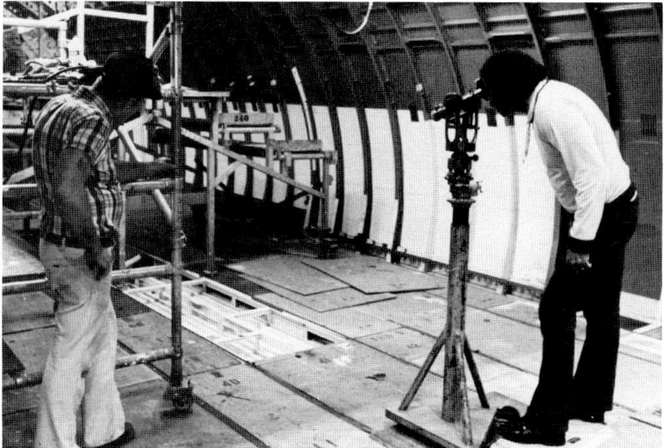

Figure 19
Large tooling fixtures for aircraft skin panels (Lockheed—California Company).

used in machining (Figure 17). A drill bushing or drill jig consists of a hardened steel sleeve through which a drill can pass freely, but with minimum clearance. The drill jig aligns the drill and the resulting hole in the workpiece is then located in its proper position. Drill jigs may be complex tools designed to locate drills for drilling many holes in the same workpiece.

Jigs may also be used to hold large fabrications in position while welding or machining occurs. Alignment of such tooling or parts of an assembly may be done by the techniques of optical tooling. In Figure 18, a jig transit is used to align large portions of an aircraft structure during assembly. Optical tooling techniques have many applications in layout, alignment and assembly of large fabrications such as ships, bridges, and aircraft.

Figure 20
Large aircraft vertical stabilizer tooling fixtures being inspected (Lockheed—California Company).

Fixtures

Fixtures are special tools designed to **hold, support, index,** and **align** a workpiece while a particular manufacturing or assembly operation is performed. An example of a simple fixture is the common machine tool vise with soft jaws that have been machined to a special shape for the purpose of holding a workpiece in a particular position.

In Figures 19 and 20 special fixtures are used to hold aircraft skin panels and the leading edge rib for a large aircraft's vertical stabilizer. Note the large size of this tooling.

Production tooling fixtures, especially in high technology manufacturing, can be complex. In Figure 21 the entire auto body is held in position while numerous holes are drilled for body panel attachment (Figure 22).

Dies

Dies include many types of special tools. They may be **cavities** into which materials are shaped by various processes. An example discussed in previous chapters is die casting where molten metal is poured or pumped into a cavity shaped to the form of the desired product. Dies are also used in machining for **thread cutting** and in **forging** processes. In punch press work, dies are used in conjunction with punches to **cut** parts to exact shapes (Figure 23). This type of tooling can produce a large variety of useful parts (Figure 24).

Dies are also used in **metal bending** applications. Figure 25 shows a press brake type bending die and Figure 26 shows the type of product produced by it. Dies are also extensively used in the plastics industry.

Molds

Molds are most often used to form products from raw material in a **fluid** state. Products formed in molds assume the shape of the mold before they change form, such as by cooling or hardening. Examples would be metal castings, fiber glass, and plastic resins. After the material solidifies by cooling or hardening, the mold is opened, removed, or broken, leaving the finished product (Figure 27).

Figure 21
An entire car body is held in a drilling fixture where body panel holes are drilled in one operation (General Motors Corporation).

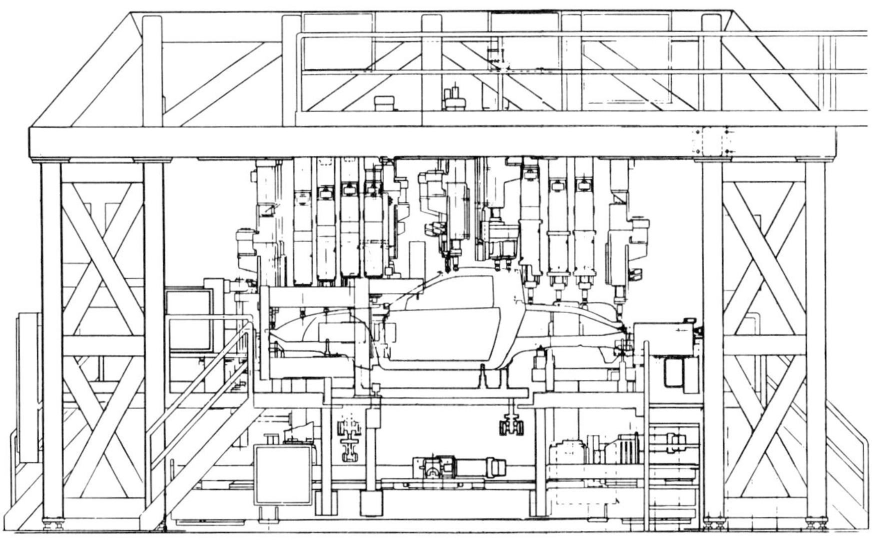

Sequence of Operations

1 *Automatic on-line load*
2 *Clamp body at eight locations*
3 *Position body per LVDT readings*
4 *Set back-ups at front, rear, top*
5 *Verify position for drilling*
6 *Pierce four slots*
7 *Drill 39 vertical holes*
8 *Machine 39 horizontal surfaces*
9 *Verify operations completed*
10 *Unclamp and release back-ups*
11 *Automatic unload to conveyor line*

Figure 22
Body panel drilling operation sequence (General Motors Corporation).

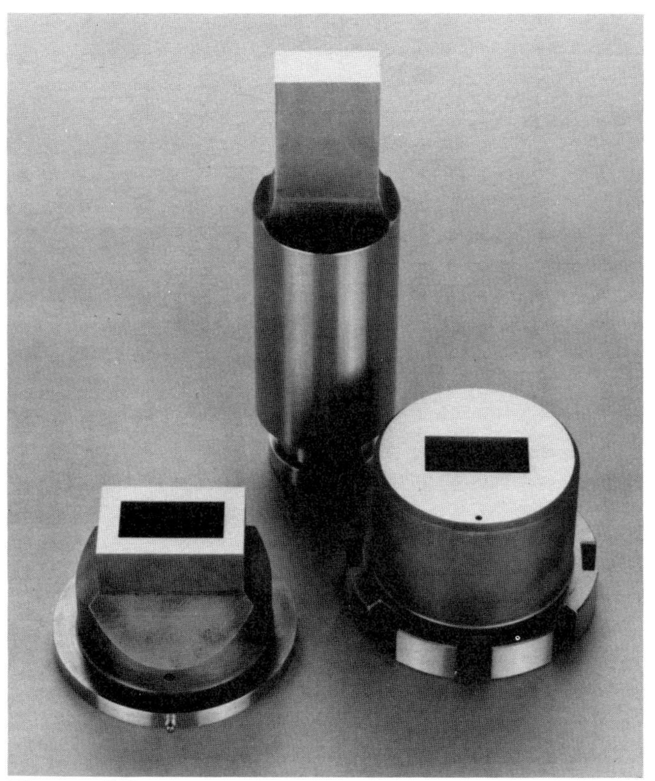

Figure 23
Punch and die tooling (Bracker Corporation/Leland J. Balber Associates).

Figure 25
Press brake tooling for sheet metal bending (Bracker Corporation/Leland J. Balber Associates).

Figure 24
Punch and die tooling can produce a wide variety of end products (Bracker Corporation/Leland J. Balber Associates).

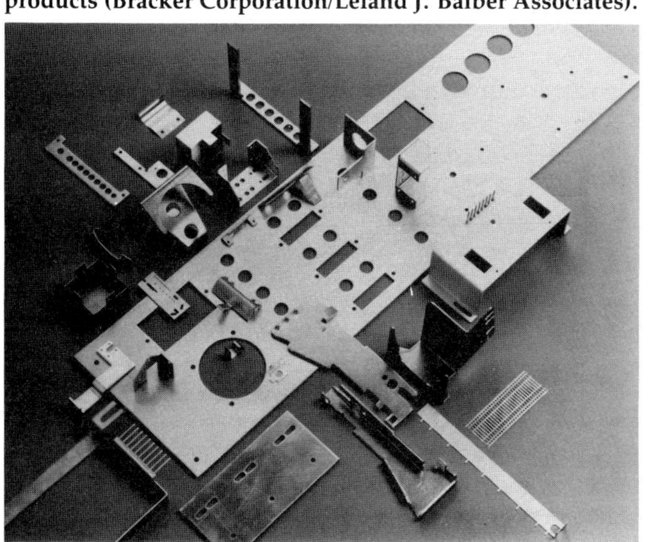

Figure 26
Example of press brake products (Bracker Corporation/Leland J. Balber Associates).

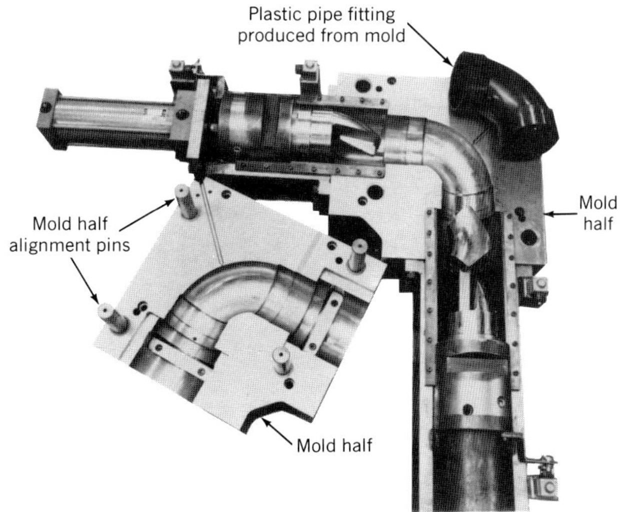

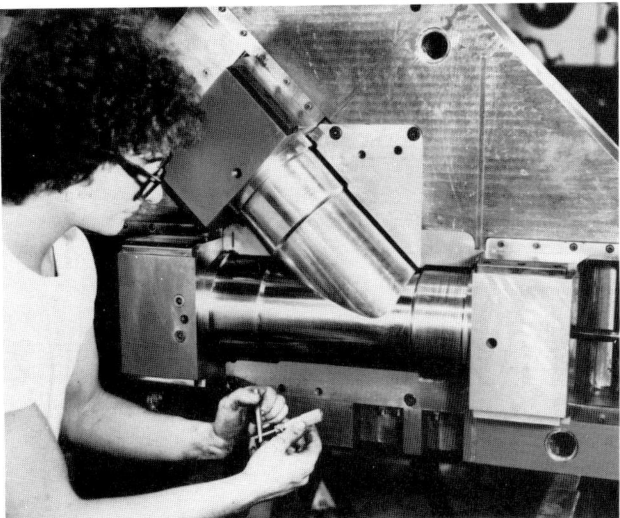

Figure 27
Mold tooling for manufacturing plastic pipe fittings (Crucible Specialty Metals Division of Colt Industries).

Figure 28
Inspecting an aircraft panel bend with a template (Lockheed—California Company).

Patterns and Templates

Patterns and templates cover a wide range of special production tooling. In Figure 28 the **shape** of the bend is checked against a template. In another example, a full sized template of a hull plate for a ship's hull may be used to trace around in order to duplicate parts of the same size and shape.

In the fabric industries, paper patterns are used in cutting materials. These may be computer generated to maximize the number of pieces that can be obtained from a given bolt of material.

Cutters

Cutters are a large and important member of the production tooling family, especially in the machining area. Cutters generally remove workpiece material by shaving off small pieces or chips. They may be rotated against the workpiece, as in the operation of milling and gear hobbing or, as in turning, the workpiece may be rotated against a stationary cutter. In machining operations such as gear hobbing, a large number and variety of cutters are required (Figure 29).

Tool Design and Toolmaking

Tool design and toolmaking are among the most important activities supporting the manufacture of almost any product. Foremost in tool design is **accuracy** that will result in the end product dimensions being as standardized as possible. Therefore special manufactured tooling will be built to exacting tolerances, often in the tenths of thousandths or smaller increments of inches. Building tools to critical tolerances requires the best quality machine tools available with experienced toolmaking personnel to operate them.

The **toolmaker** is usually an experienced machinist who has specialized in toolmaking activities. Toolmaking will often be done within the manufacturing plant in a special machine shop area dedicated to this purpose. Toolmaking personnel are frequently paid premium wages for their expertise and experience.

A typical production tooling item must meet exacting dimensional standards, but must also be convenient to set up in the manufacturing process. The pro-

Figure 29
Many styles and sizes of cutter tooling are required in gear cutting (Illinois Gear, a Household International Company).

cess operator should be able to easily load and unload the tooling with production parts during manufacturing. If the tooling is to be used in conventional machining or grinding operations, provisions should be made to clear chips easily and quickly before new parts are loaded in the fixture.

The production tooling will often have many heat treated parts so that its accuracy and durability will be preserved through much hard use. Most manufacturers will identify and catalog special tooling so that it can be located quickly for use whenever a need arises.

Design and production of tooling often involve as much effort as design and production of a manufactured product. In many cases the **cost of tooling** prior to manufacturing **represents the major cost of production.** Therefore it is often necessary to produce a minimum number of production parts just to underwrite the tooling cost. Only then does further production contribute to the earning stream of the business. In large industries with complex products, the payback process on tooling expenses can take several years.

THE MODERN PRODUCTION LINE AND MASS PRODUCTION

The culmination of design, prototyping, design review, and development of production tooling, is the successful mass production of the product. Since the earliest examples of mass production, it was soon discovered that the work can be accomplished most efficiently by setting up a series of work stations, at which individual processing or assembly operation is performed (Figure 30).

The product is then moved from one station to the next station where the next process is accomplished. This is continued until the product arrives at the end of the production line with all manufacturing operations and/or assembly operations completed. Production by this method is effective for both small and large

Figure 30
On the modern production line a portion of assembly occurs at each station (GM—Pontiac Division).

Figure 31
Production lines are equally applicable to large products (Lockheed—California Company).

products (Figure 31). Products may be moved from work station to work station by hand (Figure 32) or on assorted mechanized conveying assemblies (Figure 33). A typical system used in the automotive industry uses floor conveyers and pallets to move the product along the production line (Figure 34).

There are many advantages to production by this scheme. Each work station can be highly efficient and equipped with any specialized tools needed for the particular process occurring there. In most production accomplished by this method, only one or two processes are usually done at any one station. Hence, personnel staffing a specific station are specialists at their work. If proper supervision is applied, both efficiency and quality can result.

Machining Transfer Lines

One specialty type of production line is the machining transfer line where many repeat machining operations such as drilling, boring, and tapping must be accomplished, and it is not desirable to bring repeat product units to and from a machining center. Production under these circumstances would be inefficient as each unit would have to be mounted on the machine tool,

Figure 32
Moving products along the line by hand is also efficient (Del Manufacturing).

Figure 33
Product transportation by mechanized conveyer (Del Manufacturing).

Figure 34
Palletized product transportation on the production line (General Motors Corporation).

machined, and then off loaded to make room for the next. A much improved system is the machining transfer line where continuous product flow may be maintained while at the same time the desired multiple machining operations can be accomplished. On the transfer line, a series of machine heads are located and tooled for the desired machining operation. The product is then moved along the line to each station where required machining is accomplished. In a sense, the machining transfer line is much like the larger factory production line, but it is used for specialized manufacturing operations. Transfer lines are very popular in production of automobile engines and other high production automotive mechanical systems.

Production Lines, Present and Future

Possible disadvantages to line production are that the work is often repetitive and therefore boring for personnel. Low quality output can result. Line production of complex products and high volume output can be very labor intensive and therefore very expensive. However, the production line, even with its possible disadvantages, is likely to always remain a major institution in manufacturing. Manufacturing industries are experimenting with many techniques to make the production line workers more comfortable at their job, and thus keep quality at a high level. There is also a major trend to further automate production lines using industrial robots to replace people for repetitive or hazardous jobs. With the wide capability of the programmable computer driven robot, it is a certainty that these machines will become ever more present along the production lines of the world's manufacturing industries.

REVIEW QUESTIONS

1. What factors motivate design?
2. What are the phases of design?
3. What is CAD and how is it used in design?
4. What are the advantages of CAD?
5. Why are design review and evaluation important?
6. What are the two major types of drawings found in manufacturing?
7. What information is contained on these drawings?
8. Describe the purpose of production tooling.
9. Who is responsible for building production tooling?
10. What major items are included in production tooling?
11. What are the advantages of production lines?
12. What are the disadvantages of production lines?

CASE PROBLEMS

Case 1
Computers or Prototypes

A manufacturer of complex hydraulic equipment decides to reevaluate product design with the intent of branching into a new market requiring light-weight products (automotive and aircraft). One choice would be to build and test a large group of prototypes. This would take about one year and cost would exceed $500,000 for in-house design work and jobbed-out prototype development.

A CAD/CAM system could be purchased for about $750,000 that would evaluate designs by computer and also could be interfaced with the manufacturer's existing production capability making it more efficient and productive. Complete phase in for the CAD/CAM system to fully support manufacturing would take three years, but the design review capability could be on line almost immediately. The market potential for the new light-weight hydraulic equipment is excellent.

If the company must borrow the money for either project, which do you think would pay off in the long run from both a standpoint of efficiency and return on investment?

Case 2
Automate or Improve Working Conditions

A manufacturer is receiving poor quality work from production lines involving repetitive simple operations on simple parts. The production manager recommends that the company invest in industrial robots to replace all production line workers. The production line foreman suggests that the problem could be solved by rotating production line personnel so that their work is more varied and they take more interest in it. The company gainfully employs several hundred people, but at minimum or very low wages for production line work.

Considering all factors including costs of retooling the factory, the company's public relations image with the community, and the type of product produced, which choice would you make, purchase robots or adjust working conditions for production workers?

CHAPTER 19 AUTOMATION IN MANUFACTURING

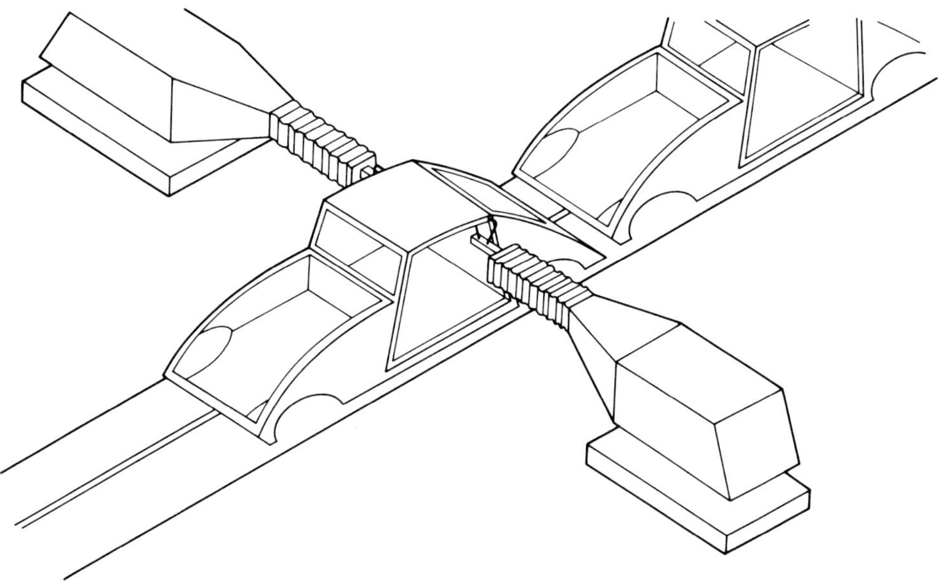

Most modern manufacturing is automated. This means that the processes used are set up in such a way that they can be repeated over and over with little or no need for manual control except to start and stop the process or to change workpieces and tooling. The manual aspects of changing the workpiece and tooling are fast becoming automated as new computerized manufacturing technology is applied in almost all phases of production. Automation trends, already at a high level in many manufacturing industries, will in the future, be refined and applied to an even higher degree. The result of this trend will, in many cases, redefine the duties of today's typical production worker or machine operator. The purpose of this chapter is to explore the techniques used to automate modern manufacturing and to look at the types of automated equipment now being applied in the high technology computer age.

OBJECTIVES

This chapter should enable you to:

1. Discuss reasons for automating manufacturing.
2. Describe the basic factors required for automation.
3. Discuss computer numerical control automation of manufacturing processes.
4. Discuss CAM, CIM, and flexible manufacturing systems.
5. Discuss the place of industrial robots in manufacturing automation.

The emphasis in most manufacturing is on mass production. In the case of most products, it is necessary to produce them in quantity so that the selling price can be low enough to make them attractive to a wide range of prospective buyers. There will always be some demand for one-of-a-kind, hand-crafted products. However, any such product that is built by hand crafting one unit at a time, will, by the very nature of the production time involved, be much more expensive compared to its mass-produced machine-made counterparts.

There is no question that machine-made goods have a sameness about them and we are not so far from the age of the individual craftspeople that we do not still think of manufacturing a product carefully by hand as being something special. It should be noted that even though the handcrafted product's quality is high, it is not necessarily any better than its machine-made counterpart. However, hand-crafted products do often represent quality, are frequently unique in design, and represent a certain dedication to manufacturing in a bygone era. In addition, as machine-made goods become more and more the standard of the industrialized world, the desire to possess something unique will always be a force for the preservation of the individualized craftspeople's dedication to their methods.

FACTORS IN AUTOMATION

Why automate manufacturing? In the opinion of some, automation is actually counterproductive as it would appear to eliminate jobs. The debate over jobs versus automation will always continue, but despite whatever negative factors there are, automation is here to stay. Manufacturing, driven by practical economic reasons, will continue to exploit automation in an aggressive manner. Primary reasons for automating manufacturing include the following. Automation

1. **Allows maximum production at minimum costs and productivity factors.**
2. **Produces the volume required.**
3. **Allows standardization, interchangeability of parts and quality control.**
4. **Meets product demand and competition.**

Costs and Productivity

Cost of manufacturing is a **major** if not the most important force behind automation. All manufacturing industries are controlled by the economic forces in play at any given time. Throughout history such factors as inflation, taxation, labor costs, material costs, as well as costs of marketing and distributing, have driven the manufacturing industry to constantly search for better and cheaper ways to accomplish production.

There has been much debate in recent years as to the long-range results of automation and its effects on employment. Advances in manufacturing technology that have occurred within the past few years have had and will have an even more profound effect on employment and it remains to be seen just what effect this will have on the production worker in the near term.

It is safe to say that in the longer term, there will be less low-skilled to semiskilled production jobs, as automated manufacturing becomes more and more the standard of industry. If economic conditions place more demands on the individuals now staffing production jobs, they in turn will place more pay demands on their employers who will compensate by further automating industries, thus reducing labor intensity.

Even though automation might appear to be counterproductive, it will, in the long run, probably create more jobs. However, these jobs will be of a higher technical level and production workers must be open minded enough to accept retraining if they expect to remain gainfully employed.

Productivity can generally be defined as the **ability to maximize the quantity and quality** of production from all phases of manufacturing including people and machinery. If plant equipment is antiquated, productivity will be low compared to modernized automated manufacturing and therefore the costs of the product may be higher. In labor-intensive manufacturing, productivity may be difficult to maintain because of the inconsistencies of the human work force. Automatic machines on the other hand do not become tired or bored, and are consistent in their quality output assuming proper adjustment and maintenance.

Automatic manufacturing equipment can also operate under conditions that humans could not. For example, robotic assembly can be done in total darkness, whereas such conditions would not be at all satisfactory for people.

Production Volume

It is simply not possible to manufacture certain high-volume products in a production environment where very many operations must be manually performed. Examples include nails, fasteners, light bulbs, electronic components, and many other parts for assemblies of all types. These products are manufactured in quantities of billions each year and the only way to meet volume requirements is to have highly automated production.

Standardization and Interchangeability

Few complex manufactured products are totally made by a single manufacturer. More likely a company will use many subcontractors to manufacture different parts for an assembly and then these will be brought together at a final assembly point. Examples are autos and aircraft. There are many advantages to manufacturing in this manner. Subcontractors become specialists in the production of individual parts or products. Their expertise develops keenly and their research and product development can be concentrated in a narrow area resulting in a product of high quality with the ability to still be mass produced.

Since a manufacturer of a general application product line such as electrical or hydraulic equipment may act as a supplier to many different prime contractors, **standardization** of the product is essential. This permits interchangeability, the key to any successful mass production of complex assemblies. The importance of standardization can not be overemphasized. Lack thereof in manufacturing will impede **interchangeability** in assembly and thus make manufacture of further complex products extremely difficult or impossible.

Once again, automation plays an important role in achieving standardization because of the ability of the machine **to duplicate products in large quantities with little or no variance in part dimensions and part feature locations.**

Product Demand and Competition

Demand for a manufactured product created by need or want will often dictate the degree of automation required in production. For complex products such as ships, off-shore oil platforms, and other large and complicated assemblies, demand will be low from a quantity standpoint. These products take time to complete, up to several months or years, and they are fairly labor intensive. Creating a highly automated production line for their production would be neither desirable nor cost effective.

On the other hand such products as light bulbs, autos, nuts and bolts, nails, home appliances, and electronics have high demand as well as wide market potential. In many cases, profit margins are very small on a per unit basis. Hence, very large production numbers are necessary, making automated production both desirable and necessary. Since products can be made quickly and efficiently using automated manufacturing, their costs can be kept low making a vast line of products available at affordable prices.

Competition is a driving force in any free enterprise economy. Rarely do we find a product only manufactured by one maker. Competition tends to weed out those producers with low quantity, too high prices or who are unable to meet the production demand for their product. If the competition to industry A is automating its manufacturing facilities in order to bring a better, cheaper and more available product to the customer, then industry B must do the same if it expects to compete in the market place. There are many examples where a manufacturer's inability or unwillingness to automate production resulted in a business failure.

AUTOMATION SYSTEMS IN MANUFACTURING

Mass production of any product always involved repetitive operations. One of the primary functions of automation is to **standardize** repeat production operations so that they may be accomplished on a continuous unvarying basis. This permits design tolerances to be met on a large number of production parts as well as there to be higher production rates. To do this, automation techniques most often replace inconsistent manual functions with consistent machine functions. This is where automation can and does eliminate production jobs.

The process of automating a manufacturing process may take many forms ranging from a rudimentary level, perhaps involving a simple mechanical device that operates part of a machine, to complex computer driven feedback systems that have considerable decision-making capacity over the process that they are controlling.

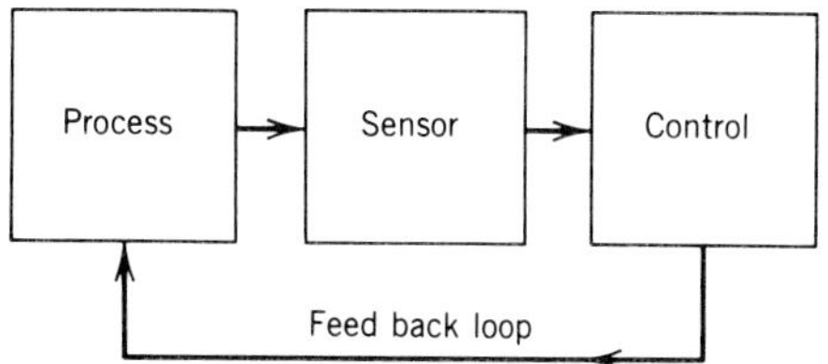

Figure 1
Factors for process automation.

In a true automatic system, the automation device has an ability to **sense** in some way factors about the process that is being accomplished and then initiate control functions.

Automation requires the following factors (Figure 1).

1. A process to automate.
2. A sensing system for making decisions about the process.
3. A control action system that operates on the sensed information and then provides a process control function.

Almost any manufacturing process can be automated. This can be done at any phase of manufacturing including the manufacturing processes themselves or systems that inspect and/or package a finished product.

Semi Automatic and Full Automatic

Consider a simple process such as the carriage feed on a lathe (Figure 2). The process must be initiated manually by the operator and disengaged manually at the end of the cut. This is not a fully automated process, but once engaged, it does serve to feed the tool automatically, without the requirement for external manual input. Thus, the process is **semi-automatic.**

Figure 2
Semi-automatic control involves manual engage and manual disengage.

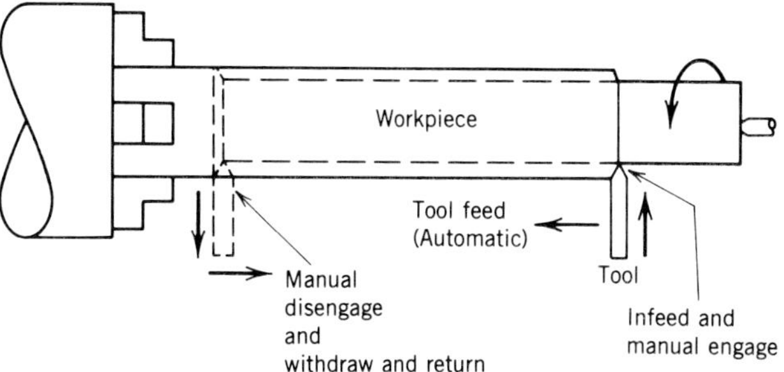

By setting the feed trip stop, the operator can further automate the process. Initiation is manual, but feed disengagement is automatically accomplished. Two automation factors are present here, the process (automatic carriage feed) and the sensor (trip stop). This simple system does not have any feedback for further action after feed disengage. The operator must take over manual control at that point to withdraw and return the tool to the starting point. Therefore, the function is semi-automatic requiring a manual initiation for each new cycle.

On a numerically controlled (NC) lathe or turning center, the feed might be engaged by computer control, disengaged, and the tool automatically withdrawn and returned to the beginning of the cut, or changed to a new tool. All three factors are present here for full automation, the process (tool feed), the sensor (computer control), and the action based on sensed information (tool withdrawal and return or tool change) (Figure 3).

Although the lathe process is automatic, initial manual inputs had to be made at some point in the system. In the case of numerical control programming for machining, the **manual data inputs (MDI)** are in the form of the **numerical control program** loaded into the machine's control computer. Such is the case with all automatic processes. They must be programmed manually or otherwise set up before they can operate with full automatic capability.

Numerical Control Automation

It would be difficult to find a device that has had more influence on manufacturing automation than the **computer.** Computer control of production processes or **Computer Aided Manufacturing (CAM)** as well as Computer Integrated Manufacturing **(CIM),** and their design counterpart **(CAD),** has profoundly altered manufacturing in dramatic ways. One very important area that has been established for many years and continues to undergo development is that of **computer numerical control** or **CNC.**

CNC is the equipment and methods used to control a process by numeric instructions through computers driving electromechanical or hydraulic actuators. Although CNC has been applied to almost every manufacturing process, **machining** certainly has been one area where its development is unparalleled. Perhaps the reason for this is that machining involves a wide variety of high precision manufacturing requirements.

All of the previously discussed factors are present in an automated numerical control machining system. The processes to be controlled include tool and workpiece positioning, tool changing, feed and speed rates, and special functions unique to the specific process or machine tool. Electromechanical and electronic sensor systems built into the computer control make decisions and initiate appropriate actions.

NC Systems NC systems include open loop and closed loop. Open loop NC systems do not make use of a feedback system to determine control action. This system was popular in the earlier days of NC before the development of more sophisticated computer servo controls.

Computer control technology permits sophisticated closed loop NC systems to be used, where sensed in-

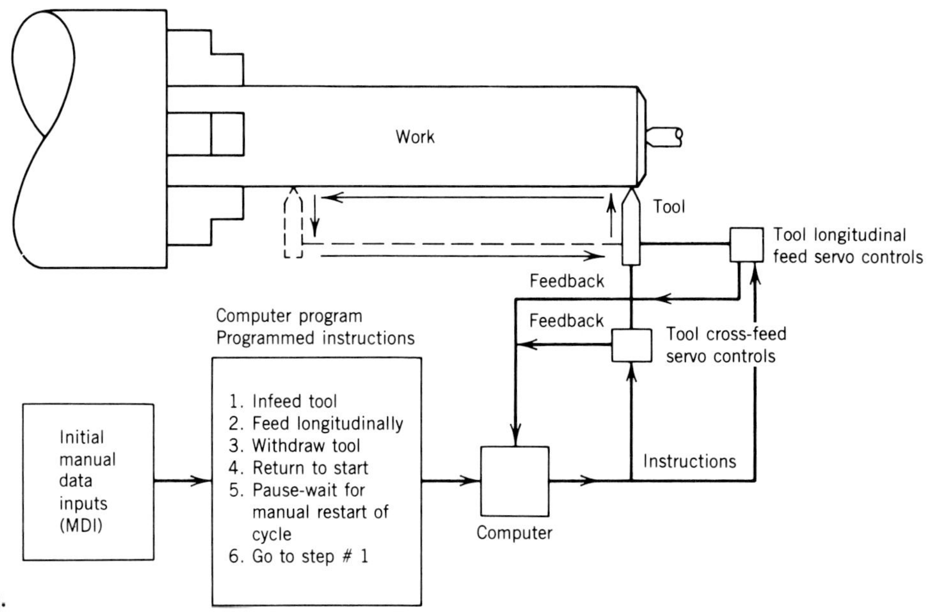

Figure 3
Full automatic programmable cycling.

formation is fed back to the control thus initiating required control action. These computer numerical control systems (CNC) are widely used and very popular in automated manufacturing.

Another version of computer numerical control places several NC machine tools or perhaps an entire production plant under a single computer control on a time sharing basis. This process, known as direct numerical control or DNC, is also popular in manufacturing automation.

NC Motion Axes and Adjunct Machine Functions

Major components of an NC machine tool have motion along fundamental machine axes. The centerline of the machine spindle is the **Z axis.** If the machine tool is a vertical machining center, the table and saddle components move at right angles to each other relative to the spindle position. Table motion is in the **X axis** and saddle motion is in the **Y axis.** If the machine tool is a horizontal spindle design, centerline of spindle is still *Z* axis whereas the *X–Y* plane turns into the vertical position (Figure 4). Turning center spindles are in the *Z* axis whereas cross slide motion is in *X* axis. Rotation about the fundamental axes is also defined permitting rotary motion of a workpiece to be specified in the NC program. Movement of the spindle, table, or saddle along each axis in either direction is specified in the program. Thus, each point within the entire volume of space in which the spindle may move or the workpiece be positioned, may be specified in the NC program. Modern CNC equipment has the capability to position the machine spindle and/or workpiece to specified locations within a few tenths of thousandths of inches. On large machine tools, this can be accurately repeated over long distances making possible extremely accurate cuts and location of holes and other features in the workpiece. By simultaneously combining motions by the machine spindle and/or workpiece, the NC machine tool can accomplish **angular and contour (continuous path)** machining that could never be done by

Figure 4
Reference axes for CNC machine tools.

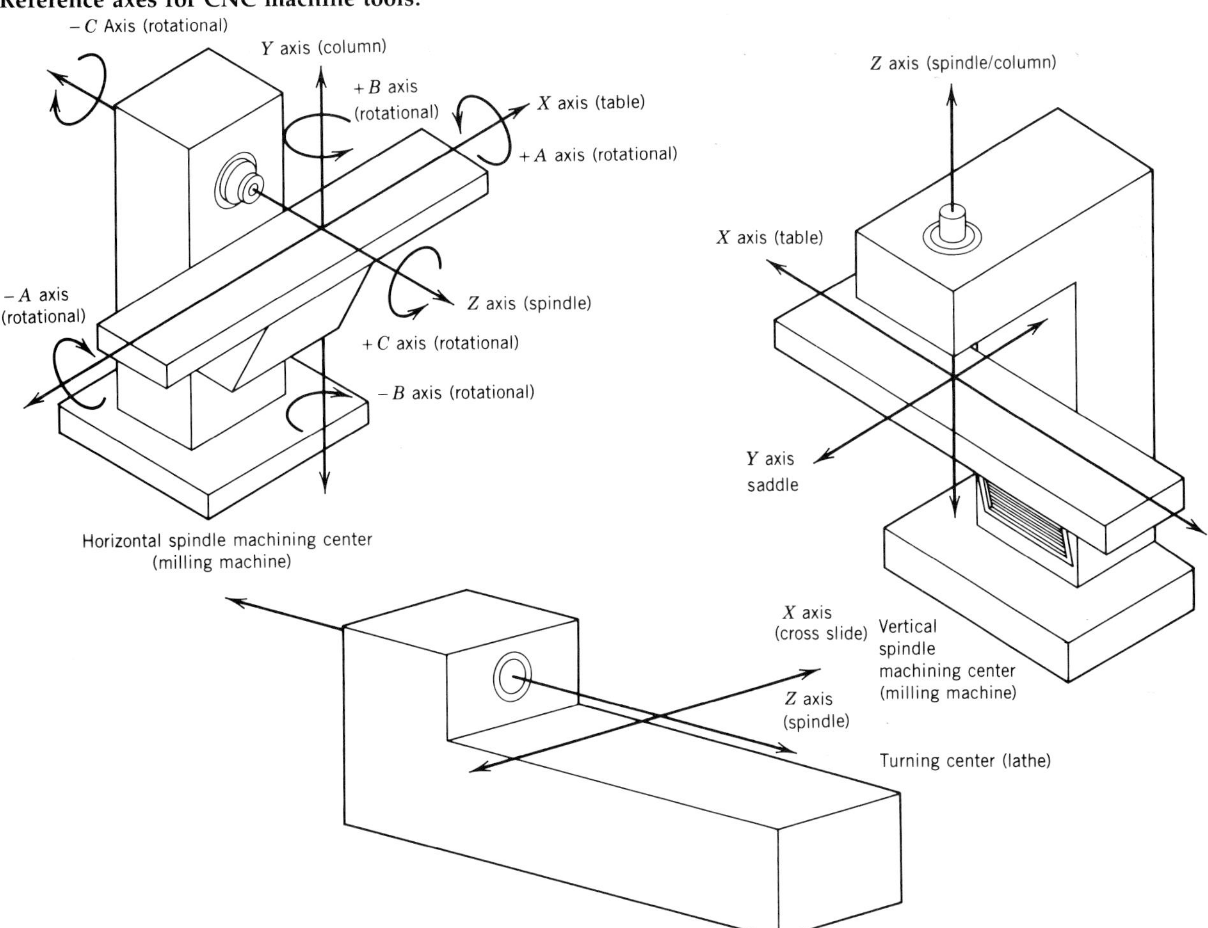

conventional hand operation methods. Thus, the NC machine tool can perform tasks that would be virtually impossible by conventional practices. Mathematics required for continuous path and angular machining is done by the control computer and the appropriate program codes are directly generated.

The NC Program To accomplish machining tasks, the control computer of the CNC machine must be programmed. This task may be done manually by the **NC programmer,** or it may be accomplished directly by the computer in a CAD station. In manual programming, the programmer must have a knowledge of the particular computer language used to direct the machine, as well as of the cutting tools required and the workholding system. NC programmers, in many cases, are experienced NC and CNC machine operators and machinists.

The CNC program consists of a set of step-by-step instruction statements describing exactly what the machine is to do regarding positioning the cutting tool or workpiece, selecting the required tool, and accomplishing the required cutting operation at the correct speed and feed. Modern CNC programming has been greatly simplified. The programmer may communicate instructions to the machine using a simple language. Such direct statements as MILL, DRILL, ARC MILL, BOLT, CIRCLE, along with the required dimensional information are understood by the machine computer and translated into actual machine actions.

Other statements in the program control adjunct machine functions such as coolant on/off, spindle stop/start, drill, tap and bore cycles, tool changes, feeds and speeds for machining, and workpiece indexing movements.

The NC program resides in the control computer memory. Programs may be entered manually at the machine control unit (MCU) or they may be fed directly from a CAD station. Some types of CNC machine tools may be programmed by manually running them through the various operations of a machining task. Each manual operation becomes a computer program step and once the first run is accomplished, the machine tool is then programmed to run the job automatically. Permanent storage of programs can be accomplished by placing the data on punched tape, magnetic tape, computer disks, or programs may simply reside in the computer's active memory, providing that the machine control unit is left energized at all times.

When a CNC machine tool is programmed, several documents may be generated if required. The machine control function codes may be fed directly to the machine from a CAD station or they may be placed on tape or disk and used at the machine as required. Generally the machine operator will need a part drawing (Figure 5a) indicating dimensions and an operation instruction sheet indicating a list of machining operations, job setup instructions, as well as a list of gages used to measure the part (Figure 5b).

The CNC programmer may generate a program definition sketch defining various points, lines, and planes that will be specified in the program (Figure 5c). The CNC program itself consists of several lines of code telling the machine where to go and what to do at the various locations where machining operations take place.

The result of the programmer's work is the CNC program (Figure 5d). It is written in a computer language that will later be translated in machine control codes by the computer numerical control on the machine tool. Figure 5d contains notes indicating what the various lines of programming code mean. If the programming is accomplished with the aid of a computer (CAD), machine control codes are generated directly (Figure 5e). The illustrations show machine control codes generated by the computer that correspond to the manual program inputs from the previous illustration. Once the machine control codes are generated, they may be fed directly to the manufacturing equipment over a data link, or they may be placed on punched tape or other storage devices. Thus the control program is available whenever needed.

Construction and Capability of NC Machine Tools Because NC machine tools are designed for high production of precision parts, they are constructed to the latest machine design available. Design effort is directed toward reliability of the mechanical and electrical/electronic components, as well as long term maintenance of machine accuracy. This makes the NC machine tool more expensive than its manually operated counterpart. However, its higher initial cost is far outweighed by its truly amazing productivity and broad capability. Numerically controlled machine tools, as well as other types of NC material processing equipment, have well established themselves as essential items in modern manufacturing automation.

With computer control, the NC machine tool has capabilities that could not be accomplished by manual operation. This has reduced or eliminated many constraints heretofore placed on the product designer. With the capability of new computer controlled production, mass produced products are appearing that would have been totally unavailable only a few years ago. Technological advances in this type of computer aided manufacturing permitting new and innovative product design, will profoundly affect us all in coming years.

Figure 5A
Part drawing for the CNC machining task.

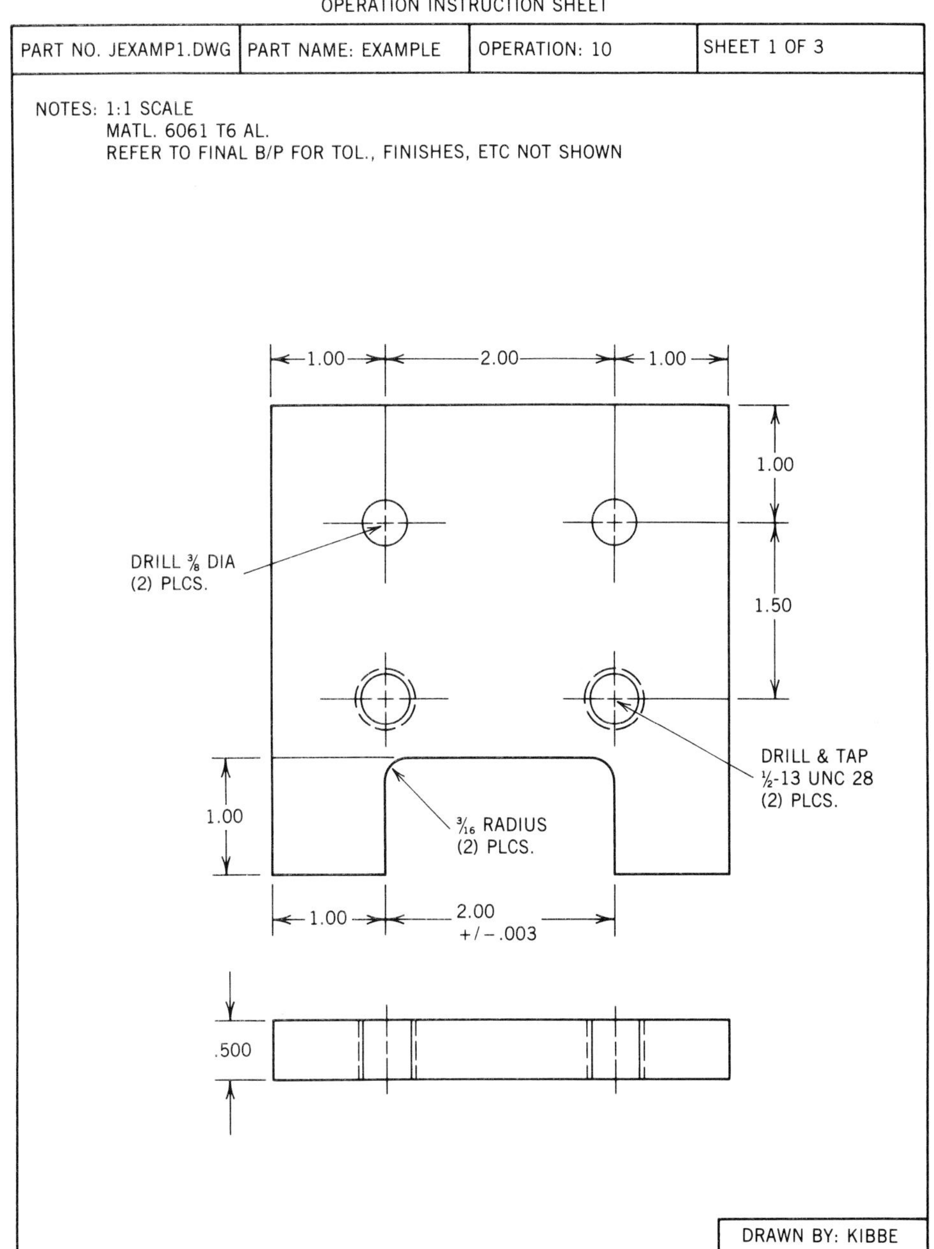

Figure 5B

CNC operations instruction sheet defines job setup requirements, program sequence, and required inspection gaging tools.

PART NO. JEXAMP3.DWG	PART NAME: EXAMPLE	OPERATION: 10	SHEET 3 OF 3		
SEQ	OPERATION DESCRIPTION		TOOL NO.	RPM	FEED
	SETUP INFO:				
	A) LOAD VISE TO NC MILL TABLE, JAWS PARALLEL TO Y AXIS.				
	B) INDICATE JAWS PARALLEL TO Y AXIS WITHIN .0005 T.I.R.				
	C) SET TOOLS IN APPROPRIATE HOLDERS PER TOOL LIST.				
	D) LOAD TAPE #001 INTO CONTROL MEMORY.				
	E) SET END LOCATING STOP ON VISE. LOAD A PART.				
	F) USING EDGEFINDER, LOCATE UPPER LEFT CORNER OF PART.				
	G) FROM THIS POINT, HANDLE OVER −10.0" IN X AXIS.				
	H) RESET AXIS READOUTS. THIS IS PROG. ORIGIN.				
	I) ZERO RETURN Z AXIS. PROGRAM BEGINS & ENDS HERE.				
	J) SET ALL OFFSET REGISTERS PER G.L.'S & T5 RADIUS.				
	K) USING STANDARD SETUP PROCEDURES (DRY RUN, MACH. LOCK,				
	SING. BLK., ETC,) WORK ALL TOOLS IN TO CUT PART.				
	L) LAYOUT 1ST PART COMPLETE & PRESENT TO INSPECTION.				
	PROG. SEQ.:				
	CENTERDRILL (4) HOLES		1	2000	16IPM
	DRILL 3/8" DIA. HOLES (2) PLCS.		2	2000	10IPM
	DRILL 7/16" MINOR DIA. (2) PLCS.		3	1200	8IPM
	TAP 1/2-13 UNC 2B THREAD (2) PLCS.		4	315	CALC
	MILL 2" POCKET WITH 3/8" E.M.		5	3000	25IPM
	GAGES REQUIRED:				
	.370/.380 DIA. PLUG GAGE				
	.430/.440 DIA. " "				
	1/2-13 UNC 2B THREAD GAGE				
	3/16" RADIUS GAGE				
	S.M.I.				
			DRAWN BY:		

Figure 5C
Program definition sketch defines part geometry that will be used in the CNC program.

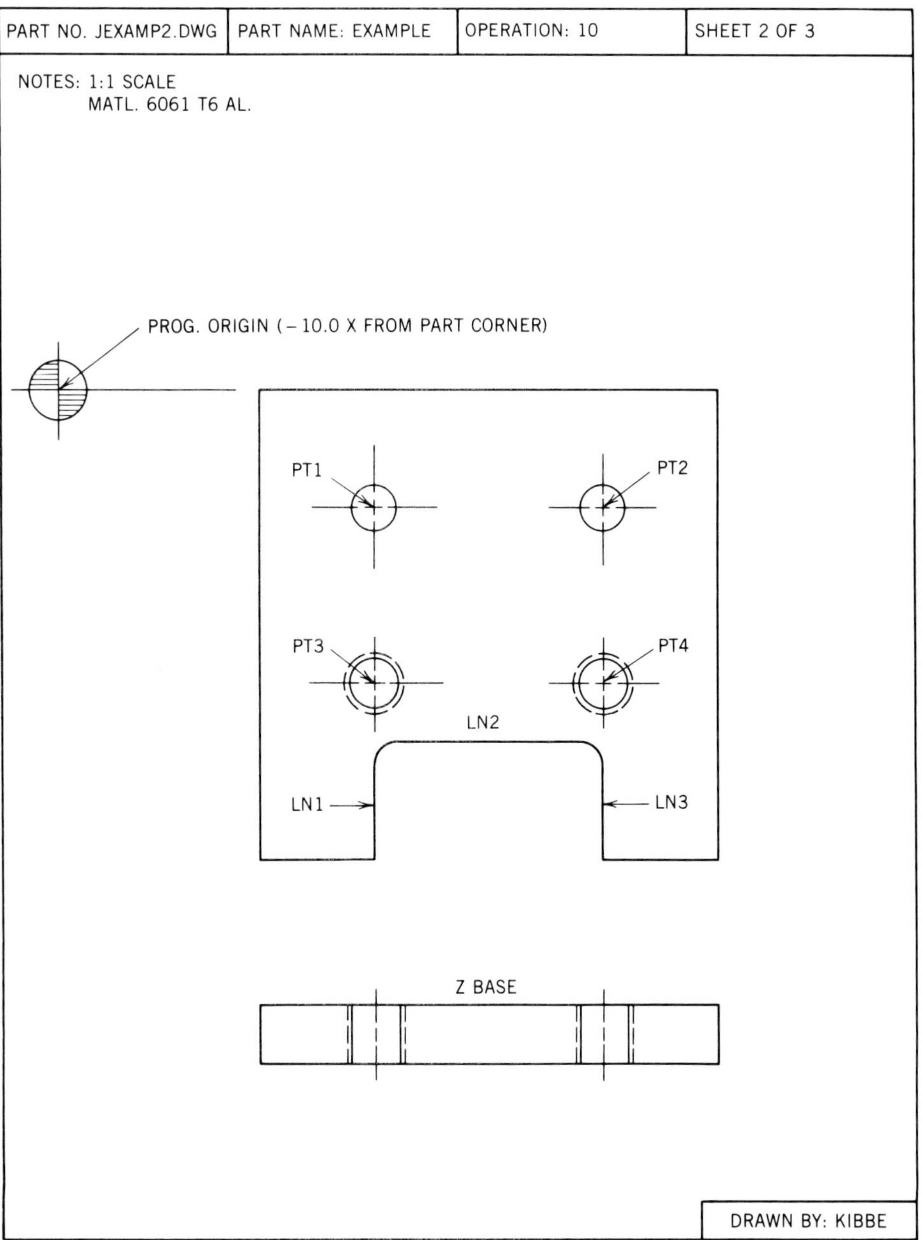

Figure 5D
The CNC program (sheet 1 of 3) contains all the codes necessary to instruct the machine tool to perform its assigned task.

Programming language: COMPACT II

All INSERT STATEMENTS are REMARK and COMMENT statements used by the programmer. These are ignored by the machine computer.

```
MACHIN, MAZVTCFAN1   <- Machine tool and link name
IDENT, V-10 JEXAMP1.DWG S/U 1 TP 1 OP. 10   <- Punched taped identification
INIT, INCH/IN, INCH/OUT   <- Program in inch measurement or optional metric
SETUP, MOD10/5/20/1,CMOD3000/1/4/5/11,TPS,RPM3150,MAGPOS99,20X,10Y,OLX,OLY   <- Setup Statement
INSRT,PART:   PARTNAME      MACHINE PER OPER. INSTRU.
INSRT,PROGRAMMED BY:   KIBBE
INSRT,DATE:  /  /  .
INSRT,QQ
BASE,OXA,OYA,8.ZA   <- Establishes home position from the machine spindle
DPT1,11.XB,-1.YB,OZB
DPT2,PT1,2.X
DPT3,PT1,-1.5Y
DPT4,PT2,-1.5Y
DLN1,11.XB,OZB
DLN2,-3.YB,OZB
DLN3,13.XB,OZB
INSRT,
INSRT,QQ
```

Geometry definitions (DPT1 through DLN3) describe parts and operation used to direct cutter path

Figure 5D ***Continued***

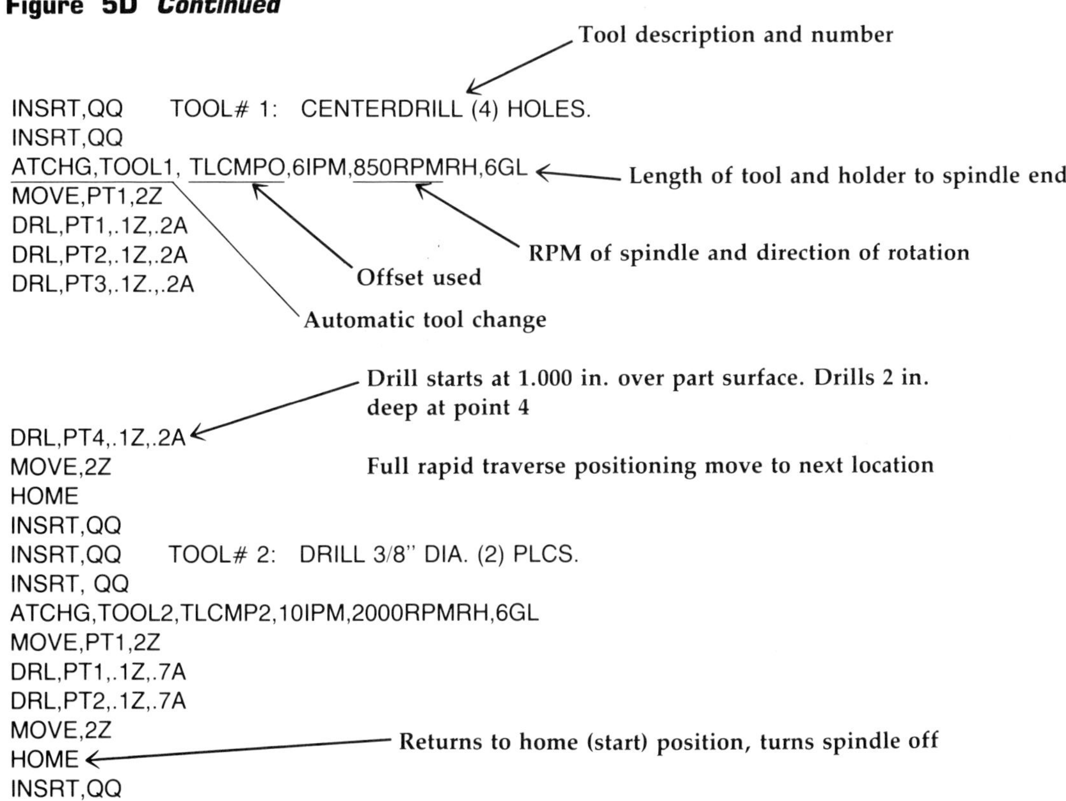

Figure 5D ***Continued***

```
INSRT,QQ      TOOL #3:  DRILL 7/16" MINOR DIA. (2) PLCS
INSRT,QQ
ATCHG,TOOL3,TLCMP3,8IPM,1200RPMRH,7GL
MOVE,PT3,2Z
DRL,PT3,.1Z,.7A
DRL,PT3,.1Z,.7A
MOVE,2Z,CUF
HOME
INSRT,QQ
INSRT,QQ      TOOL #4:  TAP 1/2-13 UNC 2B THREAD (2) PLCS.
INSRT,QQ
ATCHG,TOOL4,TLCMP4,315RPMRH,8GL
MOVE,PT3,2Z
FLT,PT3,.5Z,.1A,13PITCH
FLT,PT3,.5Z,.1A
MOVE,2Z,COF
HOME
INSRT,QQ
INSRT,QQ      TOOL #5:  MILL 2" POCKET
INSRT,QQ
ATCHG,TOOL5,TLCMP5,3000RPMRH,.0937TLR,25IPM,6GL
MOVE,OFFLN1/XL,-4.5YB,-2.Z,.010STK

CUT,-.6ZB
CUT,OFFLN2/YS,PARLN1
CUT,OFFLN2/XS,PARLN2
CUT,-1.5Y
MOVE,OFFLN1/XL,-4.5YB,OSTK,100IPM
CUT,OFFLN2/YS,PARLN1
CUT,OFFLN3/XS,PARLN2
CUT,-1.5Y
MOVE,2Z
HOME
END
```

Generates a canned (used over and over) drilling cycle

Given pitch of thread in tapped holes, the computer calculates and generates correct feed rate for the tapping operation

Generates a canned tapping cycle based on the previous data

Leaves .010 in. of material on part for finish machining

Figure 5E
The computer output for the CNC program.

```
   TOOL #1 : CENTERDRILL (4) HOLES

N001G91G28Z0T01
N002S63M03
N003T00M03
N004G92Z280000
N005G90G00X110000Y-10000Z160000
N006G81G99Z13900R141000F600
N007X130000
N008X110000Y-25000
N009X130000
N10G80Z161000
N011X0Y0Z280000

   TOOL #2 : DRILL 3/8" DIA. (2) PLCS

N012G91G28Z0T02
N013T01M06
N014G27280000
N015G70G00X110000Y-10000S73M03
N016G45Z160000H02
N017G81G99Z134000R141000F1000
N018X130000
N019G80Z161000
N02G45Z280000H02
N021X0Y0

    TOOL #3 : DRILL 7/16" MINOR DIA. (2) PLCS

N022G91G28Z0T03
N023T02M06
N024G92Z280000
N025G900G00X110000Y-25000S66M03
N026G45Z170000H03
N027G81G99Z144000R151000F800
N028X110000
N029G80Z171000
N030X0Y0

    TOOL #4 : TAP 1/2-13 UNC 2B THREAD (2) PLCS.

N032G91G28Z0T04
N033T03M06N034G92Z280000
N035G90G00X110000Y-25000S53M03
N036G45Z180000H04
N037G84G99Z164000R165000F2423
N038X110000
N039G80Z185000
N040G45Z280000H04
N041X0Y0

    TOOL #5 : MILL 2" POCKET

N042G91G28Z0T05
N043T04M06
N044G927Z2800000
N0455G90G00X111037Y-45000S76M03
N046G45Z260000H05
N047G01Z134000F2500
N048Y-31037
N049X128963
N050Y-46037
N051X110937Y-45000F1000
N052Y-30937F2500
N053X129063
N054Y-45937
N055G00Z154000
N056G45Z28000H05
N057X0Y0
N058G91G28Z0
N059T00
N060M11
N061M16
N062T05
N063M12
N064M30
=&S
```

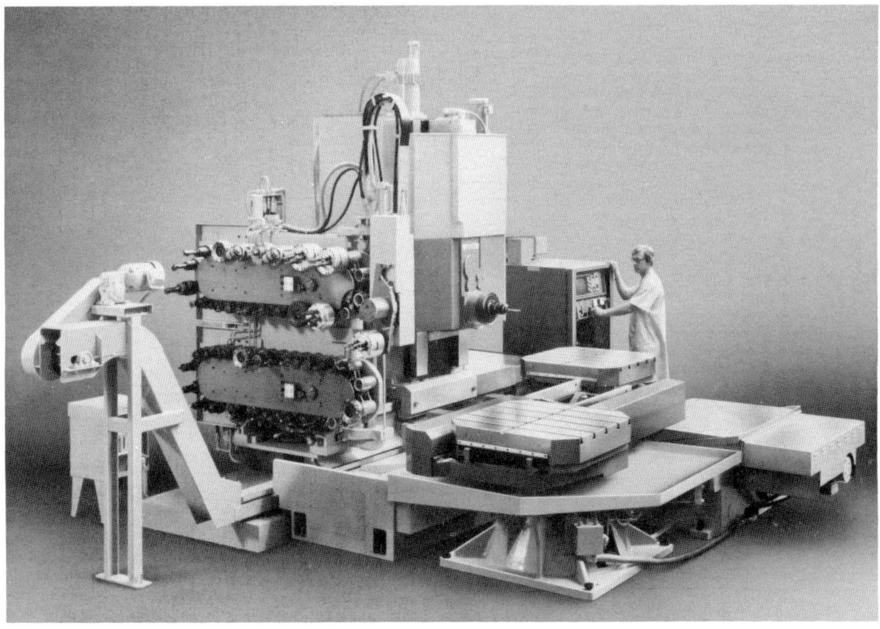

Figure 6
CNC machining center with twin tool holders and automatic tool changing (Cincinnati Milacron).

Types of NC Machine Tools Generally speaking, NC machine tools can be classified into two major groups. These include **machining centers** and **turning centers.** Machining centers are equivalent to milling machines in the conventional shop and turning centers perform lathe operations.

Machining centers CNC machining centers are one of the most popular types of automated machine tools. They can have very vertical or horizontal spindles and they can be equipped with a variety of features that can increase their versatility. These include tool changers where a number of cutting tools are mounted on

Figure 7
A face milling cutter is delivered from tool holder to spindle for a face milling task (Cincinnati Milacron).

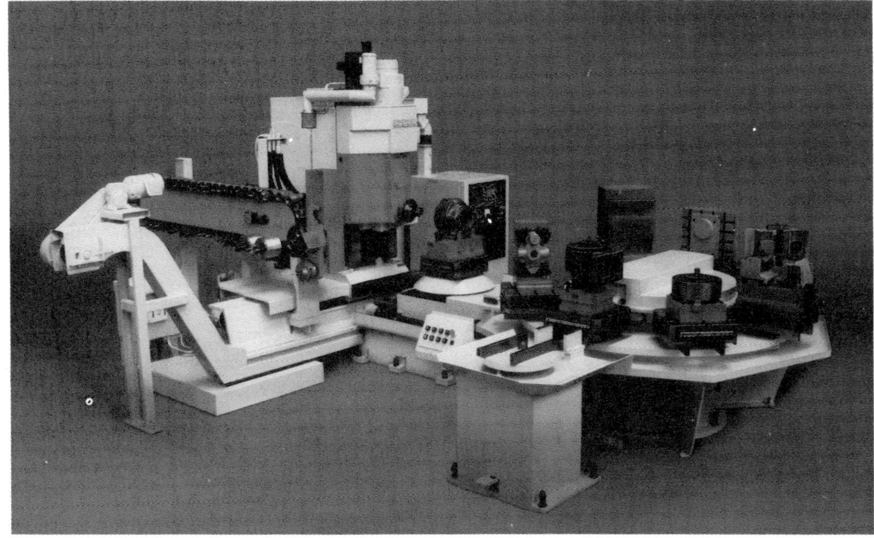

Figure 8
Multiple workpiece pallets on a CNC machining center with a variety of work-holding methods (Cincinnati Milacron).

a belt or rotary drum. Figure 6 shows a CNC machining center with twin tool holders. When the machine tool is in operation, the NC computer program selects the correct tool from the belt and mounts it in the machine spindle (Figure 7). When the cut is completed, the program causes the tools to be removed from the spindle and returned to the tool holder.

Multiple workpieces may be mounted on several pallets on the same machine tool (Figure 8). Pallets may be indexed to present different parts of the workpiece to the cutter (Figure 9). In many cases, the machining center will be used as a component in a **work cell** as part of a **flexible manufacturing system (FMS).** In this case, a provision is made to completely off-load a work-holding pallet from the machining center (Figure 10) and transport it to another machine or to a new cell or

Figure 9
With pallet indexing capability, different sides of a workpiece or different workpieces can be accessed for machining tasks (Cincinnati Milacron).

Figure 10
A workpiece pallet may be completed off loaded or on loaded when the machining center is a cell in a flexible manufacturing system (Cincinnati Milacron).

Figure 11
Duplex CNC machining centers can machine both sides of the workpiece at the same time (Cincinnati Milacron).

Figure 12
One side of the duplex machining center showing the large workpiece pallet (Cincinnati Milacron).

Figure 13
CNC turning center (Cincinnati Milacron).

Figure 14
Turning center turret mounted tooling for ID and OD machining (Cincinnati Milacron).

Figure 16
Turning center used for simultaneous OD and ID thread cutting (Cincinnati Milacron).

Figure 15
Turning center turret tooling for OD shaft turning (Cincinnati Milacron).

station in the manufacturing system. CNC machining centers can accommodate large workpieces in duplex arrangement (Figures 11 and 12), permitting both sides of the work to be machined at the same time.

Turning Centers CNC turning centers (Figure 13) are also an extremely popular and versatile member of the family of equipment used for automated machining. As on the machining center, the turning center's capability lies in its ability to bring a variety of tools into use to machine the workpiece.

Turning center tooling is often turret mounted (Figures 14 and 15) since the machine spindle holds the workpiece, all the tooling for machining on the outside diameter (OD) as well as the inside diameter (ID) of a workpiece may be installed and remain on the machine during a production run. In many cases, both ID and OD turning operations may be accomplished simultaneously. An example includes thread cutting inside and outside at the same time (Figure 16).

Figure 17
OD shaft turning with turret mounted tooling (Cincinnati Milacron).

Figure 18
CNC turning center with robot arm for loading and off-loading parts. The automated material handling system is seen at the left (Cincinnati Milacron).

Figure 19
Two views of the vertical spindle CNC profiler. The machine is equipped with three spindles and is engaged in machining complex contour aircraft parts (Lockheed—California Company).

The CNC turning center is also suitable for shaft turning operations (Figure 17) and can be equipped with a part handling robot (Figure 18) that can load and unload parts from the machine tool automatically.

NC Multispindle Profilers The NC multispindle profiler is another important member of the NC family of automated machine tools. The profiler is a multiple spindle version of the vertical or horizontal machining center. These machine tools are often quite large. Workholding tables can be 70 or more feet long. The spindles on this machine tool are carried on a gantry supported on each side of the table on a vertical spindle model. On the horizontal spindle model, the work

Figure 20
Two views of the horizontal spindle CNC profiler. Workpieces are mounted on the vertical face of the large table (Lockheed—California Company).

holding table is vertical and the spindle gantry is supported on a floor rail.

A common design of profiler is the vertical spindle model with three or more spindles (Figure 19). The horizontal spindle model may also have three or more spindles (Figure 20). Automatic tool changes are also used on multispindle profilers. These are similar to automatic tool changes commonly found on many NC machining centers. The NC profiler has numerous advantages. These include the capability to machine complex parts in multiple batches. With computer numerical control, the NC profiler is a favorite in the aircraft industry where complex contour machining of duplicate parts is routinely accomplished.

Trends and Other Applications in Numerical Control CNC is by no means limited to machining applications. It can be applied in almost any manufacturing process, including **fabric cutting, typesetting, flame cutting, punch press** operations (Figures 21 and 22), **pipe bending, welding** (Figure 23), **tool and cutter grinding, nontraditional machining processes** such as EDM, as well as **measurement, inspection,** and **industrial robotics** processes.

Numerical control is now and will be a mainstay in manufacturing automation for the foreseeable future. As computer technology advances, NC will advance as well continuing its important place in manufacturing automation. As computer technology becomes more sophisticated, it will be applied in many new and interesting automation trends in manufacturing. One such area presently undergoing development is **computer acoustic emission analysis** of machining processes. In each machining operation, characteristic acoustic emissions (sounds) are emitted by the machine and tooling

Figure 21
CNC punch press (Bracker Corporation/Leland J. Balber Associates.

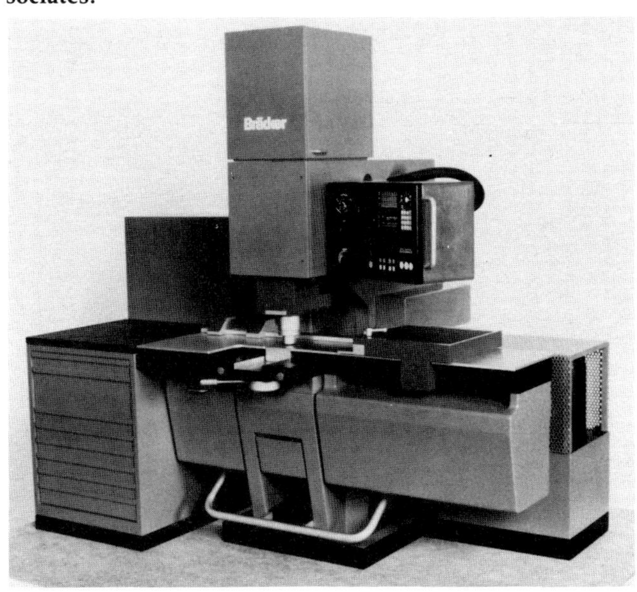

Figure 22
CNC sheet metal punching is both versatile and productive (Bracker Corporation/Leland J. Balber Associates).

during production. These sounds are excellent measures of cutting tool performance. In fact, during conventional manual machining operations, the experienced machinist or machine operator will listen to the sound of the cut as well as observe the visual action. A change in the characteristic sound will often indicate a problem such as a dull or broken tool, almost before the visual aspects of the situation appear. In electronic acoustic analysis, sounds emitted from the production processes are instantly compared to a data base in the computer representing the correct characteristic sounds. If the comparison is true, production continues. If the sounds are not comparable, the computer control will stop the machine or take other appropriate corrective action. This is another example of the high degree of sophisticated computer process control that is fast becoming the industry standard.

CAM, CADAM, ICAM, CIM, AND THE FLEXIBLE MANUFACTURING SYSTEMS FOR THE FULLY AUTOMATED FACTORY

Figure 23
Robotic welding (ESAB Robotic Welding Division).

CAM, CADAM, ICAM, and CIM

In the previous chapter, CAD or Computer Aided Design/Drafting was discussed. The counterpart of CAD in the manufacturing side is **CAM** or **Computer Aided Manufacturing.** These two developing technologies often go hand in hand to form **CAD/CAM ICAM CADAM** and **CIM** systems. **CIM** refers to **Computer Integrated Manufacturing**. *CADAM* (Computer Aided Design and Manufacturing) and *CAE* (Computer Aided Engineering) are other terms used to describe variations of CAD/CAM systems. As with CNC, these technologies have grown out of computer evolution, and they are also being fully integrated with numerical control in modern automated manufacturing.

In fully integrated CAD/CAM, CADAM, ICAD, and CIM systems, once design data have been generated, it can be used to further generate the programs needed to operate manufacturing equipment such as CNC machining centers (Figure 24), Flexible Manufacturing Systems, and/or industrial robots, or any combination of these. Steps in between such as time consuming and costly drafting can be totally eliminated. Unfortunately, jobs for people that do drafting may be eliminated as well, bringing up once again the matters of future employment under changing technology. It is likely that draftspersons of the future will have to become computer operators and programmers and they may have to involve themselves with industrial and manufacturing engineering as well.

Figure 24
Integrating CAD and CAM in machining (Lockheed—California Company).

Flexible Manufacturing Systems

The ultimate goal of the manufacturing engineer is to **completely automate** a fully integrated manufacturing plant. This concept is already in use in modern manufacturing and it is sure that the tread toward complete automation will continue. However, there are many jobs that will never be done by machine and some amount of manual labor will be necessary. It should be noted that manual labor tasks in future highly automated manufacturing will most likely be on a low skill low pay level. Major steps toward full factory automation include the flexible manufacturing system with work cells.

The flexible manufacturing system or FMS consists of manufacturing cells that are equipped with machining centers or other automated manufacturing equipment. Figure 25 shows a 10-station FMS with an automated material handling system.

Figure 26 shows a manufacturing cell equipped for vertical turning operations. Note how the cell is equipped with pallets for several workpieces. As one part is machined, another is ready for loading. This increases efficiency greatly, as the machining centers in the cell are kept constantly busy engaged in production (Figure 27). Part loading and off-loading can also be accomplished by industrial robotics (Figure 28).

The entire system of machining centers and cells may be tied together with an automated material handling system designed to move parts between cells and

Figure 25
Ten-station flexible manufacturing system or FMS with automated parts transfer between cells (Kearney & Trecker).

Figure 26
Vertical turning work cell, a component of the FMS (Giddings & Lewis).

Figure 27
CNC work cell machining center with workpiece pallets (Kearney & Trecker).

Figure 28
Loading and off-loading parts using an industrial robot (Kearney & Trecker).

cell workstations (Figure 29). Part transfers are automatically done by a track system installed in the factory floor and used to guide transport carts from station to station (Figure 30). Part inspections are also accomplished within the FMS system at the inspection cell (Figure 31). The entire FMS is under the control of a central computer that operates machining center activity as well as material transport and transfer systems (Figure 32).

The FMS has many advantages. Among these are the obvious advantages of increased productivity by automating the process in general. The FMS, because it is built up by selecting only those modules required for specific operations, is extremely versatile in its capability. Flexible manufacturing systems are extremely suited to the high production specialty machining requirement necessary to the auto industry (Figure 33). A further advantage is the flexibility of each cell and machining center to do many different jobs. Thus, the production of the system can be easily reprogrammed to handle a different part without the need for expensive and extensive retooling or the purchase of new manufacturing equipment.

Figure 29
Transporting workpieces mounted on machining pallets between cells in the FMS (Kearney & Trecker).

Figure 30
Workpiece transfer in a modern fully automated FMS (Kearney & Trecker).

INDUSTRIAL ROBOTS

Throughout the history of manufacturing, the industrial technologist and manufacturing engineer has always sought better, faster, and, especially, cheaper ways to accomplish production. In times past when human labor was easily procured at low cost, industries employed many production workers in all types of often repetitive and boring work, often at low rates of pay. In more modern times it has become more difficult and certainly more expensive to hire people to do much of the mundane work of production manufacturing. The modern production worker not only demands wages that are in line with the costs of living, they also demand safe and decent working conditions and fringe benefits, all of which add to overall labor costs and must be reflected in the final price of manufactured goods.

Although it is unlikely that automation will displace the production worker completely, it is sure that more and more of the less skilled production work will be taken over by automated equipment. There are many factors that are causing this to happen, not the least of which are complex sociological issues dealing with the desire or lack thereof on the part of the production worker to do repeat, boring operations for relatively low wages. Therefore, the trends toward replacing people with machines will continue. One important contribution toward this is the industrial robot (Figure 34).

What is a robot? A **robot** is a **fully automatic device that can accomplish a specific task** such as assembly or loading or unloading parts in a machine tool. By adding computer control, the robot may be programmed to do many different functions, thus becom-

Figure 31
The high precision multi-axis workpiece inspection cell in an FMS (Kearney & Trecker).

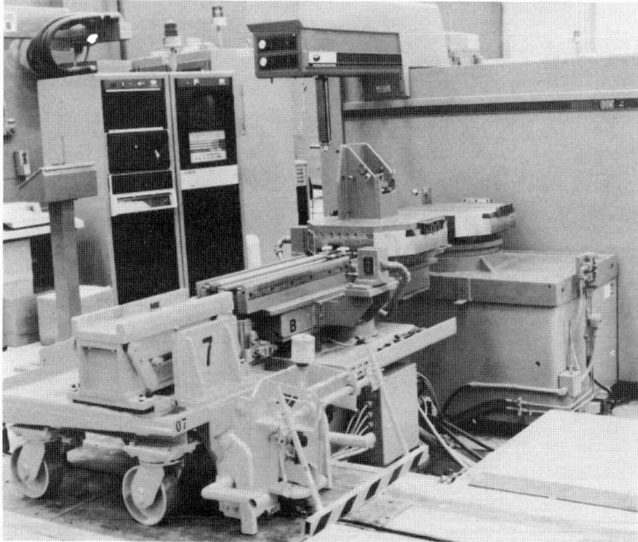

Figure 32
FMS central computer control (Kearney & Trecker).

ing an extremely versatile, untiring aid to manufacturing automation. The field of industrial robots is rapidly developing. Despite any negative aspects of their use, such as eliminating production jobs, they have already well established their place in manufacturing.

Basic Industrial Robotic Systems

A simple industrial robot may be of the **pick and place** design. These machines, also called polar robots, have limited capability and are primarily designed to pick a part from one location, turn and place the part in an-

Figure 33
Specialty FMS cell machining differential housings for the auto industry (Kearney & Trecker).

Figure 34
Programmable teachable industrial robot (Positech Corporation).

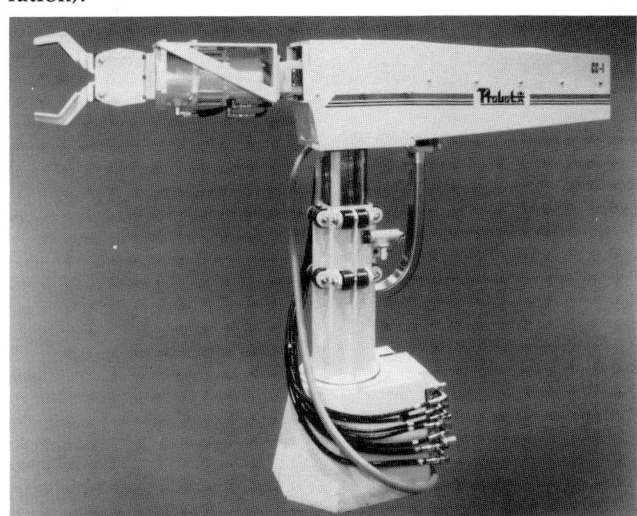

Figure 35
Industrial robot hydraulic jointed-wrist component (Bird-Johnson Fluid Power Division).

Figure 36
Motions of the jointed robotic wrist permit great flexibility in applications (Bird-Johnson Fluid Power Division).

other location or on an assembly.

Since industrial robots are built by people to accomplish tasks heretofore done by people, it is logical that they be endowed with human characteristics. A more sophisticated example of this type of robot is often of the **jointed wrist** design and can be programmed with motions in multiple axes. The jointed wrist capability (Figures 35 and 36) permits needed flexibility for complex movement necessary in many industrial applications. They have various types of grippers (fingers) to permit picking up of objects (Figure 37). The jointed wrist robot has the capability to turn up and under an assembly, and is thus very useful for such production tasks as painting, welding, and other tasks requiring the turn, bend, and multiposition capability of wrist motion. Mobility may be another important robotic characteristic depending on applications. This permits the robot to reach the work that it is accomplishing (Figure 38).

The true robotic system has a sensory system of some type and the ability to feed back information to its computer control. Through this, decisions can be made by the robot regarding its assigned task. This is to be differentiated from the abilities of a simple positioning device with only one or two functions and no real programmable capacity. It is also different from a fixed cycle automated operation where a simple task is carried out repeatedly without any variance and there is no easy way to change or reprogram the function.

Sensory capability of industrial robots may include sight, accomplished through use of a **television camera** and the ability to digitize the signal so that it may be coupled to a computer. On-going research in this area is developing robots that can **recognize** parts in a bin and then select the correct one to be placed in an assembly (Figure 39).

Figure 37
Styles of robotic finger grippers (Barrington Automation Ltd.).

Figure 38
Mobility is often essential to robotic welding applications (ESAB Robotic Welding Division).

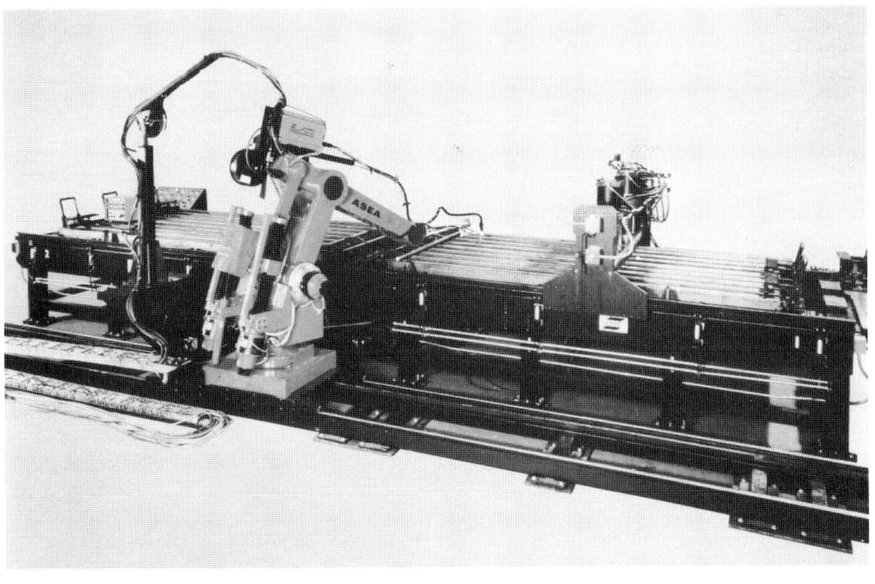

Figure 39
Sight for industrial robots, a most desirable capability (Lockheed—California Company).

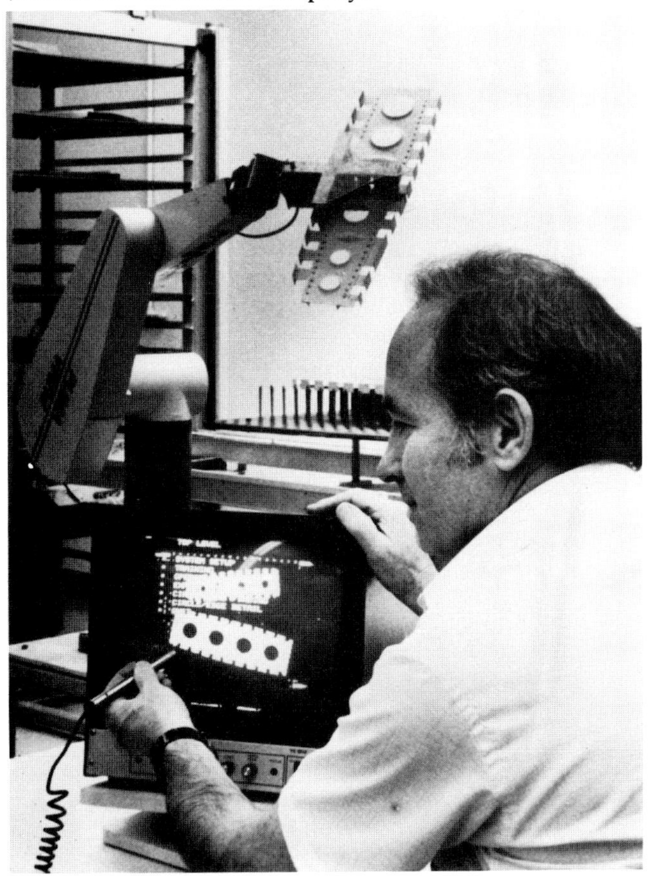

Figure 40
Using a robot to load and unload parts on a CNC turning center (Cincinnati Milacron).

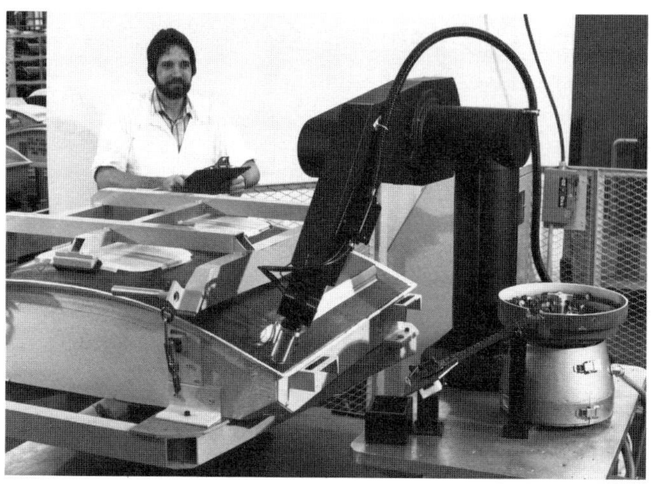

Figure 41
Using a robot for an assembly operation (Boeing photo).

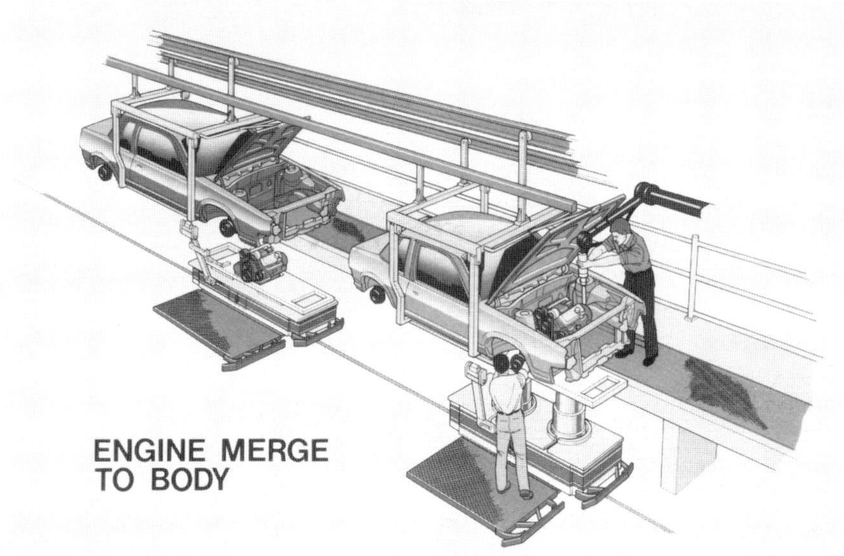

Figure 42
Robot carriers move engine assemblies to car bodies (GM—Oldsmobile Division).

Figure 43
Robot spot welders on both sides of the production line weld auto bodies with great speed and accuracy (Giddings & Lewis).

Figure 44
CNC robot welders apply 3000 spot welds to each auto body (Chrysler Corporation).

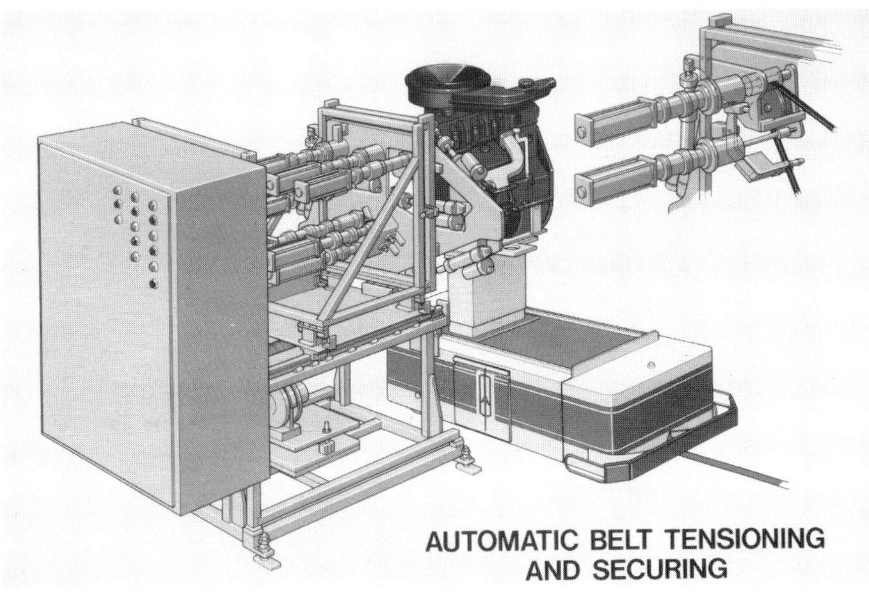

Figure 45
Belt tension adjusting, a specialty application for a specialty type industrial robot (GM—Oldsmobile Division).

The Teachable Robot Programming the teachable industrial robot is accomplished by running the machine through each of its desired positions and functions during an initial manually controlled cycle. The computer remembers each position and cycle, which is to say that it is taught what and where it is to go and do. Once this is accomplished, further duplicate cycling is repeated automatically. Reprogramming is accomplished by clearing the computer's memory of previous instructions and teaching the robot new instructions. If programming is to be saved for future repeat applications, tape and disk drives may be employed.

Industrial Robot Applications

Applications for industrial robots are numerous and expanding every day. Robots can be used to position workpieces, in a process or on a machine. They can load and unload parts from a machine (Figure 40). Many types of assembly operations are well suited to robotics (Figure 41). Robots can move parts between or to workstations in a factory. In Figure 42 robot vehicles follow a wire embedded in the floor for guidance as they carry engine assemblies to the auto assembly line.

Since the auto industry shifted to the unibody construction where the car body becomes the primary structural unit, robotic spot welding has come into wide use in auto production. Robot welders located on both sides of the assembly line (Figure 43) permit production spot welding of auto bodies to be accomplished with great accuracy and at a high production rate (Figure 44). Robots are also used for many specialty applications such as automatic belt tensioning on auto engines (Figure 45).

REVIEW QUESTIONS

1. What factors prompt manufacturing automation?
2. Are there any negative aspects to automation?
3. What factors are necessary to automation?
4. What is the difference between semi automatic and full automatic.
5. Explain in general terms what CNC automation is.
6. What are the advantages of CNC?
7. What are the common types of CNC machine tools?
8. What are CAM, ICAM, and CIM and how do they work?
9. What is a flexible manufacturing system?
10. What makes up an FMS?
11. What are the advantages of an FMS?
12. What is an industrial robot?
13. What sensory capability might an industrial robot have?
14. What is a teachable robot?
15. List some applications for robotics in manufacturing.

CASE PROBLEMS

Case 1
The Level of Automation

A manufacturer of precision mechanical equipment has many varied precision machining requirements including precision hole drilling, milling, and turning. The products produced consist of several precision subassemblies and the production requirements are often varied depending on the particular models being produced and the quantities required. Automation in machining is desirable because of its ability to meet required close tolerances. The manufacturer may equip the factory with several types of automated machine tools. These include machines that have fixed automation cycles, numerical control (NC) machine tools that are easily programmed for variable requirements, and flexible manufacturing systems that can be both programmed for different tasks and will provide a greater degree of automation by shifting parts from station to station.

Considering the primary factor of the need to machine many different close tolerance parts, which automation system would be most applicable, fixed cycle automation, numerical control, or numerical control flexible manufacturing systems? Discuss advantages and disadvantages of each.

Case 2
Robotic Applications

Given the following situations:

A. An automated assembly operation requires that the same parts be selected from a supply and placed in a location for fastening.

B. An automated assembly operation where the same parts are selected from a supply and also placed in a location for fastening. However, in this case, a visual evaluation must also be made to properly reference the part position prior to completing the assembly.

Which of these operations are suited to robotics: both, only A, only B, neither?

With common readily available robotic technology, which process would be most suited to robotics?

If a common industrial robot was used, what added equipment might make this machine more suited to handling situation B?

CHAPTER 20 QUALITY ASSURANCE AND CONTROL

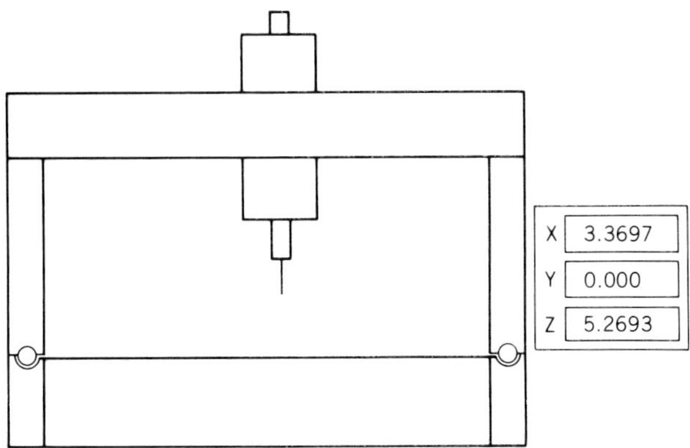

In any manufacturing, product **quality** is of paramount importance. If quality is allowed to deteriorate, then a manufacturer will soon find sales dropping off followed by a possible business failure. Customers expect quality in the products they buy and if a manufacturer expects to establish and maintain a name in the business, **quality** control and assurance **functions** must be established and maintained **before, throughout,** and **after** the production process. Quality control after manufacture and sales includes warranties and product service extended to the users of the product. Such is the purpose of quality control and assurance in manufacturing.

OBJECTIVES

This chapter should enable you to:

1. Define the purpose and intent of quality control and assurance.
2. List the major responsibilities of quality control and assurance personnel in manufacturing.
3. Discuss tolerances of size and geometry.
4. Identify common tools for dimensional measurement.
5. Define and discuss calibration.
6. Discuss aspects of nondimensional quality control.
7. Discuss quality control and assurance after manufacturing.

Generally speaking, **quality assurance** encompasses **all quality maintenance activities.** Specific quality control activities are under the quality assurance umbrella in a manufacturing industry. Quality assurance can be divided into three major areas. These include the following.

1. Source and receiving inspection before manufacturing.
2. In-process quality control during manufacturing.
3. Quality assurance after manufacturing.

SOURCE AND RECEIVING INSPECTION BEFORE MANUFACTURING

Quality assurance often begins long before any actual manufacturing takes place. This may be done at the raw materials, discrete parts, or sub-assemblies at the plants of the suppliers of the manufacturer, through source inspections. The **source inspector** travels to the supplier and inspects raw material or premanufactured parts and assemblies. Source inspections present an opportunity to sort out and reject raw materials or parts before they are shipped to the manufacturer's production facility.

The responsibility of the source inspector is to check materials and parts against design specifications and to reject the items if specifications are not met. Source inspections may include many of the same inspections that will be used again during production of the product. Included in these are:

1. Visual inspections.
2. Metallurgical testing.
3. Dimensional inspections.
4. Destructive and nondestructive inspections.
5. Performance inspections.

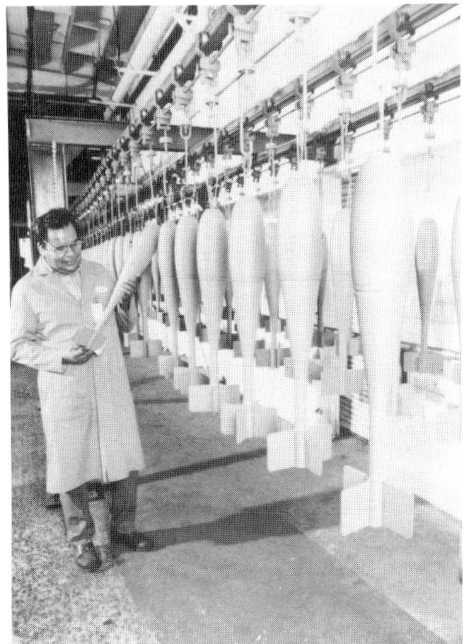

Figure 1
Making a product visual inspection (Del Manufacturing).

Figure 2
Hardness testing is an example of metallurgical inspection (MTI Corporation).

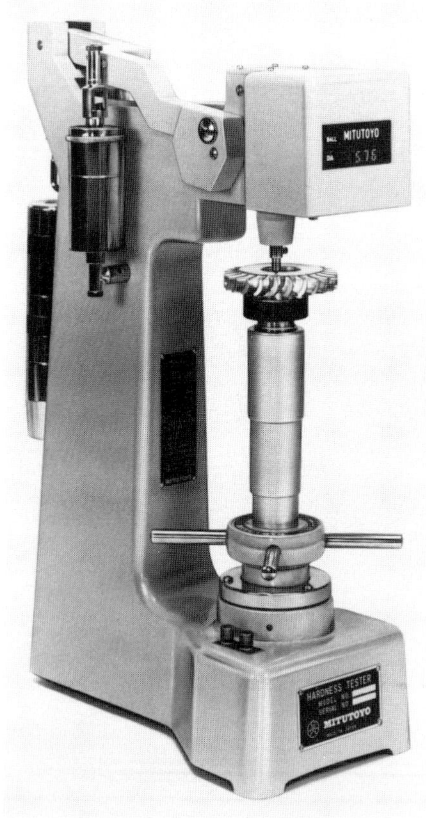

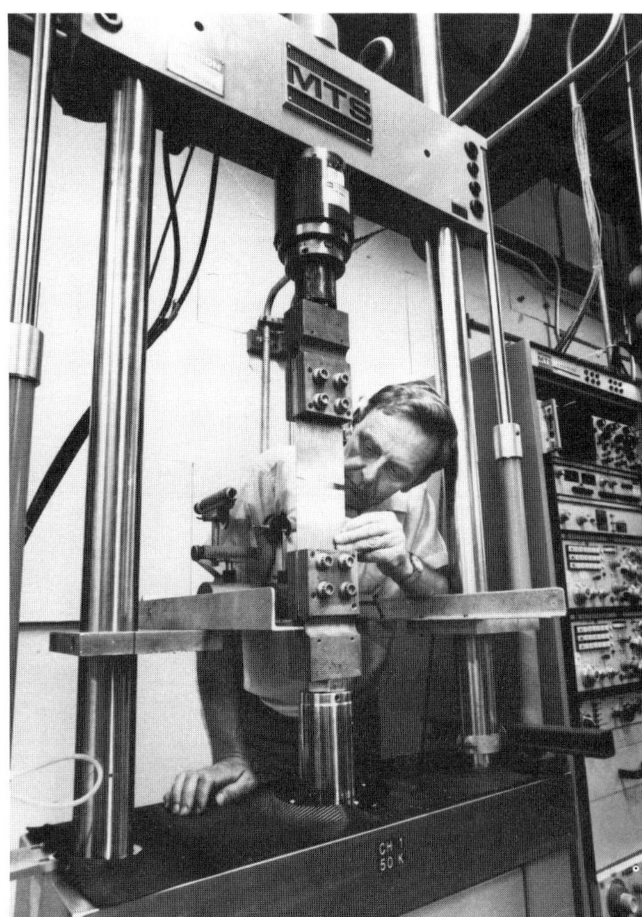

Figure 3
Tensile testing a sample (Lockheed—California Company).

Visual Inspections

Visual inspections examine a product or material for such specifications as **color, texture, surface finish,** or **overall appearance of an assembly** to determine if there are any obvious deletions of major parts or hardware (Figure 1).

Metallurgical Testing

Metallurgical testing is often an important part of source inspection, especially if the primary raw material for manufacturing is stock metal such as bar stock or structurals. Metals testing can involve all the major types of inspections including **visual, chemical, spectrographic, and** mechanical, which includes **hardness** (Figure 2), **tensile** (Figure 3), **shear compression,** and spectrographic analysis for alloy content. Metallurgical testing can be both destructive and nondestructive.

Dimensional Inspections

Few areas of quality control are as important in manufactured products as dimensional requirements. Dimensional specifications are as important in source inspecting as they are in the manufacturing process. This is especially critical if the source is parts for an assembly. Dimensional specifications are inspected at the source using all of the standard dimensional measuring tools available, plus any special fit, form, and function gages that may be required. Meeting dimensional specifications is **critical to interchangeability** of manufactured parts and to the successful assembly of many parts into complex assemblies such as autos, ships, aircraft, and many other multipart products.

Destructive and Nondestructive Inspections

In some cases it may be necessary for the source inspections to call for destructive or nondestructive tests on raw materials or parts and assemblies. This is particularly true when large amounts of stock raw materials are involved. For example it may be necessary to **inspect castings** for flaws by **radiographic, magnetic particle,** or **dye penetrant** before they are shipped to the manufacturer for final machining. Specifications calling for **burn in** time **for electronics** or **endurance run tests** for **mechanical components** are further examples of nondestructive tests.

Although it is sometimes necessary to test material and parts to destruction, because of the costs and time involved, destructive testing is most likely avoided whenever possible. However it is sometimes required. Examples include such things as pressure vessel tests to determine if safety factors are adequate in the design. Destructive tests are probably more frequent in prototype design testing than in routine inspections of raw material or parts. Once design specifications are known to be met in regard to the strength of materials, it is probably not necessary to test further parts to destruction unless they are genuinely suspect.

Performance Inspections

Performance inspections involve checking assemblies, especially those of complex mechanical systems, prior to installation in other products. Examples include electronic equipment subcomponents, aircraft and auto engines, pumps, valves, and other mechanical systems requiring performance evaluation prior to their shipment and final installations.

IN-PROCESS QUALITY CONTROL DURING MANUFACTURING

In-process quality control during manufacturing involves many of the same checks done in source inspection. A great deal of time and effort is invested by manufacturing industry in in-process quality control and the techniques and equipment used cover broad areas. Major considerations for in-process quality control include the following:

1. Receiving inspections.
2. First article inspections.
3. Manufacturing station in-process inspections.
4. Calibration.
6. Final inspections.
7. One hundred percent inspections.

Receiving Inspections

In receiving inspections, parts, material, and assemblies are checked for design specifications as they arrive at the manufacturer's production or assembly facility. If the items received are premanufactured parts requiring further assembly or processing, receiving inspection will inspect for correct dimensions and any other specifications required by the design.

Using Statistics in Quality Control Although it is sometimes necessary to check each and every part depending on the requirements of the design, this type of inspection is time consuming and therefore expensive. In most cases it is not necessary or even desirable to inspect every item in a large batch of parts that arrives at receiving inspection. Furthermore, many parts are manufactured in such large quantities that inspection of each item would be impossible. The techniques of statistical analysis can be used to assist the inspector in making an accept or reject decision.

Statistical analysis is a branch of mathematics that permits inferences to be made about a large number of items by actually measuring only a small sample of the items. In this process, a random sample is selected from a given shipment of parts. The sample is fully inspected and if sample items do not meet specifications, the entire batch is assumed to be out of specification and is rejected. Likewise, if the sample is acceptable, the batch is considered to be acceptable as well.

This might sound like an unworthy method as it could turn out that the sample selected contained bad parts and therefore the parent batch was rejected unjustly, or a truly bad lot was unknowingly accepted. When using statistics, this possibility always exists. Thus, specific sampling techniques must be adhered to. The **randomization** of the measure sample is the key to making correct inference about a large batch of a given part.

In the case of machine-made parts, however, it is quite likely that the sample taken, providing that it is random, is representative of the larger batch and therefore statistical sampling is a reasonably reliable method of inferring whether the parent batch is acceptable or should be rejected.

Difficulties can arise. If, for example, the sample drawn from the lot to be accepted or rejected is not sufficiently randomized, an unjustified rejection or acceptance could be the result. This is to say that the sample must be a random one if it is to represent the entire batch. If it is not adequately randomized, it may be badly biased and thus show tendencies that are not necessarily those of all items in the larger batch.

For example, a batch of 50,000 parts arrives at the manufacturer's receiving inspection department. The parts are in ten crates, nine of which appear to be fine. The tenth crate contains a number of rusty items. The receiving inspector draws two percent of the total order (1000 pieces) all from the rusty batch. These parts appear to be out of dimensional tolerance, but are hard to measure because of the rust. The inspector decides to reject on this sample basis.

In this case the sample was badly biased. The inspector might have been justified in rejecting only the rusty parts providing that rust rendered them unusable. However, to infer that the entire batch was dimensionally out of specification was unjustified. The inspector should have better randomized the sample so that it truly represented the larger lot.

Figure 4
In-process inspections take place constantly during production (Del Manufacturing).

First Article Inspections

Any time that a new part or product is produced for the first time or when a production machine is set up for repeat production after having been used for another job, a first article inspection is required. When the first article comes through the production process, it is inspected and if it meets specifications, further production is allowed to continue.

If the first article is not acceptable, the machine, or in some cases, the entire production line must be stopped and reset to correct any problems. Further first article inspections may be necessary before full production can resume or continue.

Manufacturing Station In-process Inspections

In-process inspection must be continuously applied during production. Each production station will have whatever tools are necessary to accomplish necessary quality checks (Figure 4). Since most manufactured goods involve the shaping of materials to specified size and geometric dimensions and the location of part features to specific dimensional specifications, part quality and acceptability are totally tied to dimensional inspections.

Precision Measurement and Dimension Tolerances in Manufacturing

Generally speaking, dimensional measurement involves evaluating quantities of length. Length, or the shortest distance between two points, appears under many different names in manufactured products and is measured in the common units of linear measurement, such as inches, meters, and fractions thereof.

Some of the names under which length appears include: outside and inside diameter, center-to-center distance of part features such as holes, wall thickness of tube and pipe, measures of eccentricity and concentricity, and surface finish (Figure 5). Dimensional measurement also involves evaluating angular relationships (Figure 6).

Tolerances An understanding of tolerances is vital to almost any successful product manufacturing. **Tolerance is an acceptable range of part size through which the part will still fit and function as intended.** It is not possible to manufacture any part to a truly exact size. However it is possible to closely approach an exact size depending on design requirements and the capability of the machinery used for manufacturing. It is also possible to measure parts reliably to very close tolerance by applying the tools of modern measurement science. However, the designer must design each part with acceptable tolerances that will accommodate the variations of the manufacturing process.

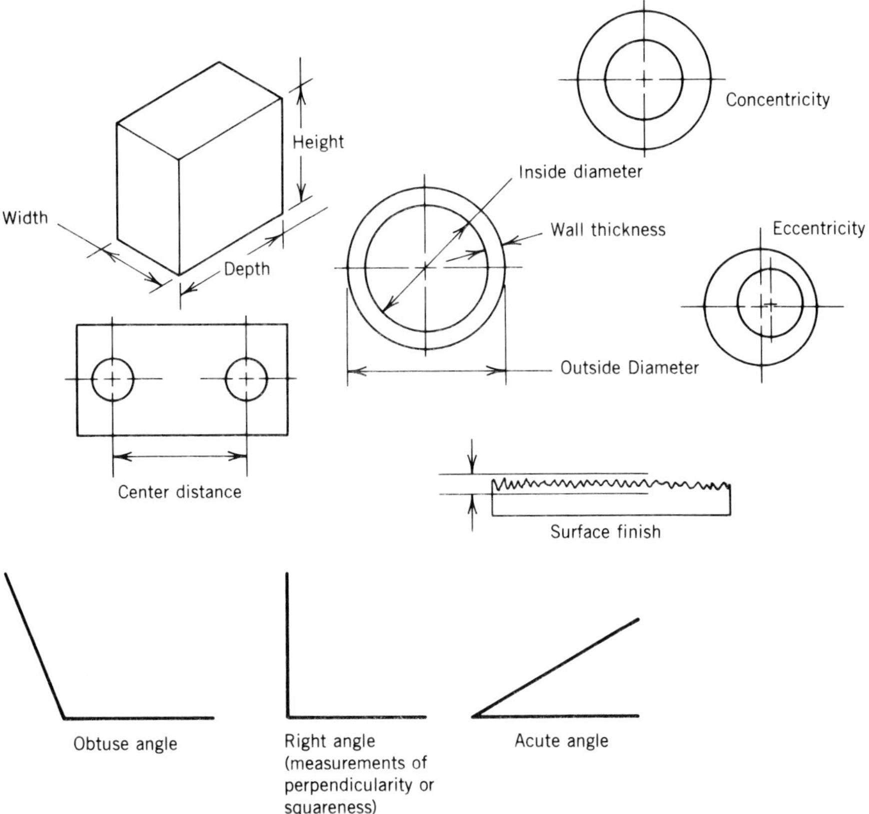

Figure 5
Many dimensional measurements evaluate the quantity of length as it appears under many different names.

Figure 6
Angular quantities must also be measured.

Tolerances generally fall into two important areas. One area deals with **tolerances of size,** that is, the acceptable range of part dimensions. Several forms of tolerance may be used (Figure 7). For example the acceptable part size may be both above and below the true exact size (**bilateral tolerance**) or it may be all above or all below the theoretical exact size (**unilateral tolerance**). Tolerances may be expressed as **high and low limits.** A typical manufacturing drawing may also contain a set of standard tolerances to be applied to all dimensions where other more specific tolerances are not specified.

The designer would, of course, prefer that the manufacturer make the parts to as near the exact specified size as possible. However, the tolerance is necessary to accommodate the variations that always exist in even the most carefully controlled manufacturing processes. Manufacturing of precision parts to tolerances of a few millionths of an inch for example will generally require more care and attention than low tolerance production. Close tolerance manufacturing will also necessitate much higher technology both in manufacturing equipment as well as measurement and inspection tools.

The other area, and one that is of equal importance, is **tolerances of form and position** of part features. This is commonly known as **geometric dimensioning and tolerancing.** Tolerances of **form** include **squareness, perpendicularity, straightness, flatness, circularity,** and **angularity** (Figure 8). **Tolerances of position** include **location** (**true position**) and **concentricity.** Almost any manufacturing drawing will contain specifications for geometric dimensions and tolerances as well as specifications for size dimensions and tolerances.

These concepts of tolerance measurement, size, and geometry should not be confused, although they are interlocked in many cases. Size and geometric specification are not necessarily the same thing although both may be vital in order for a part to meet design requirement.

In some cases, part geometry whether it be form, positional, or both, may be more important than part size. Consider the example illustrated in Figure 9. Flange A must bolt to mating part B. A dimensional specification showing the size of the bolt holes is provided on the drawing. A hole size sufficient to clear the fasteners used is naturally required. However, if the po-

Figure 7
Types of tolerances and standard tolerances.

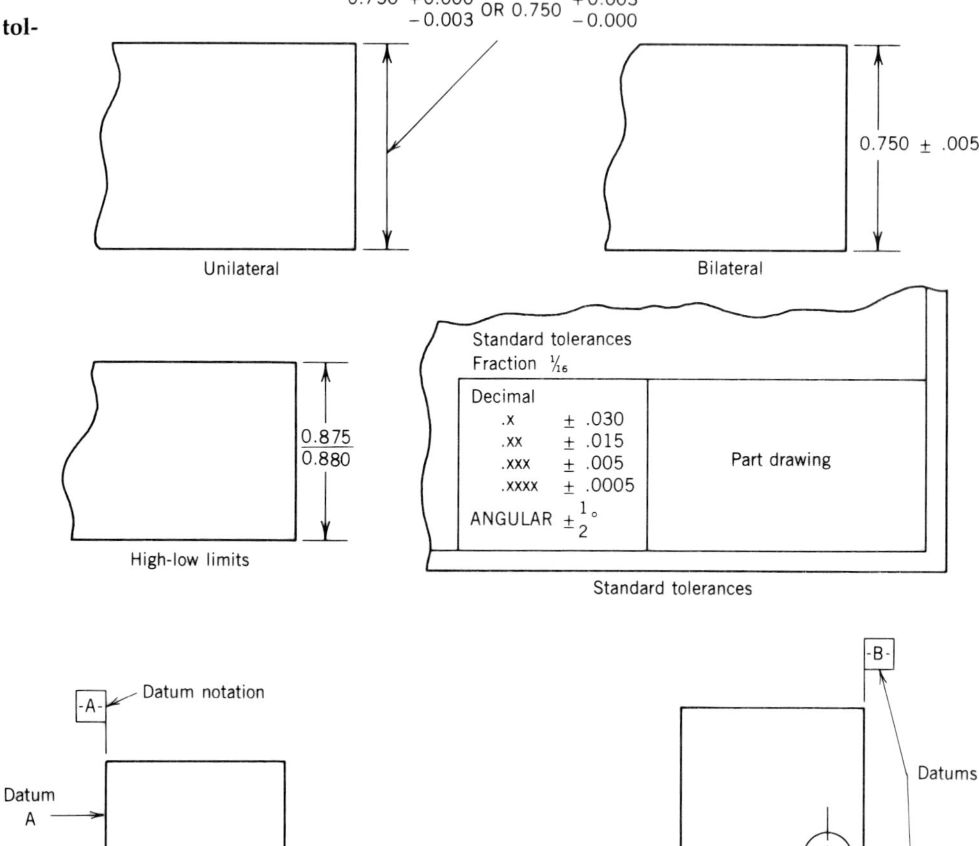

Figure 8
Examples of form and position tolerances.

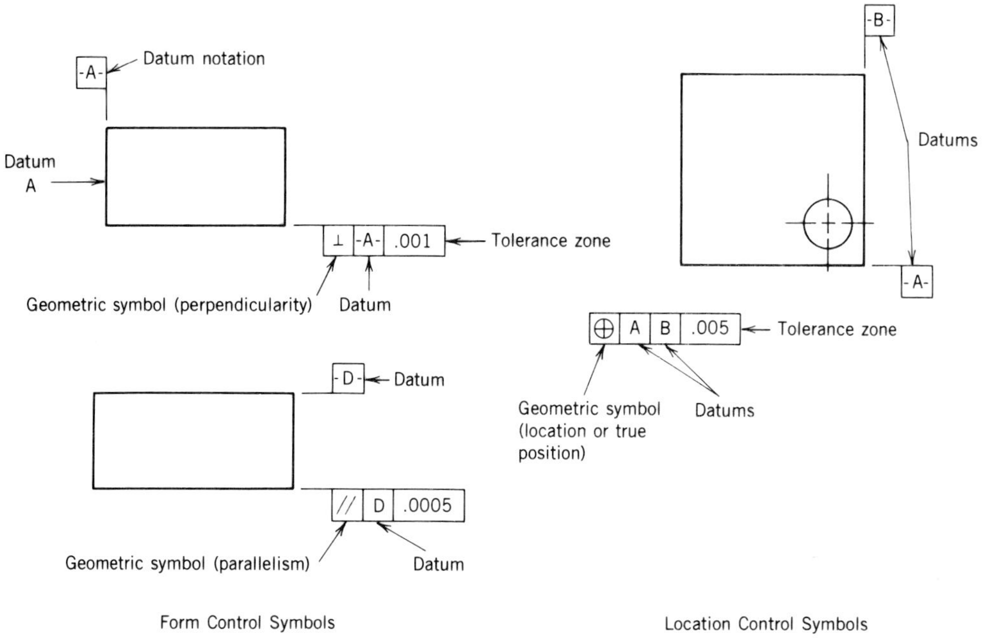

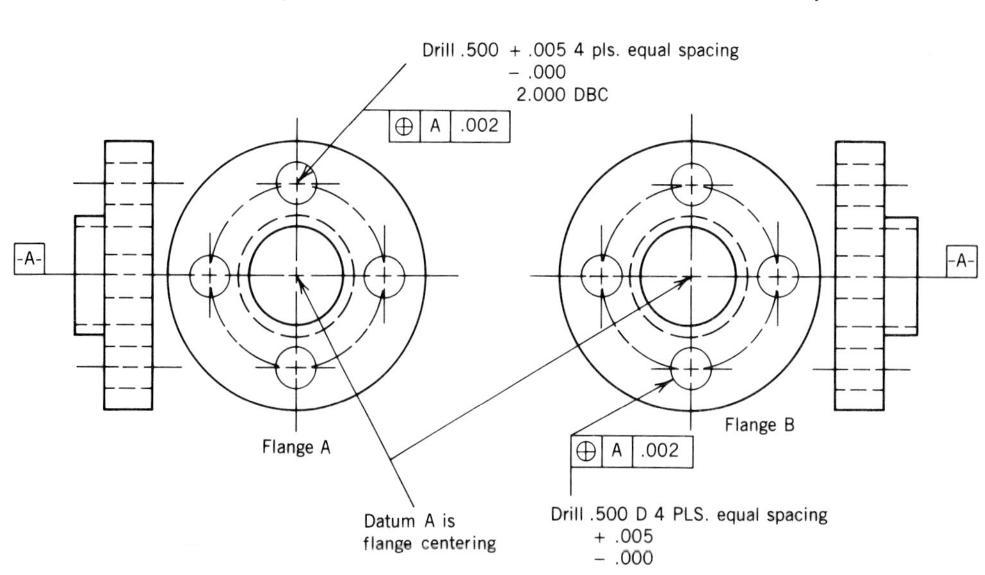

Figure 9
Accurate part feature locations are essential to successful assembly.

sition of the holes is not maintained, it will make little difference if the hole size tolerance is met, the parts will not assemble in any case. Therefore, the manufacturer must maintain sufficient quality control in the manufacturing process to both meet hole diameter specifications (size tolerances) as well as hole position specifications (geometric tolerances of location or true position).

So as to provide a place from which to take all measurement pertaining to geometry, the designer has provided a **reference point, plane,** or **line** called a **datum.** This location is a master reference point from which the manufacturer must locate and set up the part for machining and from which the critical measurements must be taken.

Understanding the relationships of size and geometric dimensions is a complex subject often requiring much study. However, it is vitally important to the successful function of a national or international manufacturing system. Few complex assemblies are made by single manufacturers. It is more likely that many different makers will participate in the process, each one contributing a part or parts toward the final product. It is easy to see how assembly problems could occur if there were no attention paid to design specifications regarding size as well as geometry of assembly parts.

Tools for Dimensional Measurement

The applications of the many tools and equipment for dimensional measurement constitute one of the largest and most interesting manufacturing support activities. As the age of high technology has progressed, the requirements for more precision measurement has moved in lock step. In today's high-precision manufacturing environment, the production of parts to **tolerances of a few millionths of an inch** is commonplace.

The primary objective of in-process measurement inspection in manufacturing is to determine if parts are within tolerance specifications. Most production manufacturing processes only accomplish one or two operations at a time, although in machining, many different cuts can be taken on a workpiece while it is in the same machine. However it is usually necessary to inspect only a few dimensions at a time and then only to inspect for dimensional tolerance requirements.

For example, if the dimension .750 ± .010 in. (Figure 10) required inspection and the production process yields parts that are all quite close to this size, it is not necessary to inspect the .750 in. portion of the dimension each time. Only the tolerance of ±.010 needs inspecting. Therefore it is not necessary to use a measuring tool that has capability to measure over the full range of the dimension .750 in. Although it would be acceptable to measure this dimension with such a tool, a production gage designed and adjusted to check only the tolerance portion would be much faster, more reliable, and easier to use.

Comparison Measurement for Inspection of Tolerances Since much production gaging is done to check tolerances only, the techniques of **comparison measurement** can be applied. In comparison measurement, dimensions or dimensional tolerances are checked by simply comparing the part to a measuring tool that is known to be the correct size. If the gage fits, then the part dimension is good. If the gage does not fit, then the part dimension is out of tolerance. It is not

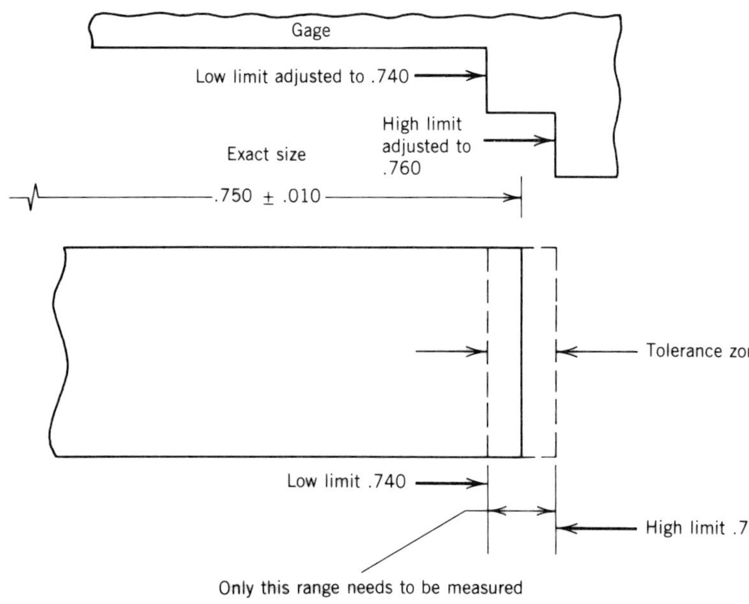

Figure 10
In-production measurement, only the tolerance portion of a dimension needs to be measured.

Figure 11
Fixed GO and NO-GO gages for checking tolerances.

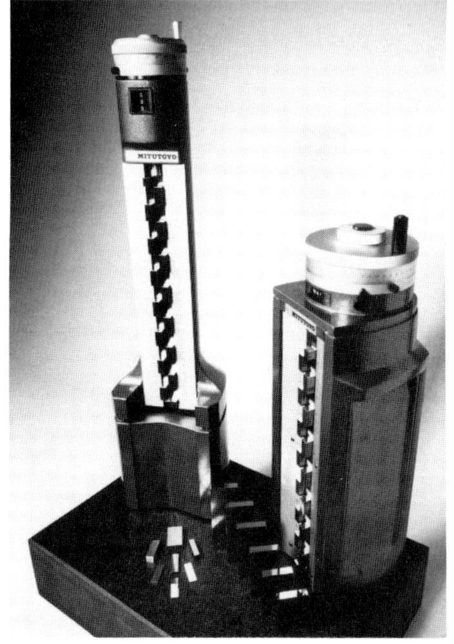

Figure 13
Precision height master (MTI Corporation).

Figure 12
Mechanical dial indicator comparator (MTI Corporation).

Figure 14
Air gages measure dimensions as a function of air flow and pressure (Federal Products Corporation).

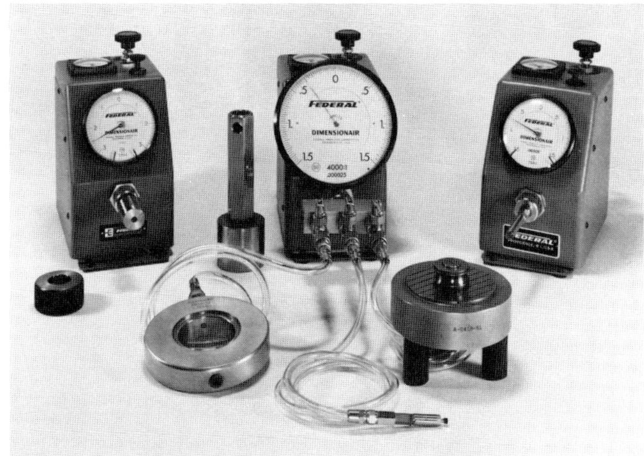

necessary to know how much out of tolerance since any amount renders the part rejected. There may be, however, a possibility of reworking many rejected parts to restore them to design specifications. This is often done in a secondary or rework operation.

In many cases, **fixed gages** are used for tolerance measurement. These measuring instruments often do not show specific numeric data regarding the measurement taken. They simply fit or do not fit. Acceptance or rejection of the part is made purely on this basis. This concept of measurement is often referred to as **GO and NO GO or GO/NOT GO.** Measurement

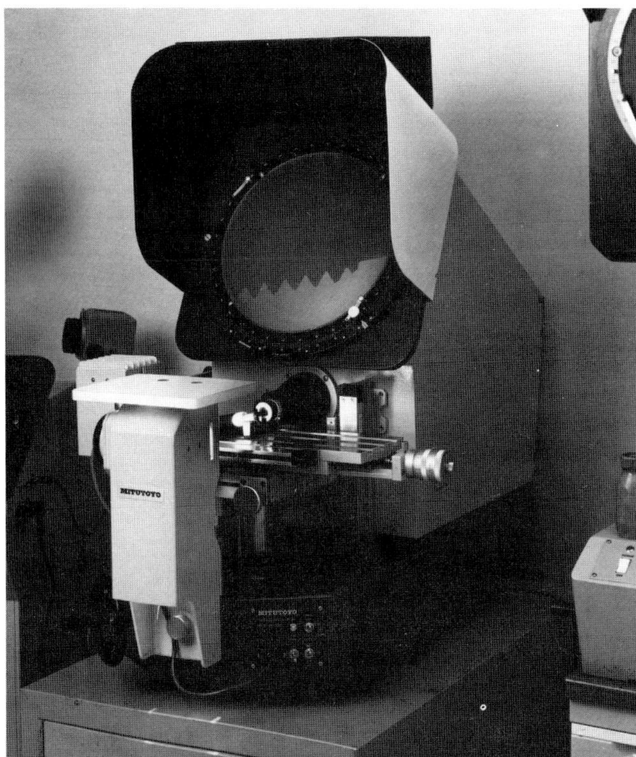

Figure 15
Dial indicator gage for fast and accurate measurement in a production situation (MTI Corporation).

Figure 16
The optical comparator projects a greatly magnified image of the part onto a large screen (MTI Corporation).

tooling used in this type of measurement is often known as **GO** and **NO GO gaging.**

Typical examples of GO/NO GO gages are **thread plug, thread ring,** and **plug gages** (Figure 11). On the thread plug gage, widely used in checking tolerances of internal threads, one end of the gage, marked the GO end, will screw into the thread providing that tolerances are correct. The other end of the gage, marked NO GO, will not screw into the thread; thus the inspector has a comparison tool that can be quickly applied to determine if thread dimensions are within tolerances.

There are several advantages to this type of measurement. Few adjustments of the measurement tool are required. Therefore variations introduced by differences in individual feel are eliminated. The operator or inspector is not required to interpret or even read any numeric data during which an error might be made. The tooling is durable and fairly easy to calibrate.

Possible disadvantages to fixed gaging are that a separate gage is required for each different measurement. This necessitates a large expense if gages must be purchased for many different dimensions.

All gaging equipment is expensive and often represents a large portion of production tooling cost. Specific production gaging designed to check the tolerance of a single dimension can only be justified if the production quantity is large enough to support its cost.

Other comparison measurement devices for checking tolerances include the **dial indicator comparator** (Figure 12), **dial test indicators and height masters** (Figure 13), **air gages** (Figure 14) where air volume and pressure are representative of part size, and the **optical comparator** (Figure 15), where a magnified image of a part or tool being inspected is projected on a screen. The optical comparator is very useful for inspecting shape on a tool or part. Many types of production comparators using mechanical dial or **electronic digital readouts** are also in wide use for measurement inspections (Figure 16).

Calibration

Interchangeability of manufactured parts depends entirely on their dimensional specifications being within design tolerances. Since different parts for the same assemblies are frequently made by many different manufacturers, maintenance of dimensional specifications is of critical importance.

To make this possible, **measuring tools used by each participating manufacturer** must correspond. Therefore calibration of measuring tools is another extremely important responsibility of quality control and assurance.

Calibration is the process of **comparing measuring tools against known standards of measurement.** The process is a continuous one in almost any manufacturing industry. The process of calibration is the responsibility of the **gage laboratory** or **tool room** in a typical manufacturing plant. Calibration equipment often performs the same kinds of measurements as produc-

Figure 19
Gage blocks assembled into stacks make versatile and highly accurate measuring tools (MTI Corporation).

Figure 17
Gage blocks, one of the most common length measurement standards in manufacturing (MTI Corporation).

Figure 18
Typical gage block sets (MTI Corporation).

Figure 20
High discrimination electronic digital readout systems may be retrofitted to existing machine tools, greatly enhancing their positioning and measurement accuracy (MTI Corporation).

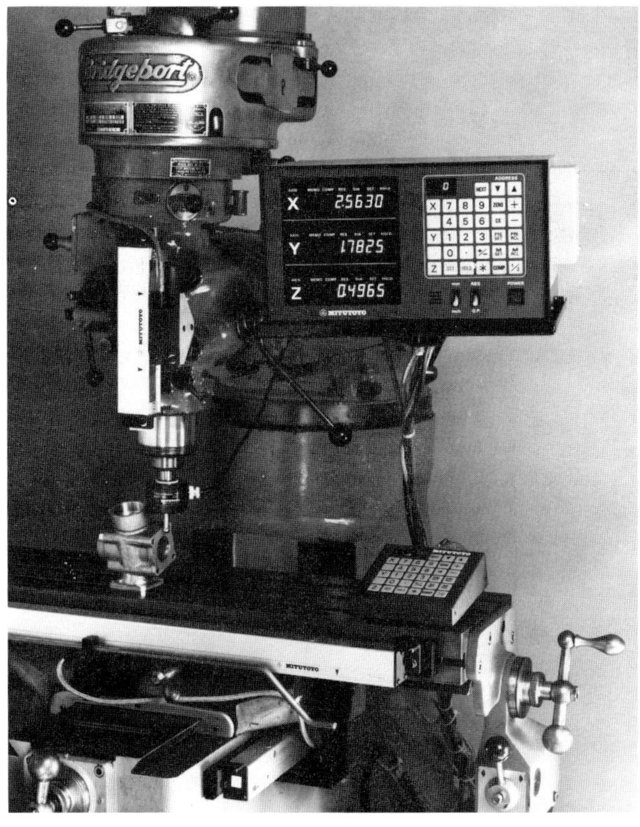

tion measuring equipment, but to a much greater degree of discrimination. Calibration equipment is very sensitive, often delicate, expensive, and must be used in controlled environments.

Gage calibration procedures are often carried out in **standard temperature clean rooms** with **filtered air** and **controlled humidity.** Measuring tools used in the production shop are rotated through the calibration facility

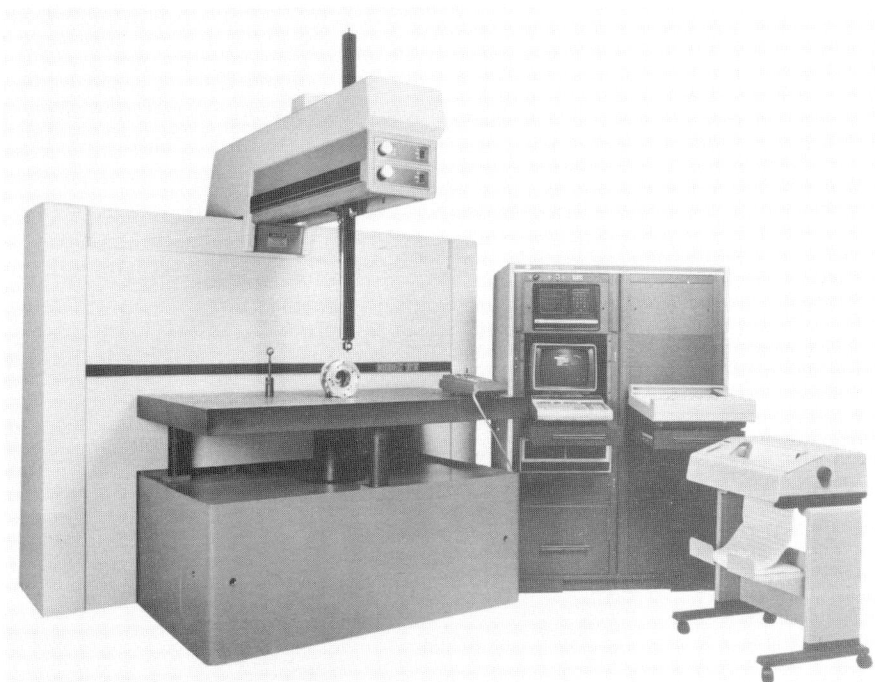

Figure 21
Three-axis coordinate measurement machine with computer readouts and printer (Sheffield Measurement Division).

on a regular basis. The tools are allowed to stabilize to gage room temperature and then they are compared to the appropriate measurement standards. Adjustments are made to bring the tools to conformity and then they are tagged with a **calibration sticker** indicating the date of calibration. When using any measuring tools on the production floor, it is the responsibility of the inspector to ensure that calibration is current.

Measurement standards Master measurement standards are most frequently established and maintained by government agencies. In the United States, the **National Bureau of Standards** (N.B.S.) has responsibility for establishment of measurement standards. Other measurement standards located in the gage labs of manufacturing industries, can have their calibration traced to N.B.S. standards. Calibration of industry standards is carried out periodically and appropriate documentation pertaining to traceability is maintained.

Gage blocks (Figure 17) are among the most common and versatile measurement standards used in manufacturing where dimensional quality control is critical to production. Gage blocks are found and used in all phases of manufacturing, production measurement, machine setup, and calibration of measurement tools.

A gage block set consists of a series of steel blocks of various sizes. Each block has been manufactured under controlled condition and is dimensionally **very close to its nominal size.** In other words, a gage block marked .500 will be, at a specified temperature (usually 68° F) within a **few millionths** of an inch of its marked size. The block may be slightly under or over nominal, but very close. Depending on the grade of gage block, tolerances vary somewhat. However, the most inexpensive gage block will usually be within .000050 in. of nominal. A laboratory grade gage block will be within .000002 in. of nominal. Block geometry including flatness and parallelism of the working surfaces will also be very close to exact. This factor of near perfect dimensions is what makes the gage block such a useful tool as a measurement standard. Because the gage block dimension is so close to exact size, several gage blocks can be stacked in any combination to make up the desired amount of length within the range of the set. These gage block stacks will also be extremely close to the exact calculated size because the block tolerances add or subtract very little to the required dimension.

The gage block set (Figure 18) is designed so that stacks of any size can be created. These can be used in many applications of **machine tool setups, production shop measurement applications** (Figure 19), and **measuring tool calibration.**

Computers and other integrated circuit electronic devices have found wide application in precision measurement. These high technology instruments have **great sensitivity** and have many advantages over their mechanical counterparts. They have made high precision measurement routine and reliable which has, in turn, let high precision manufacturing flourish and produce many of the fantastic products that we all enjoy. Examples of electronic equipment include **high resolution digital readouts** for machine tools (Figure 20) and the **three-axis coordinate measuring machine** used for inspection and calibration (Figures 21 and 22). Multi-axis coordinate measuring may be coupled to a computer that in turn will make instant calculations relative to the measurements taken. For example, the machine's measurement probe can be used to quickly touch to points on the radius or diameter of a part being inspected. Using this data, the computer will instantly calculate average diameters and radii and print out this information on a high speed printer. The process is extremely fast and reliable and it generates hard copies of inspection data. Other tools used in inspection measurement and calibration include **digital electronic height gages** for production shop precision measurement (Figure 23).

Other calibration and measurement procedures include inspecting and graphing part geometry such as **roundness** (Figure 24). The **toolmaker's microscope** has a two-axis stage allowing parts to be moved and viewed under high magnification (Figure 25).

A specific **surface finish** is vital to many parts such

Figure 22
The three-axis coordinate measuring machine used in auto body dimensional inspection (American Motors Corporation).

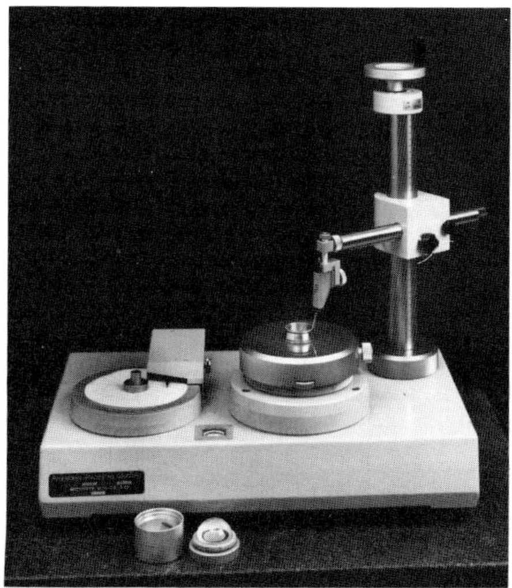

Figure 24
Measuring and graphing roundness, a geometric inspection (MTI Corporation).

Figure 23
Electronic comparators and digital height master (The L.S. Starrett Company).

Figure 25
Toolmaker's microscope, for a magnified look at dimensions (Nikon Inc., Instrument Division).

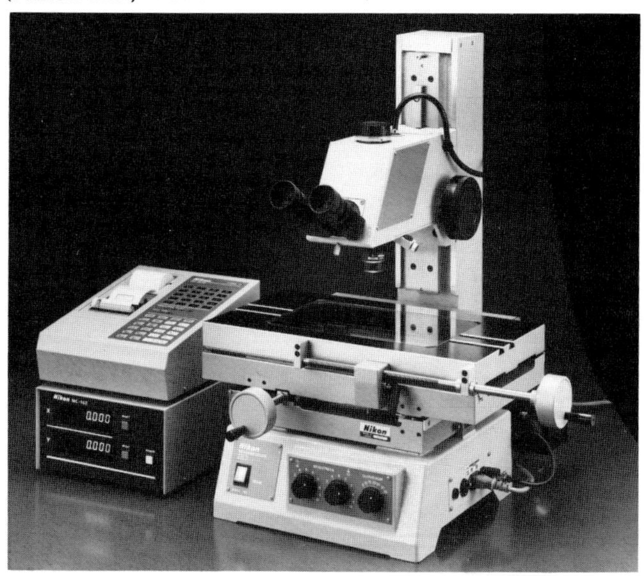

as gears, bearings, and hydraulic sealing surfaces. The surface finish measuring instrument (Figure 26) reads and records the depth of surface deviations in **microinches** (.000001 in.). Electronics can also be applied to the measurement of machined profiles (Figure 27). The electronic counterpart of the dial indicator comparator can measure to millionths of inches (Figure 28).

Calibration procedures are by no means reserved to the dimensional measurement area. Their application is, for example, equally important in many other areas such as electronics where frequency standards must be maintained and calibration of equipment for reading pressures, flow, weight, colors, surface finish, part geometry, performance, and other specifications required on manufactured products.

Figure 27
Measuring profile geometry electronically (MTI Corporation).

Final Inspections

Final inspection presents yet another opportunity to sort out parts or assemblies that do not meet specifications. Even though discrete parts are inspected at each production station, they may be inspected once again in final inspections. Parts not meeting requirements may be recycled for rework if this is possible, or in some cases, design reviews are held to determine if the specification could be adjusted and the parts accepted. It is naturally desirable to avoid scraping as many parts as possible since they often represent a large production expense.

In some cases where out of tolerance dimensions are not critical to the fit or function of an assembly, it is possible that design specifications may be temporarily set aside. This is a design engineering decision and is only done with the greatest of caution.

One Hundred Percent Inspections

One hundred percent inspection refers to the checking of **every part or assembly** coming off the production line. Even though this might appear to be a way to solve all quality control problems, it is often impractical and altogether too expensive to accomplish in most routine manufacturing. This is especially true in high volume production where it would be quite impossible to check each and every item produced.

Figure 26
Surface finish, a measurement of length in millionths of inches (MTI Corporation).

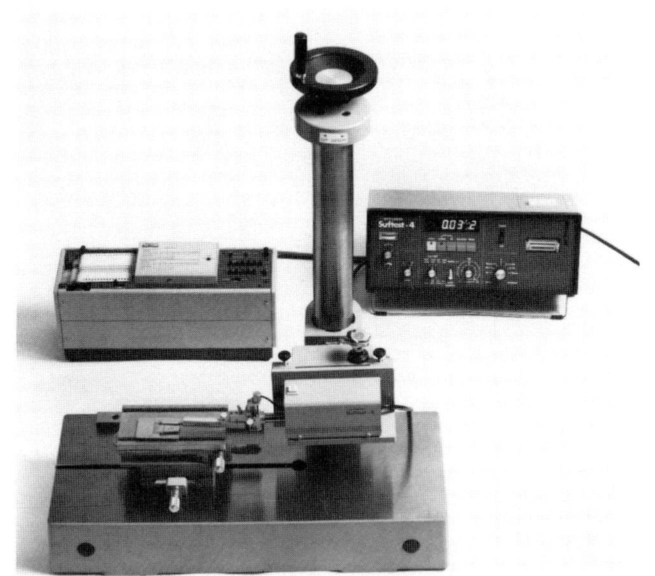

Figure 28
Digital electronic comparator (MTI Corporation).

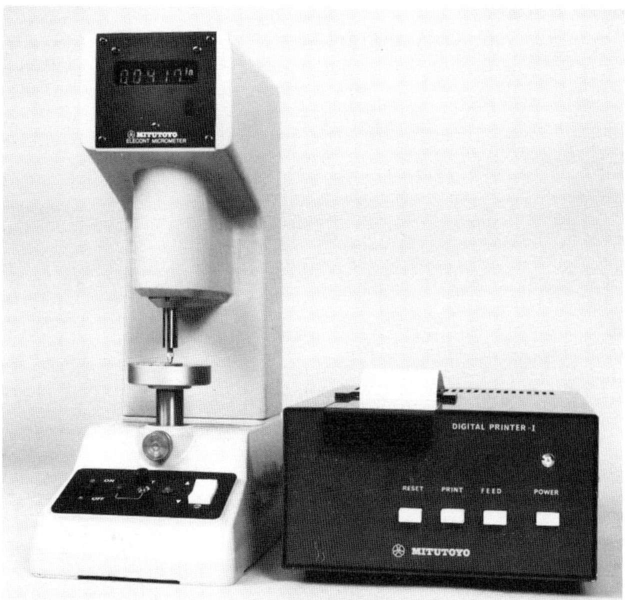

Figure 29
Automotive dynomometer testing (GM—Pontiac Division).

One hundred percent inspection is however used where very critical design requirements are specified. Certain aircraft flight critical parts may be 100 percent inspected since the possible acceptance of an out of specification part could have serious consequences.

Another place where 100 percent inspection may be applied is where out of specification parts begin to show up continually in the production process. This problem must be tracked down and immediately corrected. One hundred percent inspection may be necessary until production quality is restored, then random inspections may resume.

Figure 30
Vaporizing an engine crank case oil sample for spectrographic analysis (Allis—Chalmers).

Figure 31
Computerized spectrogaphic analysis of the oil sample (Allis—Chalmers).

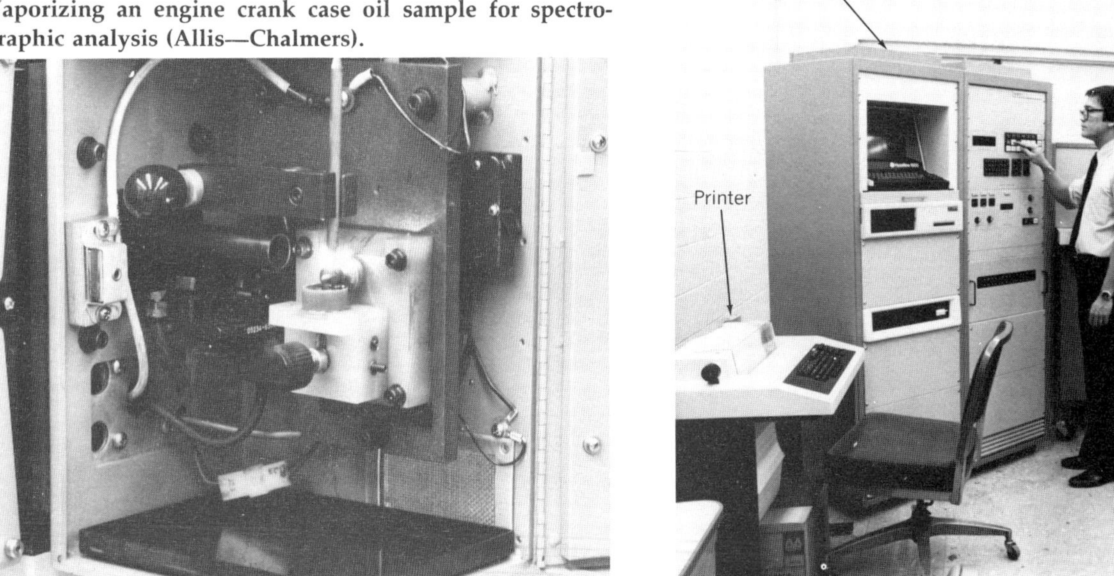

NONDIMENSIONAL QUALITY CONTROL

Quality control and assurance practices must be observed in nondimensional areas with equal dedication. This includes a wide variety of tests and inspections on all types of products especially those that take the form of a complex operating assembly.

Nondimensional quality control inspections and testing are far reaching and often involve not only state of the art technology, but many truly remarkable engineering innovations. One example is the automotive dynamometer test that reveals horsepower and emissions of engines (Figure 29). Many nondimensional quality checks require high technology to be accomplished. Computerized equipment with the ability to analyze test data with great speed plays a vital part in the overall quality control and assurance effort. In Figure 29, oil samples from lift truck engines are spectrographically analyzed to determine chemical and molecular content. The samples are first vaporized (Figure 30) and then the light energy emissions from this process are computer analyzed (Figure 31) to determine the chemical make up and content of foreign matter in the oil sample. The process can determine the type and amounts of metal particles suspended in the oil and the data can be used to make a determination of where metallic abrasion is taking place within the engine during operation.

Aircraft and spacecraft air frames are subjected to the same stresses that they will experience in service. A vehicle is enclosed in a large steel framework where

Figure 32
Space shuttle flight stress simulator subjects air frame to forces that it will encounter in actual flight (Lockheed—California Company).

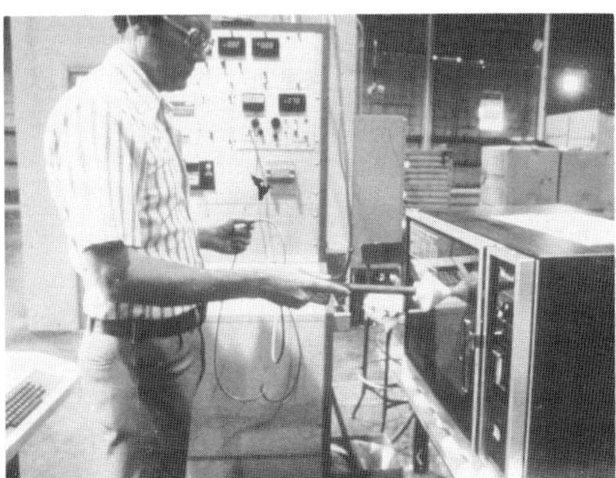

Figure 33
Computer analysis of microwave oven emissions (Whirlpool Corp.).

hydraulic jacks exert forces of up to one million pounds thus simulating what the vehicle will experience during launch, spaceflight, and reentry into the earth's atmosphere (Figure 32).

Another example of nondimensional quality control is computer analysis of microwave oven emissions (Figure 33). This type of inspection is often vital to consumer protection and safe operation of appliances for the home.

QUALITY ASSURANCE AFTER MANUFACTURING

Quality control and assurance do not end at the manufacturer's shipping dock. They continue well after manufacture in the form of **warranties** providing factory or factory authorized service, **spare parts, field service, service contracts, reliability studies, failure reports** and recommendations, and repair and **reconditioning** product support.

Purchasers of almost any product expect and are entitled to some type of warranty or guarantee that will be in force for some time after the product arrives in the hands of the end user. Warranties vary in length ranging from a few months to lifetime full replacement. Some warranties are for full repair and replacement for a short period of time and then extended warranties for certain parts only after that. Extended warranties may be purchased at extra cost in many cases. These are justified on the basis that they are probably cheaper than the cost of one significant repair.

Related to this are service contracts for which the customer pays a flat extra cost, sometimes a percentage of the purchase price, that guarantees repair service for the duration of the contract.

Many manufacturers will provide factory service on their products. So that this capability may be extended over wide geographical areas, factory authorized service through other businesses is a widely available service. Spare parts can often be purchased from the original maker or from the factory authorized service center.

Usually a large aftermarket will develop that supplies duplicate spare parts not necessarily manufactured by the original manufacturer. These are often less expensive than original equipment parts. Manufacturer field service is an important consideration as it is often impossible to return equipment to the place of manufacture.

Critical plant equipment that fails in service must often be restored to functional capability as quickly as possible. Efficient factory field service can be critical to the continuing function of other plants and businesses.

Many manufacturers, especially those of high technology products, are concerned with reliability and failure rates of their products in service. Product failure and reliability analysis is useful in determining where improvements or correction of deficiencies in the production or design need to be made. Reliability analysis can greatly improve a manufacturer's product quality. The compilation of reliability statistics can be effectively used in product advertising thus enhancing the manufacturer's standing in the business.

PRODUCT TESTING

Manufacturing industries often place a large emphasis on product testing. This may be done at the prototype stage before an actual product is manufactured for sale. Product testing may also be carried out in controlled markets where a few preproduction models are placed in selected hands for the purpose of testing and evaluation.

Product testing is often vital to successful manufacturing. In the case of aircraft, for example, a license to manufacture will not be granted until design and safety specifications have been met. The manufacturer must assume the costs of manufacturing preproduction prototypes and their testing.

Endurance testing of automobiles is another significant product testing activity. Production models are test driven through many different environments simulating what they will see in actual service. By continuous testing, several years of typical service of the vehicle can be compressed into a few months. The data collected can be invaluable in determining design problems and in predicting reliability and effective service life of the product.

REVIEW QUESTIONS

1. What is the difference between quality control and quality assurance?
2. What are the three major areas of quality assurance?
3. What are the responsibilities of the source inspector?
4. What items does the source inspector look for?
5. Give two examples of metallurgical testing.
6. How would a given metal be checked for alloy content?
7. What is involved in a dimensional inspection?
8. Why are dimensional inspections important in manufacturing?
9. What is the difference between destructive and nondestructive testing?
10. Where might destructive testing be applied?
11. What is a performance inspection?
12. What are the major points of in-process quality control?
13. How are statistics used in quality control?
14. What is the purpose of a first article inspection?
15. What are tolerances and why are they important in manufacturing?
16. Name two types of tolerance and explain them.
17. What are geometric dimensions and tolerances and why are they important in manufacturing?
18. Is there any difference between tolerances of size and tolerances of geometry?
19. Cite several examples of geometric tolerances of form.
20. Cite an example of geometric tolerances of location.
21. Which is more important, tolerance of form or tolerance of size? Why?
22. What is GO and NO GO gaging?
23. Cite several examples of tools for production gaging.
24. What is the meaning and purpose of calibration?
25. What are used as common shop measurement standards?
26. Who is finally responsible for national measurement standards?
27. What is meant by calibration traceability?
28. How are computer and electronics used in measurement?
29. What is the purpose of final and 100 percent inspection?
30. What are some examples of nondimensional quality control?
31. What quality assurance factors occur after manufacturing and shipping?
32. Why is product testing important to a manufacturer?

CASE PROBLEMS

Case 1
Calibration

A machine operator, using standard production gages, checks parts as they come out of the production process. Parts appear to be in tolerance, but are later rejected in final inspection as being out of tolerance. What could be the cause of this problem and how might it be corrected? Relate your answer to the purpose and procedures of calibration.

Case 2
Geometric Tolerances

A manufacturer of electric motors acts as a vendor to several different customers making use of several different motor models. The company ships several motors to a new customer with which it has not previously done business. A fit problem develops when the new customer attempts to attach motors to its equipment. The customer company complains that the motor manufacturer's product is not within advertised tolerances. However, the motor manufacturer indicates that the motors in question when used by other customers have caused no assembly problems.

What could be the cause or causes of this problem and what steps might be taken to ensure that there will be no assembly problems in the future. Relate your answer to tolerances of both size and geometry.

CHAPTER 21 THE MANUFACTURING INDUSTRY

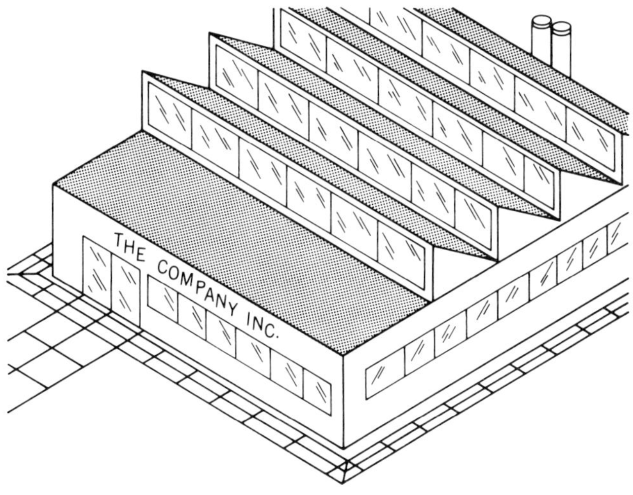

The manufacturing industry is the place where the products that make an industrialized society function are made. Manufacturing industries make up a large portion of total business and are the places of employment for a large segment of the labor force. The purpose of this chapter is to describe the structure of a typical manufacturing industry, to describe the various jobs and responsibilities of the people who fill them, and to see how a product flows through the manufacturing process.

OBJECTIVES

This chapter should enable you to:

1. Describe the structure of a typical manufacturing industry.
2. Identify general job titles and describe responsibilities of each.
3. Describe the flow of a product through the production manufacturing process.

There are many types of manufacturing industries. They range in size from one- or two-person businesses to major corporations employing several hundred thousand employees. The operations of major manufacturing corporations are accomplished on a worldwide scale, and the dollar value of sales can range from a few thousand dollars in a small business to upward of several billion dollars for a large multinational corporation. Many large companies are parts of larger groups or conglomerates. Companies that are engaged in the same or similar product manufacturing may form technology groups. In this manner, duplication of product lines is eliminated and technical expertise may be pooled for more efficient business operations, product design, and production. In other cases large manufacturers of complex products such as automobiles may own subsidiaries that supply them with component parts. This ensures that a supply of required parts is always available. The end product manufacturer may also tap much diverse expertise for all phases of business if the secondary industries are owned by the parent corporation. Regardless of the size of the business, all have a basic management structure where job responsibilities are delegated in much the same way.

BUSINESS STRUCTURE

A typical manufacturing industry of average size and having most of the typical departments and job titles will be organized along the lines shown in the organizational chart of Figure 1.

Upper Management

Upper management includes the company **president, chief executive officer (CEO), chairman of the board of directors,** and one or more **vice presidents**. The responsibilities of the CEO, president, and/or chairman of the board of directors are to oversee the entire operation of the business. The head of the company is responsible to the stockholders and, in many cases, upper management is elected by the stockholders to run and manage the business.

Vice presidents (VPs) are likely to be specialists in several areas. Examples include manufacturing, finance, advertising, research and development, engi-

Figure 1
Basic framework of a typical manufacturing industry.

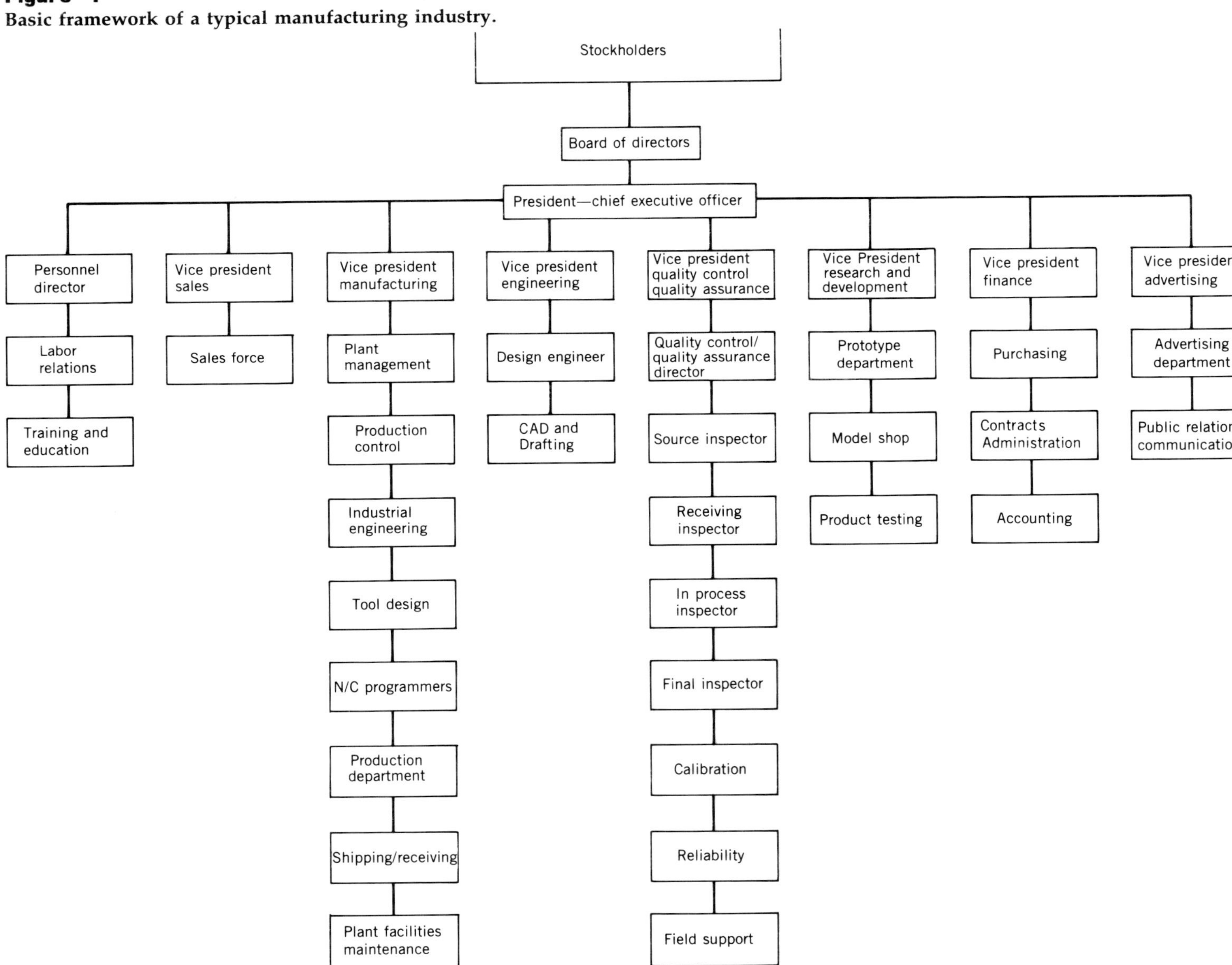

neering, quality assurance, and marketing. The VP of manufacturing is responsible for all aspects of the manufacturing operation. The VP for finance is responsible for company loans, stockholder dividends, stock and bond issues, and accounts payable and receivable. The advertising VP is responsible for advertising the company's products if this is done in house. In many cases, advertising is jobbed out to advertising agencies that specialize in product promotion.

If the company does research and development of new products or is in the business of product development for outside customers, the VP for research and development will oversee this aspect of the business. Engineering involving product design will be the responsibility of the VP for engineering.

Middle Management

Below upper management is middle management. In this area of management many job titles and far flung responsibilities are found. Middle management includes such job titles as **plant managers, department heads,** and various other titles. It is here that much of the specific management relative to production of the product is accomplished.

One important job responsibility in middle management is **industrial engineering (IE)**, sometimes called **manufacturing engineering** or **industrial technology (IT)**. Personnel that staff these jobs provide an invaluable service to the smooth running of a manufacturing industry.

The industrial or manufacturing engineer (IE) or industrial technologist is usually trained in broad areas of manufacturing. This individual knows both the business end of manufacturing as well as the processes that are used in making the product. Thus, an IE is able to communicate effectively with both upper and lower management.

Some of the responsibilities of industrial engineering include **jig, fixture, and tool design**, plant **equipment** and **facilities planning, CNC programming**, and many aspects of **quality assurance** and **control**. Although product design is the responsibility of the design engineer, industrial engineers are often the individuals who make production of the product ultimately possible.

Production Support Management

Many other individuals also contribute to the production process. Although their contributions may be indirect, they are nonetheless vital to the successful function of the business. Job titles in this area include **purchasing agents** who are responsible for purchasing tools, equipment, and raw material required for the production process and **production schedulers** who make sure that the proper material is ordered and arrives at the production facilities in a timely manner. This permits production to continue in an orderly fashion with no delays due to a lack of material. The company **personnel department** is responsible for hiring necessary individuals to meet the staffing needs of the company.

Another vital function in any manufacturing business are individuals involved in plant facilities and production equipment **maintenance**. Their job is to keep the production line moving. Production manufacturing is at best a costly business and if the production line should stop, pay for the employees does not. The line must be restored to service as soon as possible. Maintenance personnel are therefore always on call in many high production manufacturing industries. In modern computer automated manufacturing, the electromechanical and the computer numerical control electronics technicians are often directly responsible for up time on expensive production equipment. *Up time* refers to the time that a piece of manufacturing equipment is actually producing product units. *Down time* refers to times when a machine is not producing. Machine down time can be due to breakdown or required maintenance or when the machine is waiting for material to arrive and be loaded for processing. It is naturally desirable to have as little down time as possible. Thus, the mechanical and electronic maintenance technicians are often very busy in fast-paced production plants. They are often called to work during night shifts of weekend periods in order to keep production equipment operating at peak efficiency. As computer automation trends continue to predominate modern manufacturing, the maintenance technician will play an ever-increasing part in keeping manufacturing equipment in an up time mode.

Quality assurance and control personnel perform many vital tasks in all phases of manufacturing. Their responsibilities range from source and receiving inspections, first article inspections, production process inspections, final inspections, product reliability and after-production warranties.

Lower Management

Lower management is directly responsible for production worker supervision and the hour-by-hour production of the product. Job titles at this level include **line supervisors, fore, lead, and setup persons**.

The Production Worker

No discussion of business structure would be complete without mentioning those that do the actual work of production and production support. Job titles are many and varied in production work and include **machine**

operators, machinists, toolmakers, toolroom attendants, plumbers, sheet metal workers, mechanics, inspectors, electronics assemblers, helpers, electricians, electronic and **mechanical** system maintenance **technicians**.

PREPARING FOR PRODUCTION MANUFACTURING

The business of manufacturing begins when an order is received by the manufacturer or when the company's own designers create a new product that has sales potential. In many cases the order is solicited by the company's sales force as it works in the market place to generate business. The sales force is supported by the company's advertising department or an outside advertising agency. The sales force will attend trade shows to demonstrate and promote the company's products and they will make calls on prospective customers.

When an order is received, upper manufacturing management and financial management make the necessary decisions as to how production will be financed and accomplished. If necessary and justifiable, new production facilities may be built or purchased after careful study has been made to determine the profitability of the proposed production. There is always some risk. It is likely that the manufacturer will have to borrow much of the money needed to finance the production. The sales of the product will have to return this investment so that the company will be able to repay its loans while at the same time meet its daily financial requirements. These include meeting its payroll and purchasing material for the product being manufactured.

Estimating Costs

When the order (Figure 2) is received at the manufacturer's facility, it is processed and a price and time of delivery are quoted to the customer. In many cases a manufacturer will be asked to bid on a job. Under this system, the customer will shop around for the required products and ask several competing companies what their best price is for the manufacturing required. Bids are generally provided at no cost to the customer and the manufacturer must carefully prepare his bid so that it reflects actual production costs plus acceptable profits. Underbidding to get a contract may result in the manufacturer losing money on the job or being unable

Figure 2
Manufacturing requires the coordinated efforts of all the team players.

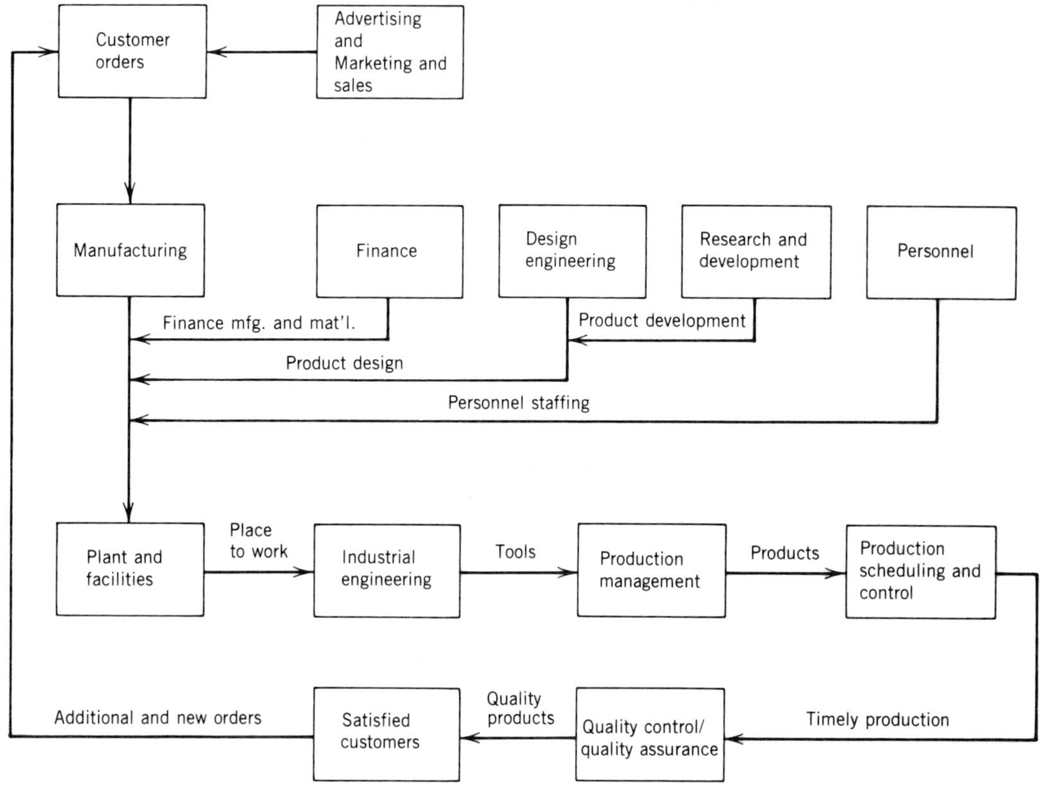

to complete the work. Both circumstances can result in a business failure. Cost estimating involves a careful analysis of all the costs involved in production of a product. Major factors influencing costs include:

1. Employee wages, salaries, and benefits.
2. Interest on borrowed money.
3. Price of material.
4. Tooling and production machinery.
5. Packaging and shipping.
6. Quality control and assurance.
7. Plant and equipment maintenance and security.
8. Utilities.
9. Product liability insurance if required.
10. Stockholder dividends and other returns to investors.
11. Design expense.
12. Rent, lease, or construction of manufacturing facilities.
13. Guarantees and services included in purchase price.
14. Company profits.

All of these factors represent a portion of the end product price. By process of careful cost accounting, each expense can be represented as a percentage of the product price. For example, if employee benefits represent 10 percent of the product price, then $10.00 of a $100.00 product represents that portion of the cost attributed to employee benefits. Product costs are also influenced by other complex business cycle factors which will all have their direct or indirect influences on the price of a product. Included in these are taxes, depreciation, investment tax credits, and projected sales, often known as sales forecasting. If a manufacturer has a strong sales position in a given market and the national economy is in an up turn, the company may be willing to risk venture capital in plant and/or product expansion, while at the same time, holding its prices in line or maybe even reducing prices, if continued strong sales justify this. However, in times of economic recession, or resulting lowered product demand, high rates of interest on borrowed funds, or foreign as well as domestic competition, a company may take a more conservative approach in new plant and equipment investments, since a large debt in a declining economy and/or market can cause a business failure and the unemployment of many workers in the labor force. Accurate cost estimating thus becomes a critical issue in any business and the manufacturer continually strives to estimate costs efficiently while reducing operating costs at all levels.

One method of cost estimating reduces all expenses to a common denominator of production hour rates or, even more simply, the price per unit of production. Production hour rates are often used by service industries or manufacturing concerns that act as incidental vendors. Job machine shops are an example. When a customer asks for a bid on a job, the price is quoted at so much per hour of production time plus any materials that are purchased for the job. The production hour rate reflects all of the costs of production with the exception of the material. The estimation of production hours requires that the estimator be generally informed of the processes involved and then be able to estimate accurately the time it would take to accomplish the task at average rates of productivity.

Example

production hour rate = $40.00 per hr
time estimate = 10 hr
cost = (time in hours) (production hour rate)
= (10 hr) ($40.00 per hr)
= $400.00

When products, particularly those that become parts of more complex assemblies, are manufactured on a continuous basis, it is simpler to estimate and quote prices simply on a cost per unit basis. This method is particularly suited to price quotations on various quantities. We have all heard the adage "cheaper by the dozen." In production manufacturing, this bit of philosophy usually turns out to be true. Once the production process is established, costs are divided by the number of products produced. The more products produced, the less the unit costs for each one.

Example

production unit cost = $1.35 per unit
production rate = 10 units per hr
cost of 20 units = ($1.35 per unit) (20)
= $27.00
hour cost = (units per hr) (cost per unit)
= (10 per hr) ($1.35 per unit)
= $13.50 per hr

Production hour rates can be converted to **cost per unit** and the reverse is also true. It is necessary to know how many product units can be produced in a production hour or how many production hours are required to produce a given number of product units. It is important that all costs be reflected in a quotation of any hour rates or costs per unit. If they are not, underbidding may occur and a cost overrun above the estimate given may result. Certain contractual arrangements between customer and vendors do not permit restructuring of estimated costs once prices have been agreed on.

To establish a production hour rate and then a cost per unit estimate, it is necessary to carefully evaluate all of the variables that will affect the final price. These expenses can be expressed in terms of **percentages of total operating budgets** in a business to determine what a customer will have to be charged per worker hour or per unit of production.

Examples

Cost per unit, knowing production rate and production hour rate.

$$\text{cost per production hour} = \$45.00$$
$$\text{production rate per production hour} = 9 \text{ units}$$
$$\text{unit cost} = \frac{\text{production hour cost}}{\text{production rate of units}} = \frac{\$45.00}{9} = \frac{\$5.00}{\text{unit}}$$

Production hour cost, knowing unit price and rate of production.

$$\text{unit price} = \$12.50$$
$$\text{production rate} = 15 \text{ per hour}$$
$$\text{production hour cost} = (\text{unit price})\frac{(\text{production rate}}{\text{hour})} = (\$12.50)\frac{(15 \text{ units}}{\text{production hour})} = \frac{\$187.50}{\text{production hour}}$$

Production hour cost breakouts as percentage of expenses. Cost of production hours = $200.00 representing:

Labor 67 percent	(.67)($200.00) = $134.00
Benefits 12 percent	(.12)($200.00) = $ 24.00
Utilities 2 percent	(.02)($200.00) = $ 4.00
Material 19 percent	(.19)($200.00) = $ 38.00
Total 100 percent	$200.00

Production hour cost as a percentage of unit cost. Unit cost = $1250.00.

Labor 67 percent	(.67)($1250.00) = $ 837.50
Benefits 12 percent	(.12)($1250.00) = $ 150.00
Utilities 2 percent	(.02)($1250.00) = $ 25.00
Material 19 percent	(.19)($1250.00) = $ 237.50
Total 100 percent	$1250.00

Facilities, Design, and Purchasing

A production facility is necessary that will accommodate the required equipment and personnel; depending on the product, a new factory may have to be built and personnel hired and/or relocated to staff it. Plant facilities and industrial engineering are responsible for the layout of the factory floor and setup of the production and/or assembly lines in such a way that product will flow smoothly through the production process.

If product design is required, it will be the responsibility of the engineering design department. Engineering design may be accomplished with the aid of the computer (CAD) in modern manufacturing; production tooling, jigs, and fixtures are developed by the industrial and manufacturing engineering staff. Construction of tooling will be done in the tool and die shop or it will jobbed out to a vendor.

The purchasing department, staffed by purchasing agents or buyers, is responsible for procurement of materials, goods, or services necessary to the support of the manufacturing operation.

PRODUCTION MANUFACTURING

When design is completed (Figure 3), manufacturing facilities are established and staffed, material is procured, and tooling is in place, production manufacturing may begin. As the major assembly components are completed, they are fitted together to make up the final

Figure 3
Designing the high technology age will be done with CAD (Lockheed—California Company).

Figure 4
Major wing subassemblies are separately constructed (Lockheed—California Company).

product (Figures 4, 5, and 6). The final product is possible if and only if all of its many parts are completed and assembled. Sequence of assembly must be properly handled so that all of the components will fit where they belong in the final assembly.

Figure 5
Major fuselage subassemblies are lowered into place (Lockheed—California Company).

Figure 6
Fuselage sections take shape in a large production facility (Lockheed—California Company).

Figure 7
All the details must be completed in proper sequence (Lockheed—California Company).

As the large components go together, work on internal parts that can be done at the same time continues (Figures 7, 8, and 9). Industrial engineering is constantly busy solving production problems while designing new tools, jigs, and fixtures. Plant and production maintenance keeps plant and production equipment functioning properly.

Figure 8
Millions of small parts such as rivets must each by installed with the same care and attention (Lockheed—California Company).

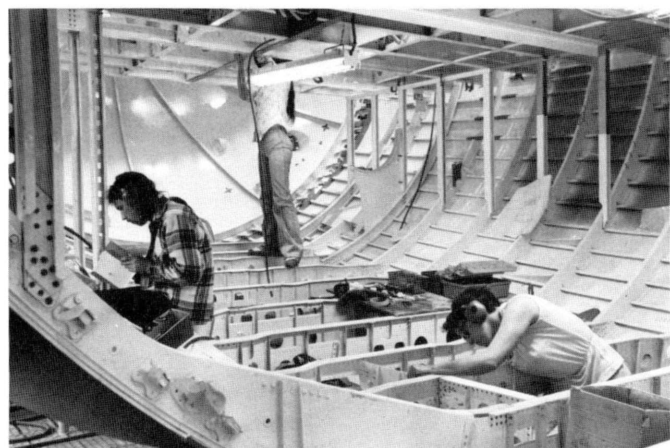

Figure 9
Internal work on major subassemblies takes place simultaneously (Lockheed—California Company).

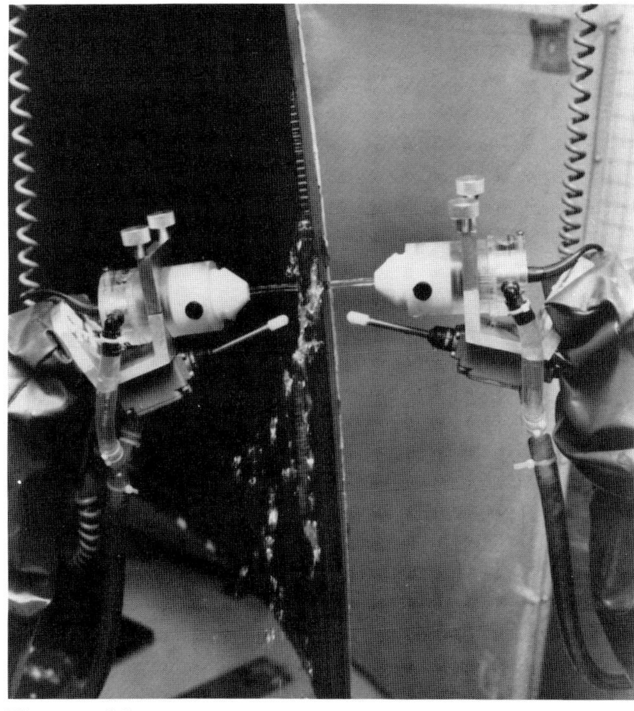

Figure 10
High technology quality control. Robotic controlled water jets inspect epoxy graphite composite bonded joints on a jet engine fan reverser (Martin Marietta Corporation).

Production control and production scheduling are responsible for the timely arrival of materials and subassemblies so that the final product may be manufactured with no delays. Quality assurance and control personnel carry out their responsibilities at every stage in the production process (Figure 10).

When major subassemblies of a large and complex

Figure 11
Completed fuselage sections prepare to join wing assemblies (Lockheed—California Company).

Figure 12
The final product takes on a familiar look (Lockheed—California Company).

product are finally ready for joining (Figures 11 and 12), the end product begins to take on a familiar shape. Behind the scenes, in the meantime, all of the many parts components and subassemblies are being manufactured by thousands of subcontractors much in a similar manner. Other complete complex subassemblies are shipped from their respective manufacturers

Figure 13
Major complex subassemblies such as aircraft engines are readied for installation on the air frame (Lockheed—California Company).

Figure 14
Gas turbine engines are hoisted for installation on the air frame (Lockheed—California Company).

for installation or assembly in the final product (Figures 13 and 14).

The product reaches its final assembly stages while at the same time new product units are only beginning their manufacture. The entire process may take place under one roof (Figure 15). End of production sees the final product off the production line ready for last de-

Figure 15
As finished product nears completion, new products begin their trip through the manufacturing process under the same roof (Lockheed—California Company).

Figure 16
Final detailing, testing, and delivery is the culmination of the manufacturing process (Lockheed—California Company).

tailing, testing, and delivery to the customer (Figure 16). The production effort has been a team effort. Thousands of individuals have each contributed a small part in the making of an incredibly complex product. Their effort has been rewarded by the production of a magnificent example of modern industrial manufacturing capability.

REVIEW QUESTIONS

1. Outline the structure of a typical manufacturing industry.
2. Discuss some of the responsibilities of the industrial engineer.
3. Who is most likely to be responsible for installing a new electrical service to a piece of production equipment?
4. The company needs 10 million rivets. Who places the order for them?
5. Who would you approach if you were seeking employment with the company?
6. Who is finally responsible for company financial affairs?
7. Who designs production tooling?
8. Who designs products?
9. Describe the flow of production in a typical manufacturing industry starting with order placement.
10. Who do you think is the most important person in the company? Why?

CASE PROBLEMS

Case 1
Estimating the Cost

Manufacturer X produces a product at a price of $.0234/production unit. The production rate for this company is 325/hour. Manufacturer Y, a competitor, sells an identical item for $.0205/production unit that can be produced at a rate of 375/hour. Calculate the cost of an order for 15,000 parts from each company, and calculate the time to produce each order. Calculate the total cost of each order and calculate the cost per hour of each company's production.

Case 2
Cost Breakouts

A manufacturer builds and sells a product costing $4500.00 per production unit. Represented in the product cost are:

Wages	— 60 percent
Employee benefits	— 11 percent
Utilities	— 2 percent
Material	— 20 percent
Other expenses	— 7 percent

Calculate each dollar cost as represented by these percentages. If the time involved is 48 worker hours, calculate the worker hour cost in this product. Manufacturing 100 units of the product will result in a 16 percent saving. Calculate the worker hour cost and product price under this condition.

GLOSSARY

Acicular Needle-shaped particles or structures as found in martensite.

Acoustic emission Machines and cutting tools emit characteristic sounds when operating at optimum efficiency. These sounds are digitized by computer and the data are compared against a data base already in the computer. The information is used to determine that the tool is operating as it should. If not, the computer may take action to stop the machine or indicate a problem to the operator.

Adhesives Materials or compositions that enable two surfaces to join together. An adhesive is not necessarily a glue, which is considered to be a sticky substance, since many adhesives are not sticky.

Aggregate Small particles such as powders that are used for powder metallurgy that are loosely combined to form a whole; also sand and rock as used in concrete.

Aging The process of holding metals at room temperature or at a predetermined temperature for the purpose of increasing their hardness and strength by precipitation; aging is also used to increase dimensional stability in metals such as castings.

AISI Abbreviation for *American Iron & Steel Institute*.

Allotropy The ability of a material to exist in several crystalline forms.

Alloy A substance that has metallic properties and is composed of two or more chemical elements of which at least one is a metal.

Amorphous Noncrystalline, a random orientation of the atomic structure.

Anisotropy A material that has specific physical properties in different directions. Rolled steel is strongest in the direction of rolling.

Anneal A heat treatment in which metals are heated and then cooled very slowly for the purpose of decreasing hardness. Annealing is used to improve machinability and to remove stresses from weldments, forgings, and castings. It is also used to remove stresses resulting from cold working and to refine and make uniform the microscopic internal structures of metals.

Anodizing To subject a metal to electrolytic action, as takes place at the anode of a cell, in order to coat it with a protective or decorative film; used for nonferrous metals.

Assembly drawing One of the two major types of typical manufacturing drawings, showing several parts in an assembly and containing pertinent information regarding the assembly, such as a bill of materials. See *Detail drawing*.

ASTM Abbreviation for *American Society for Testing Materials.*

Austempering A heat-treating process consisting of quenching a ferrous alloy at a temperature above the transformation range in a medium such as molten lead; the temperature of the quenching medium is maintained below that of pearlite and above that of martensite formation to produce a tough, hard microstructure.

Austenite A solid solution of iron and carbon and sometimes other elements in which gamma iron, characterized by a face-centered crystal structure, is the solvent.

Austenitizing The process of forming austenite by heating a ferrous alloy above the transformation range.

Autoclave A sealed chamber used to heat and cure parts assembled by adhesive bonding techniques.

Automation Processes and equipment used in mass production manufacturing and designed to accomplish repeat operations with little or no external control or inputs by people.

Bainite A structure in steel named after E. C. Bain that forms between 900° F (481° C) and the Ms temperature. At the higher temperatures, it is known as upper or feathery bainite. At the lower temperatures, it is known as lower or acicular bainite and resembles martensite.

Bill of materials A list generally appearing on an assembly drawing indicating the name of the parts in the assembly as well as their detail part drawing numbers and possibly the vendor source of the parts.

Binary data Information in a computer expressed in terms of zeros and ones called bits. Binary encoded information can then be represented as ON (1) or OFF (0) positions of electronic switches with the computer's central processor.

Blanking The operation of cutting a shape with a die from sheet metal stock. The hole material is saved and used for further operations.

Breaking point The final rupture of a material being pulled in tension, after it has reached its ultimate strength.

Bright anneal The softening process for ductile metals that have been work-hardened. The metal is enclosed in a container with inert gas to avoid scaling and then heated to its recrystallizing temperature.

Brinell hardness The hardness of a metal or alloy measured by pressing a hard ball (usually 10 mm diameter) with a standard load into the specimen. A number is derived by measuring the indentation with a special microscope.

Briquette A compacted mass of usually fine material such as metallic powder used in powder metallurgy.

Brittleness The property of materials to not deform under load but to break suddenly; for example, cast iron and glass are brittle. Brittleness is opposite to plasticity.

BUE. Built-Up Edge A condition in which a metal being cut with a tool adheres to the point or cutting face of the tool, becoming in effect a new rough cutting edge of the tool. This situation causes tearing and a rough finish in the metal being cut.

CAD Computer Aided Design or Computer Aided Drafting. Drawings are made by computer instead of pencil and paper in this system.

CADAM Computer Aided Design and Manufacturing.

CAD/CAM Computer Aided Design/Computer Aided Manufacturing. Systems where design is done by computer (CAD) and then actual machine control programs are generated directly (CAM).

Calendering A process that involves rolling of the product into sheets to achieve desired surface finishes and thickness.

Calibration The processes of comparing measuring instruments and gages to known measurement standards and then adjusting them to conform with the standards.

Carburizing A process that introduces carbon into a heated solid ferrous alloy by having it in contact with a carbonaceous material. The metal is held at a temperature above the transformation range for a period of time. This is generally followed by quenching to produce a hardened case.

Case hardening A process in which a ferrous alloy is hardened so that the surface layer or case is made considerably harder than the interior or core. Some case-hardening processes are carburizing and quenching, cyaniding, carbonitriding, nitriding, flame hardening, and induction hardening.

Casting A process of producing a metal object by pouring molten metal into a mold.

Cast iron Iron containing 2 to $4\frac{1}{2}$ percent carbon, silicon, and other trace elements. It is used for casting into molds. Cast iron is somewhat brittle.

Catalyst An agent that induces catalysis, which is a phenomenon in which the reaction between two or more substances is influenced by the presence of a third substance (the catalyst) that usually remains unchanged throughout the reaction.

Cellulose A polysaccharide of glucose units that constitutes the chief part of the cell walls of plants. For example, cotton fiber is over 90 percent cellulose and is the raw material of many manufactured goods such as paper, rayon, and cellophane. In many plant cells, the cellulose wall is strengthened by the addition of lignin, forming lignocellulose.

Cementite Fe_3C. Also known as iron carbide, a compound of iron and carbon.

Centrifuging Casting of molten metals by using centrifugal forces instead of gravity. The mold (or molds) is rotated about a center where molten metal is poured and allowed to follow sprues outward and into the mold cavity.

Ceramic Metallic oxides of metals such as silicon and aluminum.

Ceramic materials These materials demonstrate great hardness and resistance to heat and are used to make cutting tools, coatings on tools, parts subjected to very hot conditions, abrasives, and mechanical parts.

CF Cold finished.

CIM Computer Integrated Manufacturing. The system that ties design directly to production. A variation of CAD/CAM.

CIP Cold Isostatic Pressing.

Cladding The joining of one metal (usually sheet or plate) to another by using heat and pressure or by an explosive force. With this method, a thin sheet of more expensive metal or one less likely to corrode may be applied to a less expensive metal or one more likely to corrode.

CNC Computer Numerical Control. The computer is an integral part of the machine control unit in CNC.

Coining (embossing) Shaping a piece of metal in a mold or die, often creating raised figures or numbers.

Cold drawing Reducing the cross section of a metal bar or rod by drawing it through a die at a temperature below the recrystallization range, usually room temperature.

Cold finish Refers to the surface finish obtained on metal by any of several means of cold working, such as rolling or drawing.

Cold rolling Reducing the cross section of a metal bar in a rolling mill below the recrystallization temperature, usually room temperature.

Cold working Deforming a metal plastically at a temperature below its lowest recrystallization temperature. Strain hardening occurs as a result of this permanent deformation.

Comparison measurement Measurement accomplished by comparing a known standard to an unknown dimension.

Composite fibers The strands of material used as reinforcement extending through a resin or other matrix in a composite material. An example is carbon fibers in an epoxy matrix. Loads applied to the structure are carried by the fibers.

Composite material Materials exhibiting a much higher strength than the matrix or base material because of reinforcement fibers.

Compressive strength (ultimate) The maximum stress that can be applied to a brittle material in compression without fracture.

Compressive strength (yield) The maximum stress than can be applied to a metal in compression without permanent deformation.

Computer graphics Drawings, pictures, graphs, and so forth generated by a computer.

Cores In metal casting, the hollow parts that are made by using formed sand shapes that are strengthened by baking or by using epoxy.

CR Cold rolled.

Crazing Minute surface cracks on the surface of materials often caused by thermal shock.

Creep Slow plastic deformation in steel and most structural metals caused by prolonged stress under the yield point at elevated temperatures.

Crystalloid A substance that forms a true solution and is capable of being crystallized.

Crystal unit structure or **unit cell** The simplest polyhedron that embodies all the structural characteristics of a crystal and makes up the lattice of a crystal by indefinite repetition.

Cutting fluid A term referring to any of several materials used in cutting metals, such as cutting oils, soluble or emulsified oils (water based), and sulfurized oils.

Decarburization The loss of carbon from the surface of a ferrous alloy as a result of heating it in the presence of a medium such as oxygen that reacts with the carbon.

Deformation Alteration of the form or shape as a result of the plastic behavior of a metal under stress.

Dendrite A crystal characterized by a treelike pattern that is usually formed by the solidification of a metal.

Dendrites generally grow inward from the surface of a mold.

Density The density of a body is the ratio of its mass to its volume. For solids the method to determine density is to measure the buoyant force upon the specimen of liquid of known density in which it is immersed or by determining the volume of displacement from a specific gravity bottle.

Deoxidizer A substance that is used to remove oxygen from molten metals; for example, ferrosilicon in steel making.

Detail drawing One of the two major types of production drawings showing an individual part. See *Assembly drawing*.

Die casting Casting metal into a mold by using pressure instead of gravity or centrifugal force.

Diffusion The process of atoms or other particles intermingling within a solution. In solids, it is a slow movement of atoms from areas of high concentration toward areas of low concentration. The process may be (*a*) migration of interstitial atoms such as carbon, (*b*) movement of vacancies, or (*c*) direct exchange of atoms to neighboring sites.

Dimensional measurement Equipment and process used to evaluate size of a part or form and location of part features.

Dimensional tolerance Acceptable range of part size or form and positions of part features that will accommodate design specifications.

DNC Direct Numerical Control. A system in which several machines are controlled by a central computer.

Drawing **1** Reducing the diameter of wire or bar stock by pulling through successively smaller dies. **2** Reducing the wall thickness of tubing by drawing it through dies. **3** Forming sheet stock into cuplike parts by pushing it through a die. **4** Tempering of hardened tool steel.

Ductility The property of a material to deform permanently, or to exhibit plasticity without rupture while under tension.

EBM Electron Beam Machining. The process by which material is removed by an intense beam of electrons impinging on the workpiece.

Eccentricity A rotating member whose axis of rotation is different or offset from the primary axis of the part or mechanism. Thus when one turned section of a shaft centers on an axis different from that of the shaft, it is said to be eccentric or to have "runout." For example, the throws or cranks on an engine crankshaft are eccentric to the main bearing axis.

ECDB Electro Chemical Deburring. The process in which electron flow and chemical action are used to remove burrs from a workpiece.

ECM Electrochemical Machining.

EDM Electrodischarge Machining.

Elastic deformation The movement or deflection of a material when an external load is applied that is less than the elastic limit.

Elasticity The ability of a material to return to its original form after a load has been removed.

Elastic limit The extent to which a material can be deformed and still return to the original shape when the load is released. Deformation occurs beyond the elastic limit.

Elastomer Any of various elastic substances resembling rubber.

Electrolyte A nonmetallic conductor, usually a fluid, in which electric current is carried by the movement of ions.

Electromechanical Combining electrical or electronics and mechanical principles, such as a solenoid operated valve in a hydraulic system.

Electron beam welding The fusion of material by energy imparted from an intense beam of electrons.

Electroplating Coating an object with a thin layer of a metal through electrolytic deposition.

ELG Electrolytic Grinding. The process where material is removed by chemical and abrasive processes.

Embossing The raising of a pattern in relief on a metal by means of a high pressure on a die plate.

Equilibrium A condition of balance in which all the forces or processes that are present are counterbalanced by equal and opposite forces. Processes where the condition appears to be one of rest rather than of change.

Eutectic The alloy composition that freezes at the lowest constant temperature, causing a discrete mixture to form in definite proportions.

Eutectoid The alloy composition that transforms from a high temperature solid into new phases at the lowest constant temperature. In binary (double) alloy systems, it is a mechanical mixture of two phases forming simultaneously from a solid solution as it cools through the eutectoid (A_1 in steels) temperature.

Extrusion Forcing a solid metal piece (often heated)

through a shaped die by using an extremely high force in a way that is similar to that of toothpaste when it is squeezed from a tube.

Fatigue in metals The tendency of a metal to fail by breaking or cracking under conditions of repeated cyclical stressing that takes place well below the ultimate tensile strength.

Fatigue strength The amount of stress that can be applied to a metal without failure while it is subjected to ten million or more cycles of load reversals. In mild steels, the fatigue strength is about 50 percent of the tensile strength.

Feedback In an electric circuit, used as a control function. A part of the signal is fed back or returned to the controller indicating what action has or has not occurred.

Ferrite A magnetic form of iron. A solid solution in which alpha iron is the solvent, characterized by a body-centered cubic crystal structure.

Ferrous From the Latin word *ferrum*, meaning iron. Describes an alloy containing a significant amount of iron.

Fiber **1** The directional property of wrought metals that is revealed by a woody appearance when fractured. **2** A preferred orientation of metal crystals after a deformation process such as rolling or drawing. **3** Cellulosic plant cells that are used for manufacturing paper and other products. **4** Strands of materials used as reinforcement in plastic products and other materials.

Fiberglass A resin matrix reinforced with glass fibers for strength. A reinforced plastic manufacturing material with many applications.

Filament winding A composite manufacturing process where the end product is to have a hollow internal shape. A filament of the fiber is wound around a form, then bonded in place with the resin matrix.

Final inspection The last inspection done on parts at the completion of manufacturing processes.

First article inspection Inspection performed at the beginning of a manufacturing process to determine if the first parts out of production are correct.

Fixture Production tooling designed to hold and align parts during manufacturing processes.

Flame hardening A means of surface hardening steel or cast iron by applying heat, followed by a quench.

Flash The extrusion of extra material in die casting dies, forging dies, or plastic injection molding dies along the parting line.

Float glass A glass manufacturing process that produces a continuous sheet or ribbon of glass.

Flux A solid, liquid, or gaseous material that is applied to solid or molten metal in order to clean and remove the oxides.

FMS Flexible Manufacturing System. A system of material processing stations or work cells connected by automatic material handling and transfer equipment all under computer control.

Forging A method of metal working in which the metal is hammered into the desired shape, or is forced into a mold by pressure or hammering, usually after being heated to a more plastic state. Hot forging requires less force to form a given point than does cold forging, which is usually done at room temperature.

Fracture A ruptured surface of metal that shows a typical crystalline pattern. Fatigue fractures, however, often display a smooth, clam-shell appearance.

Full automatic Processes in which all phases, once started, are accomplished without the need of further manual input.

Fusion The merging of two materials while in a molten state.

Galvanic corrosion A common type of corrosion process in which a potential difference through an electrolyte causes a deplating (corroding) of the material.

Galvanizing The application of a layer of zinc to the surface of iron and steel for protection from corrosion.

Geometric dimension Dimensions that pertain to the form of a part and the position of part features rather than to size.

Geometric tolerance Tolerances pertaining to geometric dimensions of form and position.

Glue laminated beam A structural wood beam made by gluing thinner boards together until a desired dimension for beam thickness is reached. Glue laminated beams will support large loads and can span long distances with only end support.

GMAW Gas Metal Arc Welding. A shield gas welding process also known as MIG welding.

Grain Individual crystals in metals.

Grain boundary The outer perimeter of a single grain where it is in contact with adjacent grains.

Grain growth Called recrystallization. Metal grains begin to reform to larger and more regular size and shape at certain temperatures, depending to some extent on the amount of prior cold working.

Graphics In this text, a reference to pictures generated by computers.

Graphite fiber Strands of carbon in graphite form used in composite materials as the main load-bearing constituent.

GTAW Gas Tungsten Arc Welding. A shield gas welding process also known as TIG welding.

Hardenability The property that determines the depth and distribution of hardness in a ferrous alloy induced by heating and quenching.

Hardening The process of increasing the hardness of a ferrous alloy by austenitizing and quenching; also, the process of increasing the hardness of some stainless steels and nonferrous alloys by solution heat treatment and precipitation.

Hard-facing When a hard surface is desired on a soft metal, a hard material (usually another metal) is applied to the surface. Some methods employed to do this are arc welding, spray welding, and electroplating.

Hardness The property of a metal to resist being permanently deformed. This is divided into three categories: the resistance to penetration, the resistance to abrasion, and elastic hardness.

Heading Enlarging the end of a rod or bar of metal by hot or cold upset forging.

Heat sink A large mass of metal that has a high thermal conductivity used to stabilize the temperature of a part held in contact with it. In welding, the mass of the base metal often acts as a heat sink to the weld metal.

HIP Hot Isostatic Pressing.

Hot pressing Forming or forging tough metals such as alloy steel at high temperatures.

Hot rolling A process of forming metals between rolls in which the metals are heated to temperatures above the transformation range.

Hot-short Brittleness in hot metal. The presence of excess amounts of sulfur in steel causes hot-shortness.

HSLA High Strength, Low Alloy (steels).

Hydraulics **1** The dynamics of liquids (hydrodynamics), specifically, the study of the flow of water in pipes. **2** Pertaining to fluid power where mechanical actions and force are developed by the pumping of fluids (hydraulic oil) at high pressure. Hydraulics may be used to develop large mechanical advantages in many types of mechanisms, such as hydraulic presses and machine tools.

Hydrojet A manufacturing process in which a material is cut by a high-pressure jet of water often containing an abrasive material to enhance cutting action.

ICAD Integrated Computer Aided Manufacturing. A version of CAD/CAM in which computers are used to control manufacturing processes.

Impact test A test in which small notched specimens are broken in an Izod–Charpy machine. This test determines the notch toughness of a metal.

Inclusions Particles of impurities that are usually found during solidification and are usually in the form of silicates, sulfides, and oxides.

Induction hardening Heating the surface of cast iron or tool steel by means of electromagnetic currents followed by a quench.

Industrial engineer One engineering discipline where emphasis is placed on manufacturing methods, tooling, and facilities; also called manufacturing engineer, industrial technologist, or applications engineer.

Ingot A large block of metal that is usually cast in a metal mold and forms the basic material for further rolling and processing.

In-process inspections Part of quality control and assurance in which inspections occur when required during a manufacturing process.

Interchangeability The concept that parts manufactured by many different manufacturers to the same dimensional specifications may be interchanged in assemblies with no difficulty.

Interface **1** A surface that forms a boundary between two phases or systems. **2** That part of manufacturing equipment where different systems or types of equipment connect to each other to make complete functioning systems such as computer numerical control and flexible manufacturing systems.

Involute Geometry found in modern gears that permits mating gear teath to engage each other with rolling rather than sliding friction.

Ion An atom or molecule electrostatically charged by losing or gaining one or more valence electrons.

IPM Inches per minute.

IPR Inches per revolution.

Iron The term *iron* always refers to the element Fe and not cast iron, steel, or any other alloy of iron.

Isomerism **1** Atoms are said to be isomeric when their nuclei contain the same number of protons and neutrons but differ in energetics and behavior. **2** Compounds are said to be isomeric when they have the same elementary composition (i.e., their molecules contain the same numbers and kinds of atoms) but different structures, and hence, properties. It is believed that differences are due to the arrangement of atoms in each molecule.

Isothermal transformation (I–T) Transformation in some metals that takes place at a constant temperature.

Jig Production tooling designed to guide a cutting tool during production manufacturing processes.

Kaolin A fine white clay that is used in ceramics and refractories composed mostly of kaolinite, a hydrous silicate of aluminum. Impurities may cause various colors and tints.

Killed steel Steel that has been deoxidized with agents such as silicon or aluminum to reduce the oxygen content. This prevents gases from evolving during the solidification period.

Lamellar An alternating platelike structure in metals (as in pearlite).

Laminates Composed of multiple layers of the same or different materials.

Laser Light Amplification by Stimulated Emission of Radiation. A device in which heat is derived from the intense coherent beam of laser light energy. This intense, narrow beam of light is used in some welding and machining operations.

Lattice, space A term that is used to denote a regular array of points in space. For example, the sites of atoms in a crystal. The points of the three-dimensional space lattice are constructed by the repeated application of the basic translations that carry a unit cell into its neighbor.

Lignin A substance that is related to cellulose that with cellulose forms the woody cell walls of plants and the material that cements them together. Methyl alcohol is derived from lignin in the destructive distillation of wood.

Liquidus The temperature at which freezing begins during cooling and ends during heating under equilibrium conditions, represented by a line on a two-phase diagram.

Machinability The relative ease of machining that is related to the hardness of the material to be cut.

Macroscopic Structural details on an object that are large enough to be observed by the naked eye or with low magnification (about 10×).

Macrostructure The structure of metals as revealed by macroscopic examination.

Malleability The ability of a metal to deform permanently without rupture when loaded in compression.

Manufacturing outline (MO) A document that accompanies parts through the manufacturing process and indicates the steps required, types of machines, and gaging equipment required.

Martempering The process of quenching an austenitized ferrous alloy to a temperature just above or near the Ms point and maintaining until the temperature throughout the part is uniform. The alloy is then allowed to cool slowly in air through the range of martensitic formation.

Martensite An unstable constituent that is formed by heating and quenching steel. It is formed without diffusion and only below a certain temperature, known as the Ms temperature. Martensite is the hardest of the transformation products of austenite, having an acicular, or needlelike, microstructure.

Mass production Continuous production of duplicate product units such as appliances, cars, or any parts of these products.

Measurement standards Known standards to which production gages and other measurement tools are periodically compared to ensure their conformity and accuracy. Gage blocks are an example.

Metalloid A nonmetal that exhibits some, but not all, of the properties of a metal. Examples are sulfur, silicon, carbon, phosphorus, and arsenic.

Metallurgy The science and study of the behaviors and properties of metals and their extraction from their ores.

Methanol (methyl alcohol, wood alcohol) Produced by the destructive distillation of wood or made synthetically.

Microscopy The use of, or investigation with, the microscope.

Microstructure The structure of polished and etched metal specimens as seen enlarged through a microscope.

MIG Metal Inert Gas Welding. Also known as GMAW. A shield gas welding process that uses a continuous wire electrode fed from a reel.

Modulus of elasticity The ratio of the unit stress to the unit deformation (strain) of a structural material is a constant as long as the unit stress is below the elastic limit. Shearing modulus of elasticity is often called the modulus of rigidity.

Mold A general classification of tooling used to form or shape a product while the material is in a molten or liquid state. Mold tooling is found in many manufacturing processes including metal and plastic casting, injection molding, fiberglassing, and composite part manufacturing.

Monomer A single molecule or a substance consisting of single molecules. The basic unit in a polymer.

Muffled furnace A gas- or oil-fired furnace in which the work is separated from the flame by an inner lining or "muffle."

NC Numerical Control. Equipment and processes that control manufacturing processes by digital data, often involving the use of computers.

NC axes On NC manufacturing equipment, the primary paths along which major components of the machine or workpiece move.

NC program A step-by-step set of instructions written in a format understood by an NC machine and defining all of the information necessary to accomplish the manufacturing task required.

Nitriding A process of case hardening in which a special ferrous alloy is heated in an atmosphere of ammonia or is in contact with another nitrogenous material. In this process, surface hardening is achieved by the absorption of nitrogen without quenching.

Nonferrous Metals other than iron and iron alloys.

Normalizing To homogenize and produce a uniform structure in alloy steels by heating above the transformation range and cooling in air.

Notch toughness The resistance to fracture of a metal specimen having a notch or groove when subjected to a sudden load, usually tested on an Izod– Charpy testing machine.

Open-die forging Also called smith forging. In open-die forging a drop hammer delivers blows of great force to a heated metal that is shaped by manipulating it under the hammer.

Operations sketch A portion of a part drawing used in a specific manufacturing step emphasizing certain dimensional information and indicating the steps to be accomplished.

Oxidation The slow or rapid reaction of oxygen with other elements; burning. In metals, overoxidation during heating under oxidizing conditions often results in permanent damage to metals.

Oxidation–reduction A chemical reaction in which one or more electrons are transferred from one atom or molecule to another.

Part drawing A drawing of an individual part; a detail drawing.

Patterns A general class of production tooling representing the shape and/or size of the parts to be made. Patterns can be used to form molds in casting, and as measurement and gaging tools to check or to guide material cutting and preparation activities.

Pearlite The lamellar mixture of ferrite and cementite in slowly cooled iron–carbon alloys as found in steel and cast iron.

Peel load In metal, plastics, or composites, a force that acts to peel apart joined pieces.

Peening Work hardening the surface of metal by hammering or blasting with shot (small steel balls). Peening introduces compressive stresses on weld surfaces that tend to counteract unwanted tensile stresses.

Perforating Piercing many small holes close together.

Performance inspection An inspection designed to evaluate the performance of a part or product. Examples include the burn-in test for electronics and automobile engine tests.

Periphery The perimeter or external boundary of a surface or body.

Permeability In casting of metals, the term is used to define the porosity of foundry sands in molds and the ability of trapped gases to escape through the sand.

Petrochemicals Chemicals derived from petroleum; substances or materials manufactured from a component of crude oil or natural gas.

Phase A portion of an alloy, physically homogeneous throughout, that is separated from the rest of the alloy by distinct boundary surfaces. The following phases occur in the iron–carbon alloy: molten alloy, austenite, ferrite, cementite, and graphite.

Pickling A process in which metal parts are dipped in acid for the purpose of cleaning and etching prior to plating, painting, or further cold working.

Pig iron The product of a blast furnace. It is a raw iron that usually contains about 4.5 percent carbon and impurities such as phosphorus, sulfur, and silicon.

Pitch diameter For threads, the pitch diameter is an imaginary circle, which on a perfect thread occurs at the point where the widths of the thread and groove are equal. On gears, it is the diameter of the pitch circle.

Plasma An ionized gas of extremely high temperature achieved by passing an inert gas through an electric arc. Plasma arcs are used in welding, cutting, and machining processes.

Plastic deformation Deformation that occurs when sufficient stress is applied to a solid and it does not return to its original condition.

Plasticity The quality of material such that it can be deformed without breaking. Clay is a completely plastic material. Metals exhibit plasticity in varying amounts.

Plating The process of depositing a layer of one metal on another, often done electrically, for the purpose of corrosion protection, appearance, improved electrical conductivity, and other engineering requirements.

Plotter A computer-driven drawing machine used to convert computer drawings to paper drawings.

Poisson's ratio When a rod of elastic material is elongated by stretching (strain), the lateral (crosswise) dimensions will contract; Poisson's ratio is the ratio between the strain and the amount of lateral contraction.

Polymer A chemical compound or mixture of compounds formed by polymerization and consisting essentially of repeating structural units.

Polymerization A chemical reaction in which two or more small molecules combine to form larger molecules that contain repeating structural units of the original molecules.

Precast concrete Concrete products that have been precast in molds and then removed and transported to construction sites. Precast concrete walls for buildings are examples.

Precipitation hardening A process of hardening an alloy by heat treatment in which a constituent precipitates from a supersaturated solid solution while at room temperature or at some slightly elevated temperature.

Preform To bring to the approximate shape and size before the final forming takes place. Used in forging and powder metallurgy.

Process anneal See *Bright anneal*.

Production line The set of stations where various steps in the assembly or making of parts and products is accomplished in a manufacturing industry.

Production tooling Tools, equipment, and special devices necessary for the manufacture of a product.

Productivity A measure of the time spent producing a product compared to the time spent not producing a product during manufacturing.

Programming The preparation of machine control instructions for numerically controlled manufacturing equipment. Programming may be done manually or by computer (CAM).

Prototype A working model of a product to be manufactured and used for testing and design evaluation, often full sized.

PSI Pounds per square inch.

Pulltrusion A process that is the opposite of extrusion and used in composite part manufacturing.

Punching The operation of cutting a hole in sheet metal using a die. The hole material is scrapped.

Quality assurance All activities in manufacturing directed toward ensuring production of a quality product.

Quality control One segment of quality assurance often responsible for dimensional inspections during production.

Quenching The process of rapid cooling of metal alloys for the purpose of hardening. Quenching media include air, oil, water, molten metals, and fused salts.

Reciprocate A movement back and forth along a given axis.

Recovery The relaxation and reduction of locked-in internal stresses in cold-worked metals.

Recrystallization A process in which the distorted grain structure of metals that are subjected to mechanical deformation is replaced by a new strain-free grain structure during annealing.

Reduction Reaction involving oxidation–reduction.

Refractory Materials that will resist change of shape, weight, or physical properties at high temperatures. These materials are usually silica, fire clay, diaspore, alumina, and kaolin. They are used for furnace linings.

Residual stress Stresses induced within the structure of a material by cold working, machining, and heat treatments.

RIM Reaction Injection Molding. A chemical reaction between two resins generates the heat required to make the mix a liquid.

Rimmed steel A low carbon steel (insufficiently deoxidized) that during solidification releases considerable quantities of gases (mainly carbon monoxide). When the mold top is not capped, a side and bottom rim of several inches forms. The solidified ingot has scattered blow holes and porosity in the center but a relatively thick skin free from blow holes.

Robotics Systems and equipment used to operate industrial robots.

Rockwell Hardness A hardness test that uses a penetrator and known weights. Several scales are used to cover the very soft to the very hard materials. The Rockwell "C" scale is used mostly for steel.

RPM Revolutions per minute.

Runout An eccentricity of rotation as that of a cylindrical part held in a lathe chuck being off center as it rotates. The amount of runout of a rotating member is often checked with a dial indicator.

Sacrificial anode A metal slug, usually magnesium and zinc, designed to concentrate galvanic corrosion upon itself and thus save a more important structure on which the anode is attached, such as a ship's hull or a buried pipe line.

SAE Abbreviation for *The Society of Automotive Engineers*.

Scrap Materials or metals that have lost their usefulness and are collected for reprocessing.

Sealant A sealing agent that has some adhesive qualities: it is used to prevent leakage.

Semiautomatic A process in manufacturing that requires some degree of manual input, but acts without this input for at least part of the cycle.

Sensor A device used to measure some quantity and then send the information back to the control equipment.

Servo A motor or valve that can be controlled by an electrical signal, thus converting electrical inputs to mechanical action outputs.

SFM or **SFPM** Surface feet per minute. Also FPM (feet per minute).

Shearing A concentration of forces in which the bending moment is virtually zero and the metal tends to tear or be cut along a transversal axis at the point of applied pressure.

Shear load A load that tends to force materials apart by application of side slip action.

Shell molding A form of gravity casting of metal (usually a high melting temperature metal) in which the mold is made of a thin shell of refractory material.

Shield gas Usually an inert gas used to displace air from around a weld zone thus keeping the weld uncontaminated.

Shot peening A cold working process in which the surface of a finished part is pelted with finely ground steel shot or glass beads to form a compression layer.

SI Systéme Internationale. The metric system of weights and measures.

Sintering A process of fusing compacted material such as metal powders into a solid piece by applying heat sufficient to bond, but not melt, the particles.

Skelp The name of the semifinished steel of which butt-welded pipe is made.

Sketch A simple drawing made without the use of drawing instruments.

Slag (dross) A fused product that occurs in the melting of metals and is composed of oxidized impurities of a metal and a fluxing substance such as limestone. The slag protects the metal from oxidation by the atmosphere since it floats on the surface of the molten metal.

Slip planes Also called slip bands. These are lines that appear on the polished surface of a plastically deformed metal. The slip bands are the result of crystal displacement, defining planes in which shear has taken place.

Slurry A watery mixture of insoluble material, such as mud, lime, or plaster of paris.

Smelting The process of heating ores to a high temperature in the presence of a reducing agent such as carbon (coke) and of a fluxing agent to remove the gangue.

Soaking A prolonged heating of a metal at a predetermined temperature to create a uniform temperature throughout its mass.

Sodium silicate (Na_2SiO_3). Also called water glass.

Solid solution Found in metals at temperatures below the solidus. Some of the types of solid solutions are continuous, intermediate, interstitial, substitutional, and terminal.

Solidus Seen as a line on a two-phase diagram, it represents the temperatures at which freezing ends when cooling, or melting begins when heating under equilibrium conditions.

Solubility The degree to which one substance will dissolve in another.

Solute A substance that is dissolved in a solution and is present in minor amounts.

Solution heat treatment A process in which an alloy is heated to a predetermined temperature for a length of time that is suitable to allow a certain constituent to enter into solid solution. The alloy is then cooled quickly to hold the constituent in solution, causing the metal to be in an unstable supersaturated condition. This condition is often followed by age hardening.

Solvent A substance that is capable of dissolving another substance and is the major constituent in a solution.

Spalling Breaking small pieces from a surface, often caused by a thermal shock.

Specific gravity A numerical value that represents the weight of a given substance with the weight of an equal volume of water. The specific gravity for pure water is taken as 1.000.

Spheroidizing Consists of holding carbon steel for a period of time at just under the transformation temperature. An aggregate of globular carbide is formed from other microstructures such as pearlite.

Spinning Cold drawing ductile sheet metals into cylinders and other shapes having rotational symmetry in a spinning lathe.

Springback The tendency of a formed metal part to return to some extent to its former shape because of the elasticity of the metal.

Sputtering To dislodge atoms from the surface of a material by collision with high energy particles for the purpose of depositing a metallic film on a part.

Stainless steel An alloy of iron containing at least 11 percent chromium and sometimes nickel that resists almost all forms of rusting and corrosion.

Standardization The repeat manufacturing of many parts, each meeting acceptable tolerances.

Statistics Mathematically generated data derived from small samples of large batches on which inferences may be made about the specifications of a large batch without the need to inspect every part.

Steel An alloy of iron and less than 2 percent carbon plus some impurities and small amounts of alloying elements is known as plain carbon steel. Alloy steels contain substantial amounts of alloying elements such as chromium or nickel besides carbon.

Strain The unit deformation of a metal when stress is applied.

Strain hardening An increase in hardness and strength of a metal that has been deformed by cold working or at temperatures lower than the recrystallization range.

Strength The ability of a metal to resist external forces. This is called tensile, compressive, or shear strength, depending on the load. See *Stress*.

Stress The load per unit of area on a stress–strain diagram. *Tensile stress* refers to an object loaded in tension, denoting the longitudinal force that causes the fibers of a material to elongate. *Compressive stress* refers to a member loaded in compression, which either gives rise to a given reduction in volume or a transverse displacement of material. *Shear stress* refers to a force that lies in a parallel plane. The force tends to cause the plane of the area involved to slide on the adjacent planes. *Torsional stress* is a shearing stress that occurs at any point in a body as the result of an applied torque or torsional load.

Stress raiser Can be a notch, nick, weld undercut, sharp change in section, or machining grooves or hairline cracks that provide a concentration of stresses when the metal is under tensile stress. Stress raisers pose a particular problem and can cause early failure in members that are subjected to many cycles of stress reversals.

Stress relief anneal The reduction of residual stress in a metal part by heating it to a given temperature and holding it there for a suitable length of time. This treatment is used to relieve stresses caused by welding, cold working, machining, casting, and quenching.

Swaging Compacting or necking down metal bars or tubes by hammering or rotary forming.

Temper In ferrous metals, the stress relief of steels that are hardened by quenching for the purpose of toughening them and reducing their brittleness. In nonferrous metals, temper is a condition produced by mechanical treatment such as cold working. An alloy may be cold worked to the hard temper, fully softened to the annealed temper, or two intermediate tempers.

Tension load A load applied to joined parts that attempts to separate them by a pulling or stretching action.

Thermal conductivity The quantity of heat that is transmitted per unit time, per unit cross section, per unit temperature gradient through a given substance. All materials are in some measure conductors of heat.

Thermal expansion The increase of the dimension of material that results from the increased movement of atoms caused by increased temperature.

Thermal shock A stress induced on the surface of a material such as carbide tools or fire brick caused by a rapid rate of heating and surface expansion.

Thermal stress Shear stress that is induced in a material due to unequal heating or cooling rates. The difference of expansion and contraction between the interior and exterior surfaces of a metal that is being heated or cooled is an example.

Thermoplastic Capable of softening or fusing when heated and of hardening again when cooled.

Thermosetting Capable of becoming permanently rigid when cured by heating; will not soften by reheating.

TIG Tungsten Inert Gas Welding. See *GTAW*.

Tolerance The allowance of acceptable error within which the mechanism will still fit together and be totally functional.

Tool and die making The processes of building specialty production tooling to support manufacture of a product.

Tool design The process of designing cutting tools, jigs, and fixtures that will support product manufacturing.

Tool geometry The proper shape of a cutting tool that makes it work effectively for a particular application.

Tooling Generally, any machine tool accessory separate from the machine itself. Tooling includes cutting tools, holders, workholding accessories, jigs, and fixtures.

Tool steel A special group of steels that is designed for specific uses, such as heat-resistant steels that can be heat treated to produce certain properties, mainly hardness and wear resistance.

Toughness Generally measured in terms of notch toughness, which is the ability of a metal to resist rupture from impact loading when a notch is present. A standard test specimen containing a prepared notch is inserted into the vise of a testing machine. This device, called the Izod–Charpy testing machine, consists of a weight on a swinging arm. The arm or pendulum is released, strikes the specimen, and continues to swing forward. The amount of energy absorbed by the breaking of the specimen is measured by how far the pendulum continues to swing.

Transducer A device by means of which energy can flow from one or more transmission systems to other transmission systems, such as a mechanical force being converted into electrical energy (or the opposite effect) by means of a piezoelectrical crystal.

Transformation temperature The temperature(s) at which one phase transforms into another phase; for example, where ferrite or alpha iron transforms into austenite or gamma iron.

Transition temperature The temperature at which normally ductile metals become brittle.

Ultimate strength Also tensile strength. The highest strength that a metal exhibits after it begins to deform plastically under load. Rupture of the material occurs either at the peak of its ultimate strength or at a point of further elongation and at a drop in stress load.

Ultrasonic Sound frequency above that able to be heard by the human ear.

UM Universal milled (steel plate).

Uncertainty principle A principle in quantum mechanics. It is impossible to assert in terms of the ordinary conventions of geometrical position and of motion that a particle (as an electron) is at the same time at a specified point and moving with a specified velocity.

Upset forging The process of increasing the cross section of stock at the expense of its length.

Valence The capacity of an atom to combine with other atoms to form a molecule. The inert gases have zero valence; valence is determined by considering the positive and negative atoms as determined by the atoms gaining or losing of valence electrons.

Video display The television display screen of a computer. Also known as a CRT (Cathode Ray Tube).

Viscosity The property in fluids, either liquid or gaseous, that may be described as a resistance to flow; also, the capability of continuous yielding under stress.

Visual inspection An inspection made during manufacturing based on simply looking at a product to determine if any deficiencies are apparently visible.

Void A cavity or hole in a substance.

Voltage Electromotive force or a surplus of electrons at a point in an electric circuit. Voltage is equivalent to pressure in a fluid power system.

Vulcanization The process of treating crude or synthetic rubber or similar plastic material chemically to give it useful properties, such as elasticity, strength, and stability.

Weldment A unit formed by welding together an assembly of pieces.

Work hardening Also called strain hardening, in which the grains become distorted and elongated in the direction of working (rolling). This process hardens and strengthens metals but reduces their ductility. Excessive work hardening can cause ultimate brittle failure of the part. Stress relief is often used between periods of working of metals to restore their ductility.

Wrought iron Contains 1 or 2 percent slag, which is distributed through the iron as threads and fibers imparting a tough fibrous structure. Usually contains less than .1 percent carbon. It is tough, malleable, and relatively soft.

Yield point The stress at which a marked increase in deformation occurs without an increase in load stress as seen in mild or medium carbon steel. This phenomenon is not seen in nonferrous metals and other alloy steels.

Yield strength Observed at the proportional limit of metals and is the stress at which a material deviates from that proportionality of stress to strain to a specified amount.

INDEX

THERMAL RADIATION PHENOMENA

Volume 1

Radiative Properties of Air

Volume 1:

Radiative Properties of Air

Volume 2:

Excitation and Non-Equilibrium Phenomena in Air